Principles of
Occupational
Health & Hygiene

Principles of
Occupational Health & Hygiene

An introduction

3rd edition

Edited by Sue Reed, Dino Pisaniello & Geza Benke

Routledge
Taylor & Francis Group

LONDON AND NEW YORK

First published 2019 by Allen & Unwin

Published 2020 by Routledge
2 Park Square, Milton Park, Abingdon, Oxon OX14 4RN
605 Third Avenue, New York, NY 10017

Routledge is an imprint of the Taylor & Francis Group, an informa business

 A catalogue record for this
book is available from the
National Library of Australia

Index by Puddingburn Publishing Services Pty Ltd
Set in 10.75/13 pt Adobe Garamond by Midland Typesetters, Australia

ISBN-13: 9781760528508 (pbk)

 Printed and bound in Great Britain by
TJ Books Limited, Padstow, Cornwall

Contents

Foreword

The Australian Institute of Occupational Hygienists (AIOH) defines occupational hygiene as the art and science of occupational disease prevention. More specifically, occupational hygiene addresses the anticipation, recognition, evaluation, communication and control of environmental hazards in, or arising from, the workplace, which can result in injury, illness, impairment, or affect the well-being of workers and members of the community.

The importance of occupational hygiene has been increasingly recognised over the past three centuries. In 1775, surgeon Percival Pott noted an increased incidence of scrotal cancer in English chimney sweeps, documenting the first causal association between workplace exposure to hazards and the subsequent development of disease. Since this time, we have continued to see clusters of occupational diseases associated with particular industries and hazards, including the prevalence of asbestos-related lung diseases in the mid-late 20th century and the recent recognition of accelerated silicosis in manufactured stone workers in 2018.

Over this time, constantly changing technology has created constantly changing workplaces and hazards, highlighting the need to apply the fundamental principles of occupational hygiene to facilitate the anticipation, recognition, evaluation and control of new and emerging hazards.

To promote these principles, the first edition of *Principles of Occupational Health and Hygiene: An Introduction* was published in 2007. This book was based on the 1992 book *Occupational Health and Hygiene Guidebook for the WHSO (Workplace Health and Safety Officer)* written by Dr David Grantham, who generously ceded the rights and royalties for this book to AIOH in 2002.

The new book was aimed at a wide audience of health and safety professionals, and included emerging issues and topics that had become important since the original 1992 book was published. Many respected authors, all AIOH members, freely contributed their time to review the text and deliver updated chapters on their areas of expertise.

In 2013, a second edition of the *Principles of Occupational Health and Hygiene: An Introduction* was published. By this time the book had been adopted as a valued reference source by many occupational health and hygiene professionals and as a standard textbook by a growing number of educational institutions.

Building on this work, most of the second edition's authors have again donated their time to review and update their chapter of expertise to produce the current third edition. *Principles of Occupational Health and Hygiene: An Introduction* remains an ideal resource for students of occupational health and safety, and a professional reference for occupational hygienists, occupational health and safety professionals and anyone working in an occupational health and safety role.

The increased use of this book as a textbook and reference source will promote the application of the principles of occupational hygiene to ensure focus is maintained on prevention of occupational diseases in Australia and overseas.

Dr Julia Norris
AIOH President, 2019

Preface

The ambitious project by the AIOH to revise and expand David Grantham's *Occupational Health and Hygiene Guidebook for the WHSO* began in 2003 under the guidance of Dr Cherilyn Tillman. The result, published in 2007, was the 1st edition of this book, and covered a broad range of topics, which have continued to grow in importance. The 2nd edition was published in 2013, and in 2017, the AIOH initiated work on this 3rd edition.

The editors and the Council of the AIOH thank all the authors and reviewers for their professionalism and attention to detail and for their generous collaboration with us during the project. All of the 30 authors of this 3rd edition are members of the AIOH who have written chapters in their own time, without payment or favour, to support the AIOH in its role of furthering the practice and recognition of occupational hygiene in the Australasian region, and in support of the Australian Work Health and Safety Strategy 2012–2022.

We would like to thank Dr David Grantham for his interest, encouragement and assistance in updating and expanding his original material. In addition, we would like to thank the Council of the AIOH for inviting us to serve as editors of the 3rd edition. It was an honour to be entrusted with delivery of this important publication.

Where appropriate, the book emphasises issues as seen in Australian workplaces such as the coverage of thermal stress, focusing on heat rather than cold. The book refers to Australian legislation where it is appropriate but in the main legislation is not covered in depth due to differences between states. Australian standards are also only referred to where appropriate and in some areas, such as indoor air quality, international standards are discussed. Occupational exposure limits are referred to as OEL standards, and not WES, due to the broad readership of the book, and similarly OHS or H&S is used in preference to WHS.

Finally, we would like to thank our families for their patience, understanding and support during this project, which consumed far more of our lives and free time than we had envisaged.

Associate Professor Sue Reed, Professor Dino Pisaniello and Dr Geza Benke
Perth, 2019

Acknowledgements

The AIOH editorial committee gratefully acknowledges the pioneering work of Dr David Grantham, on which this book was originally based.

We also gratefully acknowledge the invaluable efforts of Dr Cherilyn Tillman, editor of the 1st edition, along with authors who decided not to be involved in the later editions but have allowed their work to be used. The relevant authors are: Dr Ian Grayson (concept of the exposure standard), Dr Steven Brown (indoor air quality), Dr Georgia Sinclair (ionising radiation), Dr Denise Elson (biological hazards).

In addition we would like to acknowledge authors from the 2nd edition who are not part of the 3rd edition, who are: Dr David Grantham AM (Chapter 1: The hazardous work environment: the occupational hygiene challenge; Chapter 3: The concept of the exposure limit strategies for workplace health hazards; and Chapter 4: Control of work-place health hazards), Dr John Edwards (Chapter 2: Occupational health, basic toxicology and epidemiology; and Chapter 10: Biological monitoring of chemical exposure), Geoff Pickford (Chapter 7: Aerosols), Dr Brian Davies AM (Chapter 7: Aerosols), Gary Rhyder (Chapter 8: Metals in the workplace), Noel Tresider AM (Chapter 9: Gases and vapours), Kate Leahy (Chapter 9: Gases and vapours), Michelle Wakelam (Chapter 13: Radiation—ionising and non-ionising), Sarah Thornton (Chapter 16: Biological hazards), Dr Howard Morris (Chapter 17: Emerging and evolving issues) and Dr Bob Rajan OBE (Chapter 17: Emerging and evolving issues).

The editorial committee would also like to thank the many people who have supported this edition by providing images; they are acknowledged individually throughout the text. Images without source citation have been provided by the chapter author/s.

The AIOH is also grateful to the AIOH members who volunteered to peer-review chapters of the book, namely Dr Barry Chesson and Dr Geza Benke (Chapter 1), A. Prof. Susanne Tepe (Chapter 2), Prof. Jacques Oosthuizen (Chapter 3), Kerrie Burton (Chapter 4), Stephen Turner (Chapter 5), Jane Whitelaw (Chapter 6), Prof. Brian Davies (Chapter 7), Gary Rhyder (Chapter 8), Prof. Sue Reed (Chapter 9), Sally North (Chapter 10), Mitchell Thompson (Chapter 11), Dr Kelly Johnstone (Chapter 12), Russell Brown (Chapter 13), Dr Vinod Gopaldasani (Chapter 14), Adelle Liebenberg (Chapter 15) and David Hughes (Chapter 16).

Finally, the editors thank the AIOH council for supporting the project and the AIOH staff for administrative support.

Author biographies

Linda Apthorpe FAIOH COH
Linda Apthorpe has a master's degree in occupational hygiene practice. She has been working in occupational hygiene since 1994, and currently lectures in occupational hygiene at University of Wollongong. She provides consultancy and laboratory services to a wide variety of workplaces, and facilitates training for occupational hygiene students and worker groups. With her analytical background in asbestos, dust and quartz, Linda contributes volunteer technical assistance to the AIOH, National Association of Testing Authorities (NATA) and Proficiency Testing Australia. She has also authored or co-authored various papers in the occupational hygiene field.

Geza Benke FAIOH COH
Geza Benke has a BSc (Physics), MAppSc (Environmental Engineering) and PhD (Epidemiology). He worked with the Victorian Environment Protection Authority for five years before joining the Victorian state government's Occupational Hygiene branch in 1985. His work in occupational hygiene mainly involved noise assessment and control, radiation safety and asbestos work. Since 1994 Geza has undertaken research in a range of occupational and environmental epidemiology studies. He is currently Senior Research Fellow with the Centre for Occupational and Environmental Health, Department of Epidemiology and Preventive Medicine, Monash University.

Jodie Britton MAIOH COH
Jodie Britton has an MSc Occupational Hygiene Practice and a Graduate Diploma in Occupational Health and Safety. She is a Full Member of the AIOH and is also a member of the American Conference of Governmental Industrial Hygienists (ACGIH). Her background in heavy industry spans 29 years and includes 26 years in the Aluminium Smelting Industry. She has extensive practical experience within this field which complements her role as a Specialist Occupational Hygienist for Rio Tinto at the Boyne Island Aluminium Smelter. Her areas of interest include the thermal work environment and coal tar pitch volatiles (CTPV).

David Bromwich FAIOH (Retired)
David Bromwich has an honours degree in physics, master's degrees in both medical physics and occupational hygiene and a research PhD in occupational hygiene. He is a Fellow of the Australian Institute of Occupational Hygienists. He was an academic at

Griffith University for 20 years in the School of Environmental Engineering and is now an adjunct Associate Professor in the School of Medicine, but is largely retired. His interest in industrial ventilation started as a mines inspector in the Northern Territory and continued with his Occupational Hygiene Master's thesis in London and his teaching where he developed an large industrial ventilation laboratory. His research interests were chemical protective clothing, industrial ventilation and exposure visualisation.

Ron Capil FAIOH COH

Ron Capil has a background in Chemistry and a Graduate Diploma in Occupational Hygiene. He has over 45 years' experience including a decade in environmental monitoring and the remainder in occupational hygiene. His occupational hygiene work has involved workplace environment assessments in aluminium smelters and alumina refineries with a major focus on asbestos removal and management. He also provided strategic guidance as principal adviser health and hygiene to a leading Australasian aluminium industry. He is currently providing occupational hygiene consulting services to the aluminium industry.

Kate Cole MAIOH COH

Kate Cole has a BSc (Biotechnology) and master's degrees in engineering and occupational hygiene. A passionate advocate for preserving the health of those who service the construction industry, her key areas of interest are complex work environments including the remediation of contaminated land and tunnel construction. She is a Winston Churchill Fellow where she investigated world's best practice to prevent illness and disease in tunnel construction workers and was named as one of the Top 100 Women of Influence by the *Australian Financial Review* for her work in proactively addressing the issue of silica dust exposure in the construction industry. She is currently the Manager of Occupational Health and Hygiene with Ventia.

Martyn Cross MAIOH

Martyn began his career as a toxicologist working in the UK chemical industry. He has an honours degree in Toxicology, a Master's of Public Health and a PhD. Martyn has considerable experience in Occupational Hygiene, as Occupational Hygienist for WorkSafe WA, the WA Department of Health, and Principal Occupational Hygienist for Minara Resources. Martyn has 35+ years as a safety professional and occupational hygienist in a variety of industries. He is currently Senior Lecturer and Principal Supervisor, supervising Masters' and PhD students at Edith Cowan University.

Maggie Davidson MAIOH

Dr Maggie Davidson MAIOH is a lecturer and researcher in the fields of occupational hygiene and environmental health based at Western Sydney University and an Adjunct Researcher with Edith Cowan University. Maggie has been undertaking research in the field of biological hazards and bioareosols for fifteen years, completing her post doctoral research on respiratory hazards associated with working in Colorado dairies at the NIOSH High Plains Intermountain Centre for Health and Safety, and PhD on the impact of smoking bans on indoor and outdoor air quality in NSW licensed clubs at UWS. Current research projects include promoting health and safety in the emergent Australian medicinal cannabis and industrial hemp industries, worker exposure to solvents and repairable crystalline silica in the manufacture of artificial stone products, and a citizen science project investigating ambient air quality in the Blue Mountains, NSW.

Ross Di Corleto FAIOH COH

Ross Di Corleto has a BSc in applied chemistry, a graduate diploma in occupational hygiene, a master's degree by research thesis in heat stress and a PhD in occupational health in the area of polycyclic aromatic hydrocarbon exposure. He has worked for over 40 years in the areas of power generation, minerals, mining, refining and smelting. This involvement has been predominantly across the chemical, health, safety and environment fields, with occupational hygiene the main emphasis. The thermal environment and its impact on the industrial employee has been a key area of Ross's interest and involvement. He is currently the Principal Consultant at Monitor Consulting Services.

Ian Firth FAIOH COH

Ian Firth has a BSc (Hons) in zoology and MSc in applied biology (toxicology). He has over 38 years' work experience, including a decade in environmental sciences and the rest in occupational hygiene. His environmental work has principally involved research on freshwater animal toxicology, environmental management and acid rock drainage management. His occupational hygiene work has involved workplace environment assessment in hard rock mining settings and in smelting in the zinc/lead and aluminium industries, and he has provided tactical and strategic advice on health management as the corporate principal adviser of a leading international resources company. He is currently a consultant providing occupational hygiene services to a variety of clients.

Garry Gately FAIOH (Retired)

Garry Gately has a Diploma of Applied Chemistry and a Master of Chemistry specialising in analytical chemistry and held a CIH. He had over 35 years of industrial experience, of which more than 30 years was in occupational health and safety. Garry was the Corporate Hygienist and Corporate Lead SH&E Auditor for Orica (formerly ICI Australia and ICI plc) and he had extensive experience in many technologies and industrial sectors.

Robert Golec FAIOH COH

Robert Golec holds a Bachelor's degree in Applied Chemistry, a Graduate Diploma in Analytical Chemistry and a Master of Applied Science in Applied Toxicology. He has been a member of the AIOH since 1985 and is a certified occupational hygienist and Fellow of the Institute. Robert has over 36 years' experience in occupational hygiene, initially with the state government and then in a consulting capacity as Principal Occupational Hygienist with AMCOSH Pty Ltd. He is a long-standing member of the AIOH Exposure Standards Committee, a member of Standards Australia Committee CH/31 Methods for Examination of Workplace Atmospheres and a member of the National Association of Testing Authorities' (NATA) Life Sciences Accreditation Advisory Committee and an Occupational Hygiene Technical Adviser to NATA as well as a Technical Assessor for Occupational Hygiene monitoring and analysis.

Terry Gorman FAIOH COH

Terry Gorman has a Master's degree in Safety Science, is a certified occupational hygienist and has been involved in workplace safety and hygiene for over 30 years. He worked in production, safety and hygiene at the Lucas Heights reactor site for 20 years before joining 3M Australia in 1999. He is a member of the Australian/New Zealand Standards Committee responsible for respiratory protection (AS/NZS 1715 and 1716) as well as the Eye/Face Protection Standards Committee (AS/NZS 1337 and 1338). He currently represents Standards Australia on the International Standards Organisation (ISO) Committee TC94/SC15, a team of international representatives creating a set of global respiratory standards.

Beno Groothoff FAIOH COH

Beno has a background in mechanical engineering, a post graduate in occupational health and safety and a Master's in health sciences, majoring in occupational health and safety. He has been working in the fields of environmental pollution control, occupational hygiene and health and safety since 1970. In Queensland he has worked in private practice and government positions in both environmental protection and work health and safety. His occupational hygiene work has involved workplace environment audits and assessment (chemical, dust, noise, acoustic shock, human vibration, radiation, heat stress, accident investigation, etc.) in a large variety of industries in both the Netherlands and Queensland. In 2015 he was the course coordinator at QUT's Occupational Hygiene and Toxicology course, updating and presenting lectures to post graduate and master's students. He is a member of the AV10 Committee of Standards Australia on occupational noise and vibration. Beno's company, Environmental Directions Pty Ltd, provides training and occupational hygiene services to a variety of clients in industry, mining, universities and government agencies.

Jennifer Hines MAIOH COH

Jennifer Hines has a BSc with Honours, a Graduate Diploma and a Master's in Occupational Hygiene, and is currently a doctoral candidate researching emissions-based maintenance of diesel engines and its benefits to the workplace. Jennifer has worked extensively in heavy industry including alumina refining, copper refining and smelting, and underground coal mining. She is the director of EHS Solutions and is passionate about teaching and encouraging others into the world of occupational hygiene. Jennifer facilitates the Australian Department of Defence Monitoring of Occupational Hygiene Course. She is a lecturer at the University of Wollongong and has served on several AIOH committees.

Martin Jennings FAIOH COH

Martin Jennings is an occupational hygienist with nearly 40 years' experience in government and in the nuclear, defence and chemical industries. His qualifications include a BSc from Surrey University, a Master of Science (MSc) in Occupational Hygiene from the London School of Hygiene and Tropical Medicine and a further Master's degree in Administration from Griffith University. He is a CIH in the United States. He is a past president of the Australian Institute of Occupational Hygienists.

Ryan Kift

Ryan Kift has a PhD in Occupational Hygiene and undergraduate degrees in Environmental Health and Occupational Health and Environment. Ryan has over 15 years of occupational hygiene and safety experience in local government, academic and industrial settings. He has been involved in occupational hygiene management on large construction projects, and worked in the oil and gas and metals industries. He is currently a senior lecture at CQ University, Australia and a Senior Consultant for GCG Health, Safety, Hygiene. Ryan has published over 50 scientific papers in the areas of biological and dust monitoring, cognitive ergonomics and safety management.

Caroline Langley FAIOH COH

Caroline Langley has a BSc, a GDip (Occup Hygiene), an MSafetySc and was a Certified Occupational Hygienist until her recent retirement. She has over 30 years of professional experience in health, safety and occupational/environmental hygiene. She held corporate OHSE positions in research organisations, developing specialist expertise in laboratory safety, before moving to the consulting sector as a director. As a consultant Caroline focused on enhancing worker health, leading a team providing a full suite of professional occupational hygiene consulting and analytical services to clients in the mining, manufacturing, food processing, aquaculture, construction, utilities, shipping, health, government and education sectors. She also continued to provide professional advice on laboratory design as well as safety in research, teaching and production laboratories. Caroline is a past president of the Australian Institute of Occupational Hygienists (AIOH) and is a member of the AIOH Foundation Board.

So Young Lee

So Young Lee has Bachelor of Business Accounting, associate degree of Optometry and Master of Public Health (Environmental Health) in South Korea. She has over 10 years' optometry, optical companies and epidemiological research experience in South Korea and is now doing a PhD in Public Health (Adelaide Exposure Science and Health) in the University of Adelaide. Her current research is about the photochemical damage from blue light exposure in the workplace and she is also interested in the health effects of workers from exposure to intense light sources.

Elaine Lindars MAIOH COH

Elaine Lindars has a PhD in geochemistry, a post graduate in industrial hygiene and an honours degree in environmental chemistry. She is a Certified Occupational Hygienist, a Chartered Professional Member of the Safety Institute of Australia, a Chartered Chemist with the Royal Australian Chemical Institute and an adjunct lecturer with Edith Cowan University. Elaine has worked with the AIOH for over ten years and is the first president of Workplace Health Without Borders (Australia). Elaine has 30 years' 'experience in occupational, environmental and analytical chemistry, with 20 scientific papers/conference publications, and is now an owner of the OHMS Hygiene, OHMS Environment and OHMS Training consultancies.

Martin Mazereeuw

Dr Martin Mazereeuw gained his PhD in analytical chemistry at the Leiden Academic Centre for Drug Research at Leiden University in The Netherlands and has since worked in several analytical and managerial roles within biotechnology, research support and large industry. He has been a GLP study director for food safety studies and implemented the first NATA accredited ISO17025 system for research within NSW. Martin joined TestSafe Australia in 2012 as the manager of the Chemical Analysis Branch of TestSafe Australia, which is part of SafeWork NSW.

Greg O'Donnell MAIOH (Retired)

Greg O'Donnell has a PhD in chemistry and an undergraduate degree in Applied Science majoring in chemistry and is a Chartered Chemist with the Royal Australian Chemical Institute. He has been working in occupational exposure chemical analysis for over 27 years with SafeWork NSW. He has been a member of the SafeWork NSW Biological Occupational Exposure Limit (BOEL) committee for 23 years and is the current chairman. Greg is also the chairman of the Standards Australia Committee CH-31 Methods for Examination of Workplace Atmospheres. He is a member of the AIOH Exposure Standards Committee and was a member of the Safe Work Australia's Health Monitoring Review Expert Working Group. He has published 14 scientific papers and over 35 conference presentations in the field of occupational health and hygiene, with particular interest in biological monitoring. He is an honorary associate of Macquarie University.

Dino Pisaniello FAIOH COH

Dino Pisaniello has a PhD in chemistry and a Master's degree in Public Health. He is a Fellow of the Safety Institute of Australia and the Royal Australian Chemical Institute. Dino has served as president of AIOH, chairman of the Congress of Occupational Safety and Health Association Presidents (2001–5) and Australian Secretary for the International Commission on Occupational Health. He is currently Professor in Occupational and Environmental Hygiene in the School of Public Health at the University of Adelaide. Dino has published over 250 scientific papers and technical reports. His research interests include chemical exposure assessment and control, intervention research, occupational and environmental epidemiology, work and vision, heat and work injury and health and safety education.

Wayne Powys FAIOH COH

Wayne Powys completed the Graduate Examination of the Royal Society of Chemistry (UK) and is a Member of the Royal Australian Chemical Institute and a Chartered Chemist. He spent the first part of his career working in the field of analytical chemistry, principally in the water sector in the UK and later the oil and gas sector in Western Australia. After obtaining a Graduate Diploma of Occupational Hygiene, he moved into the discipline and worked for well over twenty years in the oil and gas sector in Western Australia, during which time he completed an MSc degree in Occupational Hygiene.

Sue Reed FAIOH COH

Sue Reed has a Bachelor of Science, a Master of Engineering Science, a Master of Science and a PhD in Occupational Hygiene. She is a CIH in the United States. Sue has over 40 years of occupational hygiene experience in government and defence, and as an academic. She is currently an associate professor at Edith Cowan University, Perth, and director of a small occupational hygiene consultancy. Sue has published over 50 scientific papers and technical reports on subjects including chemical exposure assessment and control, noise, indoor air quality and OHS education, specifically in relation to occupational hygiene. She is a past president of the Australian Institute of Occupational Hygienists.

Roy Schmid MAIOH COH

Roy Schmid has a Bachelor of Science and honours degree in chemistry and a Graduate Diploma of Occupational Hygiene, and is a CIH in the United States. He has over 25 years' experience in occupational health and safety within the tertiary education and research sector as the hygienist and WHS manager for the Australian National University and currently with the Department of Defence. He has an extensive knowledge of conventional and novel workplace hazards.

Michael Shepherd FAIOH COH

Michael Shepherd has a Bachelor of Science in Chemistry and a Graduate Diploma in Occupational Health and Safety, is a Chartered Chemist, a Certified Occupational Hygienist and Fellow of the Australian Institute of Occupational Hygienists. Michael has over twenty-five years' experience in multidisciplinary occupational health and safety assessments and advises government on health and safety policy and legislation. Michael's specialties are recognition, evaluation and control of workplace hazards. His main expertise includes water quality, indoor air quality, asbestos, dust, mould and hazardous chemicals. He is currently the Principal Occupational Hygienist for COHLABS Pty Ltd.

Charles Steer FAIOH COH

Charles Steer has a BAppSc (Applied Chemistry), a GDip (Environmental Studies) and completed the 13-week University of Sydney post-graduate course in occupational hygiene. He has had extensive experience in health, safety, occupational hygiene, environment, risk management and chemical technology, holding senior corporate positions in these areas in the electricity industry in South Australia until mid-2000. He has run his own consultancy since that time, mainly focused on developing sustainable occupational health and hygiene management frameworks, occupational hygiene consulting and mentoring as well as exposure assessment and hands on monitoring. He is a past president of the Australian Institute of Occupational Hygienists and is currently Chair of the AIOH Foundation.

Aleks Todorovic MAIOH

Aleks Todorovic has a Master of Science (Occupational Hygiene Practice). He has over 23 years' experience in the industry as a supplier of occupational hygiene monitoring and detection equipment, and for the last 17 years has owned and managed Active Environmental Solutions (AES). Throughout this period he has been involved in the development and testing of many new types of monitoring instrumentation, focusing mainly on chemicals in the petrochemical, defence and emergency services industries. He has also been a guest lecturer and presenter at various tertiary institutions and associations. His current direction is in the advancement of connected real-time wireless detection and monitoring solutions in the occupational hygiene and safety industries, with an emphasis on total worker exposure from simultaneous multiple stressors.

Jane Whitelaw FAIOH COH

Jane Whitelaw has a Master of Applied Science (Environmental Health) and a Post Graduate Diploma in OHS and is a Certified Occupational Hygienist, Certified Industrial Hygienist and Fellow of AIOH. She had over 25 years' experience in heavy manufacturing industries before moving into teaching and research at the University of Wollongong; where she is co-ordinator of the Occupational Hygiene program. She is a member of the AIOH Professional Development and Education committee and the AIOH representative on the AOHSEAB (Australian Occupational Health and Safety Accreditation Board). Jane's research interests are in protecting worker health from chemical and physical hazards, and her major grants and research have been in evaluating the efficacy of protective equipment. She has also authored numerous papers in the occupational hygiene field.

Abbreviations and definitions

A	ampere
A/m	ampere per metre
ABC	Australian Building Code
ABCB	Australian Building Codes Board
AC	asbestos cement
ACGIH®	American Conference of Governmental Industrial Hygienists
ACIF	Australian Construction Industry Forum
ACM	asbestos-containing materials
ACPSEM	Australasian College of Physical Scientists and Engineers in Medicine
ADI	acceptable daily intake
AFOM	Australasian Faculty of Occupational Medicine
AIHA	American Industrial Hygiene Association
AIOH	Australian Institute of Occupational Hygienists
ALARA	as low as reasonably achievable
ALARP	as low as reasonably practicable
ALI	annual limit of intake
ANSTO	Australian Nuclear Science and Technology Organisation
APVMA	Australian Pesticides and Veterinary Medicines Authority
ARPANSA	Australasian Radiation Protection and Nuclear Safety Agency
ARPS	Australian Radiation Protection Society
AS	Standards Australia
AS/NZS	Australian/New Zealand Standard
ASCC	Australian Safety and Compensation Council (now Safe Work Australia)
ASHRAE	American Society of Heating, Refrigerating and Air-conditioning Engineers
ASTM	American Society for Testing and Materials
B	magnetic flux density
BCIRA	British Cast Iron Research Association
BEI®	biological exposure index
BET	basic effective temperature
BMI	body mass index

BMRC	British Medical Research Council
BOHS	British Occupational Hygiene Society
Bq	becquerel
BSE	bovine spongiform encephalopathy
°C	degrees Celsius
C/kg	coulomb per kilogram
CCA	copper-chrome-arsenic
CCT	correlated colour temperature
cd	candela
CDC	Centers for Disease Control
CEC	Commission of European Communities
CFC	chlorofluorocarbons
CFU	colony-forming units
Ci	Curie
CJD	Creutzfeldt-Jakob disease
cm	centimetre
CNF	carbon nanofibre
CNT	carbon nanotubes
COH	Certified Occupational Hygienist (COH)® status is recognition of a professionally competent, independent and ethical practitioner and an industry leader in occupational hygiene
CoP	Code of Practice
COPD	chronic obstructive pulmonary disease
COSHH	control of substances hazardous to health
CPC	chemical protective clothing
CRI	colour rendering index
CSIRO	Commonwealth Scientific and Industrial Research Organisation
CTPV	coal tar pitch volatiles
CWP	coal workers' pneumoconiosis
D	absorbed dose
d-ALA	d-aminolaevulinic acid
dB	decibel
dB(A)	decibel measured on the A-weighting scale
dB(C)	decibel measured on the C-weighting scale
DC	direct current
DFG	Deutsche Forschungsgemeinschaft
DNA	deoxyribonucleic acid
DND	daily noise dose
DNELS	derived no-effect levels
DP	diesel particulate
DPM	diesel particulate matter
E	illuminance (Chapter 15)
E	effective dose (Section 13.2.9.3)

E	electric field strength (Section 13.3.6.3)
$E_{A,T}$	A-weighted noise exposure
EAD	equivalent aerodynamic diameter
EAV	daily exposure action
EC	elemental carbon
EC	European Commission
ED_{50}	dose causing 50 per cent of maximal effect
EDTA	ethylenediaminetetraacetic acid
E_{eff}	effective irradiance
ELF	extremely low frequency
ELV	daily exposure limit (vibration)
EM	electromagnetic
EMB	eosin methylene blue
ES	exposure standard
ETS	environmental tobacco smoke
ETSI	European Telecommunications Standards Institute
eV	electron-volt
f	frequency
FCA	flux-cored arc
FTIR	Fourier transform infrared spectrometry
g	gram
G	Gauss; 10,000G = 1 T (Chapter 13.3.5.2)
GHS	Global Harmonized System
GHz	gigahertz
GM	Geiger-Müller
GM	geometric mean
GMO	genetically manipulated organisms
GSD	geometric standard deviation
Gy	gray
H	magnetic field strength
H&S	health and safety
HAVS	hand-arm vibration syndrome
HAZOP	hazard and operability study
HDM	house dust mite
HEPA	high-efficiency particulate air filter
HID	high-intensity discharge
HIV	human immunodeficiency virus
HML	high, medium-low
HP	hypersensitivity pneumonitis
HPD	hearing protective device
HPE	hearing protective equipment
HSE	Health and Safety Executive (United Kingdom)
H_T	equivalent dose

xxii   PRINCIPLES OF OCCUPATIONAL HEALTH AND HYGIENE

HTL	hygienic threshold limits
HVAC	heating, ventilation and air-conditioning
Hz	hertz
I	current
IAEA	International Atomic Energy Agency
IAQ	indoor air quality
IARC	International Agency for Research on Cancer
ICNIRP	International Commission on Non-Ionizing Radiation Protection
ICP	inductively coupled argon plasma
ICRP	International Commission on Radiological Protection
IEC	International Electrotechnical Commission
IEQ	Indoor Environment Quality
IH	industrial hygiene
IHSTAT	Industrial Hygiene Statistical Package
ILO	International Labor Office
in vitro	in an artificial environment outside the living organism
in vivo	within a living organism
IOHA	International Occupational Hygiene Association
IOM	Institute of Occupational Medicine
ipRGC	intrinsically photosensitive retinal ganglion cells
IR	infrared
IR-A	wavelengths 780 to 1400 nm and frequencies around 10^{14} Hz
IR-B	wavelengths 1400 to 3000 nm and frequencies around 10^{14} Hz
IR-C	wavelengths 3000 nm to 1 mm and frequencies around 10^{11}–10^{14} Hz
IREQ	required clothing index
ISO	International Organization for Standardization
IVC	individually ventilated cage
J	joule
K	kelvin
kg	kilogram
kHz	kilohertz
km	kilometre
kPa	kilopascals
L	litre
L	luminance (Chapter 15)
L/min	litres per minute
LAA	laboratory animal allergy
$L_{Aeq,x}$	sound pressure level, A-weighted, over time period 'x'
LCD	liquid crystal display
LCL	lower confidence limit
$L_{c, peak}$	sound pressure level, C-weighted, peak
LC_{50}	lethal concentration for 50 per cent of specific population
LD_{50}	lethal dose for 50 per cent of specific population

LED	light-emitting diode
Leq	equivalent sound pressure level
LEV	local exhaust ventilation
LLS	laser light-scattering
lm	lumen
LOAEL	lowest observed adverse effect levels
low-E	low emissivity
LPG	liquid petroleum gas
lx	lux
m	metre
m/s	metres per second
m/s^2	metres per second squared
m^2	square metres
MA	mandelic acid
MCS	multiple chemical sensitivity
MDHS	methods for the determination of hazardous substances
MDI	methylene bisphenyl isocyanate
MEA	malt extract agar
MEC	minimum emission concentration
MEPS	minimum energy performance standards
mg/kg	milligrams per kilogram
mg/m^3	milligrams per cubic metre
mGy	milligray
MHz	megahertz
ml	millilitre (10^{-3} litre)
MMA	manual metal arc
MMI	mucous membrane irritation
MMT	methylcyclopentadienyl manganese tricarbonyl
MPP	mucus-penetrating particles
MOCA	a curing agent (4,4′-methylenebis(2-chloroaniline))
ms$^{-1.75}$	metres per second to the power of 1.75
MSDS	material safety data sheet (now called SDS—safety data sheet)
mSv	millisievert
mT	Millitesla
MVOC	microbial volatile organic compounds
MVUE	estimate of the arithmetic mean (Est. AM)
mW	milliwatt
NAL	National Acoustics Laboratories
nanoparticle	particle in size range of 0 to 100 nm
NATA	National Association of Testing Authorities
NEPM	National Environment Protection Measures
nGy	nanogray
NHMRC	National Health and Medical Research Council

NICNAS	National Industrial Chemicals Notification and Assessment Scheme
NIHL	noise-induced hearing loss
NIOSH	National Institute for Occupational Safety and Health
nm	nanometre (10^{-9} metre)
NOA	naturally occurring asbestos
NOAEL	no observed adverse effect levels
NOHSC	National Occupational Health and Safety Commission
NORA	National Occupational Research Agenda
NORM	naturally occurring radioactive material
NRR	American system of determining the expected sound attenuation of a hearing protector
NVvA	Nederlandse Vereniging voor Arbeidshygiëne
OARS	Occupational Alliance for Risk Science
OCSEH	Office of Chemical Safety and Environmental Health
ODTS	organic toxic dust syndrome
OECD	Organisation for Economic Co-operation and Development
OEL	occupational exposure limits
OGTR	Office of the Gene Technology Regulator
OHS	occupational health and safety
OP	organophosphates
OSHA	Occupational Safety and Health Administration
p	pressure
P_0	reference sound pressure
Pa	pascal
PAH	polycyclic aromatic hydrocarbon
PAPR	powered air-purifying respirator
PC	physical containment
PCBU	person in control of a business or undertaking
PGA	phenylglyoxylic acid
PHS	predicted heat strain
PLM	polarised light microscope
$PM_{2.5}$	particulate mass with a median aerodynamic equivalent diameter cut-point of 2.5 µm
PM_{10}	particulate mass with a median aerodynamic equivalent diameter cut-point of 10 µm
PMF	progressive massive fibrosis
PPE	personal protective equipment
ppm	parts per million
PTFE	polytetrafluoroethylene
PVC	polyvinyl chloride
QAP	quarantine-approved premises
R	roentgen

RCS	Reuter Centrifugal Sampler
REACH	Registration, Evaluation, Authorization and Restriction of Chemicals
REL	recommended exposure limit
RfD	reference doses
RG	risk groupings
RH	relative humidity
rms	root mean square
RPE	respiratory protective equipment
RPM	respirable particulate matter
rpm	rotations/revs per minute
S	power flux density
SA	specific absorption
SAD	seasonal affective disorder
SAR	specific absorption rate
SBS	sick building syndrome
SCBA	self-contained breathing apparatus
SCOEL	Scientific Committee on Occupational Exposure Limits
SDS	safety data sheet (previously called MSDS—material safety data sheet)
sec or s	second (time)
SEGs	similar exposed groups
SEM	scanning electron microscope
Sen	Respiratory Sensitiser
SI	International System of Units
SIMPED	Safety in Mines Research Establishment
Sk	Skin
Sk:Sen	Skin Sensitiser
SLC_{80}	sound level conversion valid for 80% of the wears
SMF	synthetic mineral fibres
SMR	standardised mortality ratio
STEL	short-term exposure limit
Sv	sievert
SWA	Safe Work Australia
T	tesla
$T^{1/2}$	radiological half-life
T_a	air temperature
TCA	trichloroacetic acid
TCE	trichloroethylene
TEM	transmission electron microscope
T_g	globe temperature
TI	tolerable intakes of chemicals
TLD	thermoluminescent dosimeter
TLV®	threshold limit value (ACGIH®)
T_{nwb}	natural wet bulb temperature

TSA	tryptic soy agar
TWA	time-weighted average
TWL	thermal work limit
UCL	upper confidence limit
UKAEA	United Kingdom Atomic Energy Association
USEPA	United States Environmental Protection Agency
U_{SG}	specific gravity of urine
UV	ultraviolet
UV-A	wavelengths 315–400 nm and frequencies around 10^{14} Hz
UV-B	wavelengths 280–315 nm and frequencies around 10^{15} Hz
UV-C	wavelengths 100–280 nm and frequencies around 10^{16} Hz
V/m	volts per metre
vCJD	variant CJD
VDV	vibration dose value
VOC	volatile organic compound
VWF	vibration white finger
W	watt (equivalent to J/sec)
WBGT	wet bulb globe temperature
WEL	workplace exposure limit
WES	workplace exposure standard
WHO	World Health Organization
WHS	workplace health and safety
w_R	radiation weighting factors
ZPP	zinc protoporphyrins
β	small spatial angle
μg	microgram (10^{-6} g)
$\mu g/g$	micrograms per gram
$\mu g/L$	micrograms per litre
$\mu g/dl$	micrograms per decilitre
$\mu g/m^3$	micrograms (10^{-6} g) per cubic metre
μGy	microgray
μm	micrometre
$\mu mol/L$	micromoles per litre
μPa	micropascal
μSv	microsievert
μT	microtesla
Φ	luminous flux
ω_R	radiation-weighting factor
ω_T	tissue-weighting factor
ν	speed

1. The hazardous work environment:

The occupational hygiene challenge

Charles Steer and Caroline Langley

1.1   INTRODUCTION

Occupational hygiene is a profession grounded in science and engineering, but also requiring good communications skills, dedicated to achieving healthy workplaces. Its focus is to eliminate or, if that is not practical, to control chemical, physical or biological hazards. These hazards include hazardous chemicals, dusts, gases, vapours, mists, smokes, fumes, fibres, noise, vibration, heat and cold, ionising and non-ionising radiation, and biological hazards such as mould, fungi, bacteria and viruses.

A key characteristic of many workplace health hazards is their slow and insidious effect on health, resulting in a toll of worker illness and death that greatly exceeds the number of traumatic workplace injuries and fatalities. This is not to understate the importance of traumatic injuries and their impact. However, there is unrecognised and significant disease in our community arising from exposure at work to largely known hazards, that we can and should control.

While it would be ideal to eliminate rather than control these hazards, it is often not practical. For example, elimination of dust when mining an ore and using chemicals for which there is no safer substitute. The examples are almost endless.

Nearly all workplace activity and the associated workplace health hazards involves some aspect of occupational hygiene practice. The preventative approach of occupational hygiene, namely interventions made to the work and work environment, rather than the worker, are responsible for a large proportion of improvements in worker health. In order to be effective, occupational hygienists need a detailed understanding of the nature of hazards in the workplace, how these hazards arise and the actual extent of workers' exposure to them. They also require a sound knowledge of the risks to health which arise from exposure to each hazard, how those risks relate to exposure levels and the means to mitigate the risks.

This chapter presents an overview of procedures for workplace health investigations using the occupational hygiene principles of anticipation, recognition, evaluation and control of workplace hazards. It also lays out a framework for considering all the topics covered in this book.

This book is aimed as a general primer for occupational hygienists as well as those in other relevant disciplines who work together to achieve a healthy workplace. Figure 1.1 sets out the range of disciplines that may be involved (adapted from Harden et al, 2015).

Although the material in this book has its origins in Australia, it is also generally applicable to workplaces in the Asia-Pacific region.

1.2   HISTORICAL BACKGROUND

It is clear that humans' efforts to feed ourselves and eventually use our environment to create and build civilisations are intrinsically connected to health hazards that arise from our activities. The first mention of occupational disease is credited to Hippocrates (460–370 BCE), who documented lead poisoning in miners and metallurgists. Pliny the Elder may have been the first to document protective equipment when he noted in around 50 CE that animal bladders were used to prevent inhalation of dust and lead fumes (Blunt et al., 2011).

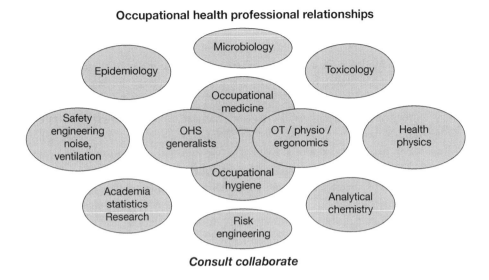

Figure 1.1 The interrelationship of occupational health and safety professions
Source: Adapted from Harden et al. (2015).

Very little occupational disease was recorded until after medieval times. Georgius Agricola described mining, smelting and refining in *De Re Metallica* (1556), including diseases and accidents, as well as the need for ventilation. Paracelsus, the medieval physician often described as the 'father of toxicology', described miners' respiratory diseases in 1567, and is said to have stated that, 'All substances are poisons . . . the right dose differentiates a poison and a remedy.' In 1700, Bernadino Ramazzini, the 'father of occupational medicine', published a book on the diseases of workers and introduced the question, 'What trade are you?'

From the late 1700s, the first Industrial Revolution accelerated workplace injuries and illness as new trades and industries such as mining and associated steam-powered factories emerged on an unprecedented scale. Awareness of workplace hazards was low or non-existent. Many who survived physical injury or escaped death in the mines became ill from dust diseases. In mines, mills and factories, workers—including children as young as six—faced injury or death from machinery and work practices designed for output, not worker health and safety. Some factory and mill owners took account of the general safety, health and welfare of their workers, but these were in the minority. Legislation in western countries became the driver for improvements to workplace health and safety but change was slow. It was not until 1833 that the first real labour laws and the Factory Inspectorate were established in the United Kingdom; this was followed by the *Coal Mines Act* in 1842.

With the development of chemical-based industries in the late nineteenth and early twentieth centuries, many new occupational diseases emerged associated with exposure to chemicals; some of them continued unchecked until relatively recent times, despite readily available evidence of the hazards. Occupational diseases commonly associated with mining and quarrying (pneumoconiosis or dust diseases), tunnelling (silicosis), fur carroting to make felt (mercury poisoning), yarn and fabric manufacture and textile weaving (byssinosis, scrotal cancer), potteries (silicosis), metal casting and finishing (silicosis and metal fume fever), chimney sweeping (scrotal cancer), matchstick manufacture ('phossy' jaw or bone necrosis from oxide of phosphorus exposure) and bridge building (caisson disease) became accepted as ordinary risks of these jobs. While many of these industries and issues have disappeared, some of these diseases remain—such as silicosis, pneumoconiosis and metal fume fever even though the causes and controls are well known.

While Georgius Agricola, Bernadino Ramazzini and Charles Thackrah made astute observations about the importance of understanding a worker's job or trade so as to assess the impacts on health, occupational hygiene as we know it made no real advances until World War I. Pioneering work done by Dr Alice Hamilton (1943) and others in the early twentieth century brought industrial disease—particularly in the war industries such as munitions—to the attention of legislators, employers and workers in the United States and Britain.

In Australia, factory and mining work health and safety laws followed the UK example, but were not enacted by the states until the late nineteenth and early twentieth centuries. Australian social reform, including an emerging labour movement and a federal arbitration system established in 1905, further improved protections for workers, including children (Damousi, Rubenstein and Tomsic, 2014).

Like its UK predecessors, early Australian legislation was industry based, with targeted prescriptive regulations to control specific hazards. These regulations set out complex requirements that were difficult to comply with and that were, by their nature, incomplete. Furthermore, with the rapid advancement of industry, including chemical manufacturing and defence, health and safety legislation was always lagging in terms of technical, medical and epidemiological knowledge.

Occupational health and safety (OHS) legislation fundamentally changed in the 1970s and 1980s in Australia, as so called Robens-style regulation took over, with its central tenets of overarching legislation featuring self-regulation and the statutory duty of employers to consult employees and to ensure their health and safety at work (Australian National University, REGNET, 2017). This approach marked the beginning of codes of practice and specific sections in the regulations to provide guidance. Australia has been moving towards consistent harmonised OHS legislation between all states and territories, but this has not yet been fully achieved. The harmonised legislation sets out clear responsibilities and requirements for employers to eliminate risks to health and safety so far as is reasonably practicable and, if not reasonably practicable, to minimise those risks so far as is reasonably practicable. The legislation is backed up by extensive codes of practice and other guidance material that is freely available. This rather legalistic approach effectively enshrines the basic human right of being safe and healthy at work.

The profession we now know as occupational hygiene (also called industrial hygiene in the United States, Malaysia, Indonesia and other countries) emerged after World War II in the United States and the United Kingdom. Scientists and engineers from government and industry began working and collaborating in this field in Australia in the 1960s, with the Australian Institute of Occupational Hygienists, Inc. (AIOH) being formed in 1980.

The AIOH website, <www.aioh.org.au>, provides details of the professional membership grades, activities such as conferences and training and university courses accredited by the AIOH. Courses at various levels are also available online through <www.OHLearning.com>.

1.3 THE PRESENT

In Australia, there is a sound, although not yet fully consistent, OHS legislative framework, with associated federal and state/territory regulators, a well-educated workforce and a cohort of well-trained OHS professionals including occupational hygienists. However, there are continuing challenges, as evidenced by the high rates of occupational disease outlined below in Section 1.4. Many of the 'older hazards' have disappeared, but many remain, along with new occupational health challenges. The following sets out the extensive range of health hazards that can be experienced in modern workplaces:

- *hazardous chemicals* (Chapter 2), including:
  - carcinogens such as arsenic and benzene
  - substances that are toxic to organs or organ systems, such as toluene, mercury, glutaraldehyde or ethanol
  - reproductive and teratogenic substances, such as lead
  - corrosive substances, whether acidic or alkaline, such as acid gases
  - sensitising agents, such as nickel, isocyanates and glutaraldehyde, which can affect the skin, respiratory system and gut
  - irritants such as ammonia and ozone.
- *aerosols* (Chapter 7), including:
  - airborne liquids and solid particulates, such as smokes, fumes, dusts, mists and fibres
  - mining and quarrying, construction and manufacturing dusts, such as coal, metal and wood dusts, respirable crystalline silica
  - vehicle exhaust emissions, including diesel particulate matter (DPM)
  - biologically active particulates in pharmaceutical manufacturing, bioaerosols from cutting fluids
  - nanoparticles in pharmaceutical and other manufacturing processes
  - acid fumes from electroplating and anodising in manufacturing and defence-related industries
  - metal fumes from welding, smelting, refining, soldering and cutting (Chapter 8)
  - acid mists from electro-refining
  - hydrocarbon mists from spray painting
  - fibres from manufacture and handling of human-made fibre products, asbestos removal, cotton and paper manufacture
  - pesticides, including insecticides and herbicides, and grain dusts found in agriculture
  - airborne contaminants from chemical manufacture, process plants, mining, construction and emergency services.
- *gases and vapours* (Chapter 9) from almost all work settings, including:
  - oxides of nitrogen, carbon dioxide, carbon monoxide, cyanides from internal combustion engines in mining, logistics and construction
  - ozone from welding
  - hydrocarbons from petroleum refining and storage, solvent degreasing
  - anaesthetic gases and other biologically active vapours in medical and veterinary settings

- hazardous and irritant chemicals such as formaldehyde from building materials in offices
- asphyxiants such as nitrogen, carbon monoxide and refrigerants.
- *noise and vibration* (Chapter 12) from almost all industries, including manufacturing, agriculture, defence, construction, mining, tunnelling and logistics.
- *thermal (heat or cold) stress* (Chapter 14) arising from outdoor and cold store work
- *biological hazards* (Chapter 16), including:
  - fungi and mould in the built environment, particularly in warm humid environments
  - biological hazards, moulds and fungi in agriculture
  - bacteria such as *Legionella pneumophila* in excess concentrations in potable water and cooling towers
  - zoonoses, including glanders, anthrax, Q fever, bird flu, lyssavirus, Hendra virus and so on in animal husbandry
  - biological hazards in health and medical settings, including accidental contact with blood and body fluids.
- *potable water quality*, where sites provide the drinking water supply, such as remote mine sites; this includes bacterial, chemical and physical quality (such as turbidity) (NHMRC and NRMMC, 2018; NHMRC, 2011).
- *ionising and non-ionising radiation* (Chapter 13) in a wide variety of industries, including mining, power, manufacturing, medical diagnostics, research and defence.
- *ultraviolet radiation* exposure is a hazard for all outdoor workers as well as from welding processes and specialist equipment.

Occupational hazards such as those listed above can exist separately or together in a range of industries and workplaces, including construction, manufacturing, defence, research, health, mining, agriculture and even office or retail environments. Although the body of knowledge on these hazards and their control is now extensive, some hazards are underestimated, some are ignored and yet others have re-emerged due to complacency. A salient example is the control of coal mine dust: the prevalence of coal workers' pneumoconiosis was as high as 27 per cent in Australia before World War II (Moore and Badham, 1931) and 16 per cent in 1948 (Glick, 1968). Unfortunately, there have been recent recurrences of pneumoconiosis in Queensland (Parliament of Australia, 2017) as well as in the United States (Blackley, Halldin and Laney, 2018), demonstrating the importance of maintaining controls and understanding the hazard.

The recent re-emergence of silicosis from working with manufactured (artificial or composite) stone kitchen benches is an example of a long-standing hazard not being recognised in a new industry (Thoracic Society of Australia and New Zealand, 2017).

A further complication is the proliferation of small workplaces, including the so-called 'gig economy', or piecemeal work where there is no union or organisational support or accumulated knowledge, almost certainly resulting in vulnerable workers. Vulnerability increases where the workforce includes migrants, disabled workers, non-English speakers and informal workers.

The challenge is significant for occupational hygienists and other professionals to provide good guidance, especially for hazards that cannot be seen, heard, felt or smelt, and that can produce health effects that may not become apparent for many years.

1.4 THE SERIOUS PROBLEM OF UNDER-ESTIMATION OF OCCUPATIONALLY RELATED DISEASE

Is work-related ill-health important enough to deserve the attention it now receives? The last 30 years in Australia have witnessed an expansion of occupational health laws and regulations, with greater government administration and new inspectorates. There are many technical guides, numerous training programs and expanded legal services. The numbers of OHS professionals and research programs are growing. But are workplace hazards—from chemicals to conditions—really detrimental to our health? After all, there are relatively few ill people in any given workplace and our workers' compensation systems report relatively few cases of occupational disease. When a disaster like a rail crash or an explosion occurs, it is easy to count fatalities. In contrast, occupational disease resulting from exposure to physical, chemical or biological agents may take years or decades to develop, so its causes are often not immediately apparent. Many similar illnesses are also the result of lifestyle or social conditions. Workers move away or change jobs or retire. A sick worker's doctor may simply be unaware of the kind of work that their patient does or has done in the past, and asking about a patient's occupation is probably unusual in typical doctor–patient interactions unless the doctor has some training in occupational health. Historical records of occupational exposures are rare, although new laws now mandate record-keeping in some cases.

Reliable data on the contributions of workplace conditions to ill-health in the community have traditionally been difficult to assemble, and this is a worldwide problem. Data on compensation for work-related ill-health is lacking, leading to a misleading impression of the true prevalence of occupational disease. Most work-related ill-health is the result not of accidents (falls, high-energy impacts, crushing or piercing injuries, etc.), but of exposure to hazardous chemicals or environmental conditions. Consider the following examples where the findings of epidemiological studies—sometimes conducted years after exposure first commenced—confirmed the need for the controls now widely demanded by law:

- *The world's worst single-event industrial disaster (with the probable exception of Chernobyl).* At Bhopal, India in 1984, the inadvertent release of methyl isocyanate gas (an intermediate of manufacturing a pesticide) killed more than 3000 people and injured some 17,000 more who lived in the chemical factory environs. Although the cause of the disaster was soon evident, its true scale was not so immediately obvious.
- *The United States' worst individual industrial accident.* Hawk's Nest Tunnel in West Virginia, built to divert a river in the early 1930s, required drilling through silica rock. As a result of inhaling the resulting dust, more than 600 men died from silicosis within two to six years. Neither the cause of the illness nor its true prevalence were obvious at the time of the work.
- *Australia's worst industrial accident.* Mining of blue asbestos at Wittenoom, Western Australia continued from 1937 to 1966. At that time, the link between asbestos fibre exposure and asbestosis, lung cancer and mesothelioma was not well defined. Workers, town residents and visitors developed these diseases over the ensuing decades and deaths are still occurring. The death toll could eventually exceed 2000. Worldwide, the death toll from asbestos exposure in the late twentieth and early twenty-first centuries could possibly reach into the millions.

A further disturbing aspect of asbestos-related disease is that the number of new meso-thelioma cases continues to be high. As the first wave of deaths from exposure during mining and processing prior to 1983 declines, a second wave of disease has emerged in tradespeople and workers in asbestos buildings (ASEA, 2016). Although these exposures were much lower, far more people were exposed, resulting in continuing high numbers of mesothelioma cases and deaths. Asbestos is covered in more detail in Chapter 7.

These examples demonstrate not only that exposure to hazardous substances can cause severe ill-health or death, but that the true impact of this exposure can take years or decades to become apparent.

In the 1980s, the US National Institute for Occupational Safety and Health (NIOSH) identified the 'top ten' workplace injuries and illnesses (US Department of Health and Human Services, 1985). The list included occupational cancers, dust diseases, musculo-skeletal injuries and hearing loss, and a number of other diseases still with us today (Safe Work Australia, 2017).

It is generally acknowledged (Schulte, 2005) that the burden of occupational disease is significant but under-estimated as the recording of long latency diseases is difficult and there may well be co-contributing causes.

The latest estimates of the global burden of occupational disease indicate that a sig-nificant challenge exists. In 2017, the Singapore Workplace Safety and Health Institute (Hämäläinen, Takala and Kiat, 2017) reported an estimated 2.78 million global deaths each year that could be attributed to work. This is approximately 5 per cent of total global mortalities and higher than the 2.33 million occupational fatalities reported in 2014. Contrary to the focus of effort by many workplace regulators, the institute states that 86 per cent of these work-related deaths are caused by occupational disease, compared with approximately 14 per cent of fatalities that were directly related to accidents. Globally, the most significant occupational diseases are respiratory disease, including chronic obstructive pulmonary disease, circulatory diseases and malignant neoplasms.

In the absence of any published Australian estimates and using the ratio of around ten fatal illnesses to one fatal injury, as has been derived in the world, US and NZ estimates as shown in Table 1.1, this results in an estimate of 1820 Australian illness fatalities for the year 2016 (Safe Work Australia, 2017). Although this result is an estimate, it serves to illustrate the size of the challenge.

An obvious question to consider is whether injury and illness rates will decrease in future as workplaces become better controlled.

On the positive side:

- Australia's injury fatality rates have been falling—from 259 in 2003 to 185 in 2017 (Safe Work Australia, 2017)—suggesting improved management commitment, awareness, design and operational improvements that are likely flowing to health hazard control.
- Australia is 'exporting' more dangerous work—for example, in manufacturing and processing industries. This is a positive for injury and illness statistics in Australia; however, it may be compounding problems elsewhere and could be deeply misleading in terms of human suffering.
- Legislation has improved (for example, banning asbestos, the use of quartz in grit blasting) along with more comprehensive codes of practice and consequent awareness.

Table 1.1 Summary of estimated fatalities by illness and accident in various jurisdictions

Jurisdiction	Estimated annual illness and accident fatalities	Ratio of fatalities from illness compared with injury
Worldwide	2.4 million deaths linked to work-related illnesses 0.38 million fatal injuries (Hämäläinen, Takala and Kiat, 2017)	6
United States	26,000 to 72,000 linked to work-related illnesses 6200 fatal injuries (Steenland et al., 2003); estimated annual cost US$128–155 billion	9
Great Britain	13,000 deaths related to work-related illness primarily linked to chemicals and dust 144 fatal injuries (HSE, 2016/17); estimated cost £9.7 billion for illness and £5.3 billion for injury	99
New Zealand	600–900 deaths related to work-related illness (WorkSafe NZ 2017) 73 fatal injuries (WorkSafe NZ 2017)	Approximately 11

- Smoking rates have declined, decreasing the chance of synergistic health effects.
- Engineering control technology has improved—for example, large-scale dust and vapour control, noise control, hand-held tools with local exhaust ventilation.
- Environmental legislation has become more stringent, coincidentally causing improvements to worker health and safety—for example, vehicle emission standards, environmental dust and noise requirements.

On the negative side:

- The so-called 'gig' economy is growing, along with increasing numbers of small to very small businesses that have limited resources and more workers in multiple transient jobs.
- New advances in technology such as nanomaterials may be outpacing health research.
- There have been reductions in the numbers of specialists employed by regulators, including occupational hygienists, occupational physicians and occupational health nurses.
- There are large areas of industry that have not experienced significant exposure to occupational hygiene principles and application, such as construction, manufacturing, agriculture and health care. This is now changing in some sectors—for example, in tunnelling, where high-level occupational hygiene expertise is starting to be applied. This is, however, by no means universal. In 2017, Australia had approximately

1.17 million persons employed in construction, 885,000 in manufacturing, 325,000 in agriculture, forestry and fishing and 1.66 million in health care and social assistance. This compares with mining, which has a relatively mature occupational hygiene culture and employed around 216,000 persons (Parliament of Australia, 2018).

- It is estimated that around 40 per cent of those in the Australian workforce are still exposed to carcinogens in their current job roles (Carey et al., 2014; Carey et al., 2017).

Future challenges include:

- The need to keep manufacturing facilities and process plants up to date to reduce noise, chemical and other exposures.
- The need for companies to ensure that lessons learnt from older facilities are applied when building new ones, including retaining a 'corporate memory'. It is also important that company return on investment considerations are made in the light of cost-effective health and safety options. Occupational hygienist need to argue for the overall benefit of higher order control measures—for example, design and engineering solutions for noise rather than supply of ear plugs.

The limitations of current workers' compensation data systems for recording and estimating occupational illness are recognised. Work-related illness is a continuing burden, with appreciable cost-shifting of long latency disease into the public health system in Australia and other countries.

1.5   THE OCCUPATIONAL EXPOSURE RISK-MANAGEMENT PROCESS

The classic occupational hygiene process involves:

- *anticipation* of problems—a vital skill; while this usually requires considerable experience, assistance is now provided by abundant resources, including safety data sheets (SDS), various databases and websites
- *recognition*—knowing the hazards and the processes by which they may affect health or identifying them through adverse health effects
- *evaluation*—measuring exposures, comparing against standards, assessing risk
- *control*—providing contaminant or hazard control. The level of protection required is based on knowledge of the toxic or other adverse effects produced by known quantitative exposures to the hazard.

These steps may need to be added to in more complex workplaces, as depicted in Figure 1.2, where it is important to characterise the hazard and, if required, carry out monitoring. Once controls are in place, it is essential to carry out regular reviews of their continuing effectiveness.

Figure 1.3 provides more detail on the major risk assessment components. These are explained in the following sections. It is important to note that this section is a summary only. The reader is encouraged to refer to the references as well as the later chapters in this book to obtain a more detailed understanding.

Risk management process – *Extra steps*

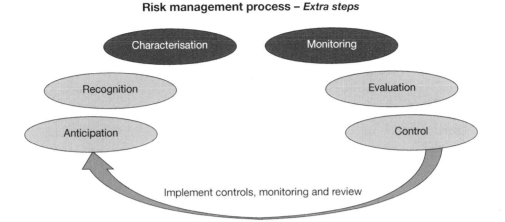

Figure 1.2 Additional steps in the risk-management process

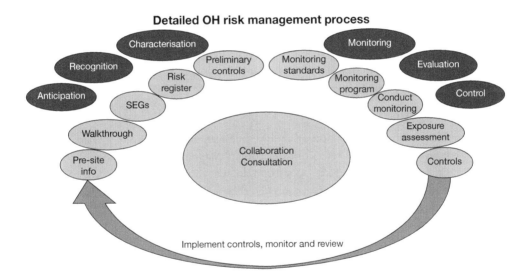

Figure 1.3 Detailed components of the occupational health risk-management process

1.5.1   *ANTICIPATION*

Anticipation involves identifying potential hazards before they are introduced into a workplace or before you visit the workplace. This can be undertaken using a HAZOP process, which would include collection and review of the following information:

- site plans and where specific chemicals are handled
- employee numbers and organisation
- site-specific legislation and regulations, including exposure standards
- any previous monitoring and analysis

- injury and illness records, taking into account the long-term nature of the health effects of some chemicals and physical agents, which may include any patterns of adverse health effects
- incident reports
- company standards and procedures
- SDS for hazardous chemicals. All SDS in Australia are now required to be in the Globally Harmonised System (GHS) format.

There are many work situations where an SDS is not available but the nature of the hazard(s) is well known—for example, asbestos exposure in building maintenance, wood dust exposure in small joinery workshops. In these cases, alternative information sources need to be used, such as codes of practice, guidelines and research publications.

There are other situations where the health effects are not so well known and further research will be required, possibly including using external specialist expertise.

1.5.2   RECOGNITION

Recognition of health hazards in the workplace is fundamental to their effective control. There are two major components in this step:

- Understanding the health effects and characteristics of the chemical, physical or biological agent.
- Being aware that the agent may hurt the worker—for example, by skin contact.

Key chemical, physical or biological characteristics include:

- The inherent toxicity or other health effects of the various physical chemical and biological agents, as set out in subsequent chapters.
- Potential exposure routes—for example, whole body, inhalation, ingestion or absorption through the skin, eyes or aurally; these are set out in subsequent chapters covering specific health-affecting agents.
- Whether the effects are acute or chronic.

With regard to chemicals and some biological agents it could include whether they are sensitisers, irritants, asphyxiants, target organ affecting, carcinogenic, corrosive to the skin or eyes, gut or respiratory tract, and whether they may react with other chemicals to produce new hazards. The physical form of the agents is also important—for example, are they aerosols, solids, liquid lipophilic or lipophobic, acidic or alkaline?

Recognition of hazards may be complicated in the following ways:

- Many health hazards cannot be seen, heard, smelt or felt.
- Some hazards, such as non-ionising radiation, are not well understood by workers or employers.
- A chemical may have no adequate warning properties such as odour.
- Workers do not know what their exposure is.
- Workers and employers may accept exposures as an unavoidable part of the job.

A *site walk-through* (or walk-through inspection or survey) is a critical step in any assessment. It is the opportunity to integrate the knowledge of the hazard with the way it is used or experienced in the workplace to gain insights into how workers can be affected.

Exposures can occur from handling, processing, transporting, packaging and storage of raw materials to by-products, intermediates, waste and finished materials. Exposure may be routine, occur intermittently (e.g. during a shutdown), be due to a non-standard operation or occur only in an emergency.

Industries that typically give rise to occupational health hazards are widespread (US National Library of Medicine, 2019). They include major employment sectors in Australia and Asia, such as the following:

- *Agriculture, fishing and farming.* These sectors involve a wide variety of planting, harvesting, crop and livestock management practices that use or produce hazardous dusts, pesticides and herbicides, and microbiological hazards associated with moulds and fungi. Dangerous gases may arise in silage pits, silos and enclosed feed sheds. Ultraviolet radiation, along with thermal stress, tool and vehicle vibration and noise, may require controls. In animal husbandry, parasites and other zoonotic agents may be present, as well as immunological sensitisers. Fin and shellfish processing involves exposure to aerosolised proteins that are immunological sensitisers. There is also extensive wet work and noise.
- *Food processing and packaging industries.* These may expose workers to zoonotic agents, wet work, toxic, corrosive, irritant or sensitising chemicals, noise and vibration, organic dusts and thermal stress. Food flavours, colourants and additives used in concentrated form in large quantities may be irritants or sensitising agents.
- *Health industries.* These industries, including hospitals, may produce a wide range of hazards, including cytotoxic, sensitising and corrosive chemicals, equipment generating ionising radiation, unsealed radioactive materials and noise. Anaesthetic administration in hospitals and veterinary clinics may pose a risk.
- *Building and construction.* This industry involves potential exposure to dusts, including crystalline silica, wood dust, asbestos and human-made mineral fibres, metal fumes, paints and coatings, solvents and adhesives, cements and fillers, noise and vibration.
- *Chemical manufacturing.* Harmful or toxic gases, liquids and solids form the basis of most areas of this industry, such as oil refining, plastics manufacturing and cyanide production. Gas reactions and gassing procedures in the production of petrochemicals, the synthesis of plastics, rubbers and fuels, and catalysis, fumigation and sterilising have the potential to give rise to serious inhalation hazards.
- *Heavy and light manufacturing and processes.* Such processes, which include welding, soldering and thermal metal cutting, produce a range of potentially toxic metal fumes, and/or irritant, oxidising and asphyxiating gases from welded material, welding rods or fluxes. Industrial plating processes pose risks of skin damage, respiratory system damage from corrosive aerosols and systemic poisoning from toxic gases and aerosols. Processes involving the manufacture, bagging and pouring of powders and dry materials produce fine dust. Industrial processes using electromagnetic operations, such as large direct current metal smelters, radiofrequency induction furnaces, microwave heaters, x-ray equipment and radar signal generators, can all produce biological

effects unless equipment is properly shielded. Radio and microwave transmission towers produce significant risks in near-field radiation zones. Smelting and hot metal handling generate metal fumes, gases and vapours from decomposition of moulds, heat and radiant energy, noise and light. Abrasive blasting typically uses steel shot, garnet, smelter (e.g. copper) slag, staurolite and heavy mineral sands (ilmenite), and produces much fine dust and very high noise levels. Vapour degreasing is a widely used industrial process posing hazards from vapour inhalation and skin and eye contact. Reinforced plastics usage may create exposures arising from various resins and human-made mineral fibres to potent skin and respiratory irritant and sensitising agents, corrosive and toxic substances. Painting and coating cover a wide number of processes using organic and inorganic powder suspensions and solutions to spray, brush and roller-coat, creating exposure to aerosols of organic materials as well as a range of solvent vapours, toxic metals, severe irritants and respiratory sensitisers. Drying ovens are widely used in manufacturing, art and craft, and industrial surface coating. Ovens produce vapours from solvents, lacquers, paints, cleaning agents and plastic resins, as well as combustion gases from fuels. Nanotechnology engineering leads to exposure to superfine particles in a wide array of processes involved in the manufacturing of products from cosmetics and fabrics to paints.

- *Mining and quarrying.* Processes in these industries may create exposures to dusts, radioactive gases, vehicle emissions, including diesel exhaust emissions, noise, vibration, risks of asphyxiation, poisoning by various gases, and injury or death from explosions.
- *Transport and logistics.* These operations, including road and rail, shipping, aircraft and warehousing, involve exposures from the materials and goods being handled, as well as dusts, vapours and solvents, vehicle exhaust, noise and vibration.
- *Laboratories.* These are locations where exposure to toxic, irritant, sensitising, corrosive or mutagenic chemicals may occur in novel operations. Physical hazards arising from equipment generating ionising and non-ionising radiation and lasers are common. Unsealed radioactive materials and biological agents may also be in use.
- *Enclosed buildings and the built environment.* Such environments, including office and retail environments, may produce hazards from volatile organic compounds arising from office equipment such as printers and photocopiers, off-gassing by building products, carbon dioxide build-up from poorly ventilated buildings and biological hazards (moulds and fungi).

Information gathered during a walk through survey could include:

- detailed site and process information and any problems (e.g. symptoms) experienced, including discussions with management, supervisors and workers. Workers may often have specific knowledge of the idiosyncrasies of machinery (and operators), processes and how well the process works or doesn't work, and non-routine operations.
- the materials used or handled and any difficulties experienced—again by specific questions
- the type and extent of worker training
- the number of workers involved and shift arrangements

- evidence of reactions, any material transformations (generated substances), by-products, intermediates and wastes. The work process itself may generate a hazard from an apparently innocuous precursor material.
- exposure times of directly involved employees as well as bystanders
- existing engineering controls that are in place, and whether they work as expected
- housekeeping at the site (e.g. spillages and cleaning methods)
- visible conditions at the site (any dusts, mists, smoke, fume, odours, deposits of material)
- possible routes of entry (inhalation, skin, ingestion, injection)
- personal protective equipment (PPE) availability, use and maintenance
- processes that expose workers to noise, vibration, heat, cold or ionising or non-ionising radiation
- processes that may contact and affect the skin.

1.5.3 PRELIMINARY CONTROLS

This is an optional step, depending on circumstances. A walk-through survey or a preliminary medical examination may identify urgent issues that require immediate controls. In these circumstances, it is important not to wait until a formal process is completed before acting to improve worker health protection.

1.5.4 CHARACTERISATION

This process involves characterising the workplace, workforce and exposures to health-affecting agents in the workplace (AIHA, 2018; Jahn, Bullock and Ignacio, 2015). Steps include:

- Use information gathered in the earlier stages to develop similar exposure groups (SEGs). An SEG is defined as a group of employees (two or more) similarly exposed to health-affecting agents. SEG development will required consultation with supervisory and workplace representatives. Note that workers may be exposed to more than one hazard at a time.
- Next, develop a site-specific risk register (Firth, Van Zanton and Tiernan, 2006) where the health risks of physical, chemical or biological agents are assessed for individual SEGs. A risk register will include existing controls, calculated inherent and residual risks and proposed controls. It provides a summary of the status of occupational health management at any workplace, and can be used to further develop action plans for additional controls and provide input into determining any medical assessment requirements. Members of the workforce, supervisory personnel and subject-matter experts need to be consulted during the development of the risk register.

1.5.5   MONITORING

Monitoring is a key aspect of the practice of occupational hygiene, and is covered in some detail in subsequent chapters. The following is a summary of the ways in which monitoring fits into the occupational exposure risk-management process.

- Once SEGs have been selected and a risk register has been developed, it is necessary to determine whether a more detailed exposure assessment is required. The main basis for this assessment is often workplace monitoring.
- The key criteria for monitoring is whether the exposure is likely to exceed an exposure standard or whether there is uncertainty about whether it is likely to affect health. Provisos include practicality and whether an appropriate occupational exposure limit (OEL) exists for the chemical, physical or biological agent.
- An optimal occupational hygiene monitoring program will provide statistically valid monitoring, where appropriate, of health-affecting agents (Grantham and Firth, 2013; Liedel, Busch and Lynch, 1977). The general requirements are sufficient sample numbers, representative sampling times within a shift and random sampling dates over a sufficient time period that takes account of seasonal variations, as well as day, night and weekend shifts as appropriate. Some monitoring may be programmed—for example, campaign monitoring of specific agents during a plant shutdown or for batch processes. There are various options in relation to monitoring equipment and techniques, including passive or active monitoring, direct reading instruments and so on (Chapters 7 and 9).
- It is generally preferable to carry out personal monitoring, but monitoring can include area, surface, biological and specific physical and biological agents, as set out in Chapter 3 as well as Chapters 7 to 16. There is a so-called hierarchy of exposure criteria (Laszcz-Davis, Maier and Perkins, 2014; Jahn, Bullock and Ignacio, 2015), with their foundation ranging from quantitative to more qualitative (see section 3.11.1: Control banding in Chapter 3).
- Exposure standards may need to be adjusted according to shift length and other factors (see AIOH, 2016; Standards Australia, 2005). Attention should also be paid to additive exposures, synergism, potentiation and ototoxic effects (Chapters 3 and 12).
- Monitoring must take account of the precision and accuracy of the sampling and analytical method if applicable. In all cases, monitoring equipment must be appropriately calibrated and maintained. Monitoring is of no value unless the results can be assured and potential errors understood.
- Wherever possible, analysis of samples must be undertaken by a laboratory accredited for the test(s) being conducted, and reported in accordance with recognised standards.
- Monitoring needs to be designed and carried out by appropriately qualified and resourced personnel with full support from site management. Expertise will depend on the agents monitored, and includes occupational hygienists for design and monitoring; it may also include analytical chemists, microbiologists, ergonomists and acoustic engineers.

- All monitoring results should be reported and made available to the workforce except where inappropriate for privacy reasons (e.g. personal biological monitoring). These results should be reported to the individual only.
- A key document summarising the monitoring program will be a SEG matrix listing health-affecting agents; a partial example is set out in Table 1.2 for a mine.

Table 1.2 Example of part of a SEG matrix

| | | | | Number of samples / measurements required | | | | | |
SEG number	SEG name	Typical tasks	Number of employees in SEG	Inhalable dust	Respirable crystalline silica	Welding fume	Thermal stress	Noise	Legionella/potable water
1	Dragline	Conduct dragline operations – operate auxiliary equipment etc.	14	11	11	Not measured	Targeted summer	8	Potable on plant Legionella in workshop
3	Field maintenance	Field maintenance of mobile plant	55	18	18	9 (opportunity basis)	Targeted summer	12	

In some circumstances, alternatives to workplace monitoring may be used, including:

- mathematical modelling (Keil and Simmons, 2009), ranging from simple models to computational fluid dynamics. This can be useful for situations including:
  - estimates during the design stage of a process or review of engineering design options for a control
  - scenario analysis
  - very large sites with many variations across processes resulting in excessive demands on monitoring and analytical resources
  - where convenient methods for monitoring are not available.
- control banding, which moves directly to the evaluation stage when monitoring data and/or occupational exposure standards may not be available (Chapter 3).

1.5.6   EVALUATION

Having measured an exposure, it is necessary to evaluate the risk associated with the hazard. This evaluation step is critical for answering the following questions:

- Is the particular risk from exposure acceptable?
- Does its existing level meet regulatory requirements?
- Will it need controlling to make it healthy and safe?
- Are there special controls for this hazard?
- How much control is needed?
- What is the most effective control mechanism for this risk?

There are, of course, many situations where evaluation will show that no further action is needed; however, a more formal and systematic approach is often required to evaluate a measured exposure.

Risks can be evaluated using a variety of methods, depending on circumstances. In the case of toxic chemicals, there are two components in the evaluation: the exposure profile and toxicity. The resulting health risk can be described as a combination of exposure and toxicity (Jahn, Bullock and Ignacio, 2015). This can be developed further to produce risk matrices to provide a qualitative risk estimate (ICMM, 2016; Jahn, Bullock and Ignacio, 2015) (see Chapter 7). These matrices are generally based on likelihood (exposure) and consequence (from toxicity). It is this integration of the exposure profile with the toxicity or other health effect that provides the overall risk. The resultant risk can then be used to revise the risk assessment in the risk register.

A range of techniques can be used as part of the risk evaluation or legal compliance assessment processes:

- If an OEL exists and statistically valid monitoring has been carried out, then formal exposure assessment methodologies can be used. These are most often used for some chemicals, aerosols, gases, vapours and noise, and are set out in various publications (BOHS and NVvA, 2011; Grantham and Firth, 2013; Jahn, Bullock and Ignacio, 2015). Techniques used include Bayesian statistics (Hewett et al., 2006) for smaller sample sizes as well as tools such as the IHStat program (AIHA, 2014). The measured exposure can then be used to provide guidance on compliance with legislation (Grantham and Firth, 2013, Chapter 3) as well as for further monitoring and controls (Jahn, Bullock and Ignacio, 2015).
- Specific risk assessment methodologies may also be used for biological monitoring, heat, vibration and radiation as set out in the relevant chapters of this book. More qualitative techniques will need to be used for monitoring such as surface wipes.
- For more straightforward workplaces simplified risk management strategies may be used (Firth et al, 2006).
- Mathematical modelling can also be used to generate an exposure profile.
- Once the exposure assessment has been completed, it needs to be considered in the context of the likelihood of the consequence—that is, the health effect. The interplay between these two components, shown using a risk matrix, will result in an assessed risk rating ranging from low to high (or catastrophic). The interpretation of the

implications of the risk rating will often require professional judgement in terms of an appreciation of the challenge of managing that particular workplace exposure. For example, one scenario may be a likely exposure (at the OEL) to an irritant such as ammonia compared with an exposure at less than 10 per cent of the occupational exposure limit of a probable human carcinogen. The risk rating may be the same according to the health risk matrix, but its management—including risk communication—will be quite different. A typical 5x5 health risk matrix is set out in Table 1.3.

Table 1.3 Example of a health risk matrix

	Consequence				
Likelihood	**1 Minor**	**2 Medium**	**3 Serious**	**4 Major**	**5 Catastrophic**
A Almost certain	Moderate	High	Critical	Critical	Critical
B Likely	Moderate	High	High	Critical	Critical
C Possible	Low	Moderate	High	High	Critical
D Unlikely	Low	Low	Moderate	High	High
E Rare	Low	Low	Moderate	Moderate	High

Likelihood or probability may be described in the context of health, as exposure to a ratio of the OEL – for example, 'Almost certain—frequent daily exposure at 10 × OEL'. Consequences may be described in the context of a health outcome—for example, 'Serious—severe reversible health effects of concern that typically would result in a lost-time illness; can include acute/short-term effects associated with extreme temperature or some infectious diseases'. Organisations may have varying thresholds and definitions for each of the categories of consequence and likelihood in the context of health.

Together, the descriptions of consequence and likelihood are used to describe the possible outcome if the event occurred. This in turn provides the risk rating from 'low' risk through to 'critical' risk. The risk rating will determine the action required, timelines for this action, management accountability and so on. For example, 'low' risk may be managed in an organisation by the adoption of routine procedures, whereas 'critical' risk would require the attention of the chief executive officer along with implementation of a detailed plan of action to reduce the risk to an acceptable level.

- The risk evaluation will inform the controls to be used, as set out in the next section, and will also signal other processes such as updating the risk register.
- In circumstances where the OEL is not known, control or hazard banding (NIOSH, 2018) can be used. This technique has a much higher degree of uncertainty, and exhibits user and model variability (Van Tongeren, 2014); however, it may be the only practical way to carry out an assessment. The outcome in this methodology is not a number; rather, it moves to the next stage of guidance on controls.

1.5.7   CONTROLS

Control remains the least understood and often the most poorly implemented component of the risk management process. Establishing controls is, quite simply, how to protect the employee from that hazard. Controls are more fully covered in Chapter 4. The following are some key aspects.

- The process of developing controls will require input from a range of employees and subject experts, ranging from occupational physicians to engineers depending on the hazard (e.g. it may require an engineering control such as ventilation, or vaccination to induce immunity). While OHS practitioners may not be required to design complex controls such as a complete ventilation system, a broad understanding of these processes is useful.
- Controls need to follow the prioritising framework known as the 'hierarchy of control' (Chapter 4).
- The recommended controls must be assessed against the regulatory requirement to control risk so far as is reasonably practicable.
- Selecting the correct type and level of control requires not only knowledge of the hazard, exposure risk and route of entry, but also of how much control is needed, the comparative effectiveness of different control processes, maintenance and testing procedures, user preferences and social impacts. The process also requires consideration of costs. Tools are available to assess the business value of implementing controls (AIHA, 2008; Jahn, Bullock and Ignacio, 2015).
- There may be specific regulatory requirements—for example, mining inspectorates have specific regulatory workplace exposure standards for inhalable and respirable dust in mines and detailed medical surveillance requirements for lead workers.
- The evaluation process can also be used to eliminate or control hazards in the design phase and as part of change management. This is a powerful prevention tool when used.
- The recommended controls can form part of any site planning process, leading to management commitment to prioritise, resource and implement controls.
- The implementation of controls is fundamental to the elimination or control of health-affecting agents. There is no point in monitoring and evaluating hazards if controls are not implemented.
- The management document for controls will be the health action plan, which will set out responsibilities and key performance indicators.

1.5.8   REVIEW

Controls should follow a continuous improvement process with reviews of all of the major elements of the occupational health management framework including SEGs, the risk register, the monitoring program, monitoring standards, medical surveillance and so on. The frequency of reviews will vary depending on the element of the occupational health management framework. The system may also include audits.

It would be beneficial for the overall occupational health management framework to fit into an occupational health and safety (OHS) management system such as those described in AS/NZS 4801 and ISO45001. The AIOH's Occupational Hygiene Monitoring and Compliance Strategies (Grantham and Firth, 2013) also provide a helpful chart (11.1) laying out the process for compliance monitoring to meet Australian harmonised OHS legislation. However, as outlined above, exposure control is broader than just compliance monitoring.

There are many elements that make up a comprehensive occupational health-management framework. It can be a complex task bringing all these aspects together. A sample occupational health-management framework is set out in Table 1.4. Review frequencies can be set by a company's own risk assessment or by specific legislative requirements if they exist.

Table 1.4 Sample occupational health-management framework

Key health-management documents	Components
Health science summary	May include legislative, data, carcinogen and SEG review.
Medical monitoring summary	May include medical monitoring data and health awareness information.
Monitoring schedule	Sets out health-affecting agents to be monitored or assessed, frequency and sample numbers. Drawn from Health Science Summary
Health action plan	Includes a review of progress against the plan.
Risk register	Sets out risks for major health-affecting agents for each SEG, identifies gaps and sets out major controls using the hierarchy of controls.
Job demands manual	Sets out major physical job demands. Used by physician as an input into medicals.
Site-specific health agent management plans	Sets out management of key health-affecting agents as determined by company risk assessments or legislative requirements. Plans may include lead, radiation, asbestos, dusts, noise, health promotion, Legionella and water quality.

Standards and procedures	Components
Contractor management	A system to ensure that contractors carry out their work in a healthy and safe way, harmonised to company requirements.
Change management	A system to ensure that any changes affecting plant, procedures or systems, as well as procurement and design, take health effects into account—for example, Buy Quiet.

Standards and procedures	Components
Legal register	Including legislative review.
First aid	Including specific site requirements, training schedule, inspection, maintenance, etc.
Personal protective clothing and equipment standards	Includes management systems for respiratory protection, hearing protection, etc.
Records system	Includes record types, retention policy and privacy considerations.
Hazard-identification and risk-assessment procedure	Includes the company risk-assessment matrix.
Hygiene-monitoring quality assurance	Sets out quality assurance requirements for providers and internal monitoring quality assurance requirements.
Medical surveillance quality assurance	Sets out quality assurance requirements for medical assessments, equipment maintenance and assessments, including respiratory and audiometry.
Manual handling task assessment	Includes task assessment methodology and risk assessment.
Hazardous substance and dangerous goods system	Incorporates systems for bringing new chemicals on site, hazardous substance-assessment procedures, including storage and disposal. May also be major hazard facility legislative requirements.
Thermal (heat and/or cold) stress-assessment procedure	Incorporates measurement, assessment and controls.
Fitness for work	Includes fatigue and hydration management, drug and alcohol, employee assistance.
Legionella-monitoring procedure	Legionella standard operating procedure sets out monitoring strategy, schedules and control strategies.
Potable water-monitoring procedure and action plan following exceedance	Sets out monitoring strategy, schedules and control strategies.
Exposure standards	Integrates company, Australian and state exposure standards into context, taking working hours/shift and other adjustments into account.

Quality Assurance (QA)	Components
Occupational hygiene monitoring QA procedures	May include particulates, vapours, gas, noise, vibration, biological, thermal, alcohol and drugs testing.
Medical systems QA procedures	Includes monitoring equipment register.
Internal or external audits of major occupational health systems	
OH training	Training needs analysis. Review of effectiveness of training, records.
Respiratory and hearing protection FIT testing	May be quantitative or semi-quantitative.

Non-routine monitoring	Components
Lighting survey	
Noise contour map of fixed plant	
Electromagnetic field survey	Focus on major likely EMF sources such as sub-stations and large electric motors.
New plant assessments— e.g. noise, vibration	As per Australian and international standards.
Ventilation	Includes local exhaust ventilation such as welding exhaust ventilation and its testing frequency, supplied air from compressors as per AS1715.

Registers	Components
Maintenance	Site-specific maintenance relating to health-affecting agents.
Signage	Includes mandatory personal protective clothing and equipment, such as hearing protection areas, pacemakers and hearing aids (electric fields), ionising radiation, dangerous goods, confined spaces, authorised access, etc.
Monitoring equipment register	Includes both occupational hygiene and occupational medical equipment as well as calibration schedules.
Training register	Training status of all employees, reviewed as part of training needs analysis. Also includes FIT testing for respirators and hearing protection.

1.6   OTHER OCCUPATIONAL HYGIENE TOOLS

One very significant aspect of occupational hygiene is the extent of the online tools and apps available to help the occupational hygienist, health professional or workplace representative to assess the workplace and improve their skills.

Due to the dynamic nature of online tool development, it is suggested that the reader consult the web pages of major OH organisations in the first instance. These include:

- AIOH—position papers, exposure assessment spreadsheets
- AIHA—exposure-assessment strategies committee; Excel-based tools including spreadsheet applications of statistical analysis, skin permeability, basic workplace exposure assessment, IH exposure scenario tool
- BOHS—links to 'Breathe Freely' for construction and manufacturing with significant material for OH risks in these industries
- Occupational Hygiene Training Association—freely available access to a comprehensive range of occupational hygiene training courses
- UK Health and Safety Executive—codes of practice and numerous computer-based tools, including vibration and noise calculators as well as the COSHH control banding system.

There are numerous occupational hygiene apps available in areas including noise exposure, octave band analysers, thermal risk, lux meters and whole-body vibration, as well as powerful apps available from instrument manufacturers providing monitoring advice and instrument catalogues.

While many apps are not validated, they can be empowering to those at the workface. Two validated apps that are likely to have longevity are:

- NIOSH Sound Level Meter app—backed up by quality NIOSH research, available for iPhone
- the Predicted Heat Strain Mobile app—developed by the University of Queensland, available for iPhone.

1.7   REFERENCES

American Industrial Hygiene Association (AIHA) 2008, 'Demonstrating the business value of industrial hygiene', <www.aiha.org/votp_new/pdf/votp_report.pdf>.
—— 2014, 'Exposure assessment strategies, tools and links: New IHStat with multi-languages', <www.aiha.org/get-involved/VolunteerGroups/Pages/Exposure-Assessment-Strategies-Committee.aspx>.
—— 2018 *Body of Knowledge Occupational Exposure Risk Assessment/Management*, eBook, <www.aiha.org/publications-and-resources/BoKs/OEA/InteractiveVersion/Pages/default.aspx>.
Asbestos Safety and Eradication Agency (ASEA) 2016, *Future Projections of the Burden of Mesothelioma in Australia,* <www.asbestossafety.gov.au/research-publications/future-projections-burden-mesothelioma-australia>.
Australian Institute of Occupational Hygienists (AIOH) 2016, *Adjustment of Workplace Exposure Standards for Extended Workshifts Position Paper*, 2nd ed., AIOH, Melbourne.

Australian National University (ANU) 2017, *REGNET Overview of Work Health and Safety Regulation in Australia*, <http://regnet.anu.edu.au/research/centres/national-research-centre-ohs-regulation-nrcohsr/overview-work-health-and-safety-regulation-australia>.

Blackley, D., Halldin, C. and Laney, A. 2018, Continued Increase in Prevalence of Coal Workers' Pneumoconiosis in the United States, 1970–2017', *American Journal of Public Health*, vol. 108, no. 9, pp. 1220–2.

Blunt, L.A., Zey, J.N., Greife, A.L. and Rose, V.E 2011, 'History and philosophy of industrial hygiene', in D.H. Anna (ed.), *The Occupational Environment: Its Evaluation, Control, and Management*, 3rd ed., AIHA, Falls Church, VA.

BOHS and NVvA 2011, 'Testing compliance with occupational exposure limits for airborne substances', developed jointly by the British Occupational Hygiene Society (BOHS) and the (Dutch) Nederlandse Vereiniging voor Arbeidshygiëne (NVvA), <www.arbeidshygiene.nl/-uploads/files/insite/2011-12-bohs-nvva-sampling-strategy-guidance.pdf>.

Carey, R., Driscoll, T., Peters, S., Glass, D., Reid, A., Benke, G. and Fritschi, L. 2014, 'Estimated prevalence of exposure to occupational carcinogens in Australia 2011–2012', *Occupational and Environmental Medicine*, vol. 71, no. 1, pp. 55–62.

Carey, R., Hutchings, S., Rushton, L., Driscoll, T., Reid, A., Glass, D., Darcey, E., Si, S., Peters, S., Benke, G. and Fritschi, L. 2017, 'The future excess fraction of occupational cancer among those exposed to carcinogens at work in Australia in 2012', *Cancer Epidemiology*, vol. 47, doi: 10.1016/j.canep.2016.12.009.

Damousi, J., Rubenstein, K. and Tomsic, M. (eds) 2014, *Diversity in Leadership: Australian Women, Past and Present*, ANU Press, Canberra.

Firth, I., Van Zanton, D. and Tiernan, G. 2006, *Simplified Occupational Hygiene Risk Management Strategies*, AIOH, Melbourne.

Glick, M. 1968 'Pneumoconiosis in New South Wales coal mines', in *Proceedings of First Australian Pneumoconiosis Conference*, Joint Coal Board, Sydney, pp. 165–77.

Grantham, D. and Firth, I. 2013, *Occupational Hygiene Monitoring and Compliance Strategies*, AIOH, Melbourne.

Hämäläinen, K., Takala, J. and Kiat, T. 2017, *Global Estimates of Occupational Accidents and Work-related Illnesses 2017*, Workplace Safety and Health Institute, Ministry of Manpower Services Centre, Singapore.

Hamilton, A. 1943, *Exploring the Dangerous Trades*, Little, Brown and Co., Boston, MA.

Harden, M., Steer, J., Harden, F., Butler, G. and Mengersen, K. 2015, 'Integrating Occupational Hygiene, Medicine and Engineering in a Mining Context', paper presented to the 10th International Occupational Hygiene Association (IOHA) London 2015 International Scientific Conference, Abstract 0150.

Health and Safety Executive (HSE) 2016/17, *Health and Safety Statistics for Great Britain 2016/2017*, <www.hse.gov.uk/sTATIstics>.

Hewett, P., Logan, P., Mulhausen, J., Ramachandran, G. and Banerjee, S. 2006, 'Rating exposure control using Bayesian decision analysis', *Journal of Occupational and Environmental Hygiene*, vol. 3, no. 10, pp. 568–81.

International Council on Mining and Metals (ICCM) 2016, *Good Practice Guidance on Occupational Health Risk Assessment*, 2nd ed., <www.icmm.com/gpg-occupational-health>.

International Organization for Standardization (ISO) 2018, ISO 45000:2018(E) Occupational Health and Safety Management Systems—Requirements with guidance for use, International Organization for Standardization, Geneva, Switzerland.

Jahn, S., Bullock, W. and Ignacio, J. 2015, *A Strategy for Assessing and Managing Occupational Exposures*, 4th ed., AIHA, Falls Church, VA.

Keil, C. and Simmons C. (eds) 2009, *Mathematical Models for Estimating Occupational Exposure to Chemicals*, 2nd ed., AIHA, Falls Church, VA.

Laszcz-Davis, C., Maier, A. and Perkins, J. 2014, 'The hierarchy of OELs', *The Synergist*, March, AIHA, Falls Church, VA.

Liedel, N., Busch, K. and Lynch, J. 1977, *Occupational Exposure Sampling Strategy Manual*, National Institute of Occupational Safety and Health, Cincinnati, OH.

Moore, R. and Badham, C. 1931, 'Fibrosis in the lungs of South Coast coal miners, New South Wales', *Health*, vol. 9, no. 5, p. 33.

National Institute for Occupational Safety and Health (NIOSH) 2018, 'Occupational exposure banding', <www.cdc.gov/niosh/topics/oeb/default.html> [accessed 26 October 2018].

NHMRC 2011, Australian Drinking Water Guidelines 6 (version 3.5), National Resource Ministerial Council, National Health and Medical Research Council <nhmrc.gov.au/about-us/publications/australian-drinking-water-guidelines>.

NHMRC and NRMMC 2018, *Australian Drinking Water Guidelines: Paper 6 National Water Quality Management Strategy*, National Health and Medical Research Council and National Resource Management Ministerial Council, Commonwealth of Australia, Canberra.

Parliament of Australia 2017, *Report No. 2, 55th Parliament Coal Workers—Pneumoconiosis Select Committee,* Commonwealth Government, Canberra.

—— 2018, *Employment by Industry Statistics: A Quick Guide*, <www.aph.gov.au/About_Parliament/Parliamentary_Departments/Parliamentary_Library/pubs/rp/rp1718/Quick_Guides/EmployIndustry>.

Safe Work Australia (SWA) 2017, *Key Work Health and Safety Statistics Australia 2017: Work Related Injury Fatalities*, <www.safeworkaustralia.gov.au/doc/key-work-health-and-safety-statistics-australia-2017>.

—— 2018, *Code of Practice: Preparation of Safety Data Sheets for Hazardous Chemicals*, <www.safeworkaustralia.gov.au/system/files/documents/1807/code_of_practice_preparation_of_safety_data_sheets_for_hazardous_chemicals.pdf>.

Schulte, P.A. 2005, 'Characterizing the burden of occupational injury and disease', *Journal of Occupational and Environmental Medicine*, vol. 47, no. 6, pp. 607–22.

Standards Australia 2001, Australian/New Zealand Standard AS/NZS 4801:2001 Occupational Health and Safety management systems—Specifications with guidance for use, Standards Australia, Sydney.

—— 2005, Occupational Noise Management Part 1: Measurement and Assessment of Noise Emission and Exposure, AS/NZS 1269.1, Standards Australia, Sydney.

Steenland, K., Burnett, C., Lalich, N., Ward, E. and Hurrell, J. 2003, 'Dying for work: The magnitude of US mortality from selected causes of death associated with occupation', *American Journal of Industrial Medicine*, vol. 43, no. 5, pp. 461–82.

Thoracic Society of Australia and New Zealand 2017, 'Peak body calls for national response following resurgence of occupational lung diseases', media release, <www.thoracic.org.au/documents/item/1194>.

US Department of Health and Human Services 1985, 'Perspectives in disease prevention and health promotion', Morbidity and Mortality Weekly Reports, vol. 34, pp. 219–22, 227.

US National Library of Medicine 2019, Haz-Map, <https://hazmap.nlm.nih.gov>.

Van Tongeren, M. 2014, 'Use of exposure models for regulatory chemical risk assessment: Can we rely on them?', paper presented to Australian Institute of Occupational Hygienists (AIOH) 32nd Annual Conference, December.

Worksafe New Zealand Feb 2017, 'Worksafe Position on Work Related (occupational) Health' <www.https://worksafe.govt.nz/laws-and-regulations/operational-policy-framework/worksafe-positions/work-related-occupational-health/>

—— Annual Report, 2016/2017 <www.https://worksafe.govt.nz/about-us/corporate-publications/annual-reports/> Wellington, New Zealand.

2. Occupational health, basic toxicology and epidemiology

Geza Benke and Martyn Cross

2.1   INTRODUCTION

This chapter provides a basic introduction to the subjects of occupational health, toxicology as it relates to the workplace and occupational epidemiology. The purpose of understanding toxicology and epidemiology is to assist in recognising hazards in the workplace with the intention of controlling them. While both these fields can require extensive expertise, most H&S practitioners will need an understanding of toxicology and epidemiology as they form the basis of many of our preventive actions. Practitioners needing in-depth toxicological or epidemiological information will need to consult further readings, some of which are identified at the end of this chapter. In addition, Chapter 10, which discusses biological monitoring and monitoring strategies, relies significantly on the toxicological principles presented in this chapter.

The chapter discusses the reasons why a systemic approach to occupational health, rather than merely a mechanistic outlook, is required. The health impacts of hazardous materials and hazardous environments on the human body will be considered. The concepts of hazard, dose and risk, their relationships and their role in workplace risk assessment will also be explained, as will the way epidemiology can provide a link between exposure and disease.

2.2   THE HUMAN BODY IN THE WORKPLACE

To make any sense of occupational health and occupational hygiene, the health and safety (H&S) practitioner needs a basic knowledge of the worker and how this human machine interacts with the work environment. The body and its responses to exposures are highly complex. The workplace is a variable mix of different factors causing differing exposures, often acting in an uncontrolled way on the human body. Hence worker–workplace interaction can be tremendously complicated.

The worker's body constantly interacts with the workplace environment and does so in a variety of ways. Some bodily reactions will be assessing the environment (such as sniffing the air), while others will be accommodating (breathing shallowly). Others may be defence mechanisms, either on a physiological level (coughing) or a cellular level (the development of antibodies). The worker interacts with the workplace through:

- *sensory input*—sight, sound, smell, feeling
- *ergonomic factors*—posture, task, energy demand
- *physical contact*—pressure, vibrations, noise, heat, radiant energy, ionising radiation

- *chemical contact*—inhalation or skin absorption
- *psychosocial factors* such as the ways in which individuals interact with the demands of their job and their work environment.

Even moderately simple workplace tasks require the use of the limbs, motor coordination, balance, vision, the lungs, the heart and circulatory system, the senses, the intellect, hearing, the voice, and body heating or cooling mechanisms. Some of the interactions are perfectly normal, and the body is designed to withstand and respond to them. If we fail to understand the factors in the workplace acting on the worker or the way the body responds to or copes with these stressors, however, we will be unable to act to control any resulting strain on the body. Damage and/or illness may then occur.

Some workplace hazards will directly affect the body system they touch; in other cases, the organ systems ultimately affected may never come into direct contact with the workplace environment. Hazardous substances entering the body through the lungs, mouth or skin may affect the brain, liver, kidneys, bones or reproductive organs, and the effects may be observed even in the offspring of exposed workers.

In order to understand damaging effects (hazards), two questions must be answered:

- What is the worker exposed to, how much and for how long (i.e. the potential exposure)?
- What is the body's response?

Answers can be difficult to establish. Sometimes the factors affecting the worker will not be apparent. For example, in the nineteenth century, illness among needle grinders was thought to be caused by the inhalation of metal splinters. In fact, the damage was done by the silica that came from their grinding wheels. X-rays and radioactivity were initially seen as curious phenomena that could sometimes be industrially useful; their carcinogenic effects were understood too late for many.

Inevitably, two more questions arise. First, are the exposure and the response linked? Second, what can be done to prevent the damage to health? This first question will be examined in this chapter; the second will be addressed throughout the remainder of the book.

2.3   ESTABLISHING CAUSAL RELATIONSHIPS

It is not always easy to recognise whether a health effect is the result of workplace exposure. We can seldom identify hazards or measure the stressing agents in isolation; there may be many hazards acting on the worker simultaneously. Consider the foundry workers in Figure 2.1.

The workers are shown loading, slag-skimming and performing other operations on the top of the furnace. In these tasks, they might be subjected to heat, noise, carbon monoxide and metal fume particles, not to mention the stress of balancing atop a vat of molten metal. When tapping the furnace into a crucible for pouring metal into sand moulds, workers may additionally be exposed to amine fumes, gaseous cyanide and/or isocyanates from mould constituents. The intensity, duration and method (e.g. breathing, skin contact) of these exposures determine the *dose* (quantity) the worker receives from

Figure 2.1 Foundry workers may be subjected to several health hazards, including various dusts, gases and heat.
Source: Courtesy R. Golec.

each hazard, which in this case will differ at various times during a single work shift, depending on the activity of the worker.

We need to know how the body responds to each of these hazards before any risk assessment can be made in this workplace. Failing to realise that carbon monoxide is a chemical asphyxiant—or, worse, failing to even recognise that it could be present—could make it difficult to understand, for example, why workers in this type of workplace could develop headaches.

Making evaluation even more complex, the amount of exposure varies from one work site and phase to the next. The body exhibits different responses depending on the dose and the rate of entry into the body. For example, breathing in relatively low doses over a protracted period of time, such as the foundry workers exposed to carbon monoxide over a work shift, may result in headache, loss of concentration and lassitude. In contrast, exposure to high concentrations over a short period of time, such as from breathing in petrol engine exhaust in a confined space, may lead to severe mental confusion, unconsciousness, coma and death. So the relationship between the hazards that exist in the workplace, a worker's actual level of exposure and the way the worker's body responds requires careful consideration in order to suggest appropriate mechanisms to control exposures. Interactions between multiple hazards and their harmful effects also need to be considered. This is especially true where interactions are additive; or even *synergistic*, meaning that the effect resulting from two hazards combined is greater than the added effects of the individual hazards.

It is especially challenging to establish *causal relationships* between workplace exposures and ill-health. Often the health effects of exposure can be delayed so it is difficult to remember the exposure or measure the amount of exposure. Humans are seldom exposed to one hazard at a time, and many workplace hazards and some lifestyle hazards may result in similar health effects or mask the effect of others. Obviously, experimenting on humans is limited by ethical considerations, so science has been forced to develop alternative ways to elicit information concerning the health hazards due to workplace exposures. One method uses experimental animals as test subjects in studies (toxicology), while another has involved long-term studies of exposed populations (epidemiology). Indeed, these studies have provided much of our understanding of the effects of metals, dusts, organic solvents, pesticides and microbiological agents. In addition, in-vitro systems and methods using cell cultures or examination of cellular changes in DNA or proteins have provided information about the mechanisms by which chemicals exert their toxic effects. Data from these sources may assist in the H&S practitioner's assessment of risks in the workplace.

Even after all these efforts to study the hazards and the body's responses, individual variability within a given work group may result in different disease outcomes. Consequently, the significance of prevention for all workers becomes apparent.

2.3.1 MULTIPLE CAUSES OF DISEASE

Another problem to surmount in studying workplace illness is that the body may respond in similar ways to exposures from lifestyle factors (such as diet and smoking) as well as workplace hazards. Foundry workers may develop asthma from exposure to workplace amines or isocyanates, or their asthma may be caused by pollens and other natural materials and not be work related at all. A headache may be the result of dehydration, excessive noise or carbon monoxide exposure.

2.3.2 LINKING A HEALTH EFFECT TO A PARTICULAR EVENT

Observation of workers exposed to thousands of different hazardous substances such as asbestos or solvents, or processes such as welding, has provided the basis for making workplaces much healthier. For these well-acknowledged problems, the H&S practitioner can confidently use the available epidemiological and toxicological information and recommended exposure standards to facilitate control strategies. Occupational exposure standards are assigned by the regulatory authorities (Safe Work Australia). An occupational exposure standard is the maximum upper limit for the exposure for a particular chemical that must not be exceeded. The concept of the exposure standard is addressed in Chapter 3.

On the other hand, many hazardous-substance exposures may lead to effects such as adverse mood or personality change, or loss of memory, which may be too subtle to be recognised easily. It is helpful for H&S practitioners to understand such phenomena. Further, materials are constantly being investigated for possible effects such as carcinogenicity (the ability to cause cancer), mutagenicity (the ability to induce mutation in cells) or teratogenicity (the ability to cause defects in offspring). However, cancer, cell mutations and birth defects occur in the general population for a variety of reasons. In order to identify occupational causes, one must show an increase in a particular disease rate following occupational exposure. For example, exposure to radon gas, which occurs naturally in underground mines, can cause lung cancer, but so can smoking. To identify any increase in the incidence of lung cancer in a group of miners beyond that experienced in a comparable non-occupationally exposed group requires specific and sensitive epidemiological methods.

The study of the health of workers and the diseases caused by their work is hence very complex. The tools used by the epidemiologist to provide better understanding of the relationship between exposure and health are presented later in this chapter.

2.4   OCCUPATIONAL HEALTH

Occupational health may be defined as the maintenance of the individual worker's state of wellbeing and freedom from occupationally related disease or injury. Occupational hygiene is the practice that achieves this.

2.4.1   *OCCUPATIONAL DISEASE COMPARED WITH OCCUPATIONAL INJURIES*

By now, the H&S practitioner will have realised that successful intervention in an occupational health issue requires a systemic approach rather than telling workers to minimise exposure. Indeed, every jurisdiction in Australia has general duty of care requirements that a workplace is safe and without risk to health as far as reasonably practicable. However, there are few regulations that prescribe detailed courses of action for specific hazards. Any successful intervention in this self-regulatory environment therefore requires knowledge of:

- the workplace process and the hazards it produces
- the equipment used in the processes
- the toxicology of hazardous materials
- the health effects of physical, chemical and biological hazards in the workplace
- the physical and organisational environment in which the task occurs
- effective risk communication and consultation
- the effectiveness and appropriateness of different control strategies
- the relative costs of implementation.

The focus of occupational health management is to protect workers in their employment from health risks, where:

- The H&S practitioner has to intervene without the benefit of being physically able to see the hazard. This may require new insights with regard to identifying problems and seeking solutions. In safety matters, one can define the hazards quite readily and also identify some of the outcomes without error. For example, unrestrained moving loads might be expected to fall, unguarded machinery can be expected to mangle limbs and unsupported excavations can be expected to collapse.
- In dealing with hazardous substances, the H&S practitioner has to be able to identify possible outcomes, even where the worker does not show immediate effects. If workers are in the way when loads fall or trenches collapse, there is a high probability of resulting injury or death. In the case of exposure to potentially hazardous materials or environments, the worker may appear not to be affected immediately afterwards, yet may fall ill or even die much later, including after retirement.
- In occupational health, there exists a clear possibility of intervention to change the course of events. In the case of accidental injury, intervention is aimed at preventing the event from occurring, since once an accident is underway, intervention is rarely possible. The use of personal protective equipment (PPE), such as helmets or steel-capped boots, does not prevent incidents but reduces the consequent harm.

Occupational health and safety practices are all based on risk assessment—that is, we aim to reduce exposure to the level where there is an acceptably small chance of harm occurring. The human body can tolerate some levels of exposure to some hazards without detriment. In general, occupational safety deals with accidents and their prevention, while occupational health deals with occupational diseases and their prevention.

A sudden injury at work can differ from disease caused by work exposures in three ways. These are the **time factor**, the **damage factor** and the **dose factor**.

In relation to the *time factor*:

- An accident and the resulting injury occur at virtually the same point in time and from a single incident. This immediacy of injury means that the link between cause and effect is obvious.
- Occupational diseases may occasionally result from single massive exposures, but usually result from exposures to the causative agent over a period of time. The disease may take some time to develop, ranging from minutes to years. Examples include:
  - carbon monoxide poisoning—minutes
  - solvent intoxication—hours
  - metal poisoning—days
  - noise-induced hearing loss—years
  - asbestos-related diseases—up to 50 years.

Some health effects appear long after exposure has ceased. Consequently, it is important to recognise that workplace exposure to hazardous substances can be equated with accidents occurring over a long timespan. The typical timespan for asbestos-related diseases, for example, may be fifteen years for asbestosis, twenty years for lung cancer and 30 years for mesothelioma.

In relation to the *damage factor*:

- An accident may injure tissues but only at the point where the energy of the accident is applied—for example, the head or the arm.
- With disease, tissue damage may or may not occur where the causative agent is applied. For example, inhaled quartz dust has a direct effect on the lungs, but inhaled solvent vapours may produce effects on the liver or on the brain (headache or drowsiness, for example).

In addition, subtle changes that are not obvious to the worker may occur to bodily functions. For example, the effects of small amounts of a toxic agent may leave the person quite well, whereas absorption of larger amounts will cause severe illness.

In relation to the *dose factor*, disease may be caused by a single large exposure or many small exposures to a workplace hazard. The likelihood of disease depends on the dose received. There may be some threshold dose below which there is no adverse effect, which will vary depending on the hazard. It will be very small for highly hazardous agents, and large for those of lower hazard.

Throughout the following chapters, the significance of this dose factor will be discussed. Different doses can provoke different responses in the case of exposure to dusts, metals, gases and vapours. The dose to which a worker is exposed affects the interpretation of risk and the kinds of controls we should use.

2.4.2   *WHAT THE H&S PRACTITIONER NEEDS TO KNOW*

Because of these three factors—**time**, **damage** and **dose**—the link between the cause and the resulting occupational disease may not immediately be obvious. Therefore, to prevent disease, or to detect minor change early before the worker becomes ill, it is essential to have some knowledge of the conditions in the workplace that make it a risk to health.

Detection and prevention of disease require an understanding of a wide array of work situations and knowledge of the effects that various hazardous agents can have on workers. It is therefore necessary for the H&S practitioner to seek good occupational hygiene knowledge of:

- those agents in use at the workplace that are hazardous
- how much exposure poses a risk
- what procedures are necessary to monitor the workplace and worker exposure
- how exposure can be controlled.

Only when armed with this information can the H&S practitioner take steps to prevent work-related illness. The following chapters provide details of workplace investigation, toxicological information on some regular workplace hazards, guidelines on how to assess the hazards and, lastly, a wide range of control techniques and methods. This information will not make the H&S practitioner an expert, but should provide a foundation upon which to build. Only practical experience will permit mastery over real work situations.

2.5   BASIC TOXICOLOGY

Toxicology may be defined as the description and study of the effects of a substance on living organisms. In terms of the workplace, we are interested only in human toxicology. Few animals—even the canary used historically in mines to warn of poisonous gas exposures—are used nowadays in workplaces. Nonetheless, animal toxicology has provided valuable insight into the potential effects of hazards upon humans. Experiments on animals involving prolonged or high exposure to various hazards, which can be carried through to the subjects' death, obviously cannot be conducted on human populations, for ethical reasons.

The H&S practitioner needs to understand basic toxicological concepts in order to:

- understand recommended or regulatory exposure standards
- recognise and prioritise chemical hazards in the workplace
- determine control measures for a particular hazard.

A foundry, for example, might provide respiratory protection such as a dust mask on its moulding line against silica dust, but fail to appreciate that polyurethane moulds release isocyanate vapours into the workplace atmosphere—vapours that are not captured by a dust mask. The H&S practitioner's knowledge of all the risk factors will be needed to provide correct protection.

2.5.1   HAZARD AND RISK

Hazard and risk are two quite different concepts, although the terms are commonly used interchangeably in discussions of hazardous substances. A hazard represents the *potential* for a substance to cause adverse effects. A hazardous substance is one that has sufficient toxicity to cause harm, given the appropriate conditions of exposure or absorption. *Risk* is a function of the *likelihood* of an adverse effect occurring in a particular situation and of the *magnitude* of the adverse effect (the severity of the outcome). If a hazardous substance is to cause harm, the worker must be exposed to a sufficiently large amount of the substance.

In short, a hazard becomes a risk when there is potential for exposure to the hazard. Potential for exposure depends on the hazard itself, the task involved in using the material with particular focus on exposure pathways, the equipment involved in the task, the people involved and their vulnerabilities, and the physical and organisational environment in which the task takes place. The likelihood and consequence of the risk are reduced by any controls put in place to reduce the potential for exposure.

Hundreds, even thousands, of substances found in workplaces have the potential to cause harm; however, it is important to differentiate between potential hazards and those hazards that present a risk of harm. Asbestos in an asbestos-cement sheet is a hazard, but may not present a serious risk because the likelihood of inhalation exposure occurring is very low if the material is not machined or abraded. However, release of asbestos fibres into the workplace air during removal of asbestos lagging will almost certainly result in an inhalation exposure, leading to a higher risk of toxic effect. Benzene is a potent leukaemia hazard, but while contained in a closed reaction vessel it presents a low risk; xylene used in open tanks or in painted coatings may pose a greater risk to health.

In mathematical terms, risk is a function of consequence or severity and likelihood. Likelihood is related to the extent of exposure, with greater exposures more likely to result in injury or disease.

Expressed as a conceptual formula:

$$\text{RISK is proportional to HAZARD} \times \text{EXPOSURE} \qquad \text{(Equation 2.1)}$$

or

$$R \propto H \times E$$

where R = risk
H = hazard
E = exposure

If exposure is zero (exposure is controlled), then risk will also be zero and risk is controlled. (Chapter 4 examines a number of ways to control exposure.) In the same manner, if the hazard is removed by replacing a hazardous material with a non-hazardous material (e.g. by replacing a solvent process with a water-based one), then both hazard and risk become zero, and exposure may not need additional control.

'All substances are poisons; there is none which is not a poison. The right dose differentiates a poison and a remedy.' This comment, by the sixteenth-century physician and philosopher Paracelsus, summarises one of the fundamental principles of modern toxicology: the concept of *threshold dose*—that is, doses of substances below the threshold for adverse effect are not harmful, and in fact some chemicals at low doses are beneficial. Essential elements such as cobalt, iron, sodium and calcium testify to the power of this observation. Doses below the threshold dose are needed for health. Doses exceeding the threshold dose will give rise to adverse effects. The threshold information is useful when considering safe levels of a toxic substance, and establishing occupational exposure standards.

2.5.2   *ABSORPTION OF HAZARDOUS SUBSTANCES*

For materials to be able to exert a toxic effect, they must have a *route of entry* into the body. In the work situation, we are generally concerned with only three routes of entry:

- inhalation
- skin contact
- ingestion.

Inhalation of a dust, fume, vapour, mist or gas is the principal route of entry in the work environment. The membranes lining the lungs provide a large area (approximately 140 square metres) across which materials may be absorbed. A moderately working male worker breathes approximately 35 litres of air per minute or 16,800 litres per working day. This equates to 16.8 square metres per working day, all washing over the 140 square metres of the lung surface to absorb the oxygen and other materials carried in the air (OSHA PEL for crystalline silica, 2016). Unless provided with special breathing apparatus, people must maintain constant contact with the air of the workplace to survive.

Many substances in the workplace readily evaporate or become airborne, and so contaminate the air. For this reason, a large part of hazardous-substance control is based on keeping the work atmosphere clean. This is why exposure standards, discussed in Chapter 3, relate specifically to inhalation of airborne substances. It is also why occupational health and hygiene investigations often focus on material that can be inhaled.

Skin contact is the second most important route of absorption. The adult body has a skin surface area of about 2 square metres. Entry through the skin applies most frequently to liquids, and some materials (e.g. oils and some pesticides) enter predominantly via the skin. Sometimes materials encountered as vapours or gases also enter through the skin, resulting in significant absorption and distribution through the body. Occasionally, substances (e.g. mercury and hydrogen cyanide) will enter the body through both the lungs and the skin, where their effects will be additive. Of course, there are also substances that have direct effects on the skin. These include corrosives (acids), alkalis and other materials (e.g. arsenic). The eye is also at risk from direct contact, particularly with biohazards and corrosives.

Skin is a very effective barrier to prevent environmental materials from entering the body; however, individual workers vary in their susceptibility to skin penetration. Those with thicker skin will be at lower risk than those with thin skin. Oily skins may provide better protection than dry skins. Some areas of skin are very thin (e.g. on the scrotum or around the eyes) and provide less of a barrier than thicker skin (e.g. on the palms or soles of the feet). Damaged skin may often present a ready route of entry for a hazardous substance, particularly a microbiological hazard. The level of skin hydration, often dependent on work rate and the extent of perspiration, may also reduce the skin's barrier characteristics, making it more freely permeable to particular chemicals.

Ingestion (swallowing)—often the result of hand to mouth contact—is usually a minor route of absorption in workplaces. Good personal hygiene, particularly attention to washing of hands, and avoidance of eating, drinking and smoking in the workplace can help reduce this route of entry. Some materials may be swallowed by virtue of internal respiratory clearance mechanisms depositing mucus at the oesophagus, or following deposition of materials in the mouth or pharynx during mouth breathing.

Other rarely encountered absorption routes include direct injection of hazards into the bloodstream or body tissues. These may be relevant to specific workplaces and processes, such as the accidental injection of infective agents from contaminated syringe needles, transdermal injection of fluids from high-pressure hoses, or puncture and laceration injuries from chemically contaminated laboratory glassware.

2.5.3   TOXIC EFFECTS OF HAZARDOUS SUBSTANCES

In the human body, the effects of hazardous materials vary greatly, as does the severity of those effects. Some are examined specifically in later chapters. In simple terms, effects can be described in terms of *where* and *when* they occur in the body.

In relation to where:

*   *Local effects* are adverse effects on the particular tissue that has been directly exposed to the hazardous substance. Examples include:
    –   damage to the eyes or skin by corrosives
    –   dermatitis caused by contact with organic solvents
    –   intense irritation of the respiratory tract by gases such as chlorine or ammonia.
*   *Systemic effects* are adverse effects on one or more body systems after absorption of the hazardous substance. In these instances, the hazardous material travels through the body to a distant, susceptible organ—for example:
    –   the nervous system, blood, kidneys and reproductive functions can be harmed by lead
    –   the nervous system can be damaged by organophosphate insecticides.

In relation to when:

*   *Acute effects* occur in a short period of time and develop during or soon after exposure. These can include irritation, acute poisoning or reproductive effects—for example:
    –   the eyes and respiratory tract immediately become irritated by gases such as ammonia
    –   narcotic effects (e.g. headache, dizziness, incoordination, unconsciousness) can rapidly follow excessive exposure to toxic organic solvents
    –   metal fume fever from high exposure to some metal fumes may result in flu-like symptoms within one to two days.
*   *Chronic adverse effects* are long lasting, if not permanent. Their onset may follow soon after exposure, or it may be delayed for many years. Some well-known chronic health effects are:

- asbestosis and silicosis following excessive exposure to asbestos and silica dust respectively
- lung cancer following exposure to dusts containing arsenic
- chronic dermatitis from exposure to chromium-containing cements
- lung cancer and mesothelioma from asbestos exposure
- damage to the DNA structure of sperm and ova, and reduced fertility (suppression of sperm production)—for example, from lead exposure.

Information about possible acute and chronic effects should appear on the safety data sheet (SDS) for the substance.

The terms 'acute' and 'chronic' are related to the duration of the exposure, rather than the duration of the toxic effect. This usage occurs in animal toxicity-testing guidelines such as those specified by the Organisation for Economic Co-operation and Development (OECD, 2019). The guidelines, as applied to laboratory-bred rats, identify acute studies (single exposures), short-term repeated-dose studies (14–28 days), sub-chronic studies (90 days) and chronic studies (six to 30 months). The generally accepted usage of these terms in occupational toxicology relates to short exposures (acute) and exposures of several years (chronic). Hence an acute *exposure* may result in a chronic *effect*—for example, a single exposure to a large dose of organophosphorus insecticides may result in delayed neurological impairment that may be long lasting.

2.5.4   THE IMPORTANCE OF DOSE

Earlier we quoted Paracelsus's observation that the dose makes the poison. A vital point to remember is that most hazardous substances are foreign to the body: they are not supposed to be there. In order to link absorption of hazardous substances and the effects they produce, we have to link the observed effects with the *amount* absorbed, stated in precise and measurable terms. It is preferable not to describe toxic doses using relative terms such as low, moderate or high exposure, but to obtain numerical data on the exact dose. All the exposure limits given in Chapter 3 are based on *measurable* quantities.

Toxicologists use two approaches to examine the relationships between dose and response. In the first, the extent of the effect is related to dose—for example, the degree of enzyme inhibition arising from chemical exposure, or the amount of skin redness (erythema) from UV exposure. When the data are plotted, this is referred to as a dose–effect curve. The expectation is that increasing dose causes increasing effect.

Figure 2.2a shows the dose–effect relationship with arithmetic axes, and Figure 2.2b shows the same data plotted on a logarithmic dose axis. A dose effect curve represented on a log axis virtually always results in a sigmoid curve (S-shaped), indicating that there are low doses that do not cause an effect, and 'high' doses that cause the effect in all subjects. The log axes enable the linear data between 20 and 80 per cent of maximal effect to be accurately plotted, and the dose causing a 50 per cent maximal effect, the ED_{50}, can be estimated. Figures 2.2a and 2.2b also show that, as dose declines, a smaller effect is observed, until we see a dose at which no effect is seen—the *threshold dose*. These two parameters— the threshold dose and the ED_{50}—together with the slope, define the location and shape of the dose–effect curve.

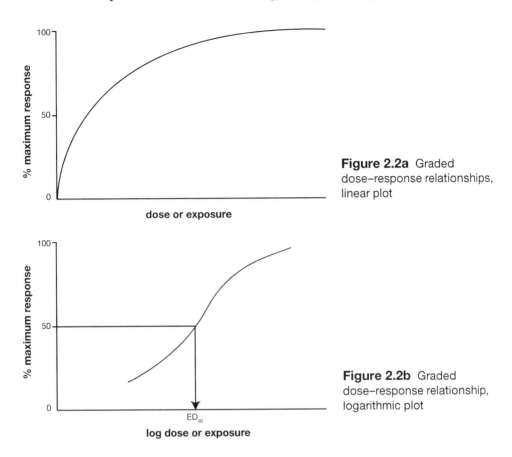

Figure 2.2a Graded dose–response relationships, linear plot

Figure 2.2b Graded dose–response relationship, logarithmic plot

The second approach describes the percentage of a population displaying a given response at a given dose, providing a dose–response curve. A defined response, such as death, is examined in groups of individuals exposed to a range of doses of a chemical. These responses may be the presence or absence of a measurable outcome—for example, at least 10 per cent inhibition of liver activity, or unconsciousness, or death. The proportion of the tested group which shows this response is plotted against the dose (or more usually log dose, as seen in Figure 2.3). From these data, toxicologists can define terms

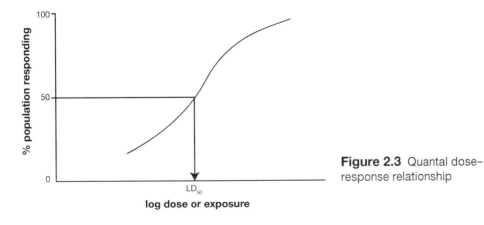

Figure 2.3 Quantal dose-response relationship

such as ED_{50} (effective dose in 50 per cent of the studied population) or LD_{50} (the dose that is lethal to 50 per cent of the studied population).

LD_{50} and related estimates are useful quantifiable concepts because they incorporate a measured dose and a measured rate of morbidity (sickness) or mortality. They represent estimates of the relative toxicity of materials by which we may compare substances encountered in the workplace. For example, the approximate LD_{50} of ethanol is 10,000 mg per kilogram body weight, whereas those of DDT and dioxin are 100 and 0.01 mg/kg respectively, making it clear that the acute dose of dioxin that is likely to kill someone is substantially less than the dose of ethanol required.

2.5.5 THRESHOLD VERSUS NON-THRESHOLD EFFECTS

So far, we have considered the proposition that, as dose declines, the effect seen or the proportion of the population affected also declines, and there is a threshold dose below which no effect is observable. This threshold, even if derived from animal experiments, can be used to estimate the safe amount of the material to which the population can be exposed. These derived safe doses may be called tolerable intakes (TIs), acceptable daily intakes (ADIs) or reference doses (RfDs). These are usually obtained by incorporating safety factors into the data based on interspecies differences (extrapolating from animals to humans), inter-individual differences (accounting for different sensitivities of individuals of the same species) and the severity of the effect observed. Thus the safety factors provide a conservative estimate of the *safe dose*. This method is usually sufficient for risk assessments of most chemical exposures.

An alternative approach suggests that at every dose above zero, there is a probability (or risk) of an adverse effect. This non-threshold approach is often applied to carcinogen exposures. For example, cancer may occur in one of every 1000 individuals with a given exposure. Reducing exposure by a factor of 100 may reduce the cancer rate to one in 10,000. Reducing exposure further may result in one person developing cancer per one million exposed, and so on. In these cases, an acceptable risk is defined (e.g. one per 100,000), which may be used as the basis for exposure standards for carcinogens, radiation and other hazards thought not to have a threshold. A more detailed examination of occupational carcinogens has been published by the Australasian Faculty of Occupational Medicine (2003).

2.5.6 FURTHER EXAMINATION OF DOSE

The dose of a hazardous substance is the amount absorbed, taking into account both the concentration and duration of exposure, as shown in Equation 2.2.

$$DOSE = \text{Concentration of exposure} \times \text{Duration of exposure} \qquad \text{(Equation 2.2)}$$

This equation disguises a few over-simplifications—for example:

- The dose may be less than the amount inhaled if most is exhaled without any absorption (e.g. many gases).
- Workers with heavy workloads breathe more air than those with light workloads, and so have larger doses.

- Dose may depend on whether the worker is a mouth or nose breather.
- Additional exposure may come from non-occupational sources (e.g. carbon monoxide from cigarette smoking).

Clearly, if we can reduce either the concentration *or* the duration of exposure, we will reduce the dose. Exposure standards therefore take both of these factors into account. Our attempts to develop safe levels for exposure standards are often based on dose studies in the workplace, volunteer studies or animal studies where various effects (even disease and death) have been identified. Acceptable or safe levels of exposure have been proposed following comparative studies of exposures where injurious effects were observed and compared with those where no effects were observed.

The dose–response curve shows that there are some susceptible individuals, the ones who succumb at low concentrations. There are also some highly resistant individuals who don't succumb even at high exposure, and average individuals somewhere in the middle.

Figure 2.4 indicates the variability of these susceptible, average and resistant individuals in their response to differing doses.

The y-axis shows relative effect, in increasing severity. The zone marked 'normality' indicates no response, 'adaptation' indicates that there are cellular, biochemical or physiological changes but these are not considered adverse, while 'disease' indicates that changes in function are adverse effects compromising the health of the individual.

- *Susceptible* individuals exhibit more severe responses, such as disease or death at doses lower than average.
- *Resistant* individuals exhibit no response, or may adapt with higher doses and may also tolerate much higher doses without disease or death.

Our determination of safe dose generally must be based on the responses of the most susceptible individuals in the working community. Exposure standards thus tend to be based on doses tolerated by 'average' subjects without ill-effect—that is, they are designed

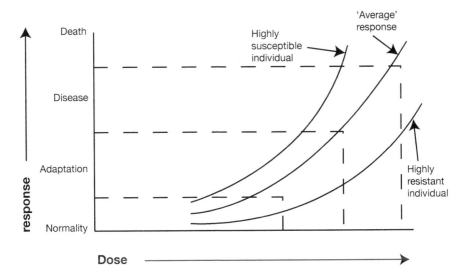

Figure 2.4 Variability of human exposure to dose

to protect nearly all workers. There may, however, be hypersensitive individuals—for example, those with a pre-existing disease or a missing metabolic enzyme—who may suffer adverse effects at doses lower than the rest of the population.

2.5.7   THE ALARP PRINCIPLE

H&S practitioners should always endeavour to reduce the exposure of the worker to as low as reasonably practicable (ALARP). This means that if it is practicable to reduce the exposure (i.e. the dose) of the worker, you should do so. Practicability takes into account the technological means of control at your disposal, physical limitations, any added health benefits and economic cost factors.

For example, hospitals use the anaesthetic nitrous oxide. Operating staff are exposed to anaesthetic gas at about 70 ppm when it is exhaled by the patient. Although the exposure standard for nitrous oxide is 25 ppm, concentrations in the workplace can be reduced to 20 ppm by increasing cross-flow ventilation, or to less than 1 ppm by using a proper scavenging mask on the patient. The choice of scavenging masks is good practice, easily achievable, economic and consistent with the ALARP principle.

At work, the H&S practitioner will be required to make decisions about three types of exposure:

- *Regular exposures that lie below the recommended exposure standards.* Here the dose (exposure × time) is likely to be within acceptable limits and the risk is low. Nonetheless, you should still aim to reduce exposure to ALARP.
- *Uncontrolled, gross over-exposures that can result from an accident, from absence or failure of control procedures, or from exceedingly poor hygiene practice*—for example, the bursting of an organic-solvent delivery line, or entry into a confined space such as a vapour degreaser. Where practicable, even in emergency situations, procedures must be available to limit exposure to ALARP.
- *Known operations producing significant exposures that could exceed exposure standards and hence exceed a safe dose*—for example, applying toluene diisocyanate-based polyurethane cork floor sealant for 30 minutes each day without proper respiratory protection. Even though the exposure is relatively short, the risk is potentially great, so control needs to be based on ALARP.

You should always consider the ALARP principle to minimise risk even further. Good basic controls to prevent exposure embody the ALARP concept because practical technical measures are readily available to control most work exposures. ALARP does not imply relentless pursuit of the reduction of exposure, although some special high-risk industries demand this approach.

2.5.8   THE CONCEPT OF HAZARD, RISK AND DOSE FOR PHYSICAL AGENTS

Not all workplace health hazards involve toxic substances. Significant health hazards are also posed by physical agents. These produce energy that is absorbed by various sense organs and by the body itself. Noise, heat and radiation, for example, are all very commonly encountered physical agents. For each of these, the strict definition of a toxic substance does not apply; however, the concept of risk as a function of exposure and consequence is equally applicable to effects of energy absorption from physical agents.

2.6   CLASSIFICATION OF WORKPLACE HEALTH HAZARDS

This book presents to the H&S practitioner some basics in occupational health and hygiene in the following areas only:

- chemical hazards (Chapters 7, 8 and 9)
- principles of biological monitoring (Chapter 10)
- indoor air quality (Chapter 11)
- physical agents (Chapters 12, 13, 14 and 15)
- biological hazards (Chapter 16).

The book does not attempt to deal with workplace ergonomics, stress-related disorders or psychosocial effects related to work. These are large areas of study—each sufficiently so to require separate treatment. The H&S practitioner should consult other texts for assistance in these specialist areas.

2.6.1   CHEMICAL HAZARDS

Chemicals give rise to a great number of health hazards in the workplace. The category 'chemicals' includes many naturally occurring substances, such as minerals and cotton, as well as both simple and complex manufactured chemical products. Chemical exposure can arise through direct use or from by-products.

Exposure to chemical hazards occurs through:

- inhalation of airborne contaminants, including:
    - dusts—silica, coal, asbestos, lead, cotton, wood, cement
    - mists—acid mists, chrome plating
    - gases—chlorine, sulphur dioxide, ethylene oxide, ozone
    - fumes—smoke, metal fumes from welding
    - vapours—chlorinated and aromatic solvents, amines, ethers, alcohols
- skin contact with hazards, including:
    - pesticides and phenol, absorbed directly through skin
    - acids and vapours, irritating eyes and mucous membranes
    - acids, alkalis and phenols, corroding skin
    - liquid toluene, methylene chloride, dissolving dermal lipids in skin (defatting)
    - creosote and bitumens, photosensitising skin
    - nickel and chromium, which have an allergenic action on skin.

2.6.2   PHYSICAL AGENTS

All workplaces expose people to physical agents that are potentially harmful, including heat, noise, vibration and light. Increasing mechanisation may decrease heat stress, but increases in industrialisation and greater use of high technology can be accompanied by new hazards. Physical agents that affect health include:

- *noise*—absorbed through the ear; some very low frequency (infrasonic) and ultrasonic sounds are absorbed directly by the body
- *vibration*—received by the body in contact with a vibrating object
- *light*—visible, ultraviolet and infrared are received by both eyes and skin; the eye is susceptible to laser energy; poor lighting may also be a workplace health hazard
- *heat*—absorbed by all parts of the body
- *cold*—experienced by whole of body; extremities in contact with cold
- *pressure*—extremes affect body tissues that contain air spaces: lungs, teeth, sinuses, inner ear
- *electromagnetic non-ionising radiation*—microwaves, radiofrequency and very low electromagnetic radiation are received directly by the body
- *ionising radiation*—x-rays and radioactive decay products—α (alpha) particles, β (beta) particles and γ (gamma) rays—received directly by the body.

Physical agents and their health effects in the workplace are discussed in detail in Chapters 12 to 15.

2.6.3   BIOLOGICAL HAZARDS

Some workers are subject to health hazards when they work with biological materials or work in environments where micro-organisms may abound. These hazards may arise from animal or plant materials, or sometimes the handling or treatment of sick persons. A few biological hazards (e.g. *Legionella*) exist more widely and affect members of the general working community. Chapter 16 looks at some of the biological or microbiological hazards of workplaces under the following classes:

- bacterial—*Legionella*, *Brucella*, tuberculosis, Q fever, etc.
- fungal—infective agents (e.g. tinea); allergenic agents (e.g. *Aspergillus*)
- viral—hepatitis B, human immunodeficiency virus (HIV), which causes AIDS.

2.7   OCCUPATIONAL EPIDEMIOLOGY

Not all the activities of the H&S practitioner will be preventative. You may need to investigate how ill-health has arisen. Sometimes the cause will be easy to recognise, particularly for hazards with well-known acute effects. If you found that older workers in a silk-screening department seemed extraordinarily clumsy, when their craft would suggest that they were capable of great care, you might investigate the extent of their exposure to solvents. Other investigations may be more difficult to carry out. If you found, for example, that none of the fifteen men working in a pesticide plant had any children, you may want to investigate their occupational exposures. Such problem situations may be the starting point for epidemiological investigations.

We have imperfect knowledge of the health outcomes associated with our industrial use of hazardous substances, although new information on the health risks of substances and processes is continually coming to light. The scientific study of disease distributions and their determinants in worker populations is known as *occupational epidemiology*, where disease outcomes are correlated with occupational exposure information. Epidemiological studies tend to be medically and statistically complex. They often require large administrative resources to conduct and large sample populations to investigate. While these occupational epidemiological studies have often disparagingly been referred to as 'counting dead bodies', reflection indicates that they are far more than this. They underpin much preventive health in the workplace, and many exposure standards are based on their findings.

H&S practitioners should be familiar with some occupational epidemiological concepts, which may help them to draw conclusions—albeit guarded ones—from meagre or unrelated facts. These concepts may also guide the H&S practitioner to put together the information that is vital for valid studies in an accessible form. Information should include demographic information about the populations at risk—for example, age, gender—and information about potential exposure histories, such as hazardous material exposure, years and duration of employment, nature of the job and any adverse health outcomes. Specifically, the data may include:

* identification of the population at risk, numbers, ages, gender and so on
* when each worker commenced and ceased employment
* complete employment histories
* nature of jobs
* hazardous-agent exposure
* medical or accident records held by employers
* details of workers' compensation claims
* smoking and possibly alcohol history for each worker and other possible confounding exposures.

One of the shortcomings of the majority of epidemiological studies conducted in the workplace is that data on the exposures of the population under study are either inadequate or absent.

2.7.1   *WHY CARRY OUT EPIDEMIOLOGICAL STUDIES?*

The purpose of epidemiological studies is to find out whether there is an association between exposure and ill-health so that appropriate action can be taken. For example, the association between mesothelioma and blue asbestos exposure has been well demonstrated through epidemiology studies. The association is so strong and the outcome so unequivocal that use of asbestos is now prohibited by law. Likewise, the increased rates of aplastic anaemia and leukaemia among workers exposed to benzene compared with members of the general population led to restrictive legislation, a reduction in benzene concentrations in petrol and substitution with far less toxic solvents in other uses.

Such dramatic links between occurrence and cause are not always evident. Many—probably most—studies suffer from deficiencies and limitations. Confounding factors such as drug or alcohol use, bias in selection of the study population and poor or missing data often prevent firm conclusions from being drawn. Old, unverifiable clinical observations may be incorrect, workplace histories may be incomplete or processes may be changed, resulting in different exposure patterns. This variability may cause the signal to become lost in the noise, rendering the findings equivocal.

Despite these shortcomings, occupational epidemiology often provides the only valid way to establish the human health hazards associated with many substances and processes.

2.7.2   HOW ARE THESE STUDIES CONDUCTED?

Two epidemiological approaches commonly employed are:

- observational studies—for example, ecological, cross-sectional, case-control, longitudinal or cohort studies
- experimental studies—for example, randomised controlled trials, field trials or community trials.

Since nearly all studies in occupational epidemiology are observational studies, we will not deal further here with experimental studies. The observational studies can be listed as follows, in order of methodological strength.

2.7.2.1   Cohort studies

Cohort studies are used to ascertain whether groups with an exposure to a hazardous agent have more disease than a non-exposed group. Studies of cohorts, or peer groups, involve:

- a known group of workers
- a fixed time period of study
- some knowledge of exposure to the agent under investigation.

The group members may all be involved in a process at a particular factory (e.g. viscose rayon workers handling carbon disulfide) or in a particular industry (e.g. all workers involved in vinyl-chloride manufacture in a country). For statistical significance, these groups usually need to be fairly large in number—several hundred or more. The assumption is that everyone within the cohort has a similar exposure to the agent that varies only with the length of time of exposure (employment).

The time period may be from some point in the past up to the present, in which case the cohort under investigation is called a *historical cohort*. Studies of such cohorts are called retrospective studies. If the study starts now and runs into the future, it is called a *prospective study*.

Knowledge of exposure to the agent requires hygiene measurement data. If this is not available, the surrogate 'years of exposure' is often used, in which the periods are split up into several classes: 1–5 years, 6–10 years, 11–15 years and so on.

Retrospective studies are comparatively cheap to conduct; however, there is little or no control over the quality of data available. Complete identification of the study population and obtaining all subjects' job and smoking histories is often problematic. The linking of subjects identified from company files to state and national cancer registry records is difficult and likely to be incomplete. Prospective studies are normally more expensive, and may take ten to twenty years to complete, but data on the cohort are easier to assemble.

These studies have one other vital element: a group to use as a yardstick. This comparison group is called the control population. The control group should differ from the population under study only in not having experienced exposure to the agent of interest. For example, in a study of the health effects of foundry work, the population under investigation cannot be compared with other foundry workers. The control population should be similar to the exposed population in age, smoking history and socioeconomic status. General population statistics are often used as the control, but special groups can be constructed when required. Investigators often have considerable difficulty matching the demographics of the control group to those of the study group.

These studies express findings as measured morbidity (e.g. incidence of a particular illness) or mortality rates compared with the control population. The relative rate (the rate experienced in the study group divided by the rate in the control group) is expressed as a ratio, such as Equation 2.3:

$$\text{Standardised mortality ratio (SMR)} = \frac{\text{observed number of deaths in the study group} \times 100}{\text{expected number of deaths in a control population}} \qquad \text{(Equation 2.3)}$$

An SMR of 100 shows that the population under investigation is dying at the same rate as the comparison population. A higher SMR shows that the population under investigation has a higher mortality rate than the comparison population.

Industrial populations often have a low SMR. This is known as the healthy worker effect. The healthy worker effect arises because, on average, workers are in better health than the general population for two reasons: first, because the general population includes people unable to work because of ill-health; and second, because workers in ill-health tend to leave work.

To express risk in cohort studies, we need to establish which subjects have experienced the exposure of interest, then identify the proportion of the exposed and unexposed subjects who have experienced the health outcome of interest. A simple example involving numbers of persons suffering from renal failure after exposure to cadmium (Cd) is given in Table 2.1.

Table 2.1 Schematic showing proportions of a cohort study displaying presence or absence of renal disease associated with cadmium exposure

	Renal failure	No renal failure	Total
Cadmium exposure	a	b	a + b
No cadmium exposure	c	d	c + d
Total	a + c	b + d	

Equation 2.4 expresses the relative risk (RR) of someone in a cohort study (such as that in Table 2.1) developing renal failure (column 1) associated with cadmium exposure (row 1) compared with the probability of a non-exposed person (row 2) developing renal

failure (column 2). If the RR is greater than 1, there is a risk that the exposure causes that disease; statistical analysis is required.

$$\text{Relative risk} = \frac{[a/(a + b)]}{[c/(c + d)]} \qquad \text{(Equation 2.4)}$$

2.7.2.2 Case-control studies

A case-control study takes a different approach: it assesses whether people with a disease of interest (a case) have more exposure to the agent of interest than a non-exposed population. The study populations comprise:

- a group of patients or workers with a particular disease, and
- a control group whose members do not have that particular disease.

The exposure histories of the cases and controls are then compared. It is possible that some cases and some controls will have had exposure to the agent under study. Being able to reach valid inferences depends on how well the exposure/non-exposure histories are recorded. This method is useful in the study of rare diseases because investigators can work with small numbers. Investigating rare diseases with a cohort study would require large numbers of participants and consequently be very expensive.

In case-control studies, epidemiologists use the measure called the *odds ratio*, which is given by the following equation, where a, b, c and d are defined as in Table 2.1.

$$\text{Odds ratio} = \frac{ac}{bd} \qquad \text{(Equation 2.5)}$$

These odds compare the probability of having the disease and exposure versus the probability of having the disease without exposure. If the number is larger than 1, the odds of having the disease and exposure are higher than having the disease and no exposure. For relatively rare diseases, the relative risk calculated in a cohort study and the odds ratio in a case-control study are approximately the same.

2.7.2.3 Cross-sectional studies

These studies are generally undertaken at a 'point in time' or over a relatively short period of time. They comprise a snapshot of the health and exposure of a particular group. As a result, causation is difficult to determine, because it is not possible to establish whether the exposure preceded or followed the disease. A major limitation of these studies is the possible influence of a survivor effect, where affected individuals have left employment and are therefore not included in the study population. For instance, a cross-sectional study investigating the prevalence of asthma in a dusty industry may result in a 'low prevalence' being reported, suggesting that the dust exposure is not related to asthma; however, the heavily exposed workers who suffered asthma may have left the industry before the study was conducted. Only those not so heavily exposed thus take part in the study, and these workers are predominantly non-diseased.

Cross-sectional studies are inexpensive and may be done quickly and easily. They are useful as a first step in identifying a suspected cause if exposure to many agents occurs in a workplace.

2.7.2.4 Ecological studies

Ecological or correlational studies are considered 'hypothesis-generating' studies. They involve the analysis of the health and exposures of groups or populations, not individuals, so the link between exposure and disease cannot be made. An example of an ecological study would be the observation that population A smokes more than population B and has higher rates of heart disease. We do not know, but we might suspect, that the smokers in population A are the ones with the higher rate of heart disease. Ecological studies can be inexpensive and simple to conduct, as they often rely on banks of data collected for other purposes, such as compensation databases, cancer registers or hospital records. These studies can easily be biased, owing to various socioeconomic factors not being controlled. They are considered the weakest in terms of study design, but are often initiated because of their low cost and quick results.

2.7.3   A FEW RELEVANT EPIDEMIOLOGICAL TERMS

Epidemiological studies express disease burden in terms of rates known as *incidence* or *prevalence*. Prevalence is the total number of cases in the population at any point in time while incidence is the rate of new cases appearing in the population during a certain time period. Incidence is expressed as new cases per 1000 people at risk per year. Determining incidence is crucial to studying changes in the rate of appearance of a disease. Prevalence relates to whether the pool of cases is growing or shrinking.

Figure 2.5 allows us to appreciate prevalence as the result of both incidence and loss. A cross-sectional study at a particular point in time will yield a point prevalence. When a study is conducted over a particular timespan, it can produce a period prevalence. Cross-sectional studies are often called 'prevalence studies' because they can measure only the prevalence and not the incidence of a disease.

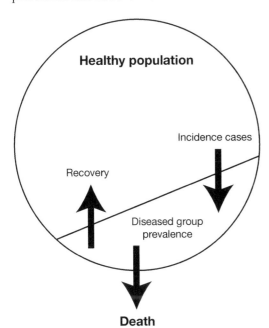

Figure 2.5 Diagram indicating the difference between incidence and prevalence

All studies suffer from bias—indeed, criticism of studies on the grounds of biased data seems to be an occupational hazard for the epidemiologist. Bias can occur in the selection of the group to study or selection of the control group, in interpreting data, in questionnaire response rates, in establishing healthy and unhealthy worker effects, in the mismatching of age distributions and in evaluating effects on survivors. It can also arise from a loss of subjects to follow-up, subjective assessments, confounders such as the effects of smoking, socioeconomic class and so on. The aim in selecting a study population is to control for these biases so that the study population is similar to the control population in all aspects except for exposure to the agent.

Epidemiological studies always refer to statistical significance in their findings. This is the probability that the association between the exposure and the outcome occurs by chance. For example, $p < 0.05$ indicates that the finding has less than a 5 per cent probability of occurring by chance.

For example, a study might look at bladder cancer among rubber workers. The epidemiologist naturally wants to attach some degree of confidence to the findings. A whole range of statistical methods are available for numerical testing of various hypotheses, but the tests generally seek one common goal: a probability of 95 per cent or more that the observations did not arise by chance—in other words, a probability of 5 per cent or less that this is a random association. This is sometimes described as 'statistically significant at $p < 0.05$'.

If the results of a study are described as statistically significant, this means that the variation in disease rates between the study and control populations is unlikely to result from chance alone. For example, a study that shows an SMR of 300 (eighteen deaths found when only six were expected) might be able to quote its findings as statistically significant if the lower 95 per cent confidence limit for the SMR is greater than 100. (See Checkoway, Pearce and Kriebel, 2004 for an explanation of these statistical terms.)

2.7.4   PRACTICAL SIGNIFICANCE

Despite the obvious difficulties in occupational epidemiology, this field of study provides the surest basis for long-range improvements in occupational health. Such studies have provided much of the information we have to date on rare disease occurrence and causative exposure. As a result, the H&S practitioner should always be receptive to the possibility that workplace factors may be implicated in ill-health, while resisting the temptation to blame them for every illness.

2.8   CONCLUSIONS

Section 2.3 introduced some of the issues associated with establishing cause in respect of workplace chemical exposures and disease. In particular, if a worker has symptoms, how can they confidently be attributed to the workplace? With an understanding of basic toxicology and epidemiology, you may be in a position to suggest a causal relationship if there are data, observations or published evidence to support the following:

- Gradients exist in space—that is, those workers remote from the source of a pollutant responsible for disease have low rates of symptoms, those near the source have high rates of symptoms and those in between have intermediate rates of symptoms.

- Distribution of cause reflects distribution of response—for example, workers at similar industrial plants without the proposed cause do not display symptoms.
- Time sequences are observed—for example, symptoms occur after exposure.
- Results are consistent with the findings of other published studies.
- The proposed mechanism for the effect is plausible.
- Experimental evidence (e.g. animal studies, human exposure studies) supports the results.
- Preventive trials or interventions reduce the effects.

In addition, to critically evaluate the extent of exposures and the veracity of symptom reports, we may ask:

- Was there a potential for exposure?
- Was the potential exposure level likely to cause symptoms?
- Are symptoms consistent with known effects of the suspected agent?
- Are there independent measurable disease/symptom indices?
- Are there independent measures of exposure?

In some cases, there may be a clear association between exposure and reported symptoms. In other cases, some of the circumstances above may be met and others not, and the H&S practitioner will need to make a judgement based on the balance of evidence. Experienced hygienists may effectively meet these challenges. However, it is also important to recognise when specialist toxicological, epidemiological or other assistance is required to reach appropriate conclusions.

2.9   REFERENCES

Australasian Faculty of Occupational Medicine (AFOM) Working Party on Occupational Cancer 2003, *Occupational Cancer: A Guide to Prevention, Assessment and Investigation*, AFOM, Sydney.

Checkoway, H., Pearce, N. and Kriebel, D. 2004, *Research Methods in Occupational Epidemiology*, 2nd ed., Oxford University Press, New York.

Organisation for Economic Cooperation and Development (OECD) 2019, *OECD Guidelines for the Testing of Chemicals*, OECD, Paris.

3. The concept of the exposure limit for workplace health hazards

Robert Golec

3.1   INTRODUCTION

The occupational exposure limit (OEL) has been the cornerstone of occupational hygiene risk assessment and risk management, and forms an important element of the control of occupational disease and the setting of policies on occupational health (Henschler, 1984; Howard, 2005; Vincent, 1999). In Australia, Safe Work Australia (2018a) has set OELs—termed workplace exposure standards (WES)—for airborne contaminants for approximately 700 workplace chemicals (Safe Work Australia, 2013a) and it is estimated that there are OELs for more than 6000 specific chemicals globally (Brandys, 2008). By way of comparison, the number of chemicals in commercial use exceeds 100 000 (Todd, 2004).

In order to understand the concept of the OEL, it is important to understand the history of its development, the thinking behind the setting and use of OELs and the strengths and weaknesses of the OEL approach, including criticisms and challenges.

3.2   HISTORY OF OCCUPATIONAL EXPOSURE LIMITS

The OEL has its basis in the toxicological concepts of a threshold of effect for exposure to chemicals and a relationship between dose and response. As discussed in Chapter 2, this concept was first theorised in the sixteenth century by the Swiss physician and chemist Phillippus Paracelsus, who declared that the dose of a chemical determines whether it exhibits toxic properties (Gallo, 2008). In the fourth century BC, Hippocrates first recorded cases of occupational disease from exposure to chemicals among mine workers with lead poisoning (Patty, 1958), and Paracelsus—who worked for ten years as a labourer in mining and smelting—later wrote an important treatise that described various miners' diseases (Patty, 1958). However, the systematic study of industrial diseases caused by exposure to contaminants in the workplace did not begin until the late seventeenth century, when the Italian pioneer of occupational medicine, Bernadino Ramazzini, showed that in the dusty trades, 'ill-health was caused by the inhalation of subtle particles that were offensive to human nature' (Oliver, 1902, p. 28). Ramazzini urged physicians to ask their patients for the details of their occupations in order to make the correct diagnoses (Hamilton and Hardy, 1949). At that time, meaningful quantification of workplace exposures to chemicals was not possible, owing to a lack of suitable sampling and analytical techniques (Paustenbach, Cowan and Sahmel, 2011). Researchers could make only qualitative assessments based on visual observations of the workplace, medical examination of workers and post-mortem findings. In the late nineteenth century, the German hygienist and toxicologist Karl Bernhard Lehmann and his pupils conducted a series of quantitative studies on 35 gases and fumes involving animal and human experiments and observations in industry. These laid the foundations for the quantitative assessment of levels of chemical exposure at which various health impacts occurred (Kober, Hayhurst and Rober, 1924). Lehmann also wrote about the quantitative analysis of workplace air for chemical and biological contaminants (Lehmann, 1893).

By the start of the twentieth century, advances in analytical chemistry and the development of occupational hygiene sampling techniques made it possible to reliably quantify exposure (dose) of chemicals to workers in the workplace. This led to estimating levels of exposure that were relatively 'safe', and those at which serious effects might be expected. One of the first attempts to formulate a set of acceptable acute OELs was that of the German researcher Rudolf Kobert, who in 1912 published a list of 20 chemicals and the doses and durations of exposure required for them to bring about effects from minimal disturbance through to rapid death (Paustenbach, Cowan and Sahmel, 2011). Kobert's list is reproduced in Table 3.1.

In 1921, the US Bureau of Mines published a technical paper listing thresholds for odour, irritation of the eyes and throat and coughing, maximum one-hour exposure levels without symptoms, and lethal exposure doses for some 33 chemicals commonly found in industry at the time (Henderson, 1921). The International Critical Tables of 1927 listed 'maximum safe concentrations' for 27 gases and vapours, a list based partly on Kobert's original 1912 data combined with data from other sources (Sayers, 1927). The former Soviet Union was the first country to establish a set of compulsory statutory OELs for fourteen chemical substances in 1933 (OSHA, 1989).

In 1941, the National Conference of Governmental Industrial Hygienists (NCGIH), since renamed the American Conference of Governmental Industrial Hygienists (ACGIH®), published a list of maximum allowable concentrations (MACs) for 63 chemicals (Baetjer, 1980; ACGIH, 2018). The MAC list was intended to be reviewed and published annually from 1946 onwards. The ACGIH® stated from the outset that these MACs were not to be taken as recommended safe concentrations (Ziem and Castleman, 1989). The ACGIH® MACs of 1946 drew heavily on a 1945 list used by various US states and the US Public Health Service and recommended by the American Standards Association compiled by the industrial hygienist Warren Cook. Some of Cook's values were based on extensive 'animal experiments and experience with workers under actual industrial conditions', while others were based on limited animal experiments or on professional judgement (Cook, 1945). Cook (1945, pp. 936–7) noted that his list was intended as a 'handy yardstick to be used as guidance for the routine industrial control of these health hazards—not that compliance with these figures listed would guarantee protection against ill-health on the part of exposed workers'. The maximum allowable concentrations were renamed *threshold limit values* (TLVs®) in 1956, and the first documentation of threshold limit values for 257 substances was published in 1962 (ACGIH, 2018). Despite the limitations and warnings by the ACGIH® that TLVs® should not be used for compliance purposes, the 1968 TLV® list was adopted by OSHA as US federal law.

Table 3.1 Kobert's 1912 list of acute exposure limits

Chemical	For human and animal rapid death	0.5–1 hour exposure serious threat to life	0.5–1 hour without serious health effects	Repeated exposure, minimal symptoms
Hydrogen chloride		1500–2000 ppm	500–1000 ppm	100 ppm
Sulfur dioxide		4000–5000 ppm	500–2000 ppm	200–300 ppm
Hydrogen cyanide	~3000 ppm	1200–1500 ppm	500–600 ppm	200–400 ppm
Carbon dioxide	30 per cent	60000–80000 ppm	40000–60000 ppm	20000–30000 ppm
Ammonia		240–450 ppm	300 ppm	100 ppm
Chlorine	~10,000 ppm	400–600 ppm	40 ppm	10 ppm
Bromine	~10,000 ppm	400–600 ppm	40 ppm	10 ppm
Iodine			30 ppm	5–10 ppm
Phosphorus trichloride	3500 mg/m^3	300–500 mg/m^3	10–20 mg/m^3	4 mg/m^3
Phosphine (phosphane)		400–600 ppm	100–200 ppm	
Hydrogen sulfide	10000–20000 ppm	5000–7000 ppm	2000–3000 ppm	1000–1500 ppm
Gasoline			15000–25000 mg/m^3	5000–10000 mg/m^3
Benzene			10000–15000 mg/m^3	~5000 mg/m^3
Carbon disulfide		10000–12000 mg/m^3	2000–3000 mg/m^3	1000–1200 mg/m^3
Carbon tetrachloride	300000–400000 mg/m^3	~150000–200000 mg/m^3	25000–40000 mg/m^3	~10000 mg/m^3
Chloroform	300000–400000 mg/m^3	70000 mg/m^3	25000–30000 mg/m^3	~10000 mg/m^3
Carbon monoxide		20000–30000 ppm	5000–10000 ppm	2000 ppm
Aniline			400–600 mg/m^3	100–250 mg/m^3
Toluidine			400–600 mg/m^3	100–250 mg/m^3
Nitrobenzene			1000 mg/m^3	200–400 mg/m^3

Source: Ripple (2010).

The ACGIH® list of TLVs® grew over time and was used as the basis for many other nations' sets of occupational exposure limits, including the German MAKs, British WELs and Finnish OELs (Henschler, 1984). Reliance on the ACGIH® list has declined in Europe since the European Union developed its own set of guidelines for setting OELs (Adkins et al., 2009a).

The European Scientific Committee on Occupational Exposure Limits (SCOEL) has defined OELs as 'limits for exposure via the airborne route such that exposure, even when repeated on a regular basis throughout a working life, will not lead to adverse effects on the health of the exposed person and/or their progeny at any time (as far as can be predicted from the contemporary state of knowledge)' (ECESAI, 2010). This definition implies that the objective of European OELs is to protect all workers, including potentially susceptible individuals such as unborn offspring. By contrast, the ACGIH® approach is to set limits that protect 'nearly all workers'. SCOEL's methodology discusses setting two types of OELs: 'health-based' OELs, which are established on the basis of a clear threshold dose below which exposure is not expected to cause adverse effects; and 'risk-based' OELs, where it is not possible to define a threshold (e.g. genotoxic carcinogens and respiratory sensitisers) and where any exposure may carry some risk.

In Australia, the influence of the ACGIH®'s TLVs® was evident by the early 1970s, when the Occupational Hygiene sub-committee of the National Health and Medical Research Council (NHMRC) published a schedule of recommended concentrations for occupational exposure based largely on the ACGIH® limits. It should, however, be noted that the setting of OELs in Australia significantly pre-dates the NHMRC schedule, as 'permissible standards for toxic gases, vapours and fumes in industry' had been published by the NHMRC Committee on Industrial Hygiene in Munitions Establishments in 1944, and a list of maximum allowable concentrations of substances was gazetted in 1945 in the Victorian Health (Harmful Gases Vapours Fumes Mists Smokes and Dusts) Regulations (De Silva, 2000). The NHMRC schedule was initially a more or less direct adoption of the ACGIH®'s TLV® list, but the Occupational Hygiene sub-committee later developed some of its own limits based on Australian experience and approaches (De Silva, 2000).

The formation of the National Occupational Health and Safety Commission (NOHSC) in 1985 saw the transfer of responsibility for the development of OELs away from the NHMRC to the NOHSC Exposure Standards Working Group, under the auspices of the Standards Development Standing Committee (SDSC), whose terms of reference were to 'Consider and recommend options to the Standards Development Standing Committee on occupational exposure standards for atmospheric contaminants based on consideration of the best available technical data from Australian and overseas sources' (NOHSC, 1995b). Initially, a list of exposure standards was adopted based on the NHMRC's schedule, but in 1995 NOHSC published a list of OELs based on the recommendations of the working group (NOHSC, 1995a). Although NOHSC had no statutory authority over the states and territories, which have ultimate responsibility for occupational health and safety, NOHSC's exposure standards became de facto legal compliance standards for workplace chemical exposure, as they were called up in the various OHS regulations in each jurisdiction. The role of OELs in Australian legislation has been reinforced with the introduction of harmonised work health and safety (WHS) regulations in a number of states and territories, in which the workplace exposure standard has been prescribed as 'the airborne concentration of a particular substance or mixture that must not be exceeded' (Safe Work Australia, 2013b, p. 6).

Since 1995, there have been only some 48 changes (amendments and additions) to the NOHSC's list of approximately 700 exposure standards. These have included lowering the limits for crystalline silica, benzene and chrysotile asbestos. Safe Work Australia's national workplace exposure standards are available in a searchable database on the organisation's Hazardous Chemical Information System website (Safe Work Australia, 2018c). At the time of writing, the methodology for setting workplace exposure standards in Australia is under review with the intent of drawing from a body of knowledge developed by OEL-setting bodies in other countries, including the American Conference of Governmental Industrial Hygienists (ACGIH®), Deutsche Forschungsgemeinschaft (DFG), the EU Scientific Committee on Occupational Exposure Limits (SCOEL), the American Industrial Hygiene Association/Occupational Alliance for Risk Science (AIHA/OARS) and the Health Council of the Netherlands (Dutch Expert Committee on Occupational Safety). More information about the proposed changes to the Safe Work Australia workplace-setting process is available from the Safe Work Australia website (Safe Work Australia, 2018c).

3.3 DEFINITION OF OCCUPATIONAL EXPOSURE LIMITS

As mentioned, Safe Work Australia's version of OELs are referred to as **workplace exposure standards** (WESs). SWA defines a WES as 'the airborne concentration of a particular substance or mixture that must not be exceeded (Safe Work Australia, 2013a).

The Australian WESs are not fine dividing lines between safe and hazardous exposures; nor are they measures of relative toxicity, as the basis on which individual standards are set differs from substance to substance.

A WES can take one or all of three forms:

- *Time-weighted average (TWA).* An eight-hour time-weighted average exposure standard is the average airborne concentration of a particular substance permitted over an eight-hour working day and a five-day working week. The TWA is the most common of the exposure standards and, except where a peak limitation has been assigned, virtually all substances are listed with a TWA.
- *Short-term exposure limit (STEL).* A short-term exposure limit is the time-weighted maximum average airborne concentration of a particular substance permitted over a fifteen-minute period. STELs are established to minimise the risk of:
  - intolerable irritation
  - irreversible tissue change
  - narcosis to an extent that could precipitate workplace incidents.
  Exposures at the STEL should not be longer than fifteen minutes and should not be repeated more than four times per day. There should be at least 60 minutes between successive exposures at the STEL.
- *Peak.* A maximum or peak airborne concentration of a particular substance determined over the shortest analytically practicable period that does not exceed fifteen minutes. Chemical substances that have workplace exposure standards are also classified as

known, presumed or suspected carcinogens, or known respiratory or skin sensitisers according to the criteria of the Globally Harmonised System of Classification and Labelling of Chemicals (GHS).

- *Skin (Sk).* Contaminants where skin absorption may be a significant source of exposure.
- *Respiratory sensitiser (Sen). A* substance that leads to hypersensitivity of the airways after being inhaled.
- *Skin sensitiser (Sk, Sen).* A substance that leads to an allergic response after skin contact.

3.3.1   *CARCINOGEN CATEGORIES*

- *Carcinogenicity Category 1A.* Known to have carcinogenic potential for humans. The classification of a chemical into this category is based largely on human evidence from studies that have established a causal relationship between human exposure and the development of cancer.
- *Carcinogenicity Category 1B.* Presumed to have a carcinogenic potential for humans. The classification of a substance into this category is based largely on animal evidence where there is sufficient evidence to demonstrate carcinogenicity in animals or where there is limited evidence of carcinogenicity in humans and animals.
- *Carcinogenicity Category 2.* Suspected human carcinogen. The classification of a chemical into this category is on the basis of evidence from human and animal studies, where the evidence is not sufficiently convincing to place the chemical into Category 1, or from limited evidence of carcinogenicity in human or animal studies.

Table 3.2 Examples of Australian workplace exposure standards

Contaminant	TWA exposure limit	STEL exposure limit	Carcinogen category	Notation
Toluene	50 ppm	150 ppm	–	Sk
Isocyanates, all (as –NCO)	0.02 mg/m³	0.07 mg/m³	–	Sen
Chromium (VI) compounds (as Cr), certain water insoluble	0.05 mg/m³	–	1A	Sen
Glutaraldehyde	0.1 ppm peak limitation	–	–	Sen
Crocidolite asbestos	0.1 f/mL	–	1A	–
Phenylhydrazine	0.1 ppm		1B	Sk, Sen

Source: Safe Work Australia (2018a).

3.4 UNITS OF CONCENTRATION

Exposure standards are concentrations in air that can be expressed as a mass per unit volume, such as milligrams of substance per cubic metre of air (mg/m^3) or parts per million (ppm). For gases and vapours, the units can be converted from one to the other using the formula:

$$\text{Concentration in ppm} = \text{Concentration in mg/m}^3 \times \left(\frac{24.45}{\text{Molecular weight of contaminant}} \right) \qquad \text{(Equation 3.1)}$$

where 24.45 is the molar volume at 25°C and 1 atmosphere (101.325 kPa)

For example, the Australian TWA-WES for formaldehyde is 1 ppm. The average concentration of formaldehyde over eight hours measured in the breathing zone is found to be 2 mg/m^3. The molecular weight of formaldehyde is 30.03. The concentration of formaldehyde in the breathing zone in ppm is:

$$2 \text{ mg/m}^3 \times \left(\frac{24.45}{30.03} \right) = 1.6 \text{ ppm (which is above the WES)}$$

For mineral fibres such as asbestos and ceramic fibres, the concentration is expressed as the number of respirable-sized fibres (f) per millilitre (ml) of air. This will be discussed further in Chapter 7.

3.5 INTERPRETATION OF EXPOSURE STANDARDS

It is important to understand that OELs/WESs:

- refer to airborne concentration of contaminants in the worker's breathing zone
- are based on current knowledge
- apply to nearly all workers.

3.5.1 WORKER'S BREATHING ZONE

Exposure standards relate to personal exposure to airborne contaminants in the worker's breathing zone, which is defined as a hemisphere of radius 300 mm extending in front of the face, measured from the midpoint of an imaginary line joining the ears (Figure 3.1). Exposure standards do not take into account exposure via other routes such as skin absorption or ingestion. It follows that valid quantitative assessment of exposure for comparison against the OEL can *only* be made via personal monitoring where the sampling device is worn within the breathing zone. Static (or area) monitoring, although useful for providing general information on contaminant release and for assessing engineering controls, does not measure the concentration of a contaminant in the worker's individual breathing zone and therefore should not be used to assess compliance with a OEL/WES (Safe Work Australia, 2013a).

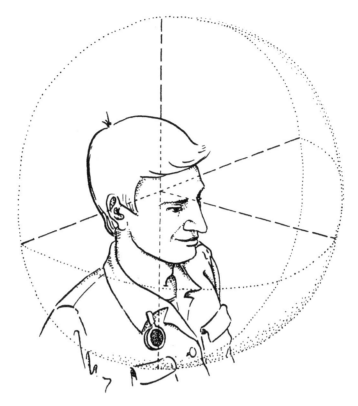

Figure 3.1 Worker's breathing zone
Source: Worksafe Victoria (1999, p. 37).

3.5.2 *ACCORDING TO CURRENT KNOWLEDGE*

The OEL-setting process involves the rigorous scientific evaluation of information from a number of sources, including:

- physicochemical data and the structure–activity relationship for the contaminant
- threshold toxicological data such as relevant no observed adverse effect levels (NOAEL) and lowest observed adverse effect levels (LOAEL), and non-threshold data from acute, chronic and sub-chronic animal experiments (Figure 3.2)
- human studies and/or literature on accidental poisonings or adverse reactions from exposure to the contaminant
- epidemiological data on exposed work populations and experience from industry
- analogy with homologues, or similar substances.

Over time, new toxicological and epidemiological data or re-evaluation of existing data may alter the basis upon which the original OEL is set. For example, some substances that were originally thought to have primarily irritant effects have since been found to cause chronic disease. Inevitably, such new information will result in a review and amendment (usually lowering) of the OEL. Figure 3.3 shows how OELs have reduced over time. For example, the ACGIH® TLV® for benzene in 1976 was 25 ppm as a ceiling value. This was

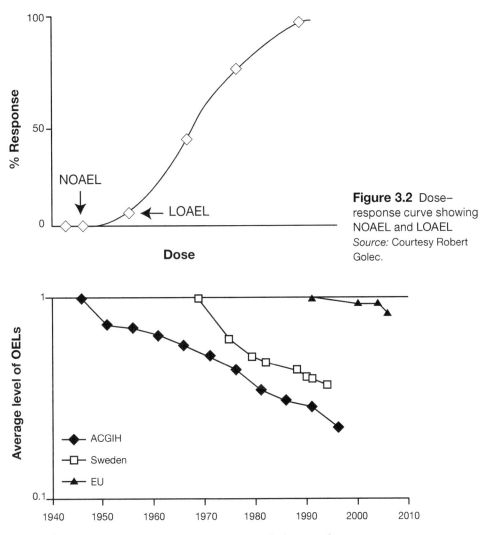

Figure 3.2 Dose–response curve showing NOAEL and LOAEL
Source: Courtesy Robert Golec.

Figure 3.3 Reduction of occupational exposure limits over time
Source: Schenk (2011, p. 17).

reduced to a TWA–TLV® of 10 ppm in 1977 and to a TWA–TLV® of 0.5 ppm in 1996 on the basis of updated toxicological and epidemiological data.

3.5.3 NEARLY ALL WORKERS

OELs should not be considered to be a fine dividing line between exposures that are safe and those that are hazardous—a fact often overlooked by those who use them. The ACGIH® states in its 'Introduction to the Chemical Substances of the TLVs®' that 'TLVs® will not adequately protect all workers. Some individuals experience discomfort or even more serious adverse health effects when exposed to a chemical substance at the TLV® or even at concentrations below the TLV®' (ACGIH, 2019, p. 3).

Factors that may make some individuals more susceptible to chemicals include genetic predisposition, hypersensitivity/sensitisation, pre-existing disease and medication, lifestyle (tobacco, alcohol and recreational drug use), age, gender and reproductive status (De Silva, 1986).

3.6   ADJUSTMENTS FOR EXTENDED WORK SHIFTS

TWA–OELs are based on the assumption that individuals are exposed for eight hours per day, five days a week (a 40-hour work week) for an entire working lifetime. Changes in work patterns mean that some employees may be exposed for periods longer than the assumed standard working hours. In these cases, it may be appropriate to adjust the TWA–OEL to take the longer exposure period and shorter recovery times into account. It should be noted that such adjustments are not appropriate in the case of OELs that are based on effects such as acute irritation and cyanosis, nor should they be used to adjust STEL and peak exposure limits.

Several methods for adjusting TWA exposure limits have been proposed, including mathematical models such as the Brief and Scala and OSHA models, and complex pharmaco-kinetic models.

The Brief and Scala method (Brief and Scala, 1975) is regarded as the most conservative model. It considers the impact of the number of increased hours worked and the recovery time between exposure periods, but not the agent's activity in the body. Using either the daily or weekly equation detailed below, a reduction factor is determined and then applied to the TWA-OEL. The Brief and Scala model reduces exposure limits according to a reduction factor calculated by the following formula:

Daily exposure:

$$RF_{daily} = \frac{8}{h} \times \frac{24 - h}{16}$$
(Equation 3.2)

Where:
RF_{daily} = daily reduction factor
h = hours worked per shift

Note that $24 - h$ represents exposure-free hours per day.

Weekly exposure for the special case of a seven-day work week is thus:

$$RF_{weekly} = \frac{40}{h} \times \frac{168 - h}{128}$$

Where:
RF_{weekly} = weekly reduction factor
h = average hours per week over full roster cycle

Note that $168 - h$ represents exposure-free hours per week.

The adjusted TWA-OEL is calculated by applying the reduction factor to the eight-hour TWA-OEL:

$$Adjusted\ TWA\text{-}OEL = 8\text{-hr}\ TWA\text{-}OEL \times RF$$

The OSHA method is based on the assumption that the magnitude of the toxic response of an agent is a function of the concentration that reaches the site of action for that agent. The model was designed to be applied to most systemic toxic substances, but not irritants, sensitisers or carcinogens.

Each agent with an exposure limit is categorised based on its toxic effect. The assigned category (these should not be confused with the carcinogen categories) is then used to determine whether any adjustment is required and, if so, what equation is to be used. The categories used are shown in Table 3.3.

The Quebec model (Drolet, 2015) is essentially based on the OSHA model and uses the most recent information on toxicological effects, including sensitisation, irritation, organ toxicity, reproductive system toxicity and teratogenicity (Verma, 2000).

Depending on the category assigned, a recommendation will be made that either:

- no adjustment is made to the TWA-OEL
- a daily or weekly adjustment is made, or
- where both apply, the most conservative of the daily or weekly adjustments is made.

The Quebec model is supported by a comprehensive technical guide and a selection tool to assist in determining the most appropriate adjustment category. Caution must be applied when using the Quebec model to ensure that exposure standards appropriate to the jurisdiction are used (e.g. Safe Work Australia's WES for use in Australia).

The adjustment of the exposure standards using this model was made much easier by the development of a downloadable tool, freely available on the website of the Institut de recherche Robert-Sauvé en santé et en sécurité du travail (IRSST). This Excel spreadsheet has a drop-down list of assessed substances that provides the adjustment category or code and computes the RF (called the adjustment factor in the spreadsheet) based on the daily and weekly average working hours. An Australian version of the tool has been developed by the AIOH, and is available from the AIOH website.

Several different pharmacokinetic models are available. These are suitable for application to exposure standards that are based on accumulated body burden. They take into account the expected behaviour of the hazardous substance in the body based on knowledge of the properties of the substance, using information such as the biological half-life of a substance and exposure time to predict body burden. The use of pharmacokinetic models can be complicated by the lack of biological half-lives for many substances. The most widely used pharmacokinetic model is the Hickey and Reist model, which requires knowledge of the substance's biological half-life and hours worked per day and per week. The Hickey and Reist model, like other pharmacokinetic models, views the body as one compartment—that is, a homogeneous mass.

Pharmacokinetic models are less conservative than the Brief and Scala and OSHA/Quebec models, usually recommending smaller reductions of the exposure standard. While pharmacokinetic models are theoretically more exact than other models, their lack of conservatism may not allow adequately for the unknown effects on the body from night work or extended shifts, both of which may influence how well the body metabolises and eliminates the substance.

Table 3.3 Reduction factors based on adjustment categories

Adjustment category	Classification	Adjustment criteria	Reduction factor (RF)
1A	Ceiling standard *Ceiling standard never intended to be exceeded at any time—independent of length or frequency of work shifts.*	None	No adjustment
1B	Mild irritants *Standard designed to prevent acute irritation or discomfort. Essentially no cumulative effects known.*	None	No adjustment
1C	Standards set by technological feasibility or good hygiene practices—independent of shift length or frequency.	None	No adjustment
2	Acute toxicants *Can accumulate during an eight-hour or longer exposure time.*	Hours/day	$RF\ daily = \dfrac{8}{daily\ exposure\ hours}$
3	Cumulative toxicants *Cumulative exposure could occur over days to even years of exposure.*	Hours/week	$RF\ weekly = \dfrac{8}{weekly\ exposure\ hours}$
4	Both acute and cumulative	Hours/day and hours/week	RF daily or weekly, whichever is lowest.

3.7 MIXTURES OF SUBSTANCES

OELs are normally applied in situations where it is assumed that there is a single substance present. In many working environments, workers are commonly exposed to mixtures of substances that could present an increased risk to health. Applying OELs in circumstances where multiple contaminants are present is a complex task which requires care. Interactions between different substances should be assessed by appropriately qualified professionals such as occupational hygienists, occupational physicians and toxicologists after specific toxicological consideration of all substances involved.

3.7.1 INDEPENDENT EFFECTS

In cases where there is clear toxicological evidence that two or more contaminants have entirely distinct effects on the body, each substance may be evaluated separately against its appropriate exposure standard. For example, wood dust affects the respiratory system, while cyclohexane vapour acts upon the central nervous system. Each of these substances may therefore be assessed individually against its appropriate exposure standard. If neither standard is exceeded, the atmosphere within the working environment is deemed to be satisfactory.

3.7.2 ADDITIVE EFFECTS

When the body is exposed to two or more contaminants, an additive effect results when the contaminants have the same target organ or the same mechanism of action. For example, ethyl acetate and methyl ethyl ketone both affect the central nervous system. In this situation, the total effect upon the body equals the sum of effects from the individual substances. For substances whose effects are purely additive, conformity with the standard results when

$$\frac{C_1}{L_1} + \frac{C_2}{L_2} + \dots + \frac{C_n}{L_n} \leq 1 \qquad \text{(Equation 3.3)}$$

where $C_1, C_2 \dots C_n$ are the average measured airborne concentrations of the particular substances $1, 2 \dots n$, and $L_1, L_2 \dots L_n$ are the appropriate exposure standards for the individual substances.

An example of an additive effect is the general effect of organic solvents on the central nervous system (narcotic or anaesthetic effect); however, the exposure standards for a number of solvents, such as benzene and carbon tetrachloride, have been assigned on the basis of effects other than those on the central nervous system. It is therefore essential to refer to the documentation for the specific substances to ascertain the basis of the standard and any potential interactions.

3.7.3 SYNERGISM AND POTENTIATION

Synergism and potentiation occur when exposure to two or more contaminants results in an adverse reaction more severe than would be expected from the sum of the individual exposures. Synergism occurs when two or more substances or mixtures have an effect individually but their total effect exceeds the additive effect. For example, exposure to both carbon tetrachloride and alcohol presents a risk of liver damage much greater than exposure to those substances one at a time. Potentiation occurs when one substance or mixture, which is sometimes of no or low toxicity, enhances the effect of another substance or mixture. For example, exposure to ototoxins such as white spirits can worsen the impact of noise on the hearing or balance functions of the inner ear.

The following table illustrates the differences between additive, synergistic and potentiation effects for exposure to substances X and Y.

Table 3.4 Differences between additive, synergistic and potentiation effects

Type of interaction	Effect of substance X	Effect of substance Y	Combined effects of exposure to substances X and Y
Additive	25%	15%	40%
Synergistic	10%	30%	60%
Potentiation	0%	15%	100%

3.8   EXCURSION LIMITS

The concentration of an airborne contaminant arising from a particular process may fluctuate significantly over time. Even where the time-weighted average concentration does not exceed the OEL over the work shift, excursions above the eight-hour TWA–OEL should be controlled. Where there is no STEL limit assigned to a substance, a process is not considered to be under reasonable control if short-term exposures exceed three times the TWA–OEL for more than a total of 30 minutes per eight-hour working day, or if a single short-term value exceeds five times the eight-hour TWA exposure standard.

3.9   OCCUPATIONAL EXPOSURE LIMITS FOR PHYSICAL HAZARDS

In addition to OELs for chemical substances, the ACGIH® sets threshold limit values for physical hazards, including noise and vibration, ionising and non-ionising radiation, heat and cold, and musculoskeletal hazards.

In Australia, OELs for physical hazards are set by a number of organisations. Safe Work Australia has set a OEL for noise exposure of 85 dB(A) as an eight-hour equivalent continuous A-weighted sound pressure level ($L_{Aeq,8h}$) and 140 dB(C) C-weighted peak limit ($L_{C,peak}$) as a part of its *Code of Practice: Managing Noise and Preventing Hearing Loss at Work* (Safe Work Australia, 2018c). This is discussed in more depth in Chapter 12.

The responsibility for setting OELs for ionising and non-ionising radiation lies with the Australian Radiation Protection and Nuclear Safety Agency (ARPANSA). The agency published the *National Standard for Limiting Occupational Exposure to Ionizing Radiation* in 2002, and the *Radiation Protection Standard for Maximum Exposure to Radiofrequency Fields—3 kHz to 300 kHz*, also in 2002. These are discussed in Chapter 13.

Exposure limits have not been set for musculoskeletal hazards or thermal stress in Australia (see Chapter 14 for more detail).

3.10 BIOLOGICAL EXPOSURE INDICES

OESs are an invaluable tool for assessing the risk posed by exposure to inhaled chemical substances in the workplace. However, since many substances are significantly absorbed through the skin, focusing on the inhalation route of entry alone provides only part of the risk estimate. For example, many organophosphate pesticides are lipid soluble, and hence readily absorbed through skin, mucous membranes and gastrointestinal tract, yet the OELs for organophosphate (OP) pesticides are based exclusively on exposure by inhalation.

Biological monitoring measures the amount of a substance, its metabolites or its biochemical effect in a suitable biological sample such as urine, blood or exhaled breath to determine exposure by all routes. It is considered complementary to atmospheric monitoring. Biological monitoring can also be used to assess both the toxicological burden on the body and the adequacy of personal protective equipment. Biological monitoring may include testing for lead in blood, mandelic acid in urine as a measure of styrene exposure, carboxyhaemoglobin in blood as a measure of carbon monoxide exposure and tertrachloroethylene in exhaled breath.

Biological exposure indices (BEIs) are guidance values for assessing biological monitoring results. They represent the levels of the analyte (chemical, metabolite or biochemical effect) in the biological sample that are likely to be observed if a person is exposed at the OEL. There are some exceptions for chemicals that might cause non-systemic effects, or where the OEL is directed towards preventing responses such as irritation or the skin is a significant route of entry. The ACGIH® establishes BEIs and publishes them in its annual threshold limit values (TLVs®) and biological exposure indices (BEIs®) booklet, as well as in documentation which supports the BEIs. Safe Work Australia (2013b) has established a number of Guidelines for Health Monitoring for a range of hazardous chemicals, which include biological monitoring action and removal levels for a number of substances including cadmium, chromium, mercury and organophosphate pesticides. Chapter 10 discusses biological monitoring in more depth.

3.11 THE FUTURE OF OCCUPATIONAL EXPOSURE LIMITS

The formal process of setting OELs has been in place for over 60 years, yet of the tens of thousands of potentially hazardous substances commonly used in industry today, organisations such as the ACGIH® and Safe Work Australia have developed limits for only about 700. Indeed, for many new chemical substances there may be little or no toxicological data available, making it impossible to develop and apply OELs in the risk-assessment process. Additionally, supposedly health-based OELs developed in different countries for the same substances can vary significantly, a fact that may reflect divergent philosophies and variations in the scientific robustness of the OEL setting processes. Table 3.5 illustrates how OELs can vary between different countries and organisations.

Table 3.5 Examples of OELs in different countries and organisations

Substance	Carbon tetrachloride	Halothane	Phenyl glycidyl ether
ACGIH	5 ppm	50 ppm	0.1 ppm
Australia	0.1 ppm	0.5 ppm	1 ppm
European Commission	1 ppm	–	–
Finland	1 ppm	1 ppm	0.5 ppm
Germany	0.5 ppm	5 ppm	No OEL
Sweden	2 ppm	5 ppm	10 ppm
Quebec, Canada	1 ppm	–	1 ppm
US OSHA	10 ppm	–	10 ppm

Source: Adapted from: Schenk (2010).

Some researchers (Ziem and Castleman, 1989) have questioned OEL development process and the adequacy of OELs to protect against adverse health effects.

These limitations have prompted some to rethink the concept of the OEL and explore alternative approaches to OEL setting and chemical risk assessment, such as setting global OELs, control banding and the development of derived no effect levels (DNELs).

3.11.1   CONTROL BANDING

Control banding is a technique that utilises the occupational hygiene approach of antici-pation, recognition, evaluation and control in cases where there is a lack of data on quantitative exposure and hazard levels for a substance in a workplace. Processes for control banding are discussed more fully in Chapter 4, but in general it is used to aid the selection of appropriate measures for controlling hazards (e.g. ventilation, containment, substi-tution) according to the extent to which the hazards are banded (that is, based on their hazard phrases) and on a subjective estimate of exposure (high to low). Control banding has been used effectively for many substances that have not had OELs assigned, such as pharmaceuticals and nanomaterials, and deployed by small to medium enterprises where lack of expertise and expensive exposure monitoring may be a barrier to implementing effective controls. A number of electronic tools have been developed by organisations such as the UK COSHH Essentials, the German REACH-CLP Help desk and the Dutch Ministry of Social Affairs and Employment to assist in conducting control banding assess-ments. The International Occupational Hygiene Association (IOHA) endorses the use of control banding, but states that 'control banding should incorporate and complement other more well-established exposure limit and assessment methodologies and control strategies and . . . should not be considered a replacement for these methodologies' (IOHA, 2010, p. 1).

3.11.2 DERIVED NO-EFFECT LEVELS (DNELs)

The European Commission (EC) brought into force the REACH (Registration, Evaluation, Authorisation and Restriction of Chemicals) regulations in June 2007. Under the REACH legislation, substances subject to registration that are manufactured, imported or used in European markets in quantities of greater than 10 tonnes a year must have chemical safety assessments undertaken by a responsible party (normally the manufacturer or supplier), including the establishment of DNELs or levels beyond which humans should not be exposed. These DNELs must be established on the basis of a number of exposure scenarios for workers, consumers/general populations and in some cases specific vulnerable subpopulations (European Chemicals Agency, 2012).

Each exposure scenario and health effect for a substance requires the establishment of a DNEL, which is calculated by dividing the dose descriptor from toxicological experimental data (e.g. NOAEL, LOAEL, LD_{50}, LC_{50}) for the health effect by an appropriate assessment factor, allowing the data to be extrapolated for human exposure situations. For non-threshold effects (e.g. genotoxicity, carcinogenicity), the derived minimal effect level (DMEL) is the level of risk below which exposure is of no concern. The lowest DNEL for a given health effect is to be documented for each exposure scenario and included in safety data sheets. Unlike OELs, DNELs take into account exposure from all routes, not just inhalation, so therefore the basis upon which they are established differs from that of OELs. Additionally, DNELs do not take account of epidemiological data or industry experience. It should be noted that OELs are generally developed by scientific specialists in occupational hygiene, occupational medicine and toxicology and/or government agencies using published data and are measures of 'acceptable risk'. DNELs, on the other hand, are 'no-effect levels' developed by manufacturers, importers or suppliers, sometimes using proprietary data. Some have speculated that DNELs may replace OELs (Paustenbach, Cowan and Sahmel, 2011), while others regard DNELs as guidance values that can coexist with OELs and believe both should be used as part of a chemical risk assessment and management process (ECESAI, 2010).

3.11.3 GLOBAL OCCUPATIONAL EXPOSURE LIMITS

Extensive efforts have been made to harmonise the classification and labelling of chemicals under the Globally Harmonized System of Classification and Labelling of Chemicals (GHS) adopted by the United Nations in 2003. However, harmonising OELs across the globe remains a major challenge. The establishment of the Scientific Committee on Occupational Exposure Limits (SCOEL) and the development of a methodology for setting European OELs are important steps towards unifying these, at least among EU members, and signal a possible diminution of the influence of the ACGIH® TLVs®. Yet it remains to be seen whether the regulatory, political, cultural/philosophical and economic factors currently impeding the progress of developing globally harmonised OELs can be overcome. Some have argued that as OELs are not absolute thresholds, but allow for 'acceptable risk' levels, 'the acceptability of the risk should ultimately be determined by the cultural body politic of the society and thus could be different for different groups' (Adkins et al., 2009b, p. 48). Nonetheless, many experienced hygienists still consider OELs to be the most effective risk assessment tools or guides for developing strategies to protect workers from exposure to chemicals (Adkins et al., 2009b).

The OEL has no utility in itself, however, unless there are reliable complementary tools for measuring workplace exposures and interpreting them against the OEL in a consistent fashion across workplaces, industries and countries. Enshrining OELs in law or state regulation has created another major challenge for regulators and users alike, since little thought has been given to how these OELs are meant to function in a regulatory fashion. Although the OEL has considerable appeal as a number against which workplace performance or compliance could be judged, in reality the myriad workplace factors, the limitations on monitoring, the differences or errors in measurement and the variety of ideas about what constitutes compliance all militate against consistent outcomes. Some workplace monitoring issues, the tools that have evolved and the requirements for decision-making about compliance with exposure regulations are all covered in the following section.

3.12 MAKING COMPLIANCE DECISIONS BASED ON REGULATORY USE OF EXPOSURE STANDARDS

Interpreting exposure standards, deciding which type to use, allowing for mixed exposures and so on are all relatively straightforward compared with making reliable decisions about compliance with regulations based on those standards.

There are several reasons for this. First, the use of a simple numerical limit in regulation does not facilitate easy decision-making for exposures which are known to vary widely depending on processes, environmental conditions, work practices, control efficacy and individual worker behaviour. Second, to accommodate all this variability, measurements need to be subject to statistical examination before informed judgements about exposure can be made. Third, even when fairly robust statistical measures of exposure can be determined, there is no common understanding on which measure should be used for making decisions on compliance with regulation, and the law itself gives no clear indication of what constitutes compliance. Finally, undertaking a valid sampling and monitoring exercise where resources are limited will still require the H&S practitioner to develop protocols for identifying differently exposed groups in a given workplace to establish a monitoring regime with adequate numbers of samples and an appropriate frequency of sampling.

One of the indeterminacies with which regulation has to deal is that 'compliance' is not a numerical condition, but the yardstick by which it is assessed is a numerical one. Further, the tools for assessing compliance are mostly numerical measurement ones and, when used, must be able to provide a sufficiently robust outcome for making a non-parametric compliance decision. Despite these obstacles, the National Occupational Health and Safety Compliance and Enforcement Policy prepared by the Heads of Workplace Safety Authorities (2008) has identified a number of goals, the first of which is 'to ensure duty holders have access to information about occupational health and safety laws and how to comply'.

3.12.1 *UNDERSTANDING REGULATORY COMPLIANCE*

Within the Australian context, regulations on hazardous chemicals are based on the principle that 'no one is exposed to a substance or mixture in an airborne concentration that exceeds the exposure standard'. Here it is clear that compliance applies not only to groups of workers, but to every individual within a given group. To encompass all the variability imposed on exposure by process, work environment, control and so on, and to manage coverage of all workers without embarking on the impossible task of sampling every worker, statistical procedures of some kind are needed for sampling and analysis. For utility, compliance decisions must be statistically based, and compliance is hinged around the most widely accepted interpretation of 'exposed'. Thus it is an exposure profile that is compared with an occupational exposure limit (OEL), not individual measurements. As a result, an individual measurement that exceeds the OEL cannot be unequivocally interpreted as evidence of non-compliance (though it may be such), other than for application of OELs utilising either peak limitations or STELs.

The basic problem is that measurements made in most exposure situations are known to vary widely for no immediately apparent reason. No individual reading represents a common value that can be compared with the OEL; it is the whole exposure profile that has to be compared with the standard. Even so, this profile is only a probability distribution of a relatively small number of measurements, and so has a good deal of uncertainty about it. Significantly reducing the uncertainty requires a very large number of measurements to be made (200–300), which is not feasible. So understanding compliance has evolved into a process of comparing probability estimates about exposure for groups and individuals with an exposure standard, and doing so on the basis of as few measurements as is possible.

To promote flexibility in approaching this problem, the Australian regulations invite the use of statistical interpretation by using the phrase 'not certain on reasonable grounds'. This encourages H&S practitioners to employ the concept of statistical probability rather than demanding absolute certainty of compliance, since even rudimentary monitoring exercises demonstrate that certainty can rarely be achieved. Guidance has been provided (Safe Work Australia, 2013a) that interpreting compliance will require a sound understanding of the nature of contaminant concentrations, together with the statistics relevant to their measurement and interpretation of measurement results. Determining that exposure of individual workers or groups of workers complies with regulations by being below an OEL now requires an accepted degree of certainty, although that degree has not been specified. It might be as low as 50 per cent or as high as 99 per cent.

However, what constitutes an accepted degree of certainty for making a compliance decision, and what tools are available to assist making those decisions, remain to be ascertained. A common convention in the social sciences is to assess statistical significance. At a 95 per cent level (usually expressed as $p < 0.05$), this indicates that an event has probably occurred for reasons other than chance. Some argue that a similar convention should be employed in occupational hygiene to aid compliance decision-making with respect to OELs.

Three tools need to be applied when making compliance decisions:

- sound knowledge of how to conduct atmospheric monitoring in the workplace
- tools for statistical analysis of monitoring data
- a compliance decision-making strategy.

3.12.2   ATMOSPHERIC MONITORING IN THE WORKPLACE

Monitoring the workplace atmosphere for airborne contaminants is a critical step in the evaluation of risks from airborne contaminants, and comprises a significant part of the work of H&S practitioners. Monitoring provides the bulk of the data about exposure that are to be compared with an ES.

Atmospheric monitoring is separated into:

- identifying the contaminants that require monitoring
- devising the most appropriate monitoring strategy for measuring exposures and making compliance decisions
- selecting the most appropriate sampling method for the contaminant in question; this has to be done in conjunction with an analytical laboratory that can provide the required analysis
- making calculations about exposures from each individual field measurement.

Recognising which contaminants require monitoring may seem obvious, but this is not always the case. Failure to recognise contaminants with significant toxicity will result in incorrect or incomplete monitoring, measurement and management regimes. For example, failing to recognise pneumotoxic oxidant gases among the fumes arising from an arc welding process will limit the value of a gravimetric monitoring of welding fumes. Discussions in later chapters on dust, gases and vapours will assist the H&S practitioner to ensure that monitoring is directed to measuring the important contaminants.

A monitoring strategy has to be tailored to a specific workplace. The monitoring strategy addresses questions of where and when to collect samples, from whom to collect samples, for how long to sample and how many samples need to be collected. H&S practitioners need to be familiar with the individual tasks performed in order to identify which workers are at risk, and also to determine whether there are similar exposed groups (SEGs) in the workplace. For example, fettlers or metal dressers in a metal foundry would be classed as a separate SEG, as would moulders or shakeout and knockout operators. The temporal pattern of exposure needs to be determined to accommodate different exposures on night shifts, maintenance procedures and seasonal variations. If there are contaminants with STELs or peak limitations, then both long-term and short-term monitoring may be needed. A monitoring strategy should observe the number of sources of contaminant release, the rate of contaminant release, the period of exposure, the type and level of maintenance of controls in place, local environmental factors and worker idiosyncrasies, which will all contribute to the exposure of the worker. Account will have to be taken of mixed exposures, and other active routes of exposure, such as skin entry. A monitoring strategy may also have to determine an appropriate re-sampling period, depending on the extent of compliance found. A monitoring strategy may also be devised for determining required protective factors of personal protective equipment where control of the working environment is not practical.

Selecting the most appropriate sampling technique requires significant attention and, as previously noted, only personal sampling should be used. There are very few sampling methods that do not require laboratory analysis. The way the atmosphere is sampled and the method of analysis are interlinked. To analyse for the thousands of possible workplace atmospheric contaminants, some hundreds of standardised and validated methods are now available (HSE, 2012; NIOSH, 2014; OSHA, 2012). It is important to match the analytical measurement range of the method selected with the expected concentration range of the workplace contaminant. Sampling and analytical constraints such as sample breakthrough, adsorption and desorption efficiency, limits of detection, interferences and measurement error should be discussed with the laboratory carrying out the analysis. Because a field measurement ultimately has to reflect the disease risk, specialised sampling protocols may be mandatory. For example, dusts that damage the lungs, such as crystalline silica, are required to be sampled with an elutriator meeting a penetration criterion matched to the human respiratory system (Safe Work Australia, 2013a). Dusts that exert systemic health effects (e.g. lead), or effects dependent on fibre morphology (e.g. asbestos), will be sampled and assessed differently. Different or more sensitive sampling techniques may be required for measuring over short periods during which STELs operate. Commercial suppliers of air sampling equipment can provide a vast array of sampling pumps, different kinds of plain and treated filters, absorption tubes and adsorption badges, as well as direct reading instrumentation for specialised purposes for gases, vapours and some aerosols. Personal sampling is often carried out in a sensitive environment, subject to some invalidation of measurement through delinquency, sample mishandling, transport loss and, occasionally, equipment failure.

Making calculations about exposure from collected samples requires correct measurement of sampling volumes, a thorough appreciation of units of measurement for conversion between gravimetric and volumetric expressions of concentration, and making corrections for temperature and pressure (if applicable). A laboratory may report a metal concentration when its ES is expressed as an oxide, so familiarity with the periodic table and chemical formulae for making the necessary conversions is useful for some applications.

3.12.3 STATISTICAL PROCESSES FOR EXAMINING MONITORING DATA

Having a numerical exposure standard in regulation is a strong invitation to make measurements. To interpret the workplace exposure measurements on individuals or groups of workers, particularly where some comparisons with an OEL are to be made, it is necessary to understand what the monitoring data represent and to select from any analysis some useful parameters that can be bases for risk-management or compliance judgements.

What typically happens with workplace measurements? Experience shows that for a group of workers doing the same kind of job but measured at different times, the results will fall across a wide range of values. Does it make any sense for all of these individual values to be compared directly with an OEL, and can risk-management protocols deal adequately with a wide range of results, some of which may exceed an OEL while others—on different workers or the same worker—do not? Probably not. Intuition suggests that it is possible to arrive at an average exposure that will somehow adequately describe the behaviour of the group; however, if the frequencies of a large number of exposure measurements are plotted against concentration, the distribution curve is not a normal one. The distribution of the log-transformed data is, however, very nearly normal. Typically, it is a lognormal probability distribution of measurements that best describes occupational exposures and provides the most useful parameters for making risk-management decisions.

A basic methodology for examining occupational hygiene data was pioneered and formalised in a NIOSH strategy document over three decades ago (Leidel, Busch and Lynch, 1977). The basic parameters produced by the lognormal distribution are the geometric mean (GM), the geometric standard deviation (GSD), an estimate of the arithmetic mean (Est. AM), the lower confidence limit (LCL), the upper confidence limit (UCL) and the 95th percentile of the distribution. While no single measure can simultaneously describe exposure and facilitate compliance judgements, a combination of a number of them can serve these purposes. The relative positions of some of these parameters are shown in Figure 3.4. Detailed explanations of methods of examining exposure data suitable for hygienists and H&S practitioners are to be found in Ignacio and Bullock (2006). A number of statistical packages, including the Industrial Hygiene Statistical Package (IHSTAT), are available to hygienists for calculating the different parameters needed for examining exposure data.

Measurements made are only representative of a probability distribution. Consequently, all calculated parameters have uncertainties associated with them, so it is not possible to know the exact value of any of them. The fewer the measurements, the larger the uncertainty or the wider is the confidence band within which the true population mean will lie. For example, the LCL is the value below which one can be 95 per cent confident that the GM will not fall, and the UCL is the value above which one can be 95 per cent confident that the true population mean will not fall. If some degree of certainty is expected, then a band of values within which the true population mean value will fall is not really a suitable measure on which to make comparisons with an OEL.

One important benefit of log-transformed data is that it provides an insight into the likely spread of data reflected in the value of the GSD. It is the spread of data that partly reflects the extent of control being exercised or, conversely, how much effort has to be put into future control. The GSD is used to calculate the confidence limits of the GM, and it is also used in more advanced methods of making compliance decisions discussed in section 3.13.4. A GSD of 2.0 is not unusual in occupational hygiene exposures, and GSDs of 2.5 to 4 may be found in some exposure scenarios.

3.12.4 A COMPLIANCE DECISION-MAKING STRATEGY—WHICH APPROACH TO ADOPT?

The regular statistical tools just discussed and used by occupational hygienists and other H&S practitioners for examining monitoring survey data provide a number of measures that might be considered useful for comparing with an OEL to make compliance decisions. The measures of interest are the Est. AM, the GM, the UCL of the GM, the GSD and the 95th percentile of a lognormal probability distribution. However, knowing how

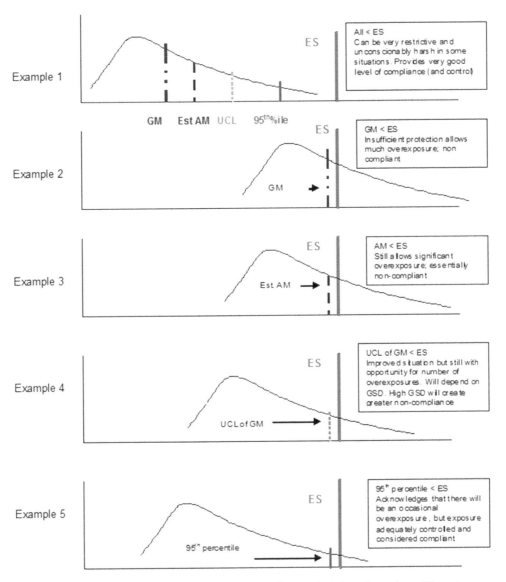

Figure 3.4 Outcomes arising from making compliance decisions based on different interpretations from monitoring data

to calculate these parameters is not the same as knowing how to use them in making a compliance decision.

3.12.4.1 What should constitute compliance?

Figure 3.4 (above) demonstrates what a compliance decision will infer about the (stylised) probability distribution of monitoring results with respect to the OEL when each of five different criteria for making a compliance decision is employed. In this example, the monitoring results exhibit a GSD of about 2.

Note that, at present, different organisations and regulators are at liberty to choose whichever one of these compliance parameters they might consider most appropriate. The five, corresponding with Examples 1–5 in Figure 3.4, are:

1   all measurements to be below the level of the regulatory OEL
2   the GM of measurements to be below the level of the regulatory OEL
3   the Est AM of measurements to be below the level of the regulatory OEL
4   the UCL of the GM to be below the level of the regulatory OEL
5   the 95th percentile of the probability distribution (represented by the few measurements that one can afford to make) to be below the regulatory OEL.

Some of the outcomes are given in terms of the practicability of meeting compliance with each option. A number of the options will permit excessive exposures to occur.

From these options for compliance, Examples 2 and 3 have to be rejected, since they permit the possibility of far too many instances of over-exposure when applied to group measurements. An example of averaging is found in the Queensland Coal Mining Safety and Health Regulation (Queensland Parliamentary Counsel, 2012) to indicate compliance, though it is not specified whether the average is a GM or an Est. AM, and these can be quite different depending on the GSD. This averaging system can confirm compliance only where all measurements relate specifically to one individual.

Example 4, using the 95 per cent UCL of the GM to be <OEL, is employed by some organisations but will still permit a significant proportion of exposure measurements to be greater than the OEL. That problem is exacerbated where the probability distribution relates to a group of workers, but some of those workers may consistently constitute the non-compliant tail. Bad performers within a group cannot be traded off against good performers. A good compromise between the stringency of Example 1 and Example 4 is to have the 95th percentile to be <ES (as in Example 5). Adoption of the 95th percentile being <OEL is a balance between the uncertainty inherent in taking a small number of measurements and the need for a practical and acceptable yardstick for making correct compliance decisions. If 95 per cent of the exposure profile represented by the probability distribution lies <OEL, then there is a high probability that the exposure profile will not exceed the exposure standard. This decision must necessarily be conservative, since the derivation of the OEL itself will also have some uncertainty associated with it; however, the need to be conservative here has been exacerbated by ascribing regulatory properties to what is a guidance value, and then concluding that compliance with the regulation will deliver the protection the OEL purports to provide. That is a perennial problem for regulators.

3.12.4.2 What other strategies are there for judging compliance against an OEL?

When making decisions about compliance where a series of measurements have been made on a group of workers, there are two considerations:

- Is the whole group compliant?
- Are all individuals in the group also compliant?

These two considerations have led to the development of several strategies to answer both questions. Each strategy aims to achieve this by taking a minimum of measurements. The two simple methods that follow have some serious limitations but can be used without the need for any complex statistical calculation. A modern robust approach (BOHS and NVvA, 2011) is also outlined, although its complete methodology cannot be fully reproduced here.

- *Simple methods—the maximally exposed worker.* One of the simplest methods proposed for evaluating compliance with regulatory standards for a group of workers all involved in a similar task is to use the NIOSH compliance strategy as follows: Typically, the maximally exposed worker in a group is identified and monitored for exposure to environmental agents. If that personal exposure is below the standard, all worker exposures are also presumed to be compliant with the regulatory exposure standard (Ignacio and Bullock, 2006, p. 425). This approach requires prior advice so that this 'worst case' or maximally exposed worker can be differentiated from all other workers. Professional input, perhaps aided by measurements made with a direct reading instrument, may assist in identifying that worker.
- *Simple methods—HSE shortcut method with measurements <OEL/3.* Similar to the NIOSH strategy, this approach (HSE, 2006) includes more of the exposed workers. Compliance can be judged as having probably been achieved if at least three-quarters of twelve or more results on the most exposed workers are below one-third of the ES. This simple method depends on the results being lognormally distributed, the most exposed workers being properly identified and the GSD not being excessive. The method is based on experience that if three-quarters of the results lie in the lower part of the distribution, then the percentage of the distribution that is >OEL is likely to be smaller than 5 per cent. However, having to make twelve measurements does not really amount to a simple shortcut. Simple methods such as these are applicable only in situations where compliance is highly probable. Where it is doubtful, more advanced decision-making methods are needed.
- *Advanced methods—group and individual compliance.* In practice, making measurements to determine compliance faces two main constraints. To keep costs low, monitoring programs need to be as economical as possible in terms of the number of samples taken. But the sampling regime must be designed to take into account all the factors that typically impart variability to its measurements. Most sampling exercises limit themselves to a cross-sectional study over one or two days, thus missing or discounting many of the variables (e.g. environment, variations in production schedule) that contribute to an exposure profile. Further, unless workers are individually sampled, most group sampling assists only in making broad compliance decisions about the group as a whole.

The technique recently demonstrated (BOHS and NVvA, 2011) meets these challenges. The process can be economical in terms of the number of measurements required, and it accommodates the variability issues through a program of structured and timed campaigns. It differs from the conventional compliance decision, which uses the 95th percentile of the probability distribution for a group, in that it can address differences both within and between workers, and it is based on the concepts of group and individual compliance with an OEL.

Use of the method depends on valid measurements and a correctly defined SEG. It is a stepwise process, with its simple first steps intended to provide:

- an indication of compliance if all results are low (<0.1 OEL), or
- immediate implementation of controls if there is evidence of possible non-compliance (one result >OEL), or
- the undertaking of two further monitoring campaigns if uncertainty remains (any results >0.1 OEL but <OEL).

Group compliance is determined by a simple calculation based on a minimum of around nine structured samples from three campaigns using the technique set out in the French law on regulatory compliance (Legifrance, 2009). However, group compliance will not necessarily imply individual compliance and may disguise individual non-compliance if there are idiosyncratic worker effects operating, even though in the same SEG. Individual compliance can be examined by using an analysis of variance (ANOVA) of the means of measurements within and between workers.

The two measures for specifying compliance (BOHS & NVvA, 2011) are:

- *group compliance*—the group complies if, with 70 per cent confidence, <5 per cent of the exposures in the SEG exceed the OEL
- *individual compliance*—the SEG complies in terms of individual exposure if there is <20 per cent probability that >5 per cent of the exposures of any individual exceed the OEL

The methodology is not reproduced here, but all analysis can be carried out conveniently using Microsoft Excel software. Calculations are based on the use of the GM, the GSD and the OEL.

It is recommended that occupational hygienists and others required to undertake compliance analysis on sampling data acquaint themselves with this method and the ANOVA tools. Procedures are also provided for making the calculations if a limited amount of data is missing because of work changes, worker absences and the like. Benefits include:

- simple protocols
- a single compliance decision-making criterion for groups
- suitability for regulatory applications
- the ability to examine for likelihood of individual non-compliance without having to sample each worker
- a considerably minimised sample set (at the expense of an increased number of campaigns)
- an immediate feedback mechanism indicating when control is required
- an approach that can also be applied when using STELs.

3.12.5   CONCLUDING REMARKS ON OELs AND THEIR USE

Occupational exposure limits or exposure standards, despite criticisms, have provided valuable service to occupational hygienists and the health and safety community for several generations. Their expression in simple numbers often belies the complexity behind their derivation, their continual refinement, their occasional differences and the myriad factors required to ensure their correct application. Their use must always be accompanied by competent best practice in occupational hygiene; however, the enshrinement of these values in regulations has bestowed on them both legality and an aura of greater authoritativeness than was ever intended.

Despite this, even good exposure standards have very limited value in law without a robust framework into which they can be inserted. Decisions in law require some acceptable degree of consistency and certainty if they are to be upheld and to achieve an intended outcome (protection of workers). The monitoring of exposure and the process for making compliance decisions based on that monitoring have rarely been attended by either complete consistency or certainty. The regulation of hazardous chemicals has established both that compliance with an exposure standard needs to be achieved, and that airborne monitoring may need to be undertaken where there is uncertainty about compliance but without identifying what is meant by compliance. Despite their necessary resort to some statistics, the monitoring programs described in section 3.12.2 and the decision-making protocols for determining compliance in section 3.12.4 have shown that it is possible to make consistent, understandable and acceptable decisions about compliance. These decisions should satisfy both the protective intent of the regulation and withstand legal scrutiny, to the extent that is possible given their basis in predictions about probability.

Properly structured monitoring programs combined with the best decision-making processes offered here will suit the needs of occupational hygienists, H&S practitioners and regulators alike. The framework of statistically based monitoring and decision-making about compliance is essential if exposure standards are to serve any regulatory function. When monitoring programs and decision-making are well integrated, exposure standards established as guides for professional use can also operate effectively as legal regulatory benchmarks.

3.13 REFERENCES

Adkins, C., Booher, L., Culver, D. et al., 2009a, *Occupational Exposure Limits: Do They Have a Future?* International Occupational Hygiene Association, Derby, <www.ioha.net/assets/files/OEL%20Green_Paper%2008%2019%2009.pdf> [accessed 3 November 2012]

—— 2009b, 'The future of occupational exposure limits—can OELs be saved?', *The Synergist*, vol. 20, no. 9, pp. 46–8.

American Conference of Governmental Industrial Hygienists (ACGIH) 2018, *History*, ACGIH®, Cincinatti, OH, <www.acgih.org/about/history.htm> [accessed September 2018]

—— 2019, *Threshold Limit Values for Chemical Substances and Physical Agents and Biological Exposure Indices*, ACGIH®, Cincinnati, OH.

American Industrial Hygiene Association (AIHA) 2018, *Tools and Links for Exposure Assessment Strategies*, AIAH, Falls Church, VA <www.aiha.org/get-involved/VolunteerGroups/Pages/Exposure-Assessment-Strategies-Committee.aspx> [accessed September 2018]

Baetjer, A.M. 1980, 'The early days of industrial hygiene: Their contribution to current problems', *American Industrial Hygiene Association Journal*, vol. 41, no. 11, pp. 773–7.

Brandys, R.C. 2008, *Global Occupational Exposure Limits for Over 6000 Specific Chemicals*, 2nd ed., Occupational and Environmental Health Consulting Services, Hinsdale, IL.

Brief, R.S. and Scala, R.A. 1975, 'Occupational exposure limits for novel work schedules', *American Industrial Hygiene Association Journal*, vol. 36, no. 6, pp. 467–9.

British Occupational Hygiene Society and Nederlandse Vereniging voor Arbeidshygiëne (BOHS and NVvA) 2011, *Sampling Strategy Guidance: Testing Compliance with Occupational Exposure Limits for Airborne Substances*, <www.bohs.org/library/technical-publications> [accessed 28 November 2012]

Cook, W.A. 1945, 'Maximum allowable concentrations of industrial atmospheric contaminants', *Industrial Medicine*, vol. 11, pp. 936–46.

De Silva, P. 1986, 'TLVs to protect "nearly all workers"', *Applied Industrial Hygiene*, vol. 1, no. 1, pp. 49–53.

—— 2000, *Science at Work: A History of Occupational Health in Victoria*, PenFolk, Melbourne.

Drolet, D. 2008, *Technical Guide T–22: Guide for the Adjustment of Permissible Exposure Values (PEVs) for Unusual Work Schedules*, 3rd edn, IRSST, Montreal.

—— 2018, Guide for the Adjustment of Permissable Exposure Values (PEVs) for Unusual Work Schedules (4th edn), IRSST, Montreal, <irsst.qc.ca/media/documents/pubirrst/t-22.pdf>.

European Chemicals Agency (ECA) 2012, 'Characterisation of dose [concentration]—response for human health', in Guidance on Information Requirements and Chemical Safety Assessment, ECHA, Helsinki, Chapter R.8, <https://echa.europa.eu/documents/10162/13632/information_requirements_r8_en.pdf/e153243a-03f0-44c5-8808-88af66223258> [accessed September 2018]

European Commission, Employment, Social Affairs and Inclusion (ECESAI) 2010, *Guidance for Employers on Controlling Risks from Chemicals: Interface Between Chemicals Agents Directive and REACH at the Workplace*, European Commission, Brussels, <https://osha.europa.eu/en/file/40569> [accessed September 2018]

Gallo, M.A. 2008, 'History and scope of toxicology', in C.D. Klaassen (ed.), *Casarett & Doull's Toxicology: The Basic Science of Poisons*, McGraw Hill, New York, pp. 1–10.

Hamilton A. and Hardy, H.L. 1949, *Industrial Toxicology*, 2nd ed., Paul B Hoeber, New York.

Heads of Workplace Safety Authorities 2008, *National Occupational Health and Safety (OHS) Compliance and Enforcement Policy*, HWSA, Gosford, <www.hwsa.org.au/files/documents/Compliance_and_Enforcement_policy.pdf> [accessed 28 November 2012]

Health and Safety Executive (HSE) 2006, *Exposure Measurement: Air Sampling*, COSHH Essentials General Guidance G409, HSE, London, <www.hse.gov.uk/pubns/guidance/g409.pdf> [accessed 28 November 2012]

—— 2012, *Methods for Determination of Hazardous Substances (MDHS) Guidance*, HSE, London, <www.hse.gov.uk/pubns/mdhs> [accessed 28 November 2012]

Henderson, Y. 1921, 'Effects of gases on men and the treatment of various forms of gas poisoning', in A.C. Fieldner, S.H. Katz and S.P. Kinney (eds), *Gas Masks for Gases Met in Fighting Fire*, Department of Interior, US Bureau of Mines, Washington, DC.

Henschler, D. 1984, 'Exposure limits: History, philosophy, future developments', *Annals of Occupational Hygiene*, vol. 28, no. 1, pp. 79–92.

Howard, J. 2005, 'Setting occupational exposure limits: are we living in a post-OEL world?', *University of Pennsylvania Journal of Labor and Employment Law*, vol. 7, no. 3, pp. 513–28.

Ignacio, J.S. and Bullock, W.H. 2006, *A Strategy for Assessing and Managing Occupational Exposures*, 3rd ed., AIHA, Fairfax, VA.

International Occupational Hygiene Association (IOHA) 2010, *IOHA Statement on Control Banding*, IOHA, Derby, <www.ioha.net/assets/files/IOHA%20Statement%20on%20Control%20Banding.pdf>.

Kober, G.M., Hayhurst, E.R. and Rober, G.M. 1924, *Industrial Health*, P. Blakiston's, Philadelphia, PA.

Legifrance 2009, 'Arrêté du 15 décembre 2009 relatif aux contrôles techniques des valeurs limites d'exposition professionelle sur les lieux de travail et aux conditions d'accréditation des organismes chargés des contrôles', *Journal Officiel de la République Française*, vol. 292, pp. 35–156, <www.journal-officiel.gouv.fr/lois_decrets_marches_publics/journal-officiel-republique-francaise.htm>.

Lehmann, K.B. 1893, *Methods of Practical Hygiene, Vol. 1*, Kegan Paul, Trench, Trübner & Co, London.

Leidel, N.A., Busch, K.A. and Lynch, J.R. 1977, *Occupational Exposure Sampling Strategy Manual*, NIOSH, Cincinnati, OH.

National Institute for Occupational Safety and Health (NIOSH) 2014, *NIOSH Manual of Analytical Methods*, NIOSH, Cincinnati, OH, <www.cdc.gov/niosh/nmam/> [accessed September 2018]

National Occupational Health and Safety Commission (NOHSC) 1995a, *Adopted National Exposure Standards for Atmospheric Contaminants in the Occupational Environment*, AGPS, Canberra, <www.safeworkaustralia.gov.au/system/files/documents/1702/adoptednationalexposurestandardsatmosphericcontaminants_nohsc1003-1995_pdf.pdf> [accessed September 2018]

—— 1995b, *Guidance Note on the Interpretation of Exposure Standards for Atmospheric Contaminants in the Occupational Environment*, 3rd ed., AGPS, Canberra, <www.safeworkaustralia.gov.au/AboutSafeWorkAustralia/WhatWeDo/Publications/Documents/238/GuidanceNote_InterpretationOfExposureStandardsForAtmosphericContaminants_3rdEdition_NOHSC3008–1995_PDF.pdf> [accessed 2 December 2012]

Occupational Safety and Health Administration (OSHA) 1989, *VI. Health Effects Discussion and Determination of Final PEL*, OSHA, Washington, DC, <www.osha.gov/pls/oshaweb/owadisp.show_document?p_id=770&p_table=PREAMBLES> [accessed 28 November 2012]

—— 2012, *Sampling and Analytical Methods*, OSHA, Washington, DC, <www.osha.gov/dts/sltc/methods> [accessed 28 November 2012]

Oliver, T. 1902, 'Dusts as a cause of occupational disease', in T. Oliver (ed.), *Dangerous Trades: The Historical, Social and Legal Aspects of Industrial Occupations as Affecting Health*, John Murray, London.

Patty, F.A. 1958, *Patty's Industrial Hygiene and Toxicology*, 2nd ed., Interscience, New York.

Paustenbach, D.J., Cowan, D.M. and Sahmel, J. 2011, 'The history and biological basis of occupational exposure limits for chemical agents', in F.A. Patty, V.E. Rose and B. Cohrssen (eds), *Patty's Industrial Hygiene*, 6th ed., Wiley, Hoboken, NJ, pp. 865–955.

Queensland Parliamentary Counsel 2012, *Coal Mining Safety and Health Regulation 2001*, Reprint no. 4A, Queensland Government, Brisbane, <www.legislation.qld.gov.au/legisltn/current/c/coalminshr01.pdf> [accessed 28 November 2012]

Ripple, S.D. 2010, 'History of occupational exposure limits', paper presented to 8th International Occupational Hygiene Association Conference, 28 September–2 October, IOHA, Rome, <www.ioha.net/assets/files/Paper2Historyoel.pdf> [accessed 28 November 2012]

Safe Work Australia (SWA) 2013a, *Guidance on the Interpretation of Workplace Exposure Standards for Airborne Contaminants*, SWA, Canberra, <www.safeworkaustralia.gov.au/doc/guidance-interpretation-workplace-exposure-standards-airborne-contaminants> [accessed September 2018]

—— 2013b, *Guidelines for Health Monitoring*, SWA, Canberra, <www.safeworkaustralia.gov.au/topics/health-monitoring>.

—— 2018a, *Workplace Exposure Standards for Airborne Contaminants*, SWA, Canberra, <www.safeworkaustralia.gov.au/system/files/documents/1804/workplace-exposure-standards-airborne-contaminants-2018_0.pdf> [accessed September 2018]

—— 2018b, *Hazardous Chemical Information System (HCIS)*, SWA, Canberra, <http://hcis.safeworkaustralia.gov.au> [accessed September 2018]

—— 2018c, *Code of Practice: Managing Noise and Preventing Hearing Loss at Work*, SWA, Canberra, <www.safeworkaustralia.gov.au/system/files/documents/1810/model-cop-managing-noise-and-preventing-hearing-loss-at-work.pdf> [accessed January 2019]

Sayers, R.R. 1927, 'Toxicology of gases and vapors', in *International Critical Tables of Numerical Data, Physics, Chemistry, and Toxicology, Vol. 2*, McGraw-Hill, New York, pp. 318–21.

Schenk, L. 2010, 'Comparison of data used for setting occupational exposure limits', *International Journal of Occupational and Environmental Health*, vol. 16, no. 3, pp. 249–62.

—— 2011, 'Setting occupational exposure limits: Practices and outcomes of toxicological risk assessment', PhD thesis, Södertörn University, Stockholm, <http://ftp.cdc.gov/pub/Documents/OEL/12.%20Niemeier/References/Schenk_2011_Thesis.pdf> [accessed 28 November 2012]

Todd, L.A. 2004, 'Occupational hygiene: Evaluation of the work environment', in *Encyclopaedia of Occupational Health and Safety*, 4th ed., ILO, Geneva, <http://ilocis.org/documents/chpt30e.htm>

Verma, D.K. 2000, 'Adjustment of occupational exposure limits for unusual work schedules', *American Industrial Hygiene Association Journal*, vol. 61, no. 3, pp. 367–74.

Vincent, J.H. 1999, 'Occupational hygiene science and its application in occupational health policy, at home and abroad', *Occupational Medicine*, vol. 49, no. 1, pp. 27–35.

Worksafe Victoria 1999, *Code of Practice for Hazardous Substances, No. 24*, WorkSafe Victoria, Melbourne, <www.worksafe.vic.gov.au/forms-and-publications/forms-and-publications/hazardous-substances-code-of-practice-no.-24,-2000> [accessed 2 December 2012]

Ziem, G.E. and Castleman, B.I. 1989, 'Threshold limit values: Historical perspectives and current practice', *Journal of Occupational Medicine*, vol. 31, no. 11, pp. 910–18.

4. Control strategies for workplace health hazards

Garry Gately and Wayne Powys

4.1   INTRODUCTION

As we saw in Chapter 1, there are four fundamental principles in occupational hygiene: **anticipation, recognition, evaluation** and **control** of health hazards. *Control* is a key objective, and is the most important of these principles since it benefits the worker. Control involves the application of technological, engineering and operational measures directed at the health hazard, work environments or workers to eliminate any adverse outcomes or reduce risk to acceptable levels. Before OHS practitioners embark on any control program, they must take the time to fully understand the workplace hazards and consider matters such as:

- how the hazardous situation or exposure arises
- what task the exposed people are conducting when they are exposed
- why they are doing the task
- whether there are any people in the area being exposed who are not involved in the process and the impact of that exposure
- what the intended outcomes are of the control program or control measures
- what the likely consequences will be if the controls are not adequate or fail to protect the worker.

The last of these considerations should be the major motivator to get it right the first time. For example, past failure to control exposures to asbestos dust in Australia has resulted in thousands of deaths (Safe Work Australia, 2014). This shows how important it is to have a full appreciation of the situation from the outset.

Invariably in any workplace, there are many and varied OHS issues that require some form of action to reduce the risks to those exposed. Consequently, there will be competing projects for the inevitably limited funds available to manage risks. It is therefore necessary to prioritise all of the identified hazards and issues—be they chemical, physical, safety, biological or environmental—and it is common practice to undertake a comparative health risk assessment for this purpose. There are many tools available to facilitate such assessments; typically, they use a risk matrix to determine the consequence of exposure and the likelihood of the consequence occurring to rank a hazardous scenario (see Appendix in Section 4.11 at the end of the chapter).

4.2   HIERARCHY OF CONTROL STRATEGIES

A full appreciation of the hazards of a particular workplace is often achieved only if a complete risk assessment is undertaken with the involvement of all parties concerned. As we saw in Chapter 1, when first investigating an issue, the H&S practitioner should first conduct a walk-through survey of the workplace, during which observations are made of such factors as the work processes used, personnel at work, hazards to which they are potentially exposed and control measures in use; this may include collecting information and data relating to workers' exposure, which could be in the form of actual exposure measurements and/or health-impact data (e.g. reports of incidence of symptoms or illness). The observations and data gathered will help the H&S practitioner to establish the current adequacy of or to ascertain what additional controls may be required.

Consider the simple example of a worker in a wine cellar who is exposed to carbon monoxide (at a concentration of 200 ppm) from a forklift exhaust. This represents a significant risk to the worker's health, being an exposure six times greater than the permitted eight-hour workplace exposure limit of 30 ppm and equivalent to the fifteen-minute short-term exposure limit of 200 ppm (Safe Work Australia, 2018a), and it must be remedied. In this workplace situation, there are several possible solutions that highlight the major principles of control include:

- Remove the forklift and shift everything by hand.
- Use an electric forklift.
- Use a catalytic converter to convert exhaust carbon monoxide to the less hazardous carbon dioxide.
- Provide ventilation to extract the exhaust gases.
- Allow the worker to work in the cellar for no more than fifteen minutes four times a day to comply with the published workplace exposure standards.
- Provide the worker with a suitable respirator to protect against carbon monoxide.

Not all of these options might be possible or practical. For example, the wine casks may not be able to be transported without a forklift, or ventilation may be impractical or too expensive because perhaps the cellar needs to be kept at natural ground temperature, or the worker may not be able to wear a respirator because they are a wine taster. Just how this challenge will best be met depends on circumstances, practicalities and economics, with a little ingenuity thrown in. As the risk associated with a particular process may change over time (e.g. the output of hazardous gas may increase), periodic review of control procedures will be needed. This will ensure that controls remain effective, as well as being efficient and economical. Simplicity is important; complicated control systems require ongoing attention (e.g. training, checking, inspections and maintenance), which increases the likelihood that they will become ineffective and their failure may possibly go undetected, thus endangering worker health.

An examination of the control strategies in the example of the wine-cellar worker above will show that controls can be applied either at the source of the hazard, in the path from the source to the receiver, or in relation to the receiver, a concept illustrated in Figure 4.1.

Further consideration of the control strategies suggested in the wine-cellar worker example above will show that some would be more effective than others, leading to an

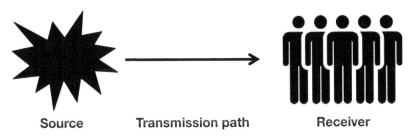

Source **Transmission path** **Receiver**

Figure 4.1 Source, transmission path, receiver model

order or hierarchy, commonly referred to as the 'hierarchy of controls' (HoC), generally set out in the following order:

1   elimination/substitution
2   engineering controls
3   administrative controls
4   personal protective equipment (PPE).

Figure 4.2 sets out a typical representation of the HoC. Based on the authors' experience, the likely degree of effectiveness of control may be:

• elimination/substitution: up to 100 per cent
• engineering: 20–90 per cent
• administrative: 0–70 per cent
• PPE: 0–50 per cent.

Elimination of a hazard from the workplace is the ultimate solution, but in practice it is often not a practical or achievable strategy. For example, zinc cannot be removed from a galvanising process, nor can lead be removed from an automotive battery factory. In such cases, the most effective, practical and economic means of control have to be established to allow the work to continue.

Frequently, two or more control methods have to be applied to reduce the risk sufficiently. For example, with abrasive blasting, the traditionally used sand has been replaced

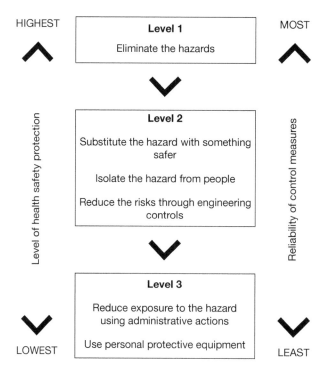

Figure 4.2
A representation of the HoC
Source: Safe Work Australia (2017).

by ilmenite and garnet, which by comparison contain very little quartz (elimination/substitution). Further, operators are trained (administrative control) in the use of the equipment and are equipped with PPE in the form of protective suits and air-supplied respiratory protection. It is generally found that a combination of controls results in a more robust and reliable system of control than would be achieved using a single, more complex method.

Experience teaches the H&S practitioner that there are always a number of possible control solutions: some work better than others; some are expensive, others less expensive; some are more acceptable to workers, others less so. Deciding which control solution to recommend will require consideration of:

- the hazard and the magnitude of the risk it poses
- the practicability of the various controls available
- the efficacy of those controls
- the consequences of failure of controls
- the relative costs of providing, operating and maintaining controls
- the likely acceptance of the controls by the workforce (if they are viewed as impractical, they will not be used).

Consideration of the above and other relevant factors constitutes in essence a cost-benefit analysis, in which risk is weighed against the effort, operational disruption, time and money needed to control it; eventually a point is reached where any further risk reduction becomes out of proportion to the resources required to reduce it further. This point is referred to 'as low as reasonably practicable' (ALARP) and represents a level to which we would expect to see workplace risks controlled.

The expectation of OH&S legislation in Australia and many other countries is that that risk is controlled 'as far as reasonably practicable' or 'so far as is reasonably practicable'. It is therefore not permissible to use controls that are lower in the HoC (e.g. administrative or PPE) if higher order controls can practicably be used (e.g. substitution or engineering); in effect, such legislation mandates the HoC.

The notion of 'reasonably practicable' is at the core of legislative expectations for risk reduction and ALARP, and while there may be a difference that must be considered when dealing with legislative issues, to all intents and purposes they can be regarded as synonymous.

When considering the trade-offs of control effectiveness against the consequences of failure and cost, the matrix shown in Figure 4.3 can be consulted; in short, where there is a high-consequence hazard (e.g. carcinogen) a more robust and effective control must be used.

The following situations are examples of best choices for workplace control of various hazards:

- replacing cancer-causing chemicals rather than controlling them, unless they are essential to the workplace (e.g. potent drug treatments in a hospital)
- not installing an expensive dust-control system for intermittent or infrequent exposures to a hazard (e.g. in a job undertaken for three hours every six months) where a combination of procedural controls and personal respiratory protection would control exposure

Primary approach to control	Expected consequence of exposure				
	First aid injury	Medically treated injury/illness	Lost work day injury	Permanent disability or illness	Single fatality
Elimination/ substitution	✗	✓	✓✓	✓✓	✓✓
Engineering	✓	✓	✓✓	✓✓	✓✓
Administrative	✓✓	✓✓	✓	✗	✗
PPE	✓✓	✓	✗	✗	✗

Key:
✓✓   benefit of implementation far outweighs cost and should be implemented
✓   benefit of implementation may or may not outweigh cost but should be considered
✗   control option not advised due to either (a) poor effectiveness or (b) cost of implementation far outweighs benefit

Figure 4.3 Control strategy related to potential consequences

- deciding against the use of half-face respirators that technically provide good protection for workers permanently employed on an acid pickling line and opting for control of the acidic aerosol by suppression and ventilation, which would provide a more acceptable long-term solution
- giving a worker requiring access to the same acid pickling line for five minutes per day the appropriate half-face respirator
- installing noise enclosures for air compressors as part of a factory design.

When considering the cost versus benefit of possible control solutions, the following sections present important guidance for selecting controls.

4.2.1   KEEP CONTROLS SIMPLE AND INVOLVE OTHERS

Controlling a hazard by elimination and substitution may in some cases be simple and effective, and should always be considered first. Applying engineering solutions, such as containment, can be somewhat expensive unless implemented at the design stage. Administrative controls and PPE can involve complex decision-making and rely heavily on worker compliance and acceptance, which can be unpredictable and uncertain. Making decisions on respiratory protection in particular requires detailed information about exposure levels, equipment performance, worker training and maintenance. Moreover, if respiratory protection is used as the primary control mechanism and fails, there is no back-up system. During each step of developing the controls, all stakeholders (e.g. process managers, exposed work group, engineers and H&S practitioners) should be involved to ensure they all take ownership of the solutions.

Despite the HoC being a well-known, understood and accepted construct, it is frequently found that the lower-order control strategies of administration and PPE are most likely to have been implemented in the workplace. The best time to implement the higher control strategies is during the design of workplaces or processes, since the cost is much less and technically easier to undertake than during the operations, as illustrated in Figure 4.4 for engineering controls.

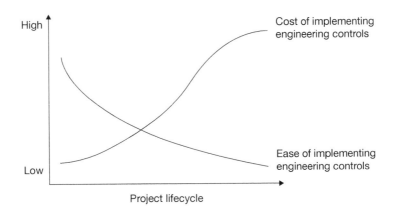

Figure 4.4 Conceptual model of the cost of implementing engineering controls against the ease of implementing them

It is important for H&S practitioners to become involved in the design of new workplaces or processes. In this way, their specific expertise, knowledge and best practices in design can be applied to eliminate hazards or at least reduce the associated risk to generally acceptable levels. Retrofitting higher-order controls is expensive and generally less effective.

Internationally, there are a number of schemes aimed at 'designing out' health and safety hazards during the course of the design development, including:

- the US National Institute for Occupational Safety and Health (NIOSH) 'Prevention through Design' (PTD) initiative, the mission of which is to 'prevent or reduce occupational injuries, illnesses and fatalities through the inclusion of prevention considerations in all designs that impact workers' health (NIOSH, 2016)
- the UK Construction (Design and Management) Regulations 2015 (CDM Regulations), which require construction designers to manage risks by applying the general principles of prevention (Health and Safety Executive 2015)
- the Safe Work Australia Good Work Design program, which addresses health and safety issues at the design stage (Safe Work Australia, 2017).

4.2.2 CHANGE MANAGEMENT

Often materials and processes find their way into the workplace without any effort being made to investigate their hazards or alternatives. In innumerable cases, materials are introduced—usually in an effort to make the task easier—without sufficient thought given to potential new hazards:

- New solvents to clean residues from parts may be toxic.
- Compressed air to blow components clean may generate a dust hazard.
- UV lamps to cure resins may expose workers to UV radiation.

A change-management system should be in place to ensure that risk assessments are conducted each time a new chemical or process is introduced, or an existing one is changed. Change-management controls are required by most OHS regulations and should involve representatives from all affected work groups.

4.3   ELIMINATION/SUBSTITUTION

Eliminating a given hazard from the workplace by completely removing a process or a substance is the definitive way to reduce risk. For example, many jurisdictions have prohibited any new uses of asbestos.

Complete elimination is often a drastic step, however, and if the process or substance is central to production, it can even shut down an industry (e.g. oil and gas production or domestic fireworks manufacturing). As a result, elimination is often rejected in favour of more practicable alternatives. The next level in the HoC is substitution—either of materials or of processes—which is directed at the source of the hazard.

4.3.1   SUBSTITUTE MATERIALS

In some workplaces, such as mines or metal smelters, it may not be possible to substitute the hazard; however, replacing one hazardous substance with a less hazardous one has occurred throughout history in many process industries—for example:

- In the painting of luminescent watch and clock dials, radium was replaced with phosphorescent zinc sulphides after it was established that the radioactivity of radium paint caused bone and tissue cancer in the painters (particularly those who licked their brush to form a fine tip).
- In the manufacture of matches, white and yellow phosphorus, responsible for a disfiguring and potentially fatal necrosis of the jawbone (termed 'phossy jaw'), was replaced by less dangerous red phosphorus. Modern safety matches incorporate a safer form of phosphorus on the striking-friction side of the box.

In fact, many of the most important developments in occupational health over the last 60 to 70 years have come from the search for less hazardous substitutes for dangerous materials. The following are some well-known examples:

- Asbestos, the fibres of which cause mesothelioma and other cancers, has been replaced by safer synthetic substitutes (glass foam, rock and glass wool). Notwithstanding this, the removal of asbestos continues to result in new exposures.
- Benzene, which causes leukaemia, has been replaced as an industrial solvent by less hazardous solvents (e.g. xylene); changed refining methods have reduced the level of benzene in gasoline.
- Beach and river sands, which have a high quartz content, have been replaced as abrasive blasting agents with low-quartz materials such as ilmenite, zircon and copper slag.

- Mercury compounds for fur carroting in felt manufacture have been replaced by less hazardous acid and peroxide mixtures.
- Flammable petroleum naphtha in dry-cleaning has been replaced, successively, by carbon tetrachloride, perchlorethylene and then chlorofluorocarbons (CFCs), with a subsequent return to perchlorethylene when CFCs became restricted as ozone-depleting substances.
- Mercury in the extraction of gold from ore has been replaced by cyanides and 'carbon-in-pulp' leaching.

These examples represent some classic advances, but there are still many workplaces where there are opportunities for substitution of less hazardous materials. Lead can be phased out, for example, as can mercury and hexavalent chromium salts. Other, far less toxic, aliphatic hydrocarbon solvents can replace the neurotoxic aliphatic solvent n-hexane, which is still used in printing. Care must always be taken, however, to ensure that the substitute does not itself pose a hazard.

4.3.2 SUBSTITUTE PROCESSES

In some industrial processes, where substitution of a less hazardous material is not possible, the risk associated with handling hazardous materials can be reduced by a change in the process. The following are some examples:

- A pelletised form of master-batch can be used instead of a dusty powder (e.g. to control lead exposure when handling stabilisers in PVC product production).
- Organic solvents can be used in gelled form to reduce the rate of vapour emissions (e.g. gelled styrene, gelled paint strippers).
- A manufacturing route may be adopted that does not give off hazardous by-products (such as dioxin in herbicide manufacture).
- A manufacturing route can be changed to obviate the need to store large quantities of extremely hazardous intermediates (such as methylisocyanate in pesticide manufacture, or hydrogen cyanide in the manufacture of sodium cyanide).
- To limit dust, a wet process can replace a dry process (e.g. damp sawdust), or sweeping can be replaced by vacuum cleaning (see Figure 4.5).
- The working temperature of a process can be lowered to reduce evaporation of volatile materials.
- In simple processes such as painting, dipping rather than spraying will greatly reduce the generation of vapours, while eliminating spillage and using low-dust cleaning methods such as vacuuming rather than dry sweeping will reduce airborne dust.

If extensive processing of materials occurs in a workplace, it is wise to examine all potential ways of reducing risks to health. For example, a task may have been done in a traditional way for years without any questions arising about its efficiency or safety. In some cases, it will be possible to substitute a ready-made product from a purpose-designed workshop rather than operate a hazardous process. The aim should be to:

- reduce the number of times a worker handles a hazardous material (e.g. the lead used in manufacturing a battery may be handled in 20 or more separate operations)

Figure 4.5 Dry sweeping generates another dust exposure route

- minimise operations producing hazards, particularly airborne hazards (e.g. restrict multiple dispensing of powders from storage to hopper to bin to bag)
- use shrouded high-pressure water-blasting equipment when cleaning to reduce the spread of contamination
- change the process to reduce fugitive dust or vapour hazards (e.g. conduct cold acid pickling processes rather than hot ones to reduce evaporation of hazardous substances and use floating ping-pong balls on the pickle liquor to reduce airborne droplets; use liquid catalysis of chemical reactions, such as in foundry moulding, rather than gaseous catalysis).

As well as chemical hazards, the strategy of substitution of processes can be applied successfully to other hazard classes—for example, noise and vibration—by:

- reducing noisy operations by replacing rivets with welds
- replacing noisy rollers with less noisy conveyor belts
- replacing high vibration hand and other tools with lower vibration models
- replacing vehicle driving seats with vibration-dampening seats.

Controls for noise and vibration, radiation, biological hazards and heat stress are discussed in more detail in later chapters.

The challenge for the H&S practitioner is to be alert to the possibility of substitution in order to reduce hazards. If the hazardous materials cannot be substituted because they are integral to a process, or the processing cannot be improved, then one of the following control strategies should be implemented.

4.4 ENGINEERING CONTROLS

Engineering controls are used to isolate people from an established or perceived hazard. Referring to Figure 4.1, engineering controls can be directed at the hazard source, the transmission path and the receiver (worker).

A wide array of engineering controls is possible depending on the hazard in question, including various types of containment and ventilation systems for hazardous chemicals, physical and biological hazards.

4.4.1 *ISOLATION*

When use of hazardous materials or processes is unavoidable, the next best procedure is often to engineer out the hazard by isolation. If the worker can be isolated completely from the hazard, the risk to health is removed. Isolation may be achieved by a physical barrier or by distance. Time (e.g. timing of activities) is also a barrier, although time may equally be considered an administrative control (see section 4.5).

The following examples illustrate the principle:

- installing noisy compressor units in sound-proof housings and well away from worker frequented areas with process plants
- using interlocked doors or barriers to prevent entry into an area while toxic substances are present
- relocating workers not directly engaged with processes that may give rise to exposure to a low-risk/hazard area
- separating materials that could create hazards if they come into contact with each other by accident (e.g. keeping oxidants and fuels in separate buildings or compartments)
- locating rest areas away from hazard sources (heat, noise, dust, etc.).

Occasionally, it is possible to use timed sequences to conduct hazardous operations when fewer workers are present. For example, burning off of plastic extrusion dies might be restricted to evening hours, allowing several hours for the air to clear before workers re-enter. Painting of a workplace could be conducted outside normal work hours to prevent unnecessary exposure to solvent vapours. Work on air-conditioning plants that may involve their shutdown should be restricted to periods outside normal working hours, particularly when duct cleaning is required. Fumigating restaurants with pesticides cannot be conducted when staff or patrons are on the premises. It is frequently a practice to conduct underground blasting at the end of the day to allow sufficient time for the ventilation systems to remove blast fumes and dust.

4.4.2 *CONTAINMENT*

Once an agent (be it chemical, physical or biological in nature) has escaped from its source, it becomes far more difficult to capture or control. A better strategy is therefore to maximise the containment by engineering means, as in the following examples:

- A whole process can be totally enclosed and coupled with an exhaust extraction system.
- Sound-proof control rooms can be installed in noisy environments.

- A glove-box or biological safety cabinet can be installed for handling infectious agents.
- A remotely controlled laboratory can be used to handle radioactive isotopes.
- Noisy machinery can be enclosed in sound-proof structures.
- Gas-tight systems can be used in chemical processing or in many sterilising or fumigation procedures.

The containment approach has some drawbacks. For example, installing totally enclosed or contained systems may carry a high initial cost and may introduce the issue of workers coming into contact with the hazard if the isolation system fails.

The complete enclosure of a hazard is usually restricted to the extreme cases where escape of the hazardous substance could have serious health consequences or may be immediately life-threatening. Containment is normally supplemented by a ventilation system to ensure complete containment. Figure 4.6 shows an example of an enclosed process; here a plexiglas enclosure is built around a corona discharge unit to contain the generated ozone. A small exhaust ventilation fan supplements the containment. Chapter 5 provides examples of the use of partial enclosures, a common and widespread method of control.

While a totally enclosed process is in operation, operator exposure will be limited. When the process is upset or maintenance is needed, however, it may be necessary for staff to enter the enclosure. This event must be treated with extreme care and utmost caution. For example, entering a toxic pesticide melting oven for maintenance or entering into a fumigation process may require the operator to use high-integrity personal protective equipment, selected according to the specific risk. Such cases are often managed via a work permit, which is an administrative control.

Figure 4.6 An example of a totally enclosed process: an ozone-generating corona discharge supplemented by an extraction fan

Even where hazardous processes are totally enclosed, their location is important. They should not be situated where users or bystanders could be harmed if the enclosure system fails. Moreover, they may need to be interlocked so the process cannot be operated unless the isolation system is operating. Lastly, primary enclosed systems should be alarmed and have secondary back-up controls. Planning for the siting and operation of totally enclosed systems is crucial to their continued safety. An extreme example of an incorrect location would be an ethylene oxide fumigation chamber, including cylinders of fumigant, in the centre of a library. A far better location would be to have the cylinders outside the normal workspace, alarmed for leakage and fitted with a continuous exhaust system to handle any accidental discharge of sterilising gas.

In any workplace, the H&S practitioner should be aware of the locations of all potentially hazardous operations and ensure that those locations will not make control difficult if the isolation system breaks down. For example, workers should not be able to stray into situations where, without warning, they could be at unnecessary risk. If new processes are being installed—particularly enclosed processes—they should not be sited in areas of constant or high worker access, and they must not block or impede emergency exit routes.

4.4.3 VENTILATION

Industrial ventilation—the control of contaminants by dilution or local exhaust ventilation—is one of the main methods of controlling airborne chemical and biological hazards. The topic has already been touched on in this chapter and is covered in more detail in Chapter 5.

4.5 ADMINISTRATIVE CONTROLS

The exposure controls examined so far work by eliminating or substituting the material or process, or by engineering out the risk. Frequently, these controls are insufficient in themselves and it becomes necessary to change work methods or systems to achieve the desired level of control. Such measures traditionally have been considered 'administrative controls'. It would be both simplistic and optimistic to suppose that a single control strategy (with the exception of complete elimination) can result in satisfactory control of exposure. Sometimes higher-level control mechanisms are found to be impractical in use and are by-passed by workers, or they cannot be made to work well enough to negate the hazard. Consider the following workplace situations:

- working inside a deep freezer
- working inside a hot oven
- working underwater at a depth of 100 metres.

In the freezer and the oven, it is rarely practicable to introduce a warm (or cool) microenvironment to compensate for the cold (or heat) in the work environment. Keeping divers safe while working at great depths for long periods—and returning to the surface—requires elaborate equipment.

The use of administrative controls in regulating workplace hazards is an alternative strategy that concentrates on work processes and systems, and worker behaviours, rather than workplace hardware. While preference must be given to higher order solutions, special attention must invariably be given to worker education, behaviour or work practices because conventional methods may be neither feasible nor adequate to control the hazard. Administrative controls should not be confused with management functions, such as responsibility, audit and review.

Documented work procedures or work permits can be an effective administrative control process for some high-risk tasks. Good work procedures that incorporate a risk assessment (i.e. hazard identification and appropriate controls) would also be considered an administrative control. For example, work in confined spaces is controlled by performing risk assessments and using work permits as described in AS 2865, Confined Spaces (Standards Australia, 2009). The hazards of restricted visibility areas can be reduced by working in pairs or groups.

Housekeeping and labelling are two administrative control measures that help limit inadvertent exposure to workplace hazards. The importance of maintaining high standards of housekeeping cannot be overstated. Dirty and untidy workplaces not only increase the likelihood of secondary exposures (e.g. by inhaling dust raised by draughts and wind, or by inadvertent skin contact with dirty surfaces and equipment), but may also send a message to personnel that poor work habits are acceptable.

To use administrative controls properly, workers have to be adequately trained so they know:

- the full nature of the hazard and potential health impacts
- why the administrative control is being used
- the exact procedures and guidelines to be followed
- the limitations of administrative control procedures
- the consequences of ignoring the administrative control.

In other words, worker involvement, participation, training and education are critical to the success of administrative control programs.

4.5.1　STATUTORY REQUIREMENTS

All OHS legislation, codes of practice and industry standards incorporate forms of administrative control. The way in which administrative controls are implemented will depend on the particular workplace. Company OH&S policies, OH&S procedures, government regulations and some industrial relations arrangements all have a role to play.

4.5.2　PROVIDE HEALTH AND SAFETY INFORMATION

Provision of information, such as the mandatory availability of a safety data sheet (SDS) for hazardous chemicals, is an administrative control mechanism. Information systems are powerful administrative control mechanisms that operate unobtrusively in most workplaces, and are often taken for granted.

4.5.3 EDUCATION AND TRAINING

OH&S legislation in Australia and many other countries imposes specific requirements for training workers and others involved in health and safety activities. Where administrative controls are instituted (such as may occur under the various health and safety or hazardous substances regulations), the law generally requires training and induction of workers.

Training programs should be formalised and administered long past the induction period and throughout the length of employment. Training should always incorporate the practical aspects of a job and include some form of competency assessment. If workers potentially exposed to a hazard are made fully aware of the consequences of over-exposure, and the routes and mechanisms of exposure, they are more likely to identify other exposure situations and act to reduce exposures in new situations. In this regard, it is useful in a training setting to visually represent exposure. For example, fluorescein dye (used by plumbers to trace drains) can be added to aqueous solutions to simulate toxic liquids. In order to show the degree of containment, or the spread of contamination, the traces of fluorescein can be made visible by using a UV lamp. Smoke generators can be used to test ventilation systems and can show the potential movement of contaminants within the workplace. Intense lighting can also be used to show dust generation, as seen in Figure 4.7.

4.5.4 WORKER ROTATION AND REMOVAL FROM EXPOSURE

A form of exposure control that involves changing a work schedule would be to rotate tasks within the work group to spread the exposure across a larger number of workers. For this to be a viable strategy, the H&S practitioner must have a reliable system to measure the exposures of all members of the work group.

Figure 4.7 Intense light illumination being used to highlight secondary dust from work clothes as a source of exposure
Source: Reproduced with permission, UK HSE.

In many circumstances, exposure to hazardous chemicals or a hazardous environment cannot be avoided. If any workers are exposed to the maximum permissible level, then they may need to be removed from exposure. The following are some examples:

- In the lead industry, workers may be removed if blood lead levels exceed a certain level, and remove themselves from further lead exposure until blood lead levels fall to an acceptable level.
- In industries where ionising radiation is involved, workers are permitted a maximum radiation dose over a specified time period (e.g. the American Conference of Governmental Industrial Hygienists' TLV® for ionising radiation (ACGIH, 2012) incorporates the concept of 'cumulative effective dose').
- Workers in excessively noisy industries who cannot be protected adequately by hearing protection should have noise exposure reduced by reassignment so that their daily equivalent noise exposure does not exceed the legal limit of 85 dB(A) L_{eq}.

Where workers have developed sensitivity to a substance, a common administrative control is to prevent any further exposure. In other instances, workers may be predisposed to experiencing effects at lower thresholds than the average worker, or they may be medically diagnosed as showing effects of exposure without actually being symptomatic. The following are some examples:

- Isocyanate-sensitised workers should be prevented from any further exposure.
- Workers with radiologically confirmed dust disease should be precluded from further work in dusty underground mining.
- Workers with certain genetic dysfunctions should not be occupationally exposed to TNT or chemicals causing haemolytic anaemia.
- Pregnant workers should not be exposed to known foetal toxins (e.g. lead, methyl ethyl ketone and other solvents).
- Asthmatics should not work with strong irritant gases.

There are examples from history where workers' exposures were controlled by compulsorily withdrawing them from exposure rather than by limiting the actual routes of exposure. This method of exposure control is suggestive of the inability of other control measures to limit exposures. In the current-day context, it would be considered unethical to continue exposing a group of workers without demonstrating the risk is ALARP. Therefore, if such a situation arises, the H&S practitioner would be advised to re-evaluate all aspects of the control systems and strategies.

4.6   PERSONAL PROTECTIVE EQUIPMENT

Personal protective equipment (PPE) takes its place at the bottom of the HoC. PPE represents the absolute last resort; beyond it is the unprotected worker and inevitable exposure if the PPE is not correctly selected, maintained and used appropriately. Even though PPE is on the bottom of the hierarchy, it is still widely used and accepted as a backup and supplement to other controls. There will also be situations where higher-level controls cannot be used and PPE will be the only practicable solution. Chapter 6 provides a detailed discussion of various aspects of the selection, care and use of PPE.

4.7 CONTROL OF CHEMICAL HAZARDS

Within Australia, hazardous chemicals regulations are based on Safe Work Australia's model regulations, *Model Work Health and Safety Regulations: November 2016* (Safe Work Australia, 2016), or its forerunner model regulations. These set out the basic obligations of manufacturers, importers and suppliers to provide information to the workplaces in which their products are used. Employers also have obligations to identify hazards (mainly based on the hazardous substances risk phrases used in SDS), provide relevant information on hazardous substances in their workplaces, assess the extent of the exposure and control risks for exposure routes to be defined, train staff, undertake health surveillance and keep records where necessary. These regulations incorporate all the HoC principles described in the various sections above.

Many published guidelines on the management of hazardous chemicals contain useful information on the hierarchy of control, but the optimal control strategy for any workplace will depend on its unique set of circumstances. Some additional regulations on asbestos, lead and carcinogens detail specific actions to be taken for their management.

4.8 CONTROL BANDING

The traditional structured approach of occupational hygiene is built around quantitative risk assessment, typically based on measurement of exposures to an agent (noise, biological hazard, toxic dust, vapours, radiation). Those measured exposures are then compared against an agreed exposure standard to determine the level of control that needs to be applied. Not all hazards encountered have exposure standards, however, thus making it difficult to do the risk assessment. In the 1990s, the pharmaceutical industry developed a control approach based on the comparison of the toxicological data combined with physical form and properties to apply one of a prescribed set of controls to a material in particular handling scenarios. This semi-quantitative approach was further refined by the UK HSE and published as Control of Substances Hazardous to Health (COSHH) Essentials, commonly referred to as control banding (CB) (HSE, 1999, 2009). The initial focus was on helping small and medium enterprises to comply with chemical control regulations. The process steps are:

1 Identify the 'hazard band' based on:
   - the toxicity (e.g. the risk phrase)
   - the ease of becoming exposed (e.g. volatility or dustiness)
   - the work processes (e.g. container filling, grinding, bag emptying)
   - the duration of the potential exposure (e.g. minutes, hours, continuous)
   - the quantity of material handled (e.g. grams, kilograms, tonnes).
2 Based on the above, a generic and conservative control approach will be identified. These are typically:
   - ventilation
   - engineering controls
   - containment.
3 Seek expert advice.

While CB seems to be an attractive system, concerns have been raised about its effectiveness. For example, it appears to be more effective when applied to dust hazards than to vapours. Attempts to apply CB to nanomaterials have not been widely accepted, one reason being the lack of benchmark exposure standards.

4.9   LEGAL REQUIREMENTS

OH&S legislative acts in Australian jurisdictions and many other countries contain requirements for a general duty of care, together with some specific directives to manage chemical and other hazardous agents, such as asbestos, silica, lead and carcinogens, noise or ionising radiation sources. There is a framework in this legislation within which the obligations of various parties, including both employers and workers, are established. Persons conducting a business or undertaking have obligations to ensure the health and safety of workers by controlling hazards at their source. Workers also have obligations to cooperate with employers to maintain their own health and safety. The law imposes obligations with regard to traditional safety matters (guarding, electrical, prevention of falls, etc.), as well as to the more difficult aspects of occupational health and occupational hygiene.

It is not defendable under law to allow hazards to persist simply because it will cost too much to control them. It is equally not defendable to leave an identified hazard in an uncontrolled state simply because workers seem prepared to tolerate it. In almost all workplaces, there is still room for improvement in the control of hazards. Along with the regulatory imperative, it has been shown that good work health and safety improves long-term business productivity (Safe Work Australia, 2018b).

4.10   REFERENCES

American Conference of Governmental Industrial Hygienists (ACGIH) 2012, *TLVs® and BEIs®*, ACGIH®, Cincinnati, OH.

Health and Safety Executive (HSE) 1999, *Control of Substances Hazardous to Health: Easy Steps to Control Chemicals*, HSE, Sudbury, UK.

—— 2009, *The Technical Basis for COSHH Essentials: Easy Steps to Control Chemicals*, HSE, Sudbury, UK, <www.coshh-essentials.org.uk/assets/live/CETB.pdf> [accessed 15/8/18]

——2015, *The Construction (Design and Management) Regulations 2015*, HSE, Sudbury, UK, <www.hse.gov.uk/construction/cdm/2015/index.htm> [accessed 15 August 2018]

National Institute for Occupational Safety and Health (NIOSH) 2016, Prevention Through Design Initiative, <www.cdc.gov/niosh/topics/ptd/default.html> [accessed 26 August 2018]

Safe Work Australia 2014, *Asbestos-related Disease Indicators*, <www.safeworkaustralia. gov.au/system/files/documents/1702/asbestos_related_disease_indicators_2014.pdf> [accessed 10 September 2018]

—— 2016, *Model Work Health and Safety Regulations*, SWA, Canberra, <www.safeworkaustralia. gov.au/doc/model-work-health-and-safety-regulations> [accessed 15 August 2018]

—— 2017, Good work design, <www.safeworkaustralia.gov.au/good-work-design>.

—— 2018, *Hazardous Chemical Information System: Exposure Standard Documentation: Carbon Monoxide*, SWA, Canberra, <http://hcis.safeworkaustralia.gov.au/Exposure Standards/Document?exposureStandardID=111> [accessed 15 August 2018]

—— 2018b, *Australian Work Health and Safety Strategy 2012–2022: Healthy, Safe and Productive Working Lives*, <www.safeworkaustralia.gov.au/doc/australian-work-health-and-safety-strategy-2012-2022> [accessed 27 June 2018]

Standards Australia 2009, Confined Spaces, AS 2865:2009, Standards Australia, Sydney.

4.11 APPENDIX SIMPLE RISK ASSESSMENT MATRIX

Consequence table

Consequences	Occupational health/safety
Notable	First aid
Significant	Medical treatment
Highly significant	Lost time
Serious	Disabling
Extremely serious	Single fatality
Catastrophic	Multiple fatalities

Risk ranking matrix

	Notable	Significant	Highly significant	Serious	Extremely serious	Catastrophic
Almost certain	B	B	A	A	A	A
Likely	C	B	B	A	A	A
Possible	C	C	B	B	A	A
Unlikely	D	D	C	C	B	A
Rare	D	D	D	D	C	B
Extremely rare	D	D	D	D	D	C

Risk Acceptability Criteria

Risk class	Risk acceptability
Class A – Very high	Risks that significantly exceed the risk acceptance threshold and need urgent and immediate attention. This represents a significant risk to health.
Class B – High	Risks that exceed the risk acceptance threshold and require proactive management. This represents a significant risk to health.
Class C – Medium	Risks that lie on the risk acceptance threshold and require active monitoring.
Class D – Low	Risks that are below the risk acceptance threshold and do not require active management

5. Industrial ventilation

David Bromwich, Elaine Lindars, Kate Cole

5.1   INTRODUCTION

Industrial ventilation is one of the most effective ways to reduce toxic airborne exposures in the workplace by engineering, and also improves the comfort of workers. To do either job, though, an industrial ventilation system must be well designed and maintained.

This chapter outlines some basic principles of industrial ventilation and should help the reader to identify poor and inefficient designs, which will assist in increasing the effectiveness of a ventilation system. While minor modifications to a system can be made without calculations, significant changes or new designs will require specialist knowledge.

5.2   PRINCIPLES AND MYTHS

Poor design, poor installation and poor maintenance can be common in industrial ventilation systems, which may be exacerbated if modifications are made without a solid grasp of their impact. The study of industrial ventilation and its application to hazard control begins with an understanding of how air moves in addition to several key concepts.

5.2.1   *MOVING CONTAMINATED AIR AWAY FROM THE BREATHING ZONE*

If contaminated air is induced to move away from a person's breathing zone (Figure 5.1), then it is likely that the person will be exposed to less of the contaminant. While this may be obvious, it is not uncommon to observe industrial ventilation systems that fail to achieve this goal.

In Figure 5.1, the worker is grinding a casting with a small wheel. Clean air is supplied through a duct and contaminated air is removed by an extraction system under the grate at his feet. The air is drawn away from his face and away from the source of contamination.

One of the most common errors in ventilation system design is to place a canopy hood directly above a worker or require the worker to operate beneath the hood, as shown in Figure 5.2. Such designs result in contaminated air being drawn past the worker's face. This is very common with welding hoods, and when the welding booth is near a wall replacement air can swirl in front of the welder as it rises, making the fume concentration inside the welding helmet higher that that outside the helmet.

5.2.2   *SWIRLING AROUND DOWNWIND OBJECTS*

When cooking on a barbecue, most people know that if they stand downwind the smoke will blow into their face. However, novice cooks may be surprised to find that when they

Figure 5.1 Contaminated air moving away from worker's face
Source: Great Britain Home Office Committee on Ventilation of Factories and Workshops (1903, p. 29).

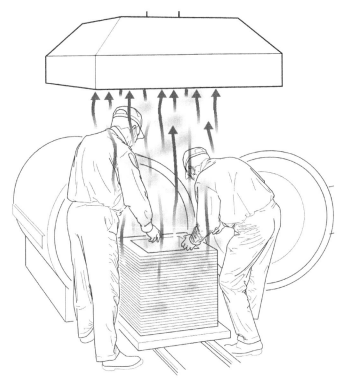

Figure 5.2 Poor canopy hood placement
Source: Health and Safety Executive (2017, p. 41).

move upwind, the smoke seems to follow them. This happens because the air swirls or eddies downwind of the body, trapping some smoke; this is sometimes referred to as the 'barbecue effect'. This phenomenon is shown in Figure 5.3. Similar effects occur in front of a fume cupboard, with a spray-painting booth or when welding.

In many workshops, a fume-extraction system is placed above the worker to capture fumes from processes such as welding. The air thus tends to flow towards and past the worker's face. This is exacerbated because the extraction system produces a local flow of replacement air that swirls around the worker, so that instead of moving directly upwards, the contaminated airflow tends to curve towards the worker's face. The concentration of fume inside a welding helmet can thus be higher than that outside the helmet.

On a larger scale, when air swirls around buildings, contaminants released from a short stack downwind of a building but still in the building's wake can be drawn back inside through windows and ventilation inlets.

5.2.3 SUCKING VS BLOWING

An understanding of the difference between sucking and blowing air is fundamental to industrial ventilation design. The difference can easily be demonstrated.

Hold your hand a distance from your mouth and *blow* towards it. The air movement is easy to feel. The air flows in a jet in the direction of your hand. With your mouth still

Downwind - Smoke lingers in inhalation zone   Upwind - Smoke still lingers in inhalation zone

Figure 5.3 The barbecue effect

open, breathe in hard. Now you are *sucking* air. The airflow is from all directions, much of it over the surface of your face: you can blow out a candle at arm's length, but if you try to suck out the candle, your lips will need to be just millimetres from the flame before you can extinguish it. This fundamental difference between sucking and blowing can be used to explain why placing a hood too far from the source of contaminated air is ineffective. It also explains why a good seal between the face and a respirator is so important: air is inhaled from all directions and so if any part of the seal is poor, the airflow will take the path of least resistance under the seal, bypassing the respirator filtration.

When a fan (or open duct) blows air, the velocity drops to roughly 10 per cent of the maximum in a distance equal to about 30 to 60 times the diameter of the fan/duct; however, when air is sucked, this velocity drop occurs within a distance equal to the diameter of the fan or duct. This can be observed with a small fan such as a desk fan (Figure 5.4). If a piece of paper is held some distance in front of the fan, the airflow ruffles it easily. If the paper is held behind the fan, it will not be disturbed until it is very close to the blades. Similarly, with a vacuum cleaner the suction operates only close to the nozzle but the exhaust plume can be felt at some distance.

Industrial ventilation systems are often designed and installed by contractors who are more familiar with heating, ventilation and air-conditioning (HVAC) systems, which condition air and blow it into buildings. The fans, ducts and some air-cleaning systems are much the same in industrial ventilation, but now the air is being sucked rather than blown, and that difference and its consequences may not be understood. As a result, industrial ventilation systems often fail to protect workers adequately from contaminated air.

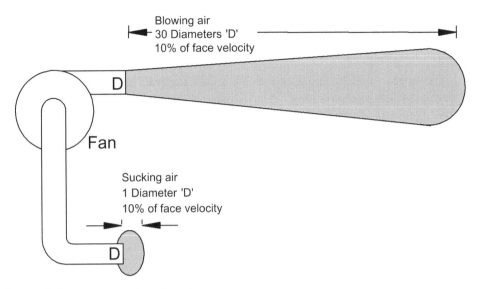

Figure 5.4 Sucking and blowing air

5.2.4   THE 'HEAVIER THAN AIR' MYTH

It is not uncommon to find industrial ventilation systems designed to remove 'heavier than air' vapours and gases from near the floor. While it is true that high concentrations of vapours and gases can be found near the floor, this usually occurs only:

- at very high concentrations these are often a result of catastrophic releases and orders of magnitude above workplace exposure standards (Figure 5.5)

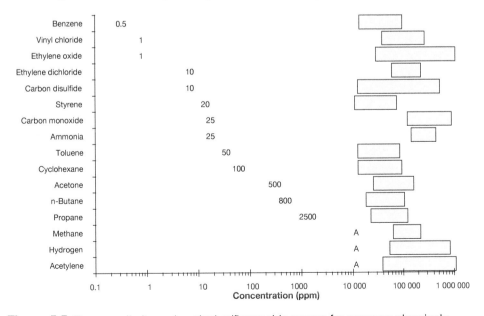

Figure 5.5 Exposure limits and explosive/flammable ranges for common chemicals

- where there are no temperature gradients (isothermal)
- where there are other air disturbances from moving people or machines.

As indicated above, these conditions may occur in the confined bilge of a boat or during a catastrophic large-volume release such as a tank rupture, but these are exceptions, because while the gas or vapour can be heavier than air, imperceptible temperature gradients can be significant in creating slow convection currents that mix the air. In Figure 5.5, exposure limits for common gases and vapours are often orders of magnitude lower than explosive limits.

A simple calculation shows that even with a concentration of 1000 ppm, a solvent with a vapour density twice that of air is unlikely to sink to the ground. Imagine a million air molecules among which are 1000 molecules of a solvent (ten times the workplace exposure standard for many solvents) whose vapour density is twice that of air (do not confuse vapour density with the density of the liquid).

Some simple calculations show the impact of molecule density vs thermal changes in a room.

The density change of a solvent vapour/air mix:

$$\frac{(\text{molecules of air} \times \text{density of air}) + (\text{molecules of solvent} \times \text{density of solvent})}{(\text{molecules of air} \times \text{density of air})}$$

$$= \frac{(999\,000 \times 1) + (1000 \times 2)}{1\,000\,000} \qquad \text{(Equation 5.1)}$$

$$= 1.001$$

$$= 0.1\% \text{ increase}$$

5.2.5   THE THERMAL CHANGE OF THE VAPOUR/AIR MIX

Now consider the ideal gas equation, $PV = nRT$ (where P = pressure, V = volume, n = moles, R = gas constant, T = absolute temperature in Kelvin). The air pressure is effectively the same at floor (subscript 1) and ceiling (subscript 2), ignoring the weight of the air column in the room. The volume of air is constant so the 'nRT' side of the equation at floor ($n_1 T_1$) and ceiling ($n_2 T_2$) is the same ($n_1 T_1 = n_2 T_2$).

$P_1 V_1 = n_1 R T_1$ and $P_2 V_2 = n_2 R T_2$
$P_1 V_1 = P_2 V_2$ and R is constant, so $n_1 T_1 = n_2 T_2$

If the air temperatures is 20°C at the floor and just 1°C warmer or 21°C at the ceiling, this gives:

$T_1 = 20°C = 273 + 20 = 293°K$ at floor
$T_2 = 21°C = 273 + 21 = 294°K$ at ceiling

$$\frac{n_1}{n_2} = \frac{T_2}{T_1} = \frac{294}{293} = 1.0034 \qquad \text{(Equation 5.2)}$$

So $n_1 = 1.0034 \times n_2$

This indicates that n_1 is 0.0034 more dense than n_2, i.e. 0.34% more dense, than n_2.

Hence, if we compare the density change as a result of 1000 ppm of a solvent (Equation 5.1) with the density change as a result of a temperature difference of just 1°C from floor to ceiling (Equation 5.2) then the result is: Equation 5.1 density change = 0.1% versus Equation 5.2 density change = 0.34%. So, thermal convection will produce greater convective forces than the density of a vapour at 1000 ppm. This means that in most workplaces, normal convective air currents will mix the air and no blanket of 'heavier than air' vapour will form on the floor. A temperature increase of a few degrees at the ceiling is quite common.

5.3   GENERAL VENTILATION DESIGN

The design of ventilation systems is well covered in texts such as the ACGIH®'s (2019) *Industrial Ventilation: A Manual of Recommended Practice for Design*, and the free booklet from Britain's Health and Safety Executive (2017), *Controlling Airborne Contaminants at Work: A Guide to Local Exhaust Ventilation (LEV)*.

5.3.1   *KEY CONSIDERATIONS*

For local exhaust ventilation, design considerations include the below.

- *The toxicity and pattern of release of contaminated air.* Is local exhaust ventilation needed or is it even the best solution?
- *Proper capture of the air and the design of hoods.* This also requires consideration of the task, movement of people past the hood, and the effect of side draughts.
- *Transport velocity of the contaminated air in ducts and sizing of ducts to make the hood operate as designed.* The roughness of the ducts' surfaces and noise need to be considered here.
- *The type of air filtration needed.* This is a specialised topic, but advice can be given by commercial suppliers.
- *Fan selection.* Issues include efficiency, power, size, noise and cost.
- *Fan location.* Ideally fans should be placed outside the building so that the low-pressure part of the system is internal and the high pressure external. That way, if there are any holes in the duct work it fails safe.
- *Ductwork.* Should have minimal (smooth) direction changes, be sufficiently strong, well supported and capable of withstanding normal wear and tear for the contaminant, and have maintenance access.
- *Ease of maintenance and monitoring of ventilation systems.*
- *Any modifications and additions and their effects.* Performance usually suffers.
- *Noise prediction and reduction.*
- *Design and location of stacks.* Some of the most expensive 'fixes' to problems involve poor design and re-entrainment of exhausts, particularly in large buildings with courtyards.

5.3.2   *PRESSURE DROPS*

In Australia, the United Kingdom and Europe, pressure drops in LEV systems are estimated with the 'velocity pressure' method, where the pressure drop is estimated for each straight run, bend or join. In the United States, the trend has been towards using the 'equivalent foot' method, where the pressure drops for bends and joins are estimated for a specific length of ducting. For complex systems, the flows are 'balanced' at the design stage so that the designed flow rates at each hood are achieved by making the pressure drop to the same extent along each 'leg'.

5.4   GENERAL VENTILATION

General ventilation is aimed at reducing the concentration of a contaminant by adding fresh air to the workplace. There are two categories: dilution ventilation and displacement ventilation.

5.4.1   *DILUTION VENTILATION*

In dilution ventilation, the contaminant is mixed with fresh air, diluting it. The air in the whole room may be mixed with the fresh air, or a stream of air from a fan or an open window may perform the task more locally.

If natural airflows are used to dilute contaminants, wind direction, wind speed and air temperatures are likely to have significant impacts on the effectiveness of this approach.

5.4.1.1   Use
Dilution ventilation may be appropriate when:

- the air contaminant has low toxicity
- there are multiple sources—that is, no single point source
- the emission is continuous
- the concentrations are close to or lower than the workplace exposure standard
- the volume of air needed is manageable
- the contaminants can be diluted sufficiently before inhalation
- comfort (or odour) is the issue, in the absence of other contaminants
- a spill has occurred, and extended airing of the workspace is needed.

5.4.1.2   Limitations
Extraction fans are sometimes mounted above a workbench to remove contaminated air. Since, in reality, the air moves towards the fan from all directions in a collapsing hemisphere (Figure 5.6), only a small amount of the contaminated air from the bench is actually extracted. Note that dilution ventilation is sometimes confused with LEV; in the case of Figure 5.6 the fan is extracting from the whole room, the table happens to be placed closest to the extraction point, however, the premise shown in the figure is true regardless of where the extraction point is set.

Figure 5.6 A fan in the wall: hope (left) and reality (right)

In many cases, poorly designed ventilation systems that aim to remove contaminated air from around a worker (Figure 5.6) only dilute the contaminant, giving little local protection to the worker but slowly cleaning the air in the workplace as a whole.

5.4.1.3 Air exchanges and mixing

Dilution ventilation of contaminated air works in much the same way as pouring clean water into a jug of coloured water dilutes the colour. If the added water forms a layer on the top and overflows the jug without mixing with the coloured water, then the colour (contaminant) persists. If the fresh water and the coloured water are well mixed, the colour gradually fades as it is diluted; however, traces of the chemical colorant remain long after all visible colour has been removed.

5.4.1.4 Decay rate calculations

In dilution ventilation, the basic formula to describe the exponential reduction in contaminant concentration (C) over time (after removal of the source) is given by the following equation, if we assume complete mixing:

$$C = C_0 e^{-Rt}$$

where:
C_0 = the initial contaminant concentration in air by volume
R = ventilation rate = Q/V
Q = airflow rate into the space (m^3 s^{-1})
V = volume of ventilated space (m^3)
t = time in seconds

As an example, in a room of 10 m^3, with 1000 ppm contaminant in the air and a diluting airflow of 0.1 m^3 s^{-1}, the concentration in the room ten minutes after the source has stopped is calculated by the following:

C_0 = 1000 ppm
V = 10 m^3
Q = 0.1 m^3 s^{-1}
R = Q/V = 0.1/10 = 0.01 s^{-1}
t = 10 × 60 = 600 s
C = $1000e^{-0.01 \times 600}$ = 2.5 ppm

This is a small fraction of the original concentration, but it is never zero.

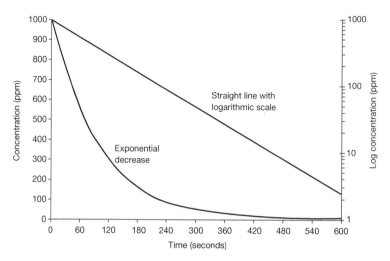

Figure 5.7 Impact of dilution ventilation represented on linear and logarithmic scales

The exponential decay of the contaminant over time is shown in Figure 5.7 on a linear plot, with the same data, plotted on a logarithmic scale, producing a straight line. The slope of the line represents the rate of contaminant decrease, which in turn is indicative of air changes, usually as air changes per hour.

This calculation would under-estimate the final concentration, as mixing is always incomplete, hence the flow should be increased by a safety factor, K, of between 3 and 10 to calculate the expected concentration at a given time. The ACGIH® (2019) also suggests that proximity, location and number of sources should be considered when selecting the 'K' factor'. K can be incorporated into the dilution ventilation equation as:

$$C = C_0 e^{-Rt/K}$$

Many texts give tables of K, terming it a 'mixing and safety factor', but this does not take into account either the variability of local concentrations or the degree of mixing (Feigley et al., 2002).

The degree of mixing of air in a room largely determines how well dilution ventilation works. For example, it is not uncommon for fresh air in an air-conditioned office to be introduced through slots in the ceiling, only to travel along the ceiling, with little effect on the air quality of the occupants. If the air inlet is directed downwards to achieve better mixing, some workers might complain of draughts. By ensuring good mixing of fresh and contaminated air while limiting draughts, the contaminated air is diluted and, so long as no more contaminant is introduced, the contaminant concentration will reduce exponentially over time.

To dilute the air after a (volatile) contaminant spill, the required flow, Q, is:

$$Q(m^3\ s^{-1}) = \frac{\text{Rate of evaporation } (mg\ s^{-1})}{\text{Density of liquid } (kg\ m^{-3}) \times \text{Workplace exposure limit } (ppm)}$$

The required flow rate would need to be multiplied by the safety factor K (3 to 10) to account for incomplete mixing.

5.4.1.5 Key considerations

Some things to watch for when considering ventilation problems are provided below:

- Is there make-up air to replace the exhausted air?
- Is the incoming or make-up air always clean?
- Where does the contaminated air go?
- If the ventilation is natural, what happens on still days or when the wind blows from another direction?

5.4.2 CONFINED SPACES

Ventilation of confined spaces is usually a special case of dilution ventilation, particularly when there is only one entrance. The emanation of toxic or flammable vapours (or absence of oxygen) and the presence of 'dead spaces' makes it difficult to predict toxic exposures within the confined space. For this reason, personal electronic monitors are advisable to warn of danger. If fresh air is provided under pressure with an air hose, a jet of air over a pool of solvent can actually increase the concentration of solvent vapour. A feed of pure oxygen can cause hair or clothing to catch alight during hot operations like welding.

5.4.3 THERMAL DISPLACEMENT VENTILATION

Thermal displacement ventilation is a mature technology in Scandinavia, and can also be found in Australia. Contaminated air is removed from the work area by the plume of warmer, more buoyant air from people and work processes. Figure 5.8 compares thermal displacement ventilation with dilution ventilation.

Air that is around 2–3°C cooler is gently introduced near floor level to fill the area to a height of 2–3 metres, in a process similar to filling a swimming pool with water. Special diffusers (often long fabric ducts) are used to ensure that the airflow does not create draughts or settle. Unlike conditioned air, which is usually cool and fed in from vents in the ceiling, displacement ventilation works with natural convection patterns and can be much more efficient than dilution ventilation because only the occupied depth of the room is cooled. Without such a temperature gradient, the whole room would have to fill to remove contaminated air.

In Figure 5.8, the contaminant concentration in the standing person's breathing zone is reduced more effectively by thermal displacement ventilation than by traditional dilution ventilation. This method not only works better but saves energy.

5.4.3.1 Use

Thermal displacement ventilation works best when:

- the contaminants are warmer than the surrounding air
- the supply air is slightly cooler than the surrounding air
- the ceiling is relatively high (more than 3 metres)
- there is limited movement in the room and no major opening.

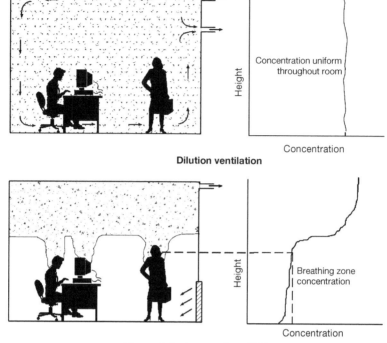

Figure 5.8 Dilution ventilation and thermal displacement ventilation
Source: Adapted from Skistad (1994, p. 7).

5.4.3.2 Limitations

If the ceiling height is less than 2.3 metres, the contaminated air is cooler than the incoming air, or there are draughts, then displacement ventilation does not work as well. Cold windows or walls can create down-draughts, and hot spots in the sun can create local updraughts. Displacement ventilation is unsuitable for operating theatres (Friberg et al., 1996) and 'clean' rooms, as it tends to act against the sedimentation of particulates.

5.5   LOCAL EXHAUST VENTILATION

Local exhaust ventilation (LEV) aims to remove air contaminants at the source before they have a chance to be inhaled. The principles of LEV have been known for more than a century, but can be poorly understood. A simple LEV system (Figure 5.9) most commonly comprises a hood to capture and remove contaminated air near the point of release, ducting to connect the hood to an air-cleaning system, a fan to move the air through the system and an exhaust stack outside the building to disperse the cleaned air.

5.5.1   GOOD DESIGN

Good LEV design can be difficult to observe. The weak point tends to be the hood, which is needed to shape the flow of contaminated air and ensure its efficient capture. To work

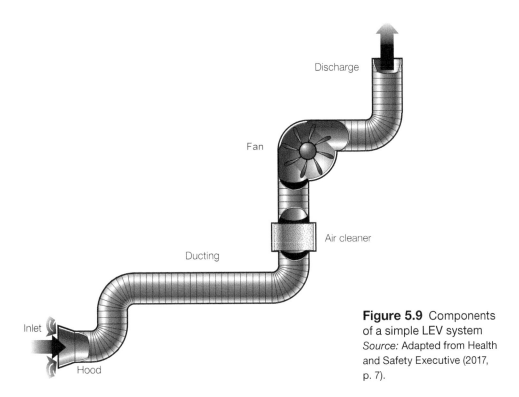

Discharge

Fan

Air cleaner

Ducting

Inlet

Hood

Figure 5.9 Components of a simple LEV system
Source: Adapted from Health and Safety Executive (2017, p. 7).

as designed, any hood needs a predetermined air inflow rate, which in turn influences the design of the rest of the LEV system; notably the maximum distance of the hood from the source can generally be no more than 25 centimetres. If the system is designed first and the hood is placed as an afterthought, air capture is unlikely to be efficient.

In Figure 5.10, workers are finishing fired earthenware with emery paper, producing dust containing silica. This 1903 example shows that modern technology is not necessary where there is good design. The hoods are ventilated enclosures with a glass top to let in light. Each hood is joined to a duct that is attached to a centrifugal fan. The exhaust from the (unguarded) belt-powered fan is cleaned by a cyclone. This is good design, but it would be even more desirable to clean the air before it reached the fan to lessen the wear on the

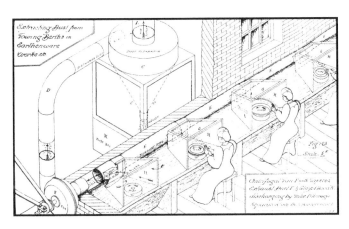

Figure 5.10 LEV system in a pottery
Source: Great Britain Home Office Committee on Ventilation of Factories and Workshops (1903, p. 58).

fan blades and limit the deposition of particulates in the ducting. This would also likely reduce the noise exposure of the workers.

Before a LEV system is designed, the potential for exposure to toxic materials should be examined so that the required degree of protection can be estimated. This assumes that an LEV system is the best approach, and that the hazard cannot be eliminated by substitution of the hazardous material or by modifications such as total enclosure of the process. If LEV is the preferred approach, then an appropriate hood is designed to capture the contaminated air, and the airflow rate needed for the hood to function is calculated. A large duct costs more and may result in settling of particulates in the duct. A small duct may result in large pressure drops or unacceptable noise. This may result in the system being turned off. The duct may bend and join other ducts and, as with merging traffic on a highway, its size should increase gradually as the airflow (traffic) increases so the velocity remains the same. The contaminated air is often filtered before being released into the atmosphere, and the air-cleaning device is usually placed ahead of the fan that sucks the air through the system, so that the fan is protected.

Once the air has passed through the fan, it is pushed through a stack under pressure. Prior to the fan, the ducting is under low pressure so not only will leaks make the system less efficient, but they can cause contaminated air to re-enter the workplace if the fan is inside the building: placing the fan outside of the building is very important where possible. In determining the height of the stack, planners should take into consideration buildings and structures, both upwind and downwind of it, to ensure that the air released does not enter other inhabited buildings. As a rule of thumb the stack should be at least 2.5 times the height of surrounding buildings or countryside to minimise turbulence and re-entrainment of exhaust into surrounding buildings.

Lastly, the type and size of the fan should be chosen to produce the desired flow into the hood and overcome any pressure drops within the system. The components of a LEV system will now be considered in their design order.

5.5.2   HOODS

Hoods are the most important part of a LEV system, as a poorly designed hood limits the performance of the whole system. It is not uncommon to see a complex system installed with little regard to how effectively it will capture the contaminated air from the process or machinery to which it is attached.

5.5.2.1   Key requirements

The most efficient hoods smoothly accelerate air from near stationary to the proper duct velocity with few eddies or changes of direction (Figure 5.11). It is difficult to capture billowing contaminants without first enclosing them in a larger volume. Similarly, high-velocity air contaminants from operations such as grinding are usually intercepted, and the residual contaminants then captured.

Simple principles that can make hoods more effective include:

- reduction of the source of emissions—such as closing the lid on a vessel, making the process wet to reduce dust or using premixed formulations to limit the dustiness of a toxic component

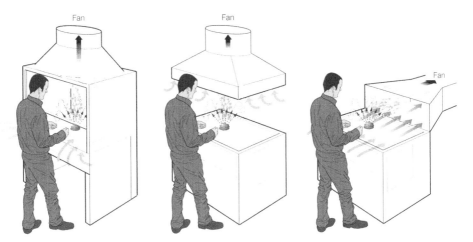

Figure 5.11 Enclosing, receiving and capturing hoods
Source: Health and Safety Excecutive (2017, p. 30).

- placing the hood as close as possible to the source, preferably enclosing it
- if the source includes fast-moving particles, positioning the hood to receive those particles (such as on a grinder or cut-off saw)
- specifying a 'capture velocity' greater than the particle velocity and
- locating the hood so that a line from the operator's face to the contaminant source leads directly towards the hood.

There are three main types of hoods: capturing, receiving and enclosing.

5.5.2.2 Capture zone and capture velocity

When installing a hood, it is first necessary to determine the zone where the air contaminants are generated and need to be captured. This in turn requires estimates of the air velocity needed at the edges of the zone to ensure that capture occurs. In Figure 5.12, the trajectories of air into a slot hood are visualised in the laboratory using a smoke tube,

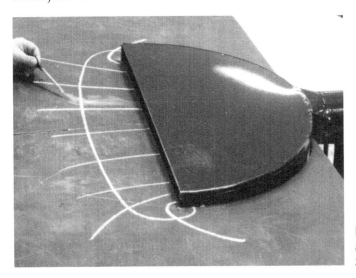

Figure 5.12 Air flow lines and 0.25 m/s^{-1} capture zone for a slot hood

the chalk lines represent the lines of airflow into the hood, and a capture zone for an air speed of 0.25 m/s^{-1} (measured with a hot wire anemometer) is shown by the line intersecting the air trajectory lines.

Air is drawn from all directions into the hood and almost as much uncontaminated air enters this unflanged hood from behind as from in front.

The 'stopping distance', S, is a useful concept for estimating how far a particle will travel before it can be affected by airflow into a hood. For particles with the density of water (unit density) that can be inhaled (<100 µm diameter) and travelling at 10 m/s^{-1} (at NTP, or normal temperature and pressure), the stopping distance varies from 13 centimetres at 100 µm diameter to 88 µm at 1 µm diameter. The stopping distance is visualised in Figure 5.13, which shows the stopping distance for given particle diameters and a density. (NTP is defined as air at 20°C [293.15K] and 1 atmosphere [101.325 kPa].)

Particles greater than 100 µm diameter tend not to be inhaled and can move some distance before the viscous drag from the air slows them. They may be more of a housekeeping problem than an inhalation issue. The stopping distance of very heavy (lead) or very light particles (glass micro-balloons or expanded styrofoam beads) has to be scaled by their density, since light particles penetrate deeper and more easily into the lungs.

5.5.2.3 Hood types

Three hoods types were shown in Figure 5.11, but flanges (mainly on receiving hoods) and baffles (mainly on enclosing hoods) can help shape the airflow and make the hoods more efficient.

Flanges inside a hood help to shape the airflow and reduce the amount of uncontaminated air entering the hood. Many hoods are poorly shaped, making them very inefficient at capturing air contaminants. The most efficient hoods avoid eddies, so that most of the suction accelerates the contaminated air to the duct velocity smoothly and with few eddies. However, a hood that is highly efficient at collecting contaminated air may obstruct the work process and be removed. Close observation and investigation of work practices are

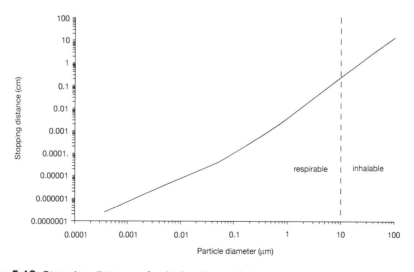

Figure 5.13 Stopping distance of unit-density particles

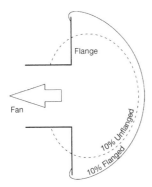

Figure 5.14 Contours for 10% of hood face velocity for a flanged and unflanged hood
Source: Adapted from Alden & Kane (1981, p. 27).

therefore needed to design a hood that is both acceptable to the workers and efficient at collecting contaminated air before it is inhaled.

In Figure 5.14, the flange prevents most clean air coming from behind the hood from being captured, and so extends the range of the hood to capture more contaminated air with the same air flow. It also decreases pressure drops associated with the hood, making it about 25 per cent more efficient.

A hood may have internal baffles to better distribute the airflow and improve its performance. Almost all fume cupboards contain a rear baffle.

5.5.2.4 Slots

Where the contaminant is not a point source—such as in a drum-filling operation or a dipping tank—long hoods with slots can make capture more efficient (Figure 5.15). The hood may be straight or curved but, most importantly, it should be very close to the source of emission, as effectiveness drops rapidly with distance. It is very common for the capture range of a slot hood to be grossly over-estimated, particularly in the case of drums, tanks and large work surfaces.

Figure 5.15 Slot hoods along the length of a tank
Source: Adapted from Alden & Kane (1981, p. 52).

For a long tank, a slot hood along each side can be effective (Figure 5.15), but this may create a dead zone down the middle of the tank where the hazardous emissions are drawn in both directions equally along the centre line of the tank and can escape into the workplace. If it is impossible to predict the direction in which contaminated air will move at any particular point, then there is a design problem.

5.5.2.5 Push–pull ventilation and other air-assisted hoods

Push–pull ventilation uses a combination of traditional hoods and jets of air to blow contaminants into the hood from a greater distance than could be managed by suction alone. This method was probably developed in the late nineteenth century and rediscovered in the 1960s. It can be very successful for large rectangular areas like tanks, where there is slot ventilation at one end and a jet of air is blown across the tank from the opposite side. It can also be used to avoid much of the swirling of air that occurs in front of a worker.

In Figure 5.16, push–pull ventilation is used to reduce a match dipper's exposure to white phosphorus. The worker is dipping matches in a basin of highly toxic white phosphorus and a curtain of air is moving over the surface of the basin towards the hood at the back of the workbench. It demonstrates the method of push–pull ventilation is the same as it was 100 years ago.

Flows of extracted air must significantly exceed those of supply air in push–pull systems, as the air jet entrains a large volume of air. The major drawback of such systems is that when the air jet is interrupted as objects are added to or removed from the tank, the objects deflect air and any contaminants into the workplace.

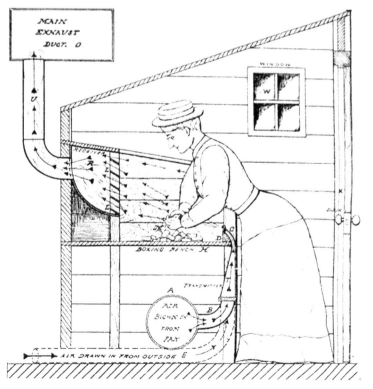

Figure 5.16 Push–pull ventilation in the manufacture of white phosphorus matches
Source: Great Britain Home Office Committee on Ventilation of Factories and Workshops (1903, p. 53).

Various attempts have been made to achieve directional airflow with hoods to extend their range. There has been some success with air assisted Aaberg-type hoods (Olander et al., 2001)

In Figure 5.17, the increase in capture range from a simple flanged circular hood (left) and Aaberg circular hood (right) is shown. The circumferential jet of air used by the Aaberg circular hood (right) entrains clean air that would otherwise be captured, leaving a tunnel in front of the hood to capture contaminated air from a greater distance. A linear version of the air jet can be used to shape the airflow into a slot hood.

Air-assisted hoods largely overcome the problem that arises with push–pull ventilation when objects are raised from dipping tanks, as there is no jet of air directed at the contaminated object. If the air jets designed to entrain clean air also entrain some contaminated air, however, they will spread this contaminated air about the workplace. Air-assisted hoods may also be unacceptably noisy.

5.5.2.6 Fume cupboards

Fume cupboards are a special type of enclosure often found in laboratories. Good designs have been available for over 50 years, but design flaws can still limit a fume cupboard's effectiveness. A respirator may reduce air contaminants by a factor of between 10 and several hundred, but even a poorly designed fume cupboard may reduce toxic exposure by a factor of 1000 or more. However, the slightest reverse airflow inside the fume cupboard will limit its effectiveness. The best designs can outperform poor designs by a factor of 10 to 100 times.

It is easy to track airflow into a fume cupboard with a smoke tube. Many technicians testing fume cupboards release smoke on the outside and watch it flow into the fume cupboard, and wrongly report that the fume cupboard works well (see section 5.6 for further detail). The question they need to answer is, 'How much smoke (representing contaminated air) released inside the fume cupboard escapes?'

Smoke released near the working surface will often move towards the front of the fume cupboard to take the place of air entrained in the flow near the front sill (see Figure 5.18 left) or just inside the sash when the incoming air cannot follow the contours of the sill,

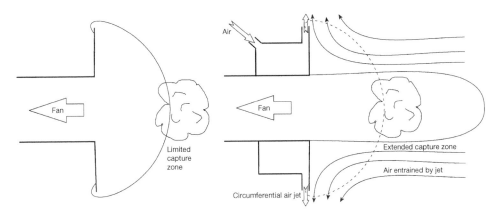

Figure 5.17 Aaberg-type hoods (right) can greatly extend the capture range of a simple flanged hood (left)

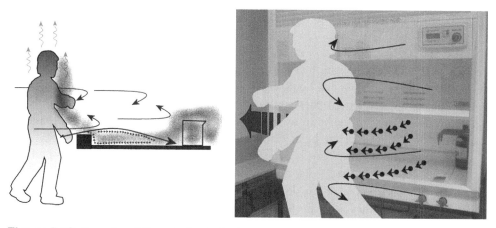

Figure 5.18 Complex airflows in front of a fume cupboard increase the exposure of the user

which is often raised. This smoke that has moved toward the sill (or the sash), like the smoke of a barbecue, becomes entrained in the air swirling in front of the worker. The problem is compounded by the air wake left by passers-by and the thermal plume of warm air generated by the fume cupboard operator (see Figure 5.18 right). This greatly lowers the degree of protection offered by the fume cupboard and allows greater toxic exposures to occur.

If the fume hood has a sash, then it must be at the operational height when the hood is in use. Many older fume cupboards currently in use in schools, industry and tertiary institutions do not have aerodynamic surfaces, and reverse airflows occur near their fronts. Working close to the sash to see inside greatly increases exposures to contaminated air.

Storage of chemicals in a cupboard connected by ventilation ducts to the fume cupboard system or underneath a fume cupboard is not allowed in AS/NZS 2243.8, so fume cupboards tend to be cluttered, reducing their effectiveness. The requirements for 10 m/s^{-1} airflow in the exhaust stack and individual stacks for each fume cupboard (to prevent fire or explosion) mean that there is a significant load on a laboratory air-conditioning system when ventilation in the fume cupboard is on but the cupboard is not in use. More efficient powered 'collecting ducts' for multiple fume cupboards are permitted in the UK but require interlocks with the collecting duct fan.

5.5.2.7   Hot processes

Designing ventilation for hot processes must take into account managing buoyant air from the source, but is complicated by:

- trying to minimise air extraction that will cool the process and result in increased energy to maintain the required temperatures for the hot process
- managing the air intake, which may take air-conditioned air that has to be replaced by additionally cooled or heated air. The sudden release of a lot of contaminated air overwhelms the collecting hood.

A canopy to collect warm, contaminated air from welding will tend to make the air move towards the worker's face. Replacement air then swirls past the worker, so contaminated air moves in exactly the wrong direction (as in Figure 5.18 left).

This additional heating or cooling of air-conditioned air or increased heating of the hot process will result in greater costs, hence the most efficient method of extracting contaminants for the breathing zone of workers is required. Some guidance for hot work is provided by the ACGIH® (2019). The key considerations are:

- ensuring contaminated air flows away from the worker's face
- minimising cross-drafts;
- whether the release is over a defined area
- whether the canopy is too high, and not capturing the buoyant air
- whether there is some sort of enclosure to aid with extraction
- the use of baffles for rectangular hoods.

The following can result in reduced capture of contaminated air.

- Nearby structural walls can cause the flow of contaminated air to become attached to the walls and not be extracted into the ventilation system. This is known as the Coanda effect.
- High-canopy hoods allow small cross-drafts to significantly affect the canopy extraction. They may also place a worker under the hood (see Figure 5.2 earlier in the chapter).
- A lack of side walls or enclosures will reduce extraction efficiency.
- Hood shapes may not match the source shape.

5.5.3   HAND TOOLS WITH LOCAL EXHAUST VENTILATION

Pneumatically powered and electrically powered hand tools are commonly designed with provision for dust extraction. This is particularly important in the construction industry, where concrete and manufactured stone benchtops produce toxic silica dust, and in the woodworking industry, where dusts from both hardwoods and softwoods have been associated with an increase in nasal cancers, occupational asthmas and dermatitis. Even with dust-extracting tools, a worker may still require properly selected and fitted respiratory protection, particularly with concrete and brick 'chasing' and other dusty grinding and cutting operations.

Often all that is required to reduce the dust is for an appropriate vacuum cleaner to be connected to the tool. This can reduce inhalable dust by up to 98 per cent, although with very fine respirable particles the method is less effective. Respirable dust is invisible to the naked eye and tends to deposit not in the nose or throat, but in the lungs. Some domestic vacuum cleaners may have high-efficiency particulate air (HEPA) filters, but they are not designed to meet industrial standards and have been found to pass up to 50 per cent of particles 0.35 µm in diameter (Trakumas et al., 2001). HEPA filters remove 99.97 per cent of particulates 0.3 µm or larger. Toxic dusts collected by a vacuum cleaner can also become a source of air contamination when they are re-suspended inside the cleaner, particularly with cyclonic and wet vacuum cleaners.

The guns of some MIG and TIG welding torches are equipped with LEV designed to capture fumes but not interfere with the shielding gas.

5.5.4   FANS

Poor LEV design may require more powerful fans, which cost more to buy and run and may do little to protect the worker. Fan design and selection is a specialised area.

The relationships between the flow rate through a fan (Q), the pressure drop (p) it can produce, the fan speed (n) and the size (d) are given in terms of fan laws in Table 5.1 (Osborne, 1977). This table also shows how small increases in flow (Q) to compensate for poor hood design can result in large and costly increases in fan motor size—and power consumption. A design that requires a doubling of airflow means an eightfold increase in fan wattage, larger ducts and higher running costs, as well as more noise.

Table 5.1 Fan laws

Fan law	Formula	Comment
Volume flow (Q)	$Q \propto d^3 n$	The flow varies with the fan speed (n), but increases with the cube of the fan diameter (d).
Fan pressure (p)	$p \propto d^2 n^2 \rho$	The pressure drop the fan can produce varies with the square of both the fan diameter and speed and linearly with the air density (ρ).
Fan power $P = p \times q$	$P = pQ \propto d^5 n^3 \rho$	Derived from the first two laws. The fan motor size and power bill will vary enormously with the fan size and fan speed.

Source: Osborne (1977).

5.5.4.1   Types

Many types of fans are used for industrial ventilation, but three types dominate in small industry: axial fans, forward-bladed centrifugal fans and radial-bladed centrifugal fans. Large axial fans are often used to move large volumes of air against a small resistance. Some axial fans are also used in LEV systems, being hidden inside a duct. In many systems, good flows can be obtained for long runs of ducts and air-cleaning devices with a forward-bladed centrifugal fan. If the air is contaminated, then a less efficient but more robust radial-bladed centrifugal fan may be required.

5.5.4.2   Which fan to choose

Once the airflow (Q in $m^3 s^{-1}$) needed to make a LEV system work is estimated, and the total pressure drop (p in Pa) in the system is calculated (these calculations are beyond the scope of this book). To know which fan to choose, the total pressure drop is required to calculate the wattage of the fan motor required, which is simply p × Q. Allowance should also be made for the actual overall efficiency of the fan and the fan motor, about 60 per cent.

For example, to move 10 $m^3 s^{-1}$ of air with a total pressure drop of 250 Pa (about 25 millimetres of water) for the whole system with a 60 per cent efficiency, the wattage of the fan motor would have to be:

$$\text{Power of fan motor} = \frac{10(\text{m}^3 \text{ s}^{-1}) \times 250 \text{ (Pa)}}{0.6 \text{ (overall efficiency)}} = 4 \text{ kW}$$

In practice, 4.5–5 kW would be chosen to allow for declines in efficiency over time.

The fan laws are used by designers and show how the flow, pressure and power usage are influenced by factors like fan speed (n) and fan diameter (d). A hood will need a design flow to make it work properly and an inefficient hood makes a huge difference to the size (initial cost and running cost) of the fan and the size of the ducting.

A ventilation system can be represented by a curve (Figure 5.19) relating the pressure drop (p) and the airflow (Q). This curve is parabolic, with the pressure drop increasing with the square of the flow. The fan's performance can be represented by a similar curve, which plots known airflow at a given speed against the resistance of the air-cleaning devices, ducts and hoods on the suction side and the ducts and stack on the exhaust pressure side. Each type of fan will have a different set of fan curves for different fan speeds; these fan curves are published by fan manufacturers. It is usual to select a duty point on the fan curve where the mechanical efficiency is high.

5.5.4.3 Common problems

After maintenance, it is not uncommon for three-phase fan motors to be reconnected backwards. With centrifugal fans, this leaves power consumption much the same but reduces efficiency by 90 per cent. When the wiring is corrected, it is not uncommon for a large amount of material that has settled in horizontal ducts to noisily hit the fan, so internal inspection and cleaning of ducts may be warranted before turning on. With radial fans, the airflow is reversed and miswiring is obvious.

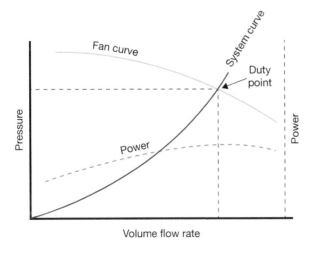

Figure 5.19 Fan and system curves
Source: Health and Safety Executive (2017, p. 57).

5.5.5   DUCTING

5.5.5.1   Key requirements

Ducts in LEV systems may be likened to highways, with merging lanes and sweeping bends. If there are unnecessary or sharp bends, greater pressure drops occur and less suction is available to power the hoods, meaning that energy is wasted and hoods under-perform in capturing contaminated air.

If sections of duct bend or join at a 90 degree angle, there will be increased turbulence where the airflows intersect. It is common for an otherwise well-designed system to be impaired when the installers introduce additional sharp bends and tortuous paths around beams that encourage the deposition of particulates and greatly increase pressure drops. Airflow is preferably smooth, with gradual corners and no abrupt changes in direction and speed. No great engineering skills are needed to recognise most badly installed ducting.

Ducts should be provided with 'cleanout ports' at strategic intervals to enable the removal of deposited particles and condensed vapours, and to permit inspection of the duct for clogging and corrosion.

Many materials may be used to form ducts, but steel sheeting and PVC pipe predominate. Abrasive or corrosive agents and clean operations may require special ducts, particularly on bends. Sometimes flexible duct supported by light spiral wire is used (and sometimes left coiled in roof spaces), but it is usually inappropriate and poorly installed. Snorkel hoods in welding shops have tough, flexible ducts.

5.5.5.2   Duct transport velocity

If the air velocity in a duct is too low, then particulates will settle in straight sections of the duct and clog them. If the air velocity is too high, large pressure drops may develop, requiring a larger fan. High air velocities lead to excessive noise, and ventilation systems will sometimes be turned off to avoid the noise. The minimum design velocities set out in Table 5.2 are based on the ACGIH® (2019) recommendations in *Industrial Ventilation: A Manual of Recommended Practice for Design*. This publication is updated regularly, and is the 'bible' of industrial ventilation.

5.5.5.3   Duct resistance

Calculations of pressure drops in ducts are beyond the scope of this book. In the United Kingdom, the velocity pressure (Vp) method is used, while in the United States the equivalent foot method is often used, where bends, junctions and cross section changes are represented by an equivalent length of ducting. Some hybrid approaches are also used.

5.5.6   AIR TREATMENT

The treatment of exhaust air to remove contaminants is becoming more important. There are many treatment technologies available; for some industrial plants, like power stations, the installations can be huge. In principle, exhaust contaminants can be considered as either particulates (solid or liquid) or gases.

Most texts on industrial ventilation will provide a good introduction to this topic, the detailed treatment of which is beyond the scope of this book. Suppliers are good sources

Table 5.2 Duct transport velocities needed to transport various contaminants through ducts

Type of contaminant	Examples	Design velocity (m/s^{-1})
Vapours, gases, smoke	All vapours, gases and smoke	Usually 5–10
Fumes	Welding fumes	10–12
Very fine light dust	Cotton lint, wood flour, expanded styrene beads, glass micro-balloons	12–15
Dry dust, powders	Fine rubber dust, wood shavings	15–20
Industrial dust	Grinding, buffing, masonry, quarry dusts	17–20
Heavy dust	Wet sawdust, heavy metallic dusts	20–22
Heavy or moist dust	Sticky dusts, heavy dusts with chips	>22

Source: Adapted from ACGIH® (2019, Chapter 5).

of information on selection of the appropriate technology, but their advice can be affected by the profit motive.

5.5.6.1 Types

To remove toxic gases and vapours, molecular processes have to be used—absorption, adsorption, condensation and incineration.

To remove particulates, a number of filtration and inertial separation technologies are used, including cyclones and inertial separators, electrostatic precipitators and fabric filters (sometimes in large 'bag houses').

5.5.6.2 Limitations

Organic vapour scrubbers are available, but limitations on their lifespan should be considered, given that the active sites can be used by water vapour as well as any other organic compounds extraneous to the contaminant. For example, they can become 'full' and need replacing before the expected volume of contaminant has been absorbed because of high humidity levels.

5.5.7 STACKS

5.5.7.1 Key requirements

A stack serves to disperse the air emitted from a ventilation system outside a workplace. Good design limits the amount of air that re-enters buildings. A stack should not incorporate structures that cap to reduce ingress of rain, as that will limit dispersion or even direct foul air back into a building, as shown in Figure 5.20. Good alternatives are shown in Figure 5.21.

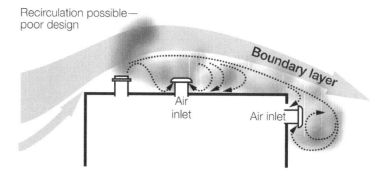

Low discharge stack relative to building height
Air inlets on roof and wall

Figure 5.20 Stack height and location are important, position here shows bad design
Source: Health and Safety Executive (2017, p. 62).

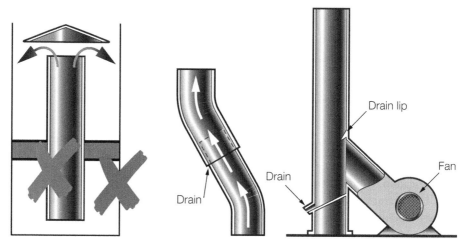

Figure 5.21 Good stack design alternatives
Source: Health and Safety Executive (2017, p. 64).

A rule of thumb is that the stack should be 2.5 times higher than the tallest building (Hughes, 1989). An alternative rule is that the stack height (H_s) should be

$$H_s = H + 1.5D$$

where:

H = the height of the building, and
D = the lesser of the building height and the maximum length of the building across the prevailing wind.

Sometimes scale wind-tunnel studies are needed, particularly if the building has a courtyard. Releasing smoke into the system from a cheap theatrical smoke generator will often reveal inadequacies in stack height and location.

Additionally, the stacks need to:

- be open at the exhaust point to use the inherent air velocity to continue taking the exhaust up above the building for local removal
- have an annular opening around the base, which allows the upper section of the stack to sit outside of the inner section so that rain will strike the outer sleeve and run off to the roof.

5.5.8 WIND ROSES

A wind rose is a diagram showing the relative frequency of wind directions at a place in a graphical form: it can show the strength, direction and frequency in a single diagram. They are used to look at potential dispersal patterns to identify air inlets that may be of concern for the stack placement, although airflows around buildings can produce reverse airflow (see Figure 5.20 above). Local airflows around buildings can negate the effect of a prevailing wind.

5.5.8.1 Common problems
Common issues include the stack exhaust being re-entrained into the building or other buildings. This is often a result of:

- the stack being too low
- incorrect placement of the stack
- a 'Chinese hat' being used on top of the stack to stop rain entering the ducting.

5.6 MEASUREMENTS

The flow at the centre of the face of a fume cupboard is often measured, but this does not reveal the limitations in containing contaminated air. Investigating the flow *into* the fume cupboard answers the wrong question, as it is the minute flows *towards the front* of the cupboard that determine its overall performance. These forward flows place contaminated air at a point where it may be entrained in the swirling air in front of a person using the fume cupboard. The front of the fume cupboard is also more exposed to room draughts, and people moving past it drag contaminated air from the fume cupboard in their wake.

The question that needs to be asked is whether contaminated air *inside* the fume cupboard is contained. Small reverse airflows near the face of the fume cupboard (at 'X' in Figure 5.24—see below) lead to significant loss of containment in almost all fume cupboards, a fact easily demonstrated with a smoke tube. These fugitive emissions are important in the case of highly toxic chemicals.

5.6.1 FACE VELOCITY

Quantitative measurement of airflows can be performed with either vane or hot-wire anemometers. Vane anemometers are mechanical and have a start-up speed around 0.2 m/s^{-1}; they are often used for measuring the face velocity of fume cupboards and for

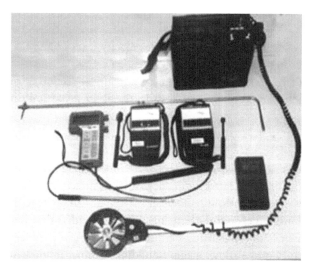

Figure 5.22 Pitot tube, hot-wire and vane anemometers

room outlets in air-conditioning systems. They can give an air speed averaged over a few seconds and over the area of the vane.

Thermo-anemometers, or hot-wire anemometers, are generally considered unsafe in explosive atmospheres, unless certified otherwise, but are useful for general industrial ventilation work and can measure lower air speeds than a vane anemometer, as shown in Figure 5.22. Generally, new hot-wire anemometers do not actually use hot wires for air flow measurements and present a low ignition hazard.

Pitot tubes give absolute measurements of air velocity by measuring 'velocity pressure' and are usually used to estimate the airflow in ducts, but they are cumbersome.

5.6.2   SMOKE TUBES

The cheapest and most useful tool for investigating industrial ventilation systems is the smoke tube. A chemical in the tube reacts with moisture in the air to form a white smoke that enables the visualisation of airflows and the estimation of both direction and magnitude of even the smallest air movements (Figure 5.23). You may not be able to take smoke tubes on an aircraft and may need to have a packet of smoke tubes sent to the destination by the supplier.

Smoke tubes are particularly useful for demonstrating the limitations of a hood in capturing air contaminants and for showing airflows from behind a hood. They can also be used to visualise airflows through windows and doors: sometimes the flow can be in one direction at the top of a door and the opposite direction at the bottom. It is also possible to make a good estimate of air speeds inside a room by estimating the time a puff of smoke takes to travel 1 metre. This makes smoke tubes particularly useful in indoor air-quality investigations.

In a fume cupboard, it is usually possible to demonstrate reverse airflows by puffing smoke just inside the sash and on the working surface near the front of the fume cupboard, as shown in Figure 5.24.

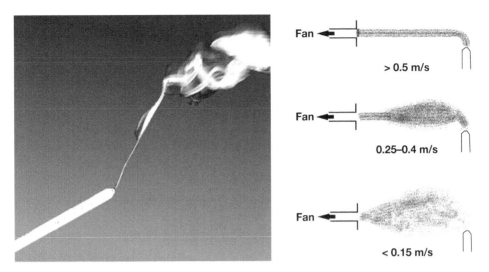

Figure 5.23 Smoke tube (left) and its use in estimating air velocity (right)

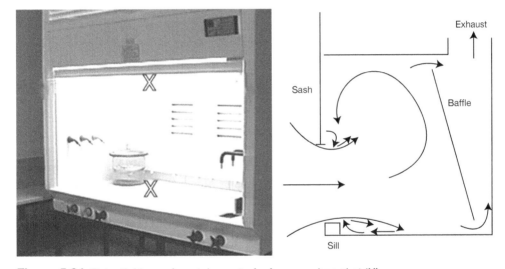

Figure 5.24 Potential loss of containment of a fume cupboard at 'X'

5.6.3 HOT PROCESSES

When measuring the flow for hot processes, that may have temperatures in excess of 1000°C, the anemometer needs to be high-temperature specific. Standard anemometers may not give the correct readings, so an anemometer capable of air velocity measurements in the region of at least 500°C must be used.

5.7 NOISE CONTROL IN VENTILATION SYSTEMS

A common reason for ventilation systems failing to provide the expected protection is that they are too noisy and are therefore switched off. Often the fan is the source of the noise, which travels along the ducting like sound through a large trumpet. It is possible to predict the levels of noise generated by industrial ventilation systems, but the calculations are complex, as the degree of turbulence, stiffness, reflections and openings all affect the amount of noise.

5.7.1 COMMON PROBLEMS

- The predominant noise problem may be the 'passing frequency' of the fan blades, which is the noise made as a result of the number of revolutions a fan makes per second multiplied by the number of blades in the fan.
- If a fan is isolated from a duct using a flexible join or some other sort of suspension that is too floppy, it will vibrate and increase the noise level.
- A fan placed too close to a bend, or wrongly sized for the job, can easily produce 10 decibels of additional noise.
- Low-frequency noise generated by air-handling systems is usually the hardest to attenuate.
- In a straight duct, the noise level varies with at least the fifth power of the air velocity (Sharland, 1972), so that for a given airflow, very small changes in duct diameter can have a large effect on noise levels.
- Small increases in duct roughness can also add 5 decibels or more to the noise.

5.7.2 FIXES

- Passing frequency noise can be greatly reduced by mechanically isolating the fan from the duct using a flexible join located at least one duct diameter away from the fan.
- If ducting is suspended from a roof or other large surface, where vibrations in the duct make the whole roof vibrate like a giant loudspeaker, a duct suspension system can reduce this 'flanking' and can have a marked effect on the noise levels in the workplace.
- Variable speed fans can reduce the sound level as the fan speed decreases.
- The use of smooth ducting and transitions can reduce sound levels.
- Reducing the number and sharpness of bends.
- Absorptive or dissipative silencers that are lined with sound-absorbing material.
- Regular maintenance of the system to ensure the system is working optimally.

5.8 REFERENCES

Alden, J.L. and Kane, J.M. 1981, *Design of Industrial Ventilation Systems*, Industrial Press, New York.

American Conference of Governmental Industrial Hygienists (ACGIH®) 2019, *Industrial Ventilation: A Manual of Recommended Practice for Design*, 30th ed., ACGIH, Cincinnati, OH.

Feigley, C.E., Bennett, J.S., Lee, E. and Khan, J. 2002, 'Improving the use of mixing factors for dilution ventilation design', *Applied Occupational and Environmental Hygiene*, vol. 17, no. 5, pp. 333–44.

Friberg, B., Friberg, S., Burman, L.G., Lundholm, R. and Ostensson, R. 1996, 'Inefficiency of upward displacement operating theatre ventilation', *Journal of Hospital Infection*, vol. 33, no. 4, pp. 263–72.

Great Britain Home Office Committee on Ventilation of Factories and Workshops 1903, *Second Report of the Departmental Committee Appointed to Inquire into the Ventilation of Factories and Workshops*, His Majesty's Stationery Office, London.

Health and Safety Executive (HSE) 2017, *Controlling Airborne Contaminants at Work: A Guide to Local Exhaust Ventilation (LEV)*, 3rd ed., HSE, London, <www.hse.gov.uk/pubns/books/hsg258.htm>.

Hughes, D. 1989, *Discharging to Atmosphere from Laboratory-scale Processes*, H&H Scientific Consultants, Leeds.

Olander, L. et al. 2001, 'Local ventilation', in H.D. Goodfellow and E. Tähti (eds), *Industrial Ventilation Design Guidebook*, Academic Press, San Diego, CA.

Osborne, W.C. 1977, *Fans*, Pergamon Press, Oxford.

Sharland, I. 1972, *Woods Practical Guide to Noise Control*, Woods Acoustics, Colchester.

Skistad, H. 1994, *Displacement Ventilation*, Research Studies Press, Taunton.

Standards Australia 2014, Safety in Laboratories: Fume Cupboards, AS/NZS 2243.8, SAI Global, Sydney.

Trakumas, S., Willeke, K., Reponen, T., Grinshpun, S.A. and Friedman, W. 2001, 'Comparison of filter bag, cyclonic, and wet dust collection methods in vacuum cleaners', *AIHA Journal*, vol. 62, no. 5, pp. 573–83.

6. Personal protective equipment

Garry Gately and Terry Gorman

6.1 INTRODUCTION

The hierarchy of hazard controls was discussed in Chapter 4. At the bottom of the hierarchy is personal protective equipment (PPE). As a control strategy, PPE represents the last resort; beyond it is the unprotected worker, who will inevitably face exposure to a toxic agent or hazard if the PPE is not correctly selected, maintained and used. Despite occupying the lowest place in the control hierarchy, PPE is still widely used and accepted as a backup and as a supplement for other controls. There will also be situations where higher-level controls cannot be used and PPE is the only practicable solution.

This chapter discusses general safety apparel, gloves and respiratory protective equipment. Refer to Chapter 12 for a detailed discussion of hearing protection.

6.2 OVER-PROTECTION

It must be borne in mind that excessive use of PPE can sometimes interfere with the worker's ability to accurately perceive the work environment and may therefore compromise personal safety; consequently, PPE should always be used wisely. For example, thick or double gloves can reduce dexterity. In some instances, workers may complain of being treated like 'Christmas trees', with pieces of PPE hung on them like ornaments, as in Figure 6.1. Considerable care should be taken when choosing basic PPE to ensure the correct level of protection. Over-protection is likely to result in the PPE not being used at all, leaving the worker totally unprotected.

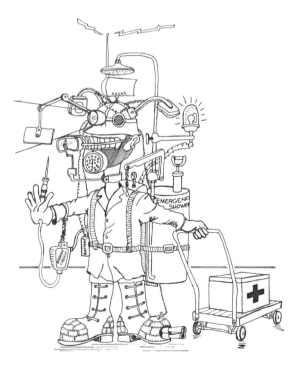

Figure 6.1 An example of over-use of PPE

6.3   PPE SELECTION STRATEGY

A thorough understanding of the hazard is essential when deciding on the appropriate PPE. The H&S practitioner needs to understand both the physical and chemical properties of the agent(s), the relevant routes of exposure and the circumstances of exposure, the what, when, where and why. The physical, chemical and relevant toxicological properties normally are gleaned from the Safety Data Sheet (SDS), but it is imperative that the H&S practitioner observes the practices and behaviours of the workplace in question before selecting the PPE. For example, the worker shown in Figure 6.2 is handling heated methylene diphenyl diisocyanate (MDI), a powerful skin and respiratory sensitiser. In this case, there is no protection from vapours and the single glove of inappropriate material provides no resistance to the chemical. Bare skin is also exposed.

Figure 6.2 Inappropriate PPE for handling heated methylene diphenyl diisocyanate (MDI)

6.4   SAFETY EQUIPMENT

Basic safety equipment is a fundamental means of protecting the body from physical hazards in the workplace. It includes hard hats, safety boots, overalls, safety spectacles or goggles, hearing protection and gloves. Each protective device is designed to isolate the worker from the hazard. It does not remove the hazard from the workplace, but reduces the risk of injury if the worker comes into contact with the hazard. For the most part, immediate protection of the eyes, hands, head and feet is relatively uncomplicated. The hazards are usually obvious and programs to use personal protective equipment can be implemented successfully with appropriate attention and resources. Measures of this basic type will not be described extensively in this text.

6.5 GLOVES

Most people use their hands all the time at work, so hand injuries are very common. They may take the form of physical trauma or chemically induced damage, as shown in Figure 6.3.

Gloves are widely used in industry to protect hands against many different hazards. The selection of the correct glove type is neither straightforward nor simple, particularly when considering chemical protection. For example, gloves that protect against methylated spirit solvents may not protect against turpentine. Knowledge of the chemical resistance and permeability of different glove materials is required before making a choice. Glove manufacturers and others have published data on the resistance of glove materials to permeation by the common solvents used in industry; however, permeation resistance data on a given material from one manufacturer may not hold true for the same type of glove material from another manufacturer. There are international standards, such as EN 374, Protective Gloves Against Chemicals and Micro-organisms (CEN, 2004), and the American Society for Testing and Materials (ASTM)'s Method F739–12E1, Standard Test Method for Permeation of Liquids and Gases Through Protective Clothing Materials Under Conditions of Continuous Contact (ASTM International, 2012), which describe test methods and terminology associated with chemical-resistant gloves. There are

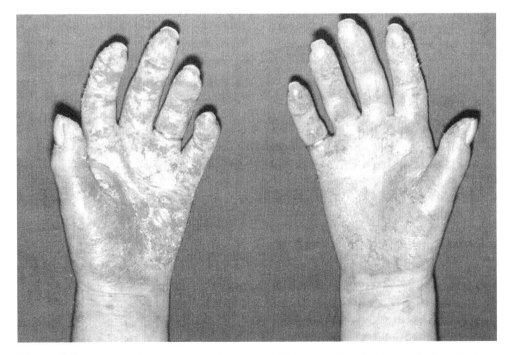

Figure 6.3 An example of severe contact dermatitis caused by direct chemical contact (e.g. MDI)

also independent online resources, such as ProtecPo (Institut National de Recherche et de Sécurité and Institut de recherche Robert-Sauvé en santé et en sécurité du travail, 2019), that the H&S practitioner can use to select the category of polymer most likely to provide protection against individual chemicals or classes of chemical. Given the rate of research in this field, it is recommended that the most current information be consulted.

Other aspects to consider when choosing gloves include the dexterity required of the worker, thermal resistance, abrasion and puncture resistance. The favoured leather 'rigger's' glove offers virtually no resistance to chemical hazards and can in fact act as a reservoir, allowing the chemical to accumulate, then slowly penetrate the glove to reach the user's skin.

The reuse and laundering of gloves can be fraught with problems, so any decision to do either should be approached with caution. The following are some potential risks:

- Gloves may contain small holes that are not detected.
- Contaminants can be translocated to the inside of the glove during washing.
- The washing process can physically damage the glove.
- The washing process may not remove chemicals that have started to migrate through the glove material.

A successful skin protection program includes the correct choice of gloves as well as training in their correct use, in the limitations to their protection and in their correct removal without causing contamination of the skin.

Some skin lotions and hand cleaners contain petroleum products that can cause swelling and degradation of the glove material. Specialist advice should be sought from the glove supplier if there are any concerns about this (Figure 6.3 above). Users of the gloves need to be alert and if any deterioration is noticed the gloves should be removed and their suitability reassessed.

Most users need to be shown how to remove gloves correctly without contaminating the unprotected skin. Loose-fitting gloves can be shaken off quite easily. Tight-fitting gloves are removed by grasping the cuff of one glove and pulling it down and inside out. While the partly removed glove is still on the hand, the second glove is pulled down and inside out, again without any contaminated glove surfaces touching the bare skin. The gloves are finally removed with one partly inside the other. It is not recommended that gloves of this type be reused because of the risk of introducing contamination onto their internal surfaces. If contamination does occur, the enclosed nature of the glove material may increase the rate of absorption of the very agent the gloves were intended to protect against.

6.6   CHEMICAL PROTECTIVE CLOTHING

All clothing—even cotton overalls—provides a degree of protection, but fabrics can act as a reservoir of contaminants, leading to continued contact with the embedded chemicals long after the wearer has left the workplace. Some form of chemical protective clothing (CPC) is required wherever hazardous chemicals are used that can be absorbed through the skin. Contaminated clothing can also act as a mechanism and transport the contaminant away from the workplace in harmful quantities. For example, there are many documented examples where toxic agents like lead, asbestos or mercury have been taken home with toxic consequences to family members.

Again, a full understanding of the what, when, where and why is required to ensure the correct protection is achieved by the CPC. Examples of types of CPC materials and their uses and limits include:

- plastic-coated woven polyolefin overalls with hood—hazardous dusts and low toxicity liquid chemical splashes
- impervious plastic pants and coat set—moderate-hazard materials, simple splash protection
- impervious plastic fully enclosed suits—high-hazard materials, such as chemical warfare agents, emergency response, spills clean-up, radiological hazards.

The selection of the correct chemically resistant protective material is only part of protecting the skin against toxic hazards. Additional measures that need to be taken include:

- sealing of gaps between boot or gloves and the suit material
- protecting the skin where respiratory protection is integrated into the suit construction
- sealing zippers or other garment closures
- using a 'dresser' or assistant, given the potential difficulty and complexity of donning and achieving the seals
- providing effective decontamination facilities so the exterior of the suit can be cleaned before it is removed
- recognising and managing heat stress during use (see Chapter 14)
- communication between staff wearing fully enclosed suits
- difficulties in manoeuvrability or limits to the range of vision when wearing enclosed suits
- operating effectively within the time constraints of the suit and/or breathing equipment.

Any situation where CPC is needed requires a detailed assessment and consideration of all the above measures. For an excellent guide to the selection and management of CPCs, refer to the Occupational Safety and Health Administration's (n.d.) *Technical Manual* (Section VIII, Chapter 1).

6.7 RESPIRATORY PROTECTIVE EQUIPMENT

Of the three principal routes of exposure, inhalation is considered the most significant (everyone has to breathe, and the absorption of airborne agents via the lungs can be very efficient). As a consequence, the provision of clean air to breathe is paramount. There are many situations in which the use of respiratory protective equipment (RPE) is an established method of protection, such as where:

- other control methods are too costly or impracticable (e.g. electrical power may not be available, or ventilation controls cannot be arranged around a large open formwork metal structure)
- the tasks may be carried out at a range of locations (e.g. a pesticide applicator providing termite treatments to buildings)

- the tasks may involve only short-term exposures (e.g. one job taking two hours a month)
- exposure may be physiologically inconsequential, not requiring elaborate controls (e.g. nuisance dust exposure)
- the agents used may have poor, or no, warning properties and over-exposure can readily occur
- RPE is required for emergency procedures, including emergency escapes (e.g. fire-fighting, escape from chemical leaks or spills)
- oxygen-deficient atmospheres exist (e.g. confined-space entries)
- RPE is necessary to supplement other control procedures (e.g. air-supplied RPE is still required for some underground tunnelling or abrasive blasting even when water suppression is used).

RPE should not be the first choice, nor should it be used merely to quell workers' concerns or complaints of over-exposure to contaminants. Before any RPE is issued, a full investigation of the task in question should be carried out to ensure that RPE is an appropriate control measure. See Chapter 4 for a full discussion on the application of the hierarchy of controls (HoC).

RPE offers relatively simple, low-cost protection from hazards that arise in many workplaces. However, it must be stressed that effective use of RPE programs is often difficult (and sometimes impossible) to achieve. Any H&S practitioner embarking upon an RPE program needs to be aware that this protective strategy requires 100 per cent commitment from both management and the worker. If there is any lack of enforcement on the part of management or lack of compliance on the part of the worker, then the protective measure will fail. RPE use requires much greater organisational effort than other hazard-control measures. If RPE is adopted, it should only be after all other options have been examined.

6.7.1   INTERNATIONAL STANDARDS

Across the globe, there are a number of national respiratory protection standards created for use in various jurisdictions—for example, CEN standards in Europe, CSA Group Standards in Canada, JIS in Japan and NIOSH standards that apply in the United States and are also used by some other countries. The equipment selected for use will depend on the relevant standards applicable in the jurisdiction.

In Australia and New Zealand, there is a pair of standards relating to respiratory protection: AS/NZS 1715 and 1716. AS/NZS1716 is the standard that lists the performance requirements and associated testing protocols for the various types of respirators used in ANZ workplaces, while AS/NZS1715 provides guidance on the selection, use and maintenance of respirators.

AS/NZS1716 recognises three levels of particle filter performance:

- P1 for mechanically generated particulates—that is, from sanding, crushing, sawing, drilling and so on
- P2 for mechanically and thermally generated particulate—as for P1, plus those particulates produced from thermal effects such as welding fume, molten metal fume, laser cutting
- P3 for all particulates, especially where high protection levels are required.

The EN respiratory protection standards use the same type of classes for particle filters.

The US NIOSH standard classifies particle filters into three types that indicate the percentage efficiency of the filter against the required test challenge aerosol. These are 95, 99 and 100.

Another filter performance characteristic tested by NIOSH is efficiency against an oil-based challenge. There are three classes here too:

- N—not resistant to oil
- R—somewhat resistant to oil
- P—strongly resistant to oil.

From the testing for these two characteristics, a particle filter is rated—for example, N95, P95, R100 or P100, etc.

For gas/vapour filters on tight-fitting masks, AS/NZS1716 uses the same class identifiers as Europe—A1, B2, E1, K1, Hg and so on, indicating the relevant chemical target groups (the letter) and capacity (the number). NIOSH has abbreviations for its specified classes—for example, OV (organic vapours), Cl (chlorine), AG (acid gas), Amm (Ammonia) and so on.

All of these standards promote the use of a comprehensive respiratory protection program to deliver the required protection to the workers.

6.7.2   AUSTRALIAN REGULATORY REQUIREMENTS

Any RPE program should comply with:

- any applicable advisory standard or code of practice on personal protective equipment, including AS/NZS 1715:2009, Selection, Use and Maintenance of Respiratory Protective Devices (Standards Australia, 2009)
- the performance requirements of AS/NZS 1716:2012, Respiratory Protective Devices (Standards Australia, 2012).

The hazardous chemical regulations mention the use of PPE, including RPE, but only as a last resort in the control of hazardous chemicals.

6.7.3   RESPIRATORY PROTECTION PROGRAMS

The implementation of a respiratory protection program needs to be based on a detailed assessment of each workplace hazard and the extent of worker exposures. For example, if the workplace uses lead, and other controls are not possible, it will be necessary to determine which type of RPE is appropriate for lead dust. The wrong choice of RPE could expose workers to excessive lead levels, resulting in possible long-term illness, as well as regulatory actions or fines, civil action by employees and other unnecessary economic costs to the business. The procedures for selecting and implementing RPE therefore require considerable attention. Where doubt exists, specific needs should be discussed with representatives from the RPE supply company.

There are some excellent respiratory protection programs operating throughout Australia; however, many purchasers of respiratory protection fail to get satisfactory performance from their RPE. Some of the possible reasons include:

- a lack of knowledge about the nature of the airborne contaminant and its concentration levels
- incorrect selection of air-purifying devices including efficiency and protection factors
- poor fitting of respirators
- a lack of scheduled maintenance
- ill-treatment and contamination of RPE in the workplace
- a lack of knowledge of filter life
- workers failing to use the RPE provided at the appropriate time.

There are many different types of respirator, and it is not the intention of this text to provide a detailed evaluation of each type. As a guide, the main types of respirator are summarised in Figure 6.4 and described in Table 6.1. For more detail, consult AS/NZS 1715:2012 (Standards Australia, 2012).

It is necessary to pay great attention to the technical detail and administration of RPE programs. There are many anecdotal examples that illustrate the incorrect use of respirators, such as the dusty respirator hanging on a nail in a workshop, fitted with a dust filter and used during spray painting with a paint containing solvents and toxic pigments by a bearded operator who could not recall when the filter was last changed but says the man who sold it to him assured him it was the correct one for spray painting.

The key requirements of an RPE program include:

- managerial capacity to administer an RPE program (it is part of the H&S practitioner's task to guide the managers)
- knowledge of respiratory hazards
- selection and purchase of the appropriate type of RPE, including the protection factor
- acceptance and appreciation of RPE by workers so it will be worn when required
- medical assessment of respirator use for some RPE users where necessary
- training in use, including correct fitting of respirator
- fit testing of close-fitting respirators
- inspection, maintenance and repair of RPE
- audit and review.

The person selecting and supervising the RPE program may also require training. Written guidelines should be prepared and followed strictly. Use of RPE must be monitored carefully to ensure the equipment is not used for purposes beyond those designated. Workers must receive adequate instruction, training and supervision to ensure they do not inadvertently use the RPE in life-threatening situations.

6.7.4   KNOWLEDGE OF RESPIRATORY HAZARDS

It is not possible to proceed with an RPE program unless there is a clear understanding of why RPE is being used. Many factors need to be considered. For instance, if choosing LEV as a preferred control, it may not be necessary to know a lot about the nature of the hazard, its concentration in the workplace, which of the workers are at risk or how far the contaminant spreads. LEV will simply exhaust the contaminant away from the work zone and remove the problem.

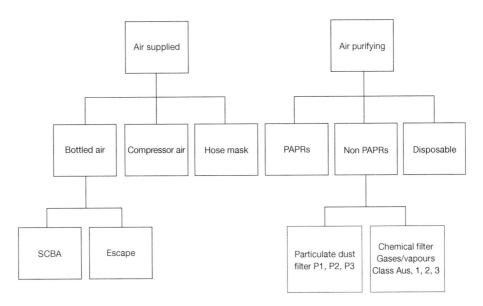

Figure 6.4 Classification of respiratory protective equipment

Table 6.1 Main types of respirators and filters

Filter types

Particulate filter types/classes	Function or purpose
P1	Protection against mechanically generated particulates
P2	Protection against mechanically and thermally generated particulates
P3	Protection against all particulates including highly toxic materials

Note: 'handyman' dust masks that do not carry the Australian Standards claim or mark do not provide protection against toxic or hazardous dusts. Extreme caution should be exercised when selecting this type of respiratory protection.

Gas/vapour filter classes
The larger the class number, the larger the capacity of the gas filter—that is, it will last longer for a given gas or vapour concentration.

Class AUS	Low-absorption capacity filters
Class 1	Low- to medium-absorption capacity filters
Class 2	Medium-absorption capacity filters
Class 3	High-absorption capacity filters

Gas/vapour filter types

For specific chemicals, confirm suitability of filter types with the manufacturer.

Type A	Organic gases and vapours
Type B	Acid gases and vapours as specified, excluding carbon monoxide
Type E	Sulphur dioxide and other inorganic gases and vapours as specified
Type G	Low vapour pressure organic compounds for vapour pressures less than 1.3 Pa (0.01 mm Hg) at 25°C as specified. These filters incorporate a particulate filter at least equivalent to P1.
Type K	Ammonia and organic ammonia derivatives
Type AX	Low-boiling-point organic compounds (less than 65°C), e.g. methanol
Type NO	Oxides of nitrogen
Type Hg	Metallic mercury
Type MB	Methyl bromide

Types of respirator

Air purifying, powered type	Powered air purifying respirator (PAPR)—uses a battery-driven fan to force air through a filter assembly and deliver cleaned air to a helmet, hood or face mask. Positive pressure.
Air purifying, replaceable filter type	A facepiece (full face or half face) to which a replaceable filter assembly is connected. The user's lung power is used to draw the air through the filter. Negative pressure.
Air purifying, disposable	A respirator with the filter as an integral part of the facepiece and the filter is not replaceable. When exhausted, the whole assembly is discarded. Negative pressure.
Air hose— natural breathing	A wide diameter hose located outside the contaminated zone connects to the respirator facepiece. The user's lung power draws air to the facepiece (can be low pressure fan assisted). Air hose systems supply air at or near atmospheric pressure.
Supplied air, compressed air line	A small-bore line connected to a compressed air source supplies clean air for respiration. This may use a compressor, or compressed air cylinders located a distance from the work location. Positive pressure.
Supplied air, self-contained	The respirator facepiece is connected by a breathing tube to a cylinder of breathable gas that is carried by the wearer. Often referred to as SCBA, or self-contained breathing apparatus. Positive pressure.

Note: AS/NZS 1715 specifies air purity for air supplied systems including SCBA systems.

Source: AS/NZS 1716:2012 (Standards Australia, 2012).

In contrast, to provide proper respiratory protection, the following steps must be taken:

1   Determine the identity of the hazardous contaminants to be controlled (i.e. dust, mist, fume, gas, vapour, asphyxiant, or any mixture of these) and the expected range of concentrations in various work situations/areas.

2   Determine if the workplace contaminant(s) concentration exceeds the published IDLH (Immediately Dangerous to Life or Health) level and the potential consequences of RPE failure.

3   Determine which workers require RPE. Not all workers in a particular process may be at risk.

4   Distinguish whether RPE is the major control method, or a backup or secondary control.

5   Determine whether there are any additional routes of entry other than inhalation, or other effects. Skin absorption and effects on the eyes are the most common considerations.

6.7.5   PROTECTION FACTORS

Selection of correct RPE should not be attempted by the inexperienced or untrained worker. Some regulations or codes of practice—for example, Safe Work Australia's Model Code of Practice: How to Safely Remove Asbestos (Safe Work Australia, 2018) spell out

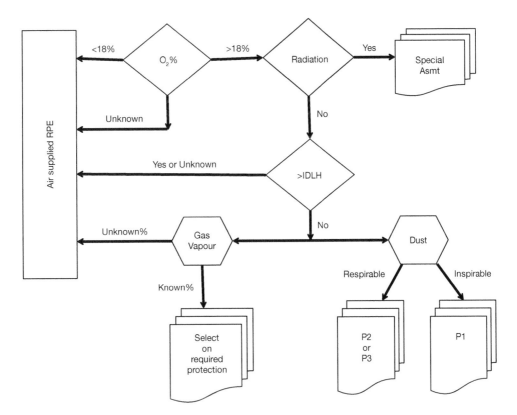

Figure 6.5 Simplified RPE selection decision tree

the minimum RPE for situations where it is required. On occasions, trials of different types and brands will be necessary to determine the RPE that is most suited to a particular workplace. Suppliers of RPE are an excellent source of information and guidance.

Two pieces of numerical information are crucial to applying RPE successfully:

- the concentration of contaminant in the workplace
- the target concentration inside the respirator.

Based on these data, it is possible to calculate the required minimum protection factor needed to reduce exposures to an acceptable level. The following formula should be used to determine the required minimum protection factor:

$$\text{Required minimum PF} = \frac{\text{concentration in workplace}}{\text{OEL* (or other target concentration)}} \qquad \text{(Equation 6.1)}$$

Note: Both the concentration and the *OEL (Occupational Exposure Limit) should be in the same units.

The required minimum protection factor is a crucial piece of information when selecting RPE. It relates to the capacity of the respirator to adequately reduce concentrations of the contaminant between the outside and the inside of the respirator. No respirator prevents all the contaminant from entering the breathing zone. AS/NZS 1715:2009 assigns maximum use limits for different types of respirators, summarised in Table 6.2.

It is notable that the Australian Standard uses the term 'acceptable exposure level/ standard' to refer to the concentration inside the respirator. It is the recommendation of this author that to simply apply the OEL may not provide any margin of safety to accommodate errors in the various estimations. As an example, let us say the cyclohexane concentration in a workplace is measured at 1800 ppm. Cyclohexane can be removed with filtering-type respirators with Class AUS, 1, 2 or 3 filters. The required protection factor to reduce the 1800 ppm concentration to below, say, 50 per cent of the OEL target concentration of 100 ppm will be 1800/50 = 36. Table 6.2 indicates that only full facepiece respirators with

Table 6.2 Required protection factors for selected respiratory devices

Required minimum protection factor	Respirator type
Up to 10	Disposable facepiece respirator, half-face respirator for particulates and gas/vapour concentrations <1000 ppm
Up to 50	PAPR or full facepiece P2 for particulates, and Class AUS or 1 gas filters for vapour concentrations <1000 ppm
Up to 100	Full facepiece P3 for particulates or Class 2 gas filters up to 5000 ppm (there are no Class 3 filters)
100+	Full facepiece positive pressure airline or SCBA positive pressure demand

Source: data based on AS/NZS 1715:2009.

Class 2 or 3 filters will have the required minimum protection factor to be used above 1000 ppm. (There are, however, other types available and suitable, as outlined in Table 4.5 of AS/NZS 1715:2009—half-face, air-line, continuous flow; PAPR 2 in full-face, etc.) AS/NZS 1715:2009 also contains an excellent flowchart/decision tree (Figure 6.5 above) to identify the most suitable RPE for most situations. This standard deals with respirator selection under the simplified headings of 'contaminant', 'task' and 'operation'.

6.7.6   *FILTER SERVICE LIFE AND BREAKTHROUGH*

When introducing air-purifying RPE to control exposure, attention must be directed to usage patterns and exposures in order to estimate the service life of protective filters. This is a complex issue and depends on many factors, including the filter construction, the concentration of the contaminant, temperature and humidity, the worker's breathing rate, the physical and chemical composition of the contaminant and general respiratory competence. Service life estimates must include some unexpended reserve capacity as a safety margin. Importantly, filters should not be used beyond the expiry of their shelf life—this should be marked on the filter or its packaging.

A recurring question is how long the filter in the respirator will last. There is no simple answer, so it is better to rely on outside experience or advice from manufacturers rather than chance unsafe practices. For any type of respirator, wearer acceptance is also an important factor in adoption of RPE. For particulate filters, filtration efficiency usually also increases with use as dust particles gradually block the filter. This causes increased inhalation resistance. Similarly, the higher filter efficiency of P3 filters comes with increased inhalation resistance. This may have adverse effects for the wearer when RPE is used for continuous work, or if the wearer has some respiratory ailment that makes the use of a respirator difficult. The service life of such filters is ended when the wearer has trouble breathing through the filter.

For gas and vapour filters, minimum service life can be calculated only if there is reliable data on exposure conditions. Otherwise, scheduled maintenance and replacement programs with a reasonable margin for safety must be followed scrupulously. Several RPE manufacturers have software to perform service-life calculations. It is essential to check the compatibility of the software with parameters such as the OELs, the RPE specifications on flow or breathing rates and relative humidity in the situation under consideration. In short, the calculation of service life is fraught and should be approached with caution.

Breakthrough of the contaminant, as indicated by odour, taste or irritation, is an unreliable means of determining the end of service life (exhaustion of capacity) of a respirator gas/vapour filter. Some contaminants have no odour (carbon monoxide); others have workplace exposure standards below their odour threshold (isocyanates). Some individuals become insensitive to odour or develop nasal or lung irritation. Some individuals have no sense of smell. Some chemicals rapidly deaden the sense of smell (carbon disulfide).

6.7.7   *TRAINING IN USE AND CORRECT FITTING*

Tests for the correct fitting of tight-fitting RPE can only be achieved using the proper equipment. To qualitatively test for facial leaks, the wearer typically is subjected to an aerosol (Bitrex or saccharin) as illustrated in Figure 6.6. The respirator user has a hood over their head and an aerosol of saccharin or Bitrex is aspirated into the hood. The user is asked to perform a series of exercises involving talking and head movements. Any significant failure of the face seal will be detected by the taste of saccharin or Bitrex. This method is suitable only to test face seals on half-face respirators fitted with particulate filters, as the saccharin or Bitrex particles can pass through gas and vapour filters. A similar test can be conducted for organic-vapour air-purifying respirators using isoamyl acetate vapour as the challenge agent. Fit testing is described in AS/NZS 1715:2009 (Standards Australia, 2009). If the wrong test agent is used it will penetrate the filter, producing misleading results. RPE suppliers can provide assistance in fit testing.

The more complex quantitative face-fit tests use a modified respirator facepiece and a particle detector to measure the inward leakage of an aerosol. Some tests of this type use ambient aerosols that occur naturally in the environment, while others use a generated sodium chloride aerosol or oil mist. These tests require both specialist equipment and training.

Both types of test (qualitative and quantitative) help to ensure that the selected respirator properly fits the wearer's face and therefore delivers effective protection. The results indicate satisfactory fit, not workplace hazard-protection levels. The level of protection is highly dependent on how the device is used in the workplace. Results of such tests should not be used to increase the protection factors above those recommended in codes of practice or standards.

The fit test should be performed when the respirator is first issued and whenever the respirator type is changed. It serves the added purpose of reminding users of the need to

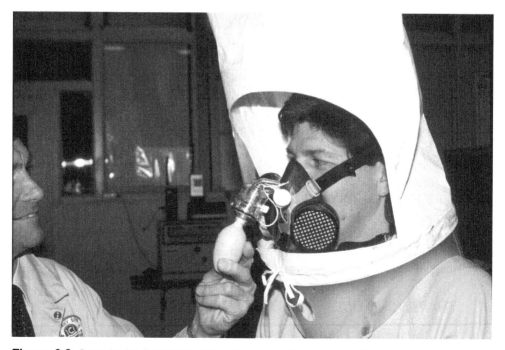

Figure 6.6 Quantitative facial fit testing using a saccharine aerosol

always put on their respirator with care, and can also demonstrate the effect of damaged parts or facial hair, including moustaches, stubble and beard growth, on the completeness of the face seal. All wearers of close-fitting respirators should be clean shaven in the area of the face seal.

Each time the mask is put on, a gross negative-pressure fit check should be done to check for leaks by blocking off the filter inlets using the palms and inhaling (Figure 6.7). Inward leakage can be detected, though this test is not very sensitive because the negative pressure within the facepiece causes the respirator to be drawn onto the face by the external atmospheric pressure, improving the seal on the face. A similar positive pressure test can be done by exhaling while covering the filters or exhalation valve and trying to detect leakage. These checks are not substitutes for fit testing, but are used each time after donning respirators to ensure there are no gross seal failures.

6.7.8 INSPECTION, MAINTENANCE AND REPAIR

With the exception of disposable-type respirators, the use of RPE in the workplace requires a constant program of inspection, maintenance and repair. Maintenance includes washing, cleaning, disinfecting where necessary, inspecting for wear and tear, checking for leaks, replacement of worn components and replacement of worn parts and filters. It is essential to store RPE properly between uses. Gas and vapour filters can continue to absorb

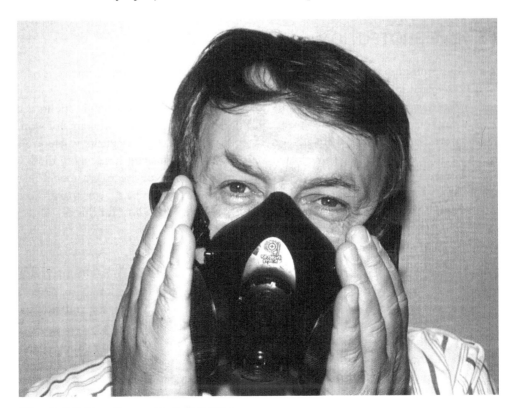

Figure 6.7 Negative pressure fit testing

contaminants when they are not in use, further exhausting their capacity. Plastic sealable food storage containers or zip-lock plastic bags are ideal for between-use storage. Suppliers will be able to provide details of suitable maintenance programs for their particular equipment. Where possible, each wearer should be provided with individual RPE. Where air compressors are used for air-supplied respirators, there should be a maintenance, inspection and testing program in place for the compressors as well, to ensure the air quality. See AS/NZS 1715:2009 (Standards Australia, 2009) for guidance on compressed air quality testing.

6.7.9   MEDICAL ASSESSMENT OF RPE USERS

Wearing RPE is not without its physiological and psychological demands. There are a number of medical conditions, including diabetes, asthma, emphysema, skin sensitivity and chronic airways disease, that can prevent a worker from using RPE. Some workers also feel claustrophobic wearing a normal filtering respirator, but may find a powered air-purifying respirator (PAPR) more acceptable. Always seek expert medical advice from an occupational physician if prospective wearers have problems wearing respirators.

In some cases, biological testing (e.g. blood lead testing) can be used as a supplementary means to assess the overall effectiveness of exposure control programs. While respirators provide respiratory protection, other uncontrolled routes of intake of the toxic agent may remain—for example, ingestion or skin absorption—leaving the worker less protected than they believe.

6.8   EYE PROTECTION

Eye protection is a common requirement in many working environments. The hazards are associated mostly with flying particles or sparks, but chemical exposure is also significant in certain industries and protection from splashes, sprays and gases and vapours is needed. Workplace experience is that many eye injuries occur to workers who are not wearing eye protection at all, but there are also a significant number of incidents where eye injuries occur to those wearing some form of eye protection. This largely occurs to those wearing protective eyeglasses with no side shields or those who are wearing a model that leaves large gaps between the spectacles and the face. Style can also be a factor to gain acceptance from workers, so a range of products should be offered to allow for this as well as getting a suitable fit outcome.

The crucial factors to prevent eye/face injuries are:

- *Effective eye protection that complies with the local standards.* To be effective, the eyewear or face shield must be of an appropriate type and rating for the eye/face hazards encountered and should be fitted properly to reduce any gaps and fit securely in position.
- *Potential interference with other PPE.* The eye or face protection needs to be compatible with any other PPE being used. Safety spectacles with wide or stiff arms may cause a significant loss of the attenuation provided by a pair of ear muffs, or the frame may interfere with the correct fit of a half-face mask respirator. This issue needs to be assessed for each individual and their selected PPE.

- *Better training and education.* Workers injured while not wearing protective eyewear have indicated that they did not believe it was required by their situation. Many employers furnish eye protection at no cost to employees, but commonly many of these workers receive no information on the kind of eyewear they should use and for what tasks.
- *Maintenance.* Eye protection devices (spectacles, goggles, face shields) must be maintained properly. Scratched and dirty devices reduce vision, can cause glare and may contribute to accidents. Scratched or damaged eyewear should be replaced immediately.

6.9 A FINAL WORD ON PERSONAL PROTECTIVE EQUIPMENT

If PPE is used as the primary method of protection, then it is essential that robust systems be in place to ensure that the correct level of protection continues to be achieved. Such systems may include workplace behavioural safety programs, in-house inspections and audits, routine training or 'toolbox talks' or biological monitoring (mentioned above). The H&S practitioner should resort to personal protective equipment only when other means of exposure control are impracticable. Use of any protective equipment places restrictions upon workers, reduces the flexibility of their operation and affects their performance. Others have suggested that if an H&S practitioner wants a worker to work permanently in PPE, they should personally try wearing the same protection for a week to judge its real suitability as a long-term measure.

Running a robust PPE program requires great attention to detail and careful record-keeping to track who has received what training and fit testing for which types of respirators and for which specific tasks.

6.10 REFERENCES

American Society for Testing and Materials (ASTM) International 2012, Standard Test Method for Permeation of Liquids and Gases Through Protective Clothing Materials Under Conditions of Continuous Contact, ASTM F739–12E1, ASTM International, West Conshohocken, PA, <www.astm.org> [accessed 12 March 2018]

CEN 2004, Protective Gloves Against Chemicals and Micro-organisms, EN 374:2004, European Committee for Standardisation, Brussels.

Institut National de Recherche et de Sécurité and Institut de recherche Robert-Sauvé en santé et en sécurité du travail 2019, *ProtecPo Software for the Selection of Protective Materials*, INRS and IRSST, <http://protecpo.inrs.fr/ProtecPo/jsp/Accueil.jsp> [accessed 17 August 2019]

Occupational Safety and Health Administration n.d., *OSHA Technical Manual*, Section VIII: Chapter 1 chemical protective clothing, US Department of Labor, Washington, DC, <www.osha.gov/dts/osta/otm/otm_viii/otm_viii_1.html> [accessed 12 March 2018]

Safe Work Australia 2018, Model Code of Practice: How to Safely Remove Asbestos, Safe Work Australia, Canberra, <www.safeworkaustralia.gov.au/doc/model-code-practice-how-safely-remove-asbestos> [accessed 17 August 2019]

Standards Australia 2009, Selection, Use and Maintenance of Respiratory Protective Devices, AS/NZS 1715:2009, SAI Global, Sydney.
—— 2012, Respiratory Protective Devices, AS/NZS 1716:2012, SAI Global, Sydney.

7. Aerosols

Linda Apthorpe and Jennifer Hines

7.1 INTRODUCTION

This chapter considers aerosols from the perspectives important to the H&S practitioner, and covers:

- information on different kinds of aerosols
- work situations where aerosols are commonly encountered
- what happens to inhaled aerosols in the human respiratory system
- how to assess various kinds of workplace aerosol hazards
- an historical review, so that the H&S practitioner knows how important the aerosol problem has been in the past.

7.2 WHAT ARE AEROSOLS?

The term 'aerosol' applies to a group of liquid or solid particles usually in the range of 0.001 to 100 μm in size (100 μm = 0.1 mm), suspended in a gaseous medium. Naturally and artificially produced aerosols are found in ambient and industrial air environments and vary greatly in size, density, particle shape and chemical composition. For example, the shapes of particles include spheres (water or oil droplets and welding fume), cylinders (asbestos and glass fibres), crystals (crystalline silica), regular and irregular particles (fly-ash and road dust). The term 'particle' implies a small, discrete object, and 'particulate' indicates that a material is made up of particles or has particle-like characteristics.

There is no simple system for classifying aerosols found in the workplace based on the nature of the aerosol, its toxic effect or the particle size, although all of these are important for different reasons. Some aerosols exert their toxic effects in the nose, throat and upper airways and others in the lung, while for some the lung is merely the route of entry, not the ultimate target organ. Further, some inhaled aerosols can cause more than one effect, depending on the amount to which a worker is exposed (the dose).

7.2.1 DUST

Dust usually comprises solid particles, generally greater than 0.5 μm in size, formed when a parent material is crushed or subjected to other mechanical forces. Dust is found everywhere and remains one of the most intractable of workplace problems. Operations carried out in today's industrial workplaces, such as mining, crushing, sieving, milling, grinding, planing, sawing, sanding, machining and pouring, all contribute to dust generation. Particles generated in one workplace can become airborne and be carried into other working locations. While many of these dusts are relatively harmless, causing only transient irritation, some give rise to lung fibrosis, and others to carcinoma, bronchitis, asthma or other lung disorders.

7.2.2   FIBRES

There are many different types of fibres. They can be naturally occurring—for example, asbestos and cotton—or man-made, such as man-made vitreous fibres (MMVF, formerly known as synthetic mineral fibres, or SMF), kevlar and carbon fibres. Fibres generally are defined by their aspect ratio—that is, the ratio of their length to width. The aspect ratio is important, as it defines the ability to become airborne and where the fibre deposits in the human respiratory system. Fibre composition is also important in considering impact to health upon exposure.

7.2.3   FUME

Fume is produced when vaporised materials—usually metal—condense to become solid particles. These are less than 0.05 µm in size, and generally agglomerate. Smelting, thermal cutting and welding operations all produce fume. The measurement of welding fume is conducted using AS 3853.1 (Standards Australia, 2006a). The Australian Standard describes the use of sampling devices suitable for insertion behind protective face shields normally worn by welders. The companion standard AS 3853.2 (Standards Australia, 2006b) deals with the measurement of gases relating to welding. 'Diesel fume', as it is commonly and incorrectly known, is a particulate and is discussed in section 7.7.

7.2.4   MIST

Mist or fog is an aerosol composed of liquid droplets. The droplets may be generated mechanically, such as by spraying, or by condensation of vapour, such as in fog. The particles of mists and fogs are commonly greater than 10 µm in size. Examples of mists are oil mists, paint spray, chemical mists and sprays containing admixtures of agricultural chemicals in water.

7.2.5   NANOPARTICLES

Nanoparticles can be naturally occurring in by-products of heating/combustion such as diesel emissions, fume and smoke. They can also be present in the form of engineered (that is, man-made) nano-sized particles. The International Organization for Standardisation (2015) indicates that the size range for nanoparticles or sub-micrometre particles is 1 nm to 100 nm (i.e. 0.001 µm to 0.1 µm). Measurement of airborne nanoparticles and the subsequent interpretation of the results are highly complex and require considerable experience, as the particles exhibit different behaviour and toxicity characteristics.

7.2.6 *SMOG*

Smog is an aerosol consisting of solid and liquid particles, generally created by the action of sunlight on various vapours (see Chapter 9). Particles in smog are generally less than 1 µm in size.

7.2.7 *SMOKE*

Smoke is an aerosol containing solid and liquid particles that result from incomplete combustion, and most smoke particles are less than 1 µm in size.

7.3 AEROSOLS AND THE WORKPLACE

Sources of aerosol includes naturally occurring and synthetic manufactured materials of **inorganic mineral origin**. Coal, quartz-bearing rock and metal dusts (e.g. lead) commonly associated with mineral extraction and processing industries are all significant sources of workplace dust. Many metal manufacturing and heavy industries give rise to dusts and fumes of **metals** (lead, zinc, copper, arsenic). Today, many workers are potentially exposed to dust from **construction and demolition activities**.

Asbestos can still be present in some 'asbestos-free' friction materials manufactured overseas (e.g. brake linings), despite the Australian ban on importation of all types of asbestos. Workers are also potentially exposed to asbestos fibre in the asbestos-removal industry. MMVF used in insulation and fire protection provide another source of workplace dust exposure; however, they are not toxic like asbestos, with the exception of some types that are classified as 'possibly carcinogenic to humans'. Refer to section 7.9.1 for more detailed information.

Naturally occurring organic dusts are common in some workplaces. Rural workers are exposed to natural dusts of grain. Sugar-mill workers may be exposed to dust from bagasse, the waste cane that remains after sugar has been extracted. Downstream processing industries expose workers to dusts containing wood, flour, cotton, paper, felt, fur, feathers and pharmacologically active plant materials.

Within industrial manufacturing processes, the H&S practitioner can encounter *manufactured dusts* from a range of plastic polymers, including epoxies, polyvinyl chloride, acrylates, polyesters and polyamides. They are found in workplaces as diverse as foundries, plastic pipe manufacture, packaging, surface coating industries, manufacture of composite materials and dental laboratories. The wide spectrum of dusts that can occur in different workplaces provides a constant challenge to the H&S practitioner to devise various strategies for their control.

From the point of view of the H&S practitioner, the two factors important for assessing the impact of inhaled aerosols in the workplace are:

- chemical composition of the aerosol, and
- particle size.

Both **composition** and **particle size** are important in terms of how an inhaled aerosol affects the worker, because together they govern how much of a material actually enters the body, where it is finally deposited and what sort of toxic effect it can exert.

7.3.1   CHEMICAL COMPOSITION OF THE AEROSOL

It is known that different kinds of aerosol can have different effects on health, thus the composition of the aerosol—principally its chemical composition—is important. In some cases, the toxic effect caused by the inhaled particles occurs quickly (e.g. within a few minutes) and this acute effect is easy to associate with exposure to the inhaled aerosol.

In other cases, the effect of inhaling the aerosol may not appear for many years following exposure. It then becomes difficult to make an association between the inhaled aerosol and its chronic (or long-term) effect. This long period, or latency, between exposure and disease has often made it difficult to establish causal links between particular aerosol exposures and disease. This can also lead to a false sense of safety in dealing with such aerosols. To assess any likely health impact of the aerosol in workplace air, the H&S practitioner first needs to know the composition of the aerosol.

The following examples highlight the importance of correct identification of the aerosol. Some particulates are acute respiratory hazards and some are chronic respiratory hazards; some are sensitisers that can cause increasing symptoms on further contact; and some are not particularly hazardous at all.

- Welding fume may consist mostly of relatively non-toxic iron oxide or it may be combined with a small proportion of highly toxic cadmium or other toxic metals, which have acute effects (refer to Chapter 8).
- Airborne asbestos fibres present significant chronic and life-threatening respiratory hazards; airborne MMVF do not. Both types of fibre can be present in some workplaces.
- Airborne quartz-containing dusts are far more hazardous than limestone dusts, because quartz (i.e. crystalline silica) causes silicosis, a chronic lung disease, whereas limestone is more a 'nuisance' dust with little effect on the body.
- Many wood dusts, along with sap, latex and lichens associated with wood, can lead to skin irritation, sensitisation, dermatitis and respiratory effects such as rhinitis, nose bleeds, asthma and other allergic reactions. Toxic activity, however, is determined by the species of tree, with some being more toxic than others, even leading to nasal cancer in the case of some hardwood dusts.

In some cases, the identity of an aerosol can be obtained directly from a safety data sheet (SDS), particularly where no chemical transformation occurs during processing. For example, in manufacturing lead accumulator batteries, it would be reasonable to expect to find lead-containing dust in the workplace atmosphere, yet the identity of the hazardous particulate is often not clear. In cases of uncertainty, expert advice and analysis should be sought from a suitable laboratory. On occasions, workplace aerosols will contain more than one hazardous component (e.g. refractory ceramic fibre and cristobalite, or coal dust and alpha-quartz). In these circumstances, each component may need to be monitored. Cost-effective control solutions can be arrived at only if the identities of hazardous materials are known. Otherwise, costly solutions may be based on incorrect information.

Depending on dose, most particles that can be inhaled are toxic to organs other than the lungs. Uptake may be directly via the bloodstream in the lungs, or by secondary absorption in the gut. Solubility in the gut may be low, while absorption in the lungs is very high. For example:

- Soluble salts such as nicotine enter via the lung and gut and target the brain.
- Metals such as arsenic, zinc, cadmium or lead may enter primarily via the lung and target various other organs.

7.3.2   PARTICLE SIZE OF AEROSOLS

Aerosols generally found in the workplace vary widely in size, as shown in Figure 7.1. Workplace particulates of interest range in diameter from around the width of a human hair (approximately 30–100 µm) to less than 0.1 µm (i.e. nanoparticles). Particle size is important for two reasons: first, it determines how long a particle remains airborne and hence how far an aerosol cloud will disperse in a workplace before settling. This factor may influence the choice and impacts the effectiveness of control strategies.

Second, the effects on health caused by many aerosols depend on their site of deposition in the respiratory tract. The site of deposition depends largely on the size of a particle or, more correctly, its aerodynamic settling velocity. Large particles (up to several hundred micrometres in diameter) settle in the nose and throat and may exert their effects at these sites. Smaller particles are deposited in the upper airways (the bronchi and bronchioles), from where they can be cleared by the mucociliary escalator (a mucus layer moved by cilia, or fine hairs, beating in an upward direction) to ultimately be swallowed and eliminated through the gastrointestinal tract. The very small particles, termed respirable and generally smaller than around 3–5 µm in size, penetrate to the alveolar gas-exchange region of the lung, from where they are cleared only very slowly. Particles in the sub-micrometre range (e.g. nanoparticles) may also be transported through the alveolar cell walls directly into the bloodstream.

Consequently, any assessment made in a workplace of an aerosol that presents a disease risk must consider particle size. Different sampling methods are available to achieve this.

Figure 7.2 shows, in simplified form, the basic structure of the human respiratory system and each of its components. It leads us to understand the concepts used in the study and measurement of dust deposition as follows:

- *inhalable mass*—for materials that are hazardous when deposited anywhere in the respiratory tract
- *thoracic mass*—for materials that are hazardous when deposited anywhere in the lung or gas-exchange region
- *respirable mass*—for materials that are hazardous when deposited in the gas-exchange region.

The primary agent measured for the thoracic area is sulphuric acid. In Australia, inhalable and respirable fractions are of major significance, and both of these fractions are examined in more detail in section 7.6. Note that the Australian Standard AS 3640 (Standards Australia, 2009c) uses the term 'inhalable dust' (formerly 'inspirable dust').

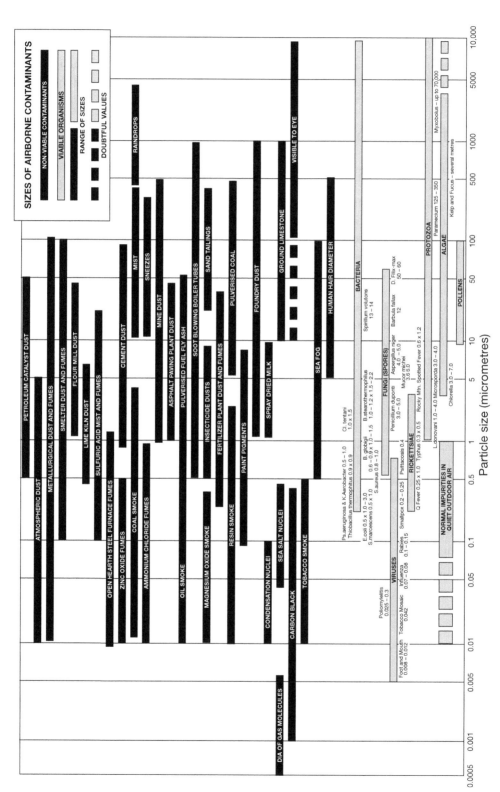

Figure 7.1 Sizes of various particles

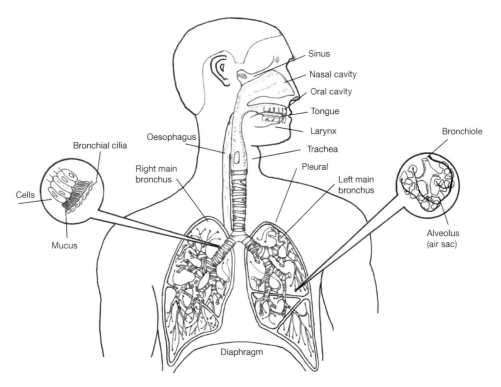

Figure 7.2 Basic structure of the respiratory system

7.4   AEROSOL MEASUREMENTS IN THE WORKPLACE

Aerosol sampling (i.e. collecting samples from workers) is not a highly complex task; however, obtaining reliable and valid results does require significant attention to detail, appropriate skills and the use of relatively costly equipment.

Four different classes of aerosol measurements important to the H&S practitioner are discussed here. They involve the sampling and analysis of:

- inhalable particles
- respirable particles
- diesel particulate
- fibrous dust.

Different devices are used to sample each of these. Sampling with the wrong device can lead to an over- or under-estimate of the risk, depending on the nature of the contaminant. This can lead to incorrect or costly control procedures or inadequate protection of workers.

Differences between inhalable and respirable particulate sampling are outlined below, together with typical applications and sampling procedures. The laboratory analysis for some applications is also described.

7.4.1 GENERAL APPROACH

The sampling of aerosols to estimate disease risk has taken centuries to understand properly, and just as long to develop into a practical form. The types of aerosol clouds found in the workplace have complex characteristics in terms of particle size and composition. Numerous different sampling procedures and extensive chemical analytical techniques may be necessary. If H&S practitioners understand the basic principles of aerosol measurement, then with adequate equipment and training they will be able to perform some workplace monitoring for aerosols.

For example, with dusts, the H&S practitioner will first need to determine:

* whether a dust should be assessed as a respirable or an inhalable dust
* whether a dust can produce acute effects on exposure or long-term chronic disease.

Such information provides the basis for proper assessment and control. It will also allow the H&S practitioner to give specific guidance to the employer or a consultant engaged to undertake workplace dust monitoring. Not all dusty situations will require monitoring. If a competent assessment is made based on knowledge of the dust, it is possible to recommend control procedures, including ventilation or personal protection, without the need for dust monitoring.

7.4.2 DEVELOPMENT OF WORKPLACE AEROSOL MONITORING

Before the current methods of aerosol sampling are examined, it is worth looking at the development of dust sampling over the last century. With the very earliest dust-sampling equipment, those interested in examining the dust conditions in a workplace—typically a mine or a tunnel—reasoned that the ill-health arising from inhaling the dust was probably correlated in some way with the amount the worker inhaled. Dust disease in miners had been known since the days of King Solomon, yet it did not attract much compassionate concern in the days when most mine workers were slaves or prisoners. In fact, it was not until the fifteenth and sixteenth centuries, when there was an unacceptable toll among the silver miners in Europe (who were responsible for maintaining the coffers of the kings) that some interest (although not much assistance) was taken in their respiratory health.

It was not until the early 1900s that methods of quantifying dust levels became available, and while they were crude, they were ingenious. Mines of the time were not places of great technological development—for example, ventilation methods included building a roaring fire above an upcast shaft to induce air movement. One of the earliest methods of quantifying mine dust involved pumping a large quantity of air through a tube containing sugar, dissolving the sugar out, then weighing the trapped dust that remained.

Such methods were not very sensitive, and only one sample could be collected per shift. In the United States in the 1920s, air samplers were developed that trapped dust particles in a liquid impinger and impacted the dust onto a cold glass slide (by inertial or thermal force). These portable samplers allowed an inspector to take many samples per shift. Impingement samples were assessed by means of a microscope, giving rise to the term 'dust count'. In Australia, these and similar methods were employed by the NSW Joint Coal Board until as late as 1984, when they were replaced by gravimetric sampling methods.

Particle-counting methods, while far more sensitive than the previous methods of weighing dust, had numerous drawbacks. Sensitivity is important because some inhaled particles cause respiratory disease even at very low levels. One of the fundamental problems found by medical researchers was there was no relationship between the number of particles inhaled and severity of the disease they caused. This was because the size-selecting characteristics of the human respiratory system were largely ignored.

Over the 25 years following World War II, the largest ever health study of a group of workers was undertaken in the British coal-mining industry to establish the relationship between inhaled dust and dust disease. By examining the dusts deposited in the lungs of deceased miners, the British Medical Research Council was able to propose a size distribution as one of the fundamentally important parameters (i.e. 'respirable' dust). Instruments have since been devised to capture particles fitting this size profile.

Current methods used for aerosol sampling consider the size-selective nature of the human respiratory system, and measurements are expressed in terms of the mass of particles, rather than the number of particles, entering different parts of the respiratory system.

7.5   TYPES OF AEROSOL-SAMPLING DEVICES

Two different types of samplers are used in workplace monitoring, namely direct reading devices and filtration samplers. Both have advantages and disadvantages, but for most applications the filtration types are recommended as the more versatile and valid. Any H&S practitioner involved in workplace monitoring should seek the advice of an experienced occupational hygienist before spending money on any one type of instrument.

7.5.1   DIRECT-READING DEVICES

Despite their apparent advantages, direct reading devices for aerosol monitoring are not commonly used. Quality instruments are expensive, while cheaper instruments lack the ability to discriminate adequately between respirable and inhalable particulates. Calibration is a problem with instruments not fitted with primary separating elutriators (devices used to separate and classify particles by allowing them to settle under gravity in a moving airstream—see section 7.6.3).

Various devices, such as the TSI DustTrak™ aerosol monitor range (Figure 7.3), are used in some dusty industries.

The DustTrak™ is a laser photometer that detects light scattered by the presence of particles. This is one type of instrument that provides direct real-time readouts as well as data-logged results that can identify peaks, means, averages and other aggregated data. Direct-reading instruments are very useful in evaluating control procedures in the workplace; however, it is important to consider the advantages and disadvantages of this type of instrumentation for each sampling situation, including appropriate calibration to the specific type of aerosol being monitored (as these instruments are normally calibrated against 'Arizona Road Dust').

Figure 7.3
DustTrak™ direct-reading aerosol monitor

7.5.2   *FILTRATION SAMPLING INSTRUMENTS*

The majority of aerosol-sampling devices use filtration techniques. Advantages include relatively low cost, robustness and the ability to sample different types of airborne particles (e.g. wood, crystalline silica, coal, metals, grains, organic dusts, powders, acid mists).

Common personal dust-sampling equipment contains the following five elements, as seen in Figure 7.4:

- filter
- filter holder incorporating a size-selective device
- suction pump
- connecting tubing
- flow meter—used to set up the sampling equipment.

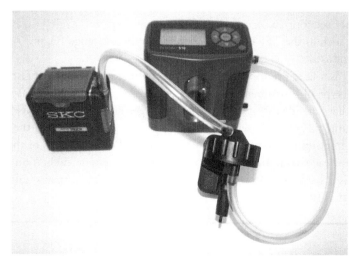

Figure 7.4 Typical personal monitor showing pump, sampling head and electronic flow meter in a calibration train

In the absence of a belt, a harness or backpack can be used to conveniently attach the sampling equipment to the person wearing it.

Collecting particles on a filter and measuring subsequent concentration occurs via the following steps:

1 A known volume of particle-laden air is drawn by the sampling pump.
2 A size-selection device (first stage) may be used to reject some of the larger dust particles.
3 The finer particles of interest are collected on the filter (second stage).
4 The collected deposit is weighed on a microbalance or analysed using special techniques in a laboratory.

7.5.2.1 Filters

Many different kinds of filters are useful for workplace monitoring. Filters are chosen depending on the aerosol being sampled, the analysis to be carried out and the environment being sampled. Some filters have grids, some are acid resistant and some are transparent in certain light wavelengths. Filters may be made from polyvinyl chloride (PVC), polycarbonate, glass fibres, polytetrafluoroethylene (PTFE), mixed cellulose esters etc., and have pore sizes ranging from 0.3 to 5 μm. Specialist advice on appropriate filter type and pore size should be sought for any new sampling task.

7.5.2.2 Sampling pumps

Different types of sampling pumps are available for aerosol monitoring in the workplace. Some operate on mains power, most are small, convenient, battery-powered pumps that can be worn by the worker. These pumps are designed to provide flow rates of between 0.5 and 5 litres per minute (L/min), with most aerosol sampling carried out between 2.0 and 3.0 L/min.

Good dust-sampling pumps have four important features:

- They are controlled so that the volume of air can be accurately measured.
- They are pulsation dampened so that any size-selective device connected may operate correctly (i.e. constant flow).
- They have the ability to set flow rates over a wide range.
- They have the capacity to operate at a reasonable suction pressure (e.g. up to 10 kPa).

Two other features important when choosing a sampling pump are continuous operating time and intrinsic safety. A fully charged pump in good operating condition should be able to operate for at least twelve hours. Intrinsically safe pumps, designed to be incapable of generating sufficient energy to ignite a flammable atmosphere, are mandatory for use in workplaces with potentially explosive atmospheres (e.g. coal mines, petroleum refineries and grain-handling facilities).

Most of the flow meters, or flow displays, built into sampling pumps should not be relied upon because they can indicate incorrect and variable flow rates. Any flow meter used to measure the flow rate of a pump must be calibrated with a primary flow meter, which is a flow meter with key properties that are traceable to national measurement standards of length, mass and/or time. One example of a primary flow meter is the 'soap film flow meter' used in a calibration train similar to that shown in Figure 7.5.

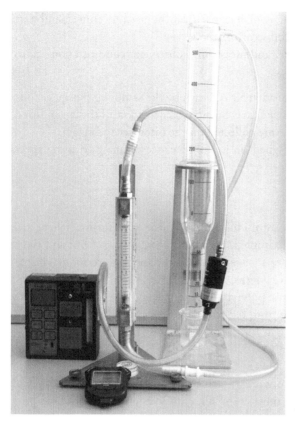

Figure 7.5 Flow rate calibration train and soap film flow meter

The soap film flow meter uses the principle of timing the movement of a soap bubble along a transparent tube of known volume as a result of air flow produced by a pump. There are several types of portable flow meters (e.g. electronic soap film and dry piston) that have been shown to provide consistent and reliable results. Although the makers of some of these instruments claim that they are primary meters, they do not meet the Australian National Measurement Standard in this respect and should be checked against a bona fide primary flow meter.

7.5.2.3 Connecting tube and harness

The connecting tube used to connect the pump to the filter holder needs to be made of high-quality PVC (e.g. Tygon) or polyethylene so that it does not readily crush and retains its elasticity.

Sampling pumps typically are worn on the belt, and a stout belt or harness will be needed to suspend the pump. Pumps are generally too large to be carried in a pocket and may be prone to accidental blockage if carried inside clothing.

7.6   SIZE-SELECTIVE SAMPLING DEVICES

Several size-selecting devices for particulate sampling are available commercially. It is important that the correct device is selected for each sampling task. Details of the recommended samplers are given on the following pages. Only samplers with the correct performance can be used to make measurements that are in accordance with the exposure standards (ES).

While most sampling commonly involves dusts, similar principles underpin sampling for other aerosols.

7.6.1 INHALABLE DUST SAMPLING

Dust hazards will be assessed mostly on the basis of the dust a worker can inhale from the workplace air. This is commonly known as inhalable dust; it is similar to, but does not produce the same sampling results as, 'total dust' (the total-dust sampler was developed in North America, and is not used in Australia or Europe). Many particles in a visible dust cloud are aerodynamically too heavy to be captured by the respiratory system, which means that only a fraction of them are inhaled. It is important to bear in mind that inhalable dusts represent a wide size range and include respirable particles. (Respirable dust is discussed in section 7.6.2.)

Some of these inhalable dusts cause their effects at the site of deposition in the upper airways. They include:

- wood dusts causing nasal cancer (e.g. oak, beech, birch, mahogany)
- cement dusts causing airway irritation
- sensitising wood dusts causing asthma (e.g. western red cedar)
- proteolytic enzymes, which attack cell structure.

Inhalable mists can also contain oils, acids or alkalis, which require special analysis and expert advice for sampling.

There is no universally accepted size criterion for inhalable dusts, because inhalability varies with the density of the dust. It may also depend on a number of workplace and human factors, such as wind speed and whether the worker is a nose or mouth breather. Australian Standard AS 3640 (Standards Australia, 2009c) provides definitions for two different samplers, now based on the common standard defined by the International Organization for Standardization (ISO, 1995), published as ISO 7708 and harmonised with the definition used by the American Conference of Governmental Industrial Hygienists (2018). The ISO criteria are seen in Figure 7.6, which shows that the collection efficiency of any inhalable-dust sampler is above 90 per cent for particles below 4 µm in diameter and drops to around 50 per cent efficiency for particles of diameter greater than 100 µm. In other words, as with the human nose and mouth, not all particles are caught by the sampler.

Two inhalable-dust samplers are recommended by AS 3640 for use in conjunction with the exposure standards. Figure 7.7a shows the IOM dust sampler that has been designed to overcome problems of different wind speeds past the wearer and sampler orientation (i.e. sampler facing horizontal or downwards). This sampler is useful for dusts that are reliably known to contain only small particle sizes (e.g. many thermally generated metal fumes).

Figure 7.7b shows the modified United Kingdom Atomic Energy Association (UKAEA) seven-hole sampler, which can over-sample in environments where large

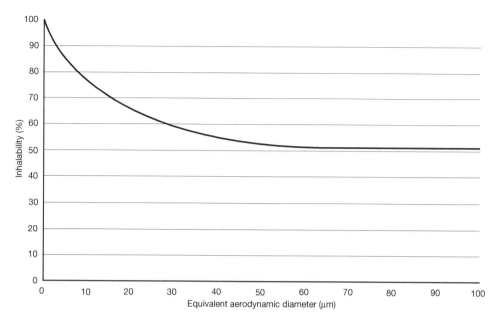

Figure 7.6 Inhalable-dust sampler performance (ISO 7708)

particulate sizes are directed towards the face of the sampler. This is not commonly used in Australia.

A detailed practical procedure for measuring inhalable dust in the workplace is given in AS 3640. The field sampling procedure specified in this standard can be conducted by H&S practitioners with appropriate sampling equipment and training. The laboratory analysis requires special equipment and techniques (e.g. microbalance, static eliminating equipment and various analytical devices). Analysis should be conducted by specialist laboratories accredited by the National Association of Testing Authorities (NATA).

On completion of sampling, the concentration in mg/m³ of inhalable dust is calculated by dividing the net weight gain (after taking into account weight change of a blank filter) by the total volume of air sampled.

Gravimetric weighing of the dust is not the only method of evaluating its concentration in the workplace. When there is a mixed dust, other types of laboratory analysis will be needed to determine the concentration of the contaminant of interest. Examples include:

- measuring metal contaminants in the presence of smoke and other fumes
- measuring polycyclic aromatic hydrocarbons (PAHs) in foundry dust
- measuring benzene-soluble fraction in coke oven emissions.

7.6.2   DUST DISEASES AND THE CONCEPT OF RESPIRABILITY

A typical aerosol in a mine, quarry or factory contains particles ranging in size from 100 μm diameter to as little as 0.1 μm. So far, we have examined how to measure that part of an aerosol that can be breathed in (inhaled).

The other fraction of the aerosol which is important to workplace health is the respirable fraction. Its recognition, its economic and health importance and how it is now

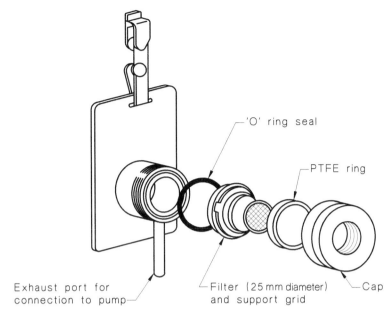

Figure 7.7a Exploded view of IOM inhalable-dust sampler

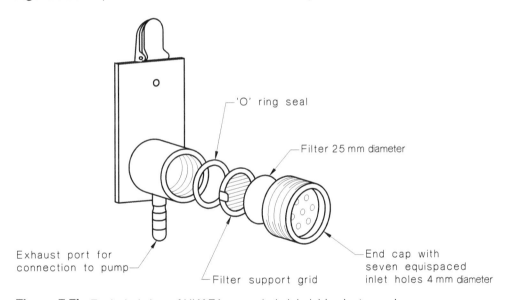

Figure 7.7b Exploded view of UKAEA seven-hole inhalable-dust sampler

measured form part of a classic industrial detective story centred on solving the mystery of the most prominent of all occupational diseases.

Historically, a number of dust diseases such as silicosis and coal workers' pneumoconiosis (also known as 'black lung') have been known to occur in miners, quarry workers, tunnellers, stonemasons and others in dusty trades. Exposure to both crystalline silica and coal dusts can cause pneumoconiosis, which results in a permanent alteration of the lung structure. Exposure to crystalline silica dust can lead to silicosis, a fibrosis of the lung.

Crystalline silica in the lung initiates a reaction involving macrophages (scavenging cells), which die, triggering an inflammatory reaction. This reaction leads to the development of fibrotic nodules, which grow to about 1 mm in diameter and can be identified by x-ray imaging of the lung. These small nodules grow and coalesce as the disease progresses, even after the exposure ceases.

Figure 7.8a shows the radiograph of the lungs of a miner with no dust disease; Figure 7.8b shows a case of advanced silicosis. The normal air space of the lung shows up as dark shadow.

Silicosis may progress from a relatively benign form in which there is little impairment to lung function to a severe form if exposure continues. Permanent disability or death may ensue. In some cases of very heavy exposure, the risk of lung cancer may be increased. Early diagnosis and cessation of exposure are therefore very important. The parts of the lung affected by fibrosis are rendered incapable of transferring oxygen into the bloodstream.

The reason that crystalline silica promotes fibrogenic disease of the lung is not entirely clear, although it may be linked to the surface structure of the crystalline particles. Silicon dioxide that is bound up in complex silicates (e.g. basalt, ilmenite) produces little or no lung fibrosis. When mined, if the crystal surface is contaminated by other minerals, it shows lower initial fibrogenicity than would be expected from the amount of quartz present. However, if the surface layers are dissolved away, the quartz demonstrates its usual toxicity.

Early researchers found these dust diseases generally slow to develop and more prevalent among miners who had worked for longer durations. Often—although not always—the dustier the mine or workplace, the greater the chance of developing dust disease. The severity and progression of disease also appeared to be related to the material inhaled. Gold miners working through quartz seemed particularly susceptible, as were stonemasons hewing and working granite. It also became clear that the disease was centred in the lung itself, not in the airways leading to the lung. This meant that not all the dust

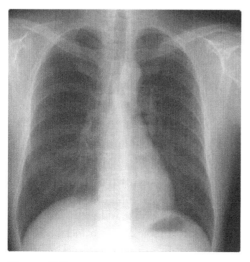

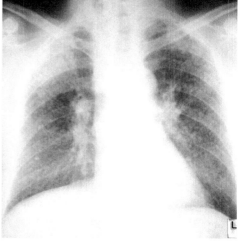

Figure 7.8a Radiograph showing healthy lungs

Figure 7.8b Radiograph showing advanced silicosis

a worker breathed in was implicated in causing the illness. The researchers also concluded that because dust disease often appeared many years after exposure ceased, some of the dust must have remained in the lung.

This evidence indicated the need for a technique to measure the concentration of dust particles within the size range reaching the critical parts of the lung: the alveolar, non-ciliated regions (i.e. those without the hair-like structures that help to transport secretions). These particles are very small (less than 5–10 µm).

Through various animal experiments, autopsy studies of the lungs of deceased miners and theoretical calculations, it was found that not all particles small enough to penetrate the tracheo-bronchial tree and deposit in the lung actually do so. In the tortuous journey to the alveolar region of the lung, only the very smallest particles stay in the airstream. Successively larger particles impinge to an increasing extent on the airway walls and are deposited there. Disease development was found to be related to the mass of minute particles reaching the alveolar region.

7.6.3 RESPIRABLE DUST SAMPLING

These considerations led to the adoption by the British Medical Research Council (BMRC) of a definition of 'respirable dust' as recommended at the Pneumoconiosis Conference held in Johannesburg in 1959. The respirable fraction, which applied to the pneumoconiosis-producing dusts—namely coal and some other minerals—was defined in terms of the free-falling speed (i.e. terminal velocity) of 'unit density particles'—that is, particles with the same density as water. The small particles (less than 1 µm in diameter) all penetrate to the alveolar region. At 5 µm, penetration is only 50 per cent, and particles larger in diameter than 7.1 µm do not penetrate to the alveolar region. This gradation is sometimes referred to as the 'Johannesburg curve'.

The BMRC recommended that two important rules be followed in measuring dust samples:

- The dust should be measured by mass (or surface area), and
- If any compositional analysis has to be undertaken on a dust cloud, it should be undertaken only on the respirable fraction.

From this research, conducted over many decades, two practical instruments for measuring respirable dust were developed: the horizontal plate elutriator and the cyclone elutriator. Both were **two-stage devices** consisting of an aerodynamic separator (first stage) followed by a collection filter (second stage). The **horizontal plate elutriator**—for example, the MRE 113-A Dust Sampler—was a primary reference device. It was quite heavy and was required to remain horizontal for correct sampling, so its use was restricted to sampling at fixed locations.

Today, the commonly employed practical device meeting the respirable dust size-selection criterion is the miniature cyclone elutriator, a small, portable device well suited to personal sampling.

Originally, the Higgins and Dewell miniature cyclone was developed by the British Cast Iron Research Association (BCIRA) and known as the BCIRA cyclone. It was soon followed by a lighter cyclone developed by the Safety in Mines Research Establishment at

Casella as a personal dust sampler. This modified version was known as the Safety in Mines Personal Equipment for Dust Sampling (SIMPEDS) cyclone. Other manufacturers—for example, SKC—also produced their versions of the Higgins and Dewell miniature cyclones. The Casella and SKC versions are now known as 'modified' Higgins and Dewell cyclones. The aluminium cyclone, commonly used in North America, was originally designed to meet the ACGIH® sampling curve, which is different from the BMRC curve.

Figure 7.9a shows the miniature cyclone sampler, with its filter cassette and sampling filter. Figure 7.9b displays both the aluminium cyclone and the SKC modified Higgins and Dewell cyclone sampling devices.

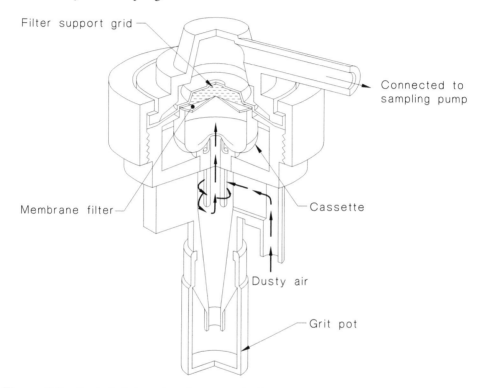

Filter support grid

Connected to sampling pump

Membrane filter

Cassette

Dusty air

Grit pot

Figure 7.9a Exploded view of cyclone sampler for respirable dust sampling

Figure 7.9b SKC modified Higgins and Dewell cyclone (left) and aluminium cyclone (right) samplers for respirable dust sampling

In 1995, the International Organization for Standardization's (ISO) technical report *ISO 7708 Air Quality: Particle Size Fraction Definitions for Health-related Sampling* (ISO, 1995) modified the BMRC definition of respirable dust so that, for the first time, the same definition was used worldwide. This definition was adopted in 2004 and continues in the 2009 version of AS 2985 *Workplace Atmospheres: Method for Sampling and Gravimetric Determination of Respirable Dust* (Standards Australia, 2009b). Respirable-dust sampling in Europe, Australia and North America can now be conducted on a common basis, even when using sampling equipment originally designed to perform to different definitions of respirable fraction. Examples of flow rates used for different cyclone samplers are provided in Table 7.1.

Table 7.1 Flow rates for size-selective cyclone samplers

Size-selective sampler	Flow rate (L/min)
Casella modified Higgins and Dewell cyclone	2.2
SKC modified Higgins and Dewell cyclone	3.0
Aluminium cyclone	2.5

If set at the correct flow rate, the samplers in Table 7.1 meet the requirements of the sampling efficiency curve for respirable dust as per the ISO 7708 respirable-dust convention (ISO 1995), as seen in Figure 7.10.

These flow rates satisfy the criteria for size-selective sampling whereby unit-density particles less than 2 μm in diameter are collected at greater than 97 per cent efficiency, 5 μm particles are collected at 34 per cent efficiency and particles larger than 18 μm are

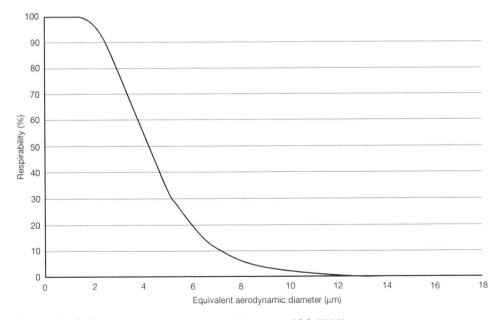

Figure 7.10 Respirable dust sampler performance (ISO 7708)

not collected at all. As the specific gravity (density) of particles increases, smaller particles become aerodynamically equivalent to larger 'unit-density' particles. For example, 3 µm quartz particles are equivalent to 5 µm unit-density particles.

The H&S practitioner involved in respirable-dust monitoring must ensure that sampling is conducted using appropriate equipment and with the correct flow rate in order to sample respirable dust that meets the ISO 7708 (1995) sampling efficiency curve. A detailed practical procedure for measuring respirable dust in the workplace is given in AS 2985 (Standards Australia, 2009b). Monitoring procedures for respirable dust permit very little latitude. Flow control is critical for correct adherence to size-selective criteria. The field sampling procedure specified in this standard can be conducted by H&S practitioners who have appropriate sampling equipment and training.

The laboratory analysis requires special equipment and techniques (e.g. micro-balance, static eliminating sources, various analytical devices). This usually makes it necessary to utilise specialist laboratory services accredited by NATA for the analysis.

On completion of sampling, the concentration (mg/m^3) of respirable dust is calculated by dividing the net weight gain (after taking the weight gain or loss of a blank filter into account) by the total volume of air sampled.

It is strongly recommended that the H&S practitioner starting out on dust measurements should first contact an experienced occupational hygienist or a laboratory experienced in this field.

7.6.4   MATERIALS REQUIRING RESPIRABLE DUST MEASUREMENT

Most dusts producing pneumoconiosis are assessed by monitoring respirable dust. Table 7.2, though not exhaustive, lists the most important of these. Alpha quartz and coal dust are the most prevalent in the workplace. The significance of coal dust arises due to the number of workers exposed.

Fibrous dusts (not to be confused with dusts that produce fibrosis of the lung) are examined in sections 7.8 and 7.9.

Table 7.2 Some pneumoconiosis-producing dusts

- Respirable crystalline silica:
  - alpha quartz
  - cristobalite
  - tridymite
- Tripoli
- Fused silica
- Fumed silica
- Microcrystalline silica
- Coal dust
- Graphite (natural)

7.6.5 SILICA DUSTS

Silicon is the most abundant element in the Earth's crust, and a number of geological silicates are toxic to humans. Silicon is ubiquitously distributed, occurs in most rocks and soils and is widely found in many workplaces, including mining, construction and manufacturing industries.

Silicon dioxide can be found in two main forms, crystalline and amorphous, each with the same chemical formula SiO_2. Both crystalline and amorphous forms can be found as a naturally occurring mineral. The crystalline form of (alpha) quartz is also known by the general term 'crystalline silica'. Alpha quartz in its original state is fibrogenic to the lungs, and may be transformed by heat to two of its other forms, cristobalite and tridymite, both of which are also fibrogenic to lung tissues.

The following examples are drawn from typical workplaces in which respirable dust exposures are important, and in which H&S practitioners may need to undertake respirable dust measurements or respirable crystalline silica (i.e. quartz) measurements.

7.6.5.1 Abrasive blasting

Abrasive blasting on large steel structural components, processing plants and exposed aggregate concrete products produces extremely high concentrations of respirable dust. Australian abrasive blasting regulations prohibit the use of crystalline silica for dry abrasive blasting, and all sands are excluded from use. Only a few abrasives—ilmenite, copper slag, aluminium oxide and some other specialty minerals—meet the stringent standards required.

7.6.5.2 Concrete and masonry work

Chasing is the process where concrete or masonry is cut to allow the laying of services such as electrical conduit into a wall, floor or ceiling. Masonry is the skill of building with brick and stone (i.e. natural and man-made) and involves processes such as cutting and finishing tasks. This includes working with sandstone and engineered stone (e.g. kitchen benchtops), and mechanical removal of grout from brickwork.

Traditional sculptors and craftspeople can be exposed to crystalline silica from working with sandstone and granite. The use of newer mechanical tools (chipping hammer and air bottle), which creates more dust from grinding and sanding than occurred with traditional hand methods, can lead to greatly increased risks unless properly controlled.

Without good dust suppression and/or control, very high concentrations of both respirable dust and respirable crystalline silica (up to 50 or more times the ES) may be generated (Alamango, Whitelaw and Apthorpe, 2015).

7.6.5.3 Pottery- and brick-making industries

In these traditional industries, the handling of clay dusts containing up to 20 per cent quartz has produced many cases of respiratory disease. A secondary hazard exists because some of the quartz can be converted to cristobalite in the furnace or kiln. Art studios and pottery schools also produce respirable dusts, but on a much smaller scale than occurs in industrial processes.

7.6.5.4 Foundry industry

Moulds in metal foundries traditionally are made from sand and various binders. Exposure to quartz is a hazard for moulders, as well as for the fettlers who clean down the poured castings. Shot blasting and working in sand reclaim (i.e. where sand is recycled from moulds already used) are also hazardous tasks.

7.6.5.5 Beverage production

A common material used in the filtering and clarification of beverages is diatomaceous earth (amorphous silica). In its calcined (roasted) form, this material may contain 40–50 per cent cristobalite, and dusts from this material pose a potential hazard unless well controlled.

7.6.5.6 Sugar cane farming

There is a rare occurrence of biogenic silica (i.e. silica formed by the action of biological organisms) in sugar cane, to which workers can be exposed during harvesting. Exposure has led to nasopharyngeal and broncho-pulmonary symptoms.

7.6.5.7 Tunnelling

With an increasing population, there is a need for large infrastructure projects for services, road and rail. Construction works include extensive earthworks and tunnelling. These activities are likely to expose workers to respirable crystalline silica (i.e. quartz). Historically, silicosis is the common disease of tunnellers; however, chronic exposure may also lead to chronic obstructive lung disease, emphysema and lung cancer.

7.6.5.8 Quarrying

Many quarries for building materials and road-base rock produce respirable dusts containing crystalline silica in their crushing plants and screening operations.

7.6.5.9 Coal mining

Coal is a combustible mineral, formed predominantly of carbon, although other lesser quantities of impurities are often found within it. During mining for coal, contaminants such as quartz (i.e. crystalline silica) can be present, which create further health risks.

Exposure to coal generally affects the lungs, and exposure to respirable coal dust (RCD) is the most significant.

The three most common health outcomes from exposure to RCD are:

- coal workers' pneumoconiosis (CWP) (also known as 'black lung')
- progressive massive fibrosis (PMF)
- chronic obstructive pulmonary disease (COPD)

Each of these affects the lungs in different ways; however, common themes include impaired lung function, emphysema and reduced life expectancy.

CWP is caused by the inhalation of coal dust, and possibly crystalline silica. In its simpler form, CWP will usually be asymptomatic, even though detectable on lung x-rays. It can progress following further exposure to show extensive but discrete lesions on an x-ray. Advanced simple CWP is significant because it may lead to the most severe form, PMF. In PMF, the lesions coalesce to show extensive large opacities on the x-ray. The lung in these areas becomes a hard, black mass, severely reducing breathing capacity, which leads to disability and likely premature death.

In addition to RCD, coal mines and coal-handling facilities may also provide exposure to dust in the following forms:

- inhalable fraction of coal dust
- respirable crystalline silica (RCS), and
- dusts not otherwise specified (NOS) (AIOH, 2018).

RCD is formed during coal mining by the action of coal-cutting machinery, from transfer and handling (e.g. conveyors, transfer points), stockpile building and reclamation, train loading and unloading.

Coal miners worldwide were plagued with dust diseases until the end of World War II. As many as 50 per cent of miners developed some form of disabling respiratory disease or died prematurely. The cost in human suffering and loss of productivity was staggering, and it came during a period when coal mining was a key industry in the West. Occupational hygiene research and the resulting control strategies brought about a dramatic and exemplary reversal in the industry's prosperity, augmented by the benefits flowing from a healthy workforce. This decline in incidence of CWP and PMF in coal miners results from the relentless application of better ventilation and other controls in response to more stringent respirable dust standards.

Unfortunately, coal mining was not as well controlled as it was thought to be, and although it was often used in examples of a well-controlled process where black lung and other lung diseases had been eliminated, this was not the case. A change was noticed in 2015–17, when more than 20 cases of CWP were diagnosed in Queensland, and one in New South Wales (AIOH, 2018). This was a disturbing revelation as the latency period is long, and the indications are that more cases may emerge. The controls were not maintained as well as they should have been in all cases, and the result is a resurgence in dust diseases.

Control technologies have been developed that can be, and have been, successfully implemented across parts of the coal mining industry. The hierarchy of controls must be applied when determining the appropriate controls to be utilised. The main exposure controls are:

- remote-controlled mining
- effective dust extraction systems on continuous miners
- enclosing coal transfer operations
- minimising drop distances at stockpiles
- providing the cleanest possible air to underground workers and using ventilation appropriately
- provision of enclosed ventilated cabins for workers in higher dust areas (if possible)
- maintaining water sprays on cutting and crushing equipment
- ensuring sharp picks on cutting head
- avoiding overloading shuttle cars to reduce spillage
- keeping coal and roadways clean from spilled material

- administrative controls, including limiting time in dusty areas
- 'positioning' of workers to avoid high exposure situations
- worker education
- implementation of a comprehensive respiratory protection program if dust exposure remains high (including being clean shaven).

It should be understood that there are multiple sources of dust and every workplace is different, so more than one control strategy will likely be required to reduce worker exposures to acceptable levels. Whatever strategy is adopted, it should be supported by an effective maintenance program on coal cutting and handling equipment and ventilation systems in particular, so that dust control effectiveness is sustained (AIOH, 2018).

Apart from underground mining, the H&S practitioner will also find coal dust associated with coal-fired furnaces, train loading stations and unloading ports, and in a few laboratory facilities that crush coal in preparation for analysis.

7.6.6   *LABORATORY ANALYSIS OF RESPIRABLE CRYSTALLINE SILICA*

Most particulate samples obtained by air sampling will contain a mixture of materials, and laboratory analysis may be required to determine their composition.

In practice, laboratory analysis for crystalline silica (i.e. quartz) is conducted on most respirable dust samples. This analysis should be conducted by specialist laboratories accredited by NATA using one of two National Health and Medical Research Council (NHMRC) methods adopted and published unchanged by NOHSC: infrared spectrometry and x-ray diffractometry (NHMRC, 1984).

Infrared or x-ray analysis can therefore be conducted on the sample filters; however, for infrared spectrometry, if interfering minerals are present—for example, kaolinite, amorphous silica or cristobalite—x-ray analysis will be necessary. Cristobalite can be present naturally or can exist where quartz has undergone heat treatment above 450°C.

Figure 7.11 shows the infrared spectrum, or 'fingerprint', of alpha quartz. The amount of quartz can be determined directly from the spectral intensity of the double peak at 798 and 779 cm^{-1} wavenumbers. The respirable concentration of alpha quartz in the air can then be calculated once the volume of air sampled is known.

7.7   DIESEL PARTICULATE

Diesel exhaust is an ever-changing and complex mixture of gases, adsorbed organics and particulate components. The complete mixture is of interest to the occupational hygienist with respect to exposure to workers; however, the particulate component is the focus of this section.

Diesel machinery is used widely in industry, due to its reliability, power and seemingly simple operation. As a result, large numbers of workers are exposed to the by-products of exhaust. There are several health effects, including the risk of increased lung cancer if exposed to diesel engine exhaust. Australia does not have a national exposure standard for diesel particulate matter, although some jurisdictions within Australia have adopted their own exposure standards.

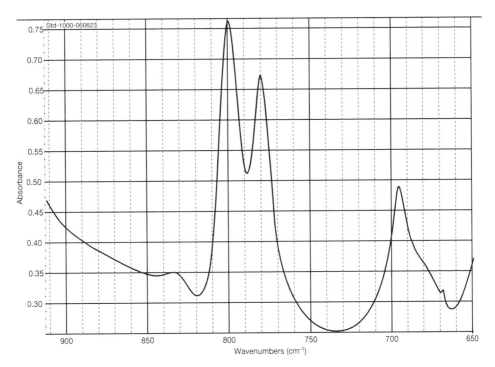

Figure 7.11 The infrared spectral doublet used for determination of respirable quartz by infrared spectrophotometry

7.7.1 COMPOSITION OF DIESEL PARTICULATE

Amman and Siegla (1981) summarised the early research into the composition of diesel particulate matter (DPM), defining it as 'consisting principally of combustion-generated carbonaceous soot with which some unburned hydrocarbons have become associated'. Using photomicrographs of the exhaust from a diesel passenger car, they demonstrated that diesel particulates were made up of a collection of spherical primary particles, termed 'spherules', which formed aggregates resembling in appearance a range of forms from a cluster of grapes to a chain of beads. Subsequent researchers have confirmed that the spherules vary in diameter from 10 to 80 nanometres (nm), with most in the 15–30 nm range.

High-resolution electron microscopy (Figure 7.12a) indicates that the basic spherule consisted of an irregular stacked graphitic structure—so-called elemental carbon (EC)—shown schematically in Figure 7.12b (Rogers and Whelan, 1996; World Health Organization, 1996).

The graphitic nature and high surface area of these very fine particles (typically <1 μm in diameter) means they can absorb significant quantities of hydrocarbons (the organic carbon fraction) originating from unburnt fuel, lubricating oils and compounds formed in complex chemical reactions during the combustion cycle. Traces of inorganic compounds have also been found in the particulates (e.g. sulphur, zinc, phosphorus, calcium, iron, silicon and chromium). These are believed to have arisen from the fuel, additives and lubricating oil used in diesel engines.

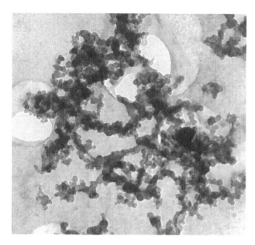

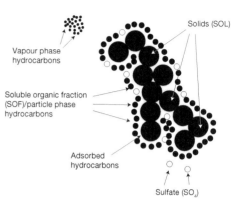

Figure 7.12a Electron micrograph—mine diesel particulate showing spherules, chains and agglomerates
Source: Rogers and Whelan (1996).

Figure 7.12b Schematic—mine diesel particulate showing spherules, chains and agglomerates

Improvements in fuel quality and emissions controls have led to a change in the chemical fingerprint of diesel exhaust. This change is defined as pre- and post-2007, as this is when the United States in particular started regulating an improvement in emissions. Post-2007, new engines emit much less quantities of carbon particles and less sulphur (due to improved fuel quality), and the consequence is there are more detectable numbers of ultrafine droplets of semi-volatile organic compounds. This is due to fewer carbon surfaces for these to adsorb onto.

7.7.2   HEALTH EFFECTS

The potential for adverse health effects arising from occupational exposure to diesel particulate has been the subject of intense scientific debate for many years. Part of this debate is attributed to the way diesel fuel and its subsequent exhaust composition have changed over the years. It is challenging to predict health effects of a substance that has changed so intensely over time. A detailed critical review of this research is provided in the AIOH Position Paper *Diesel Particulate Matter and Occupational Health Issues* (AIOH, 2017). This document should be consulted for detailed information on the health effects of diesel particulate. However, the following general statements about adverse health effects can be made:

- Diesel particulate has the potential to irritate the mucous membranes of the eyes and respiratory system, and cause neurophysiological symptoms such as headaches, light-headedness, nausea and vomiting.
- There is evidence of non-malignant respiratory disease, including increased levels of coughing and phlegm, and some evidence of altered pulmonary function.
- Animal and epidemiological studies link exposure to DPM from traditional diesel engines (pre-2007) with an increased risk of lung cancer, and to a lesser extent bladder cancer.

- The combined gaseous and particulate emissions, as 'diesel engine exhaust', have been classified as carcinogenic to humans (Group 1) by the International Agency for Research on Cancer (IARC) (IARC Working Group on the Evaluation of Carcinogenic Risks to Humans, 2012).
- The level of potency of DPM has been subject to intense scientific discussion without a definitive answer currently being available.

7.7.3   EXPOSURE STANDARDS

The development of workplace exposure standards for diesel particulate continues to be in a state of flux. This is a result of the paucity of data on dose–response, differing approaches to sample collection and analysis methodology, and differing attitudes of various industry segments, advisory groups and regulatory authorities.

Although there is no Australian workplace exposure standard set by Safe Work Australia that considers the dose–response, the mining industry within some Australian jurisdictions—New South Wales, Queensland and Western Australia—has adopted the AIOH guidance exposure value of 0.1 mg/m^3 (as submicron EC). This standard comes from research completed in the 1990s (Pratt et al., 1997; Rogers and Davies, 2001) and was proposed by the NSW Minerals Council (1999). It was determined that if EC is kept to 0.1 mg/m^3 or below, the level of irritation in the eye and upper respiratory tract is significantly reduced.

The Minerals Council at that time acknowledged that, although compliance with such a standard would offer substantial improvement in worker comfort, there was insufficient evidence to suggest that it would prevent the development of diseases such as cancer and that exposure levels to diesel particulate should be reduced to as low as reasonably practicable through effective control strategies.

The AIOH has recently reviewed its guideline and added an action level of 0.05 mg/m, designed to trigger investigation into the source of exposure and implementation of suitable control strategies. Overall, the AIOH supports the principle that diesel emissions should be controlled to levels that are as low as reasonably practicable (ALARP) (AIOH, 2017).

The United States, Canada and Europe have legislation in place to control DPM exposures in mining and tunnelling. From 2007 to 2008, the permissible exposure limit (PEL) was reduced from 0.31 mg/m^3 as EC to 0.12 mg/m^3 EC. This change was partly due to the risk of lung cancer.

Germany has a DPM exposure limit in workplace air of 0.1 mg/m^3, measured as EC diesel particulate, based on what can be achieved using technically available measures (TRGS 554, 2008).

In summary, the promulgation of a dose–response workplace exposure standard linked to sound epidemiological or dose–response evidence does not appear likely in the near future.

The effectiveness of such a standard in reducing the potential risk of cancer is unknown, owing to the uncertainties surrounding published epidemiological studies.

7.7.4   MONITORING METHODS

EC is used as a surrogate for DPM as it provides the best fingerprint of particulate in diesel emissions. It is relatively free of interferences and is chemically stable, unlike the adsorbed organic carbon fraction.

In the early 2000s, there was a shift from measuring the mass of DPM to elemental carbon concentration (by thermal analysis).

The internationally accepted monitoring and analytical method for DPM as sub-micron EC is NIOSH Method 5040 Diesel Particulate Matter as Elemental Carbon (NIOSH, 2016). Diesel particulate is collected using either a single use impactor cassette or a three-piece cassette with a specialised quartz filter. Where size selection is required, such as in situations that are inherently dusty (i.e. mines), or where contamination from other carbonaceous materials is likely (i.e. breakdown of conveyor material), an impactor and cyclone sampler should be used to remove the larger particles. Refer to Figure 7.13 for two types of DPM samplers. Analysis involves carbon speciation for EC using thermal-optical instrumentation techniques.

At this stage, there is no evidence that methods which rely on measuring the number of ultrafine droplets of semi-volatile organic compounds provide a suitable alternative or better method of defining exposures for health assessment purposes.

Real-time monitoring instruments currently on the market can be useful indicative instruments in helping to identify DPM sources, and to manage and reduce overall DPM concentrations. These do, however, need to be adequately calibrated against traceable primary standards such as for total carbon (TC) or EC (AIOH, 2017).

7.7.5   CONTROL TECHNOLOGIES

Experience has shown that no single simple solution exists to control diesel exhaust exposure and individual organisations need to explore which of the following control technologies best fit their circumstances. Often more than one control strategy may be required

Figure 7.13 DPM SKC modified size-selective sampler and sampling cassette (left) and three-piece cassette (right)

to reduce worker exposures to ALARP levels. It must be remembered that strategies that work at one location may not necessarily work at another due to differing ventilation, equipment and other workplace factors. Whatever strategy is adopted, it should be underpinned by an effective maintenance program so that emission reductions are sustained (AIOH, 2017).

Although great improvements are mandated by regulatory authorities in both the United States and Europe requiring engine manufacturers to produce cleaner engines, many older diesel engines remain in service and will be around for some time to come. The control of emissions from these engines presents unique challenges, and experience has shown that while there is no single simple method of controlling particulate levels, a range of options in one or more configurations can be effective. In some cases, this may be as simple as redirecting an exhaust away from personnel; in others, it may involve retrofitting the workplace with one or more sophisticated control technologies.

Control options for DPM include:

- low emission fuel, delivered to the engine in a clean state
- tailpipe ventilation in workshops where vehicles are required to idle constantly to maintain a state of readiness or where they are being maintained while operating
- underground mine ventilation taking into consideration numbers and size of vehicles
- vehicle restrictions—limiting the number of diesel engines operating in an area
- exhaust treatment devices such as wet scrubber systems, regenerative ceramic filters, disposable exhaust filters and exhaust dilution/dispersal systems. Exhaust filters can be permanently or temporarily fitted. Filters can be fitted to the exhaust to operate either while the vehicle is idling or when it is operating under load. Ensuring the effectiveness of the filter on its own, as well as within the filter system, is vital to the success of this strategy.
- new-generation low-emission engines, such as Euro V and Euro VI
- maintenance programs targeted at minimising exhaust emissions (Hines et al., 2017)
- well-sealed, filtered and maintained air-conditioned operator cabins
- driver and workforce education
- personal protective equipment (PPE). In all situations, other control technologies should always be explored in preference to personal protective equipment. Care should be exercised when selecting PPE to ensure a respirator is chosen which has been validated against challenge particles of comparable size and make-up to that of DPM (Burton et al., 2016).

Comprehensive diesel exhaust management plans assist the site to reduce:

- diesel engine emissions at the source
- diesel exhaust transmission throughout the mine and, importantly,
- worker exposures.

7.8   FIBROUS DUSTS: ASBESTOS

Of all the hazards, asbestos will probably have the highest profile and involve the most debate, emotion and worker concern. Asbestos is one of the most widely evaluated workplace health hazards, and a significant amount of literature has been published regarding this fibrous mineral silicate. 'Indestructible' asbestos has been used since antiquity in lamps, pottery and woven garments; however, it is only recently that its less fortunate legacy has been widely recognised. Asbestos was prized for its resistance to heat and other important qualities of strength and friction resistance. Sadly, these properties today are overshadowed by the grim toll of deaths caused by the inhalation of the fine fibres. In the past, commercial mining and use of asbestos in manufacturing have often occurred under conditions that could only be described as horrific.

7.8.1   PUBLIC CONCERN ABOUT ASBESTOS

Most of the recent concern has arisen since asbestos was shown to be a human carcinogen and singled out for banning and public paranoia. Assessment of the actual risks posed by asbestos has sometimes been affected by emotion.

7.8.1.1 Physical characteristics of asbestos

For the H&S practitioner, the term 'asbestos' is limited to the commercially used fibrous minerals from one serpentine rock and the amphibole series. The fibrous forms of these minerals have qualities of flexibility, good tensile strength, and some are able to be woven. They show good resistance to heat, they are non-conductive and the amphiboles especially show good acid resistance. Many other minerals occur in fibrous form (e.g. wollastonite, brucite and erionite); however, they are not asbestos. Fibrous minerals may also be present in some mineral deposits worked for their metals (e.g. nickel, iron ore).

7.8.2   SOURCES AND TYPES OF ASBESTOS

Many years ago, asbestos was mined in Australia at locations in New South Wales and Western Australia. The mineral continues to be mined and/or extensively used in many other countries because of its useful properties. Unlike Australia, these countries still have significant numbers of workers involved in the manufacture of asbestos-containing products.

The main types of asbestos commonly found in workplaces are:

- chrysotile, belonging to the serpentinite family (also commonly known as 'white' asbestos)
- amphiboles, including:
  - amosite, commonly known as 'brown' asbestos
  - crocidolite, or 'blue' asbestos.

Other members of the fibrous amphibole series include actinolite, tremolite and anthophyllite; however, these forms were not commonly used in commercial products.

It is useful to know about the different types of asbestos fibre in order to understand the various diseases they can cause and the risks associated with each fibre type in the workplace.

7.8.3   HEALTH EFFECTS OF ASBESTOS

Thousands of workplaces have many hazardous materials in them—the risk materialises only when workers are exposed to the hazard. As with crystalline silica and coal, exposure to asbestos causes diseases directly in the respiratory system. Where asbestos is present and a process generates airborne fibres, the risk arises if these fibres are inhaled.

Today, asbestos may be present in the workplace as part of the fabric of a building (e.g. insulant, construction materials etc.). Asbestos poses a risk solely when fibres become airborne. Occupational disease is not related to skin contact; however, ingestion is suspected to cause disease in a few sites (liver, prostate) in cases of heavy ingestion of fibres. The occupational diseases caused by asbestos are primarily respiratory diseases, related to fibres of respirable size.

Fibres of respirable size are those <3 μm in diameter, usually longer than 5 μm, and with length-to-width ratios of more than 3:1. These are known as 'respirable fibres'.

The three major occupational lung diseases of interest caused by asbestos are:

- asbestosis
- lung cancer
- mesothelioma.

Not all the types of asbestos listed above have strong associations with these diseases. Section 7.8.6 examines the risk factors and the likelihood of any of these diseases occurring from present occupational exposure to asbestos.

7.8.3.1   Asbestosis

Pulmonary and pleural asbestosis is found only in asbestos workers who have been exposed to high-fibre concentrations over a long period. It is the classic disease of asbestos miners, millers, weavers and those involved in processing fibre in large quantities (e.g. manufacturing brake linings or asbestos-cement products). Chrysotile and the amphiboles (amosite and crocidolite) have all caused asbestosis. In pulmonary asbestosis, the inhaled fibres penetrate to the alveolar region of the lung, where a fibrotic (scarring) reaction takes place. There are no well-defined nodules as seen in silicosis. Pleural asbestosis, also known as pleural plaques, presents in the form of calcification of the outer and generally top surface of the lung. It does not in itself progress to pulmonary asbestosis and is usually not debilitating.

Mild cases of pulmonary asbestosis are usually asymptomatic; however, they may progress, particularly with further exposure, to cause increasing breathlessness. Onset of asbestosis typically occurs after 15 to 40 years of substantial exposure to airborne asbestos fibres. As fibrosis progresses, the oxygen-exchange capacity of the lungs can decrease drastically, leading to associated heart failure.

7.8.3.2   Lung cancer

Historically, an increased incidence of lung cancer was observed among workers heavily exposed to any type of asbestos, and a greatly increased risk of lung cancer occurred when these workers were also heavy smokers.

Recent research has shown that where cigarette smokers with lung cancer have had only brief exposure to airborne asbestos, their cancer is completely or almost completely caused by smoking (Hodgson and Darnton, 2000). Only very small numbers of non-smoking asbestos workers not showing signs of asbestosis have died from lung cancer. Latency periods are around twenty years or more, and quitting smoking eventually reduces the risk.

7.8.3.3 Mesothelioma

Mesothelioma is a malignancy of the cells in the mesothelium, or lining, of the pleura surrounding the lung (pleural mesothelioma) or abdomen (peritoneal mesothelioma). Radiologically, a mesothelioma presents as a large mass of tumour protruding into the lungs. The disease is usually fatal, generally within one to three years after diagnosis. Mesothelioma is a rare disease, occurring most often among people exposed to asbestos but also—albeit very seldom—in some unexposed people. Its latency period is usually 30 to 40 years and it typically follows substantial exposures. In some cases, however, it is thought to have been caused by brief yet intense exposures over a few months or less. Adults are reported to have developed the disease in their twenties after being exposed as young children to dust carried home on a parent's clothing.

On the other hand, in Australia some two to four people per million per year develop mesothelioma with no known exposure to asbestos. It is believed that in these cases the disease may either not be related to asbestos at all, or be caused by 'environmental' concentrations of airborne asbestos from natural and artificial sources.

It is well established that exposure to crocidolite (blue asbestos) causes this malignancy, with some cases also attributable to amosite (brown asbestos) or to mixtures of these fibres with chrysotile (white asbestos). There is a weak association (some orders of magnitude lower than that of crocidolite) between exposure to chrysotile asbestos and mesothelioma. Crocidolite asbestos was mined and milled at Wittenoom, Western Australia until 1966, and was later found to have caused mesothelioma and other asbestos diseases in many of the workers.

The greatest risk of mesothelioma may be associated with the ability of an asbestos mineral or product to produce biodurable and long (>10 μm) fibres. The evidence for mesothelioma among asbestos workers being caused by exposure to chrysotile alone is less convincing. However, the extensive industrial use of chrysotile means that more cases of chrysotile-only related mesothelioma may appear during the next few decades.

7.8.4   OCCURRENCE OF ASBESTOS IN THE WORKPLACE

Any use of crocidolite or amosite in new applications in Australia was banned in the 1990s. It still exists in insulation in older buildings and equipment, and in some asbestos-cement (AC) materials manufactured before 1966. Industrial use of asbestos diminished rapidly throughout Australia during the 1980s. In the building materials industry, its use in the manufacture of AC sheeting and piping was phased out completely by 1984. It is still encountered, however, during maintenance, refurbishment and demolition.

Asbestos can still to be found in Australian workplaces in:

- insulation on boilers and pipes used for steam, where the asbestos can occur as raw fibre lagging, or in a cementitious form combined with magnesite (magnesium silicate) as a trowelled plaster on steam pipes, calorifiers, outer furnace skins, etc. Some ships have extensive asbestos insulation for fire and heat purposes.
- fire-retarding insulation on steel-framed supports in buildings, particularly high-rise buildings
- fire stoppings in buildings between floors, on tops of walls, in cable risers
- decorative finishes and acoustic attenuation on ceilings and walls in auditoria, public halls, schools, hospitals
- space insulation in buildings, particularly beneath metal-sheeted roofs and in the ceiling spaces of homes and other buildings. Asbestos has also been used to prevent condensation in the risers which carry air-conditioning ducts in large buildings.
- AC building products in flat or corrugated sheet form and pipes for water reticulation
- friction brake and clutch products
- some gaskets and valve packing
- millboard for air-conditioning heater banks
- asbestos fabric as fireproof rope, gloves, mats and hoses
- older-style vinyl flooring materials
- bituminous felt used on roofs and around oil and petrol pipelines
- asbestos-containing materials (mainly AC) in soil as the result of demolition or dumping
- mines and quarries, as a naturally occurring mineral.

While processing was restricted to chrysotile asbestos until December 2003, the use—that is, new use and reuse—of any type of asbestos is now prohibited throughout Australia.

Due to the durability of asbestos, it will still be present in walls, ceilings, floors, roofs, pipes and other parts of many buildings and structures for many decades to come. The ubiquitous use of AC in domestic and commercial premises causes significant problems when previously used land is contaminated and remediation is necessary for redevelopment.

7.8.5 ASBESTOS EXPOSURES IN THE WORKPLACE

The current Safe Work Australia exposure standard for all forms of asbestos is 0.1 fibres per millilitre of air (fibres/mL) in the breathing zone of an exposed person, averaged over a full work shift. Past industrial procedures involving handling of asbestos fibre on a large scale generated relatively large risks compared with those occurring today. Manufacture of AC building products often yielded fibre concentrations in the range of 1–10 fibres/mL or more in uncontrolled situations. Preparation of boiler reinsulation by hand mixing produced around 10–50 fibres/mL. The popular form of limpet or sprayed asbestos for fire insulation produced fibre concentrations of tens to hundreds of fibres/mL. Some individual operations involving cleaning of baghouse filters are believed to have produced fibre concentrations of hundreds to thousands of fibres/mL.

The largest use of asbestos in Australia was in the production of AC building products. Since the early 1980s, asbestos use has been superseded by an asbestos-free cellulose technology. Imported products labelled 'asbestos free' sometimes contain a small proportion of asbestos fibre (<1–5 per cent). Exposure can still occur in the friction material industry, particularly in the remanufacturing plants that strip and replace brake linings. The following processes that generate fibre in the workplace air may still be encountered, even though the presence of asbestos may be unknown to the worker:

- grinding, drilling, sanding and sawing (using power tools) of building materials containing asbestos (2–50 fibres/mL), including high-pressure water blasting (up to 1 fibre/mL)
- blowing down brake drums with compressed air during repair (believed to be up to 10 fibres/mL for less than one minute). It is important to note that the overall airborne asbestos concentration during brake repair of one car (approximately 1.5 hours) is significantly less because of the small amount of compressed-air use and is approximately a time-weighted average of 0.09 fibres/mL.
- asbestos-stripping operations (up to 100 fibres/mL).

The last category, the asbestos removal industry, can be a potential source of exposure to significant amounts of asbestos fibres. Asbestos-removal programs are discussed in sections 7.8.5.3 and 7.8.5.4.

7.8.5.1 Asbestos in soil

The major asbestos 'problem' of today and the future (Pickford et al. 2004) is *not* the removal of asbestos from buildings or structures; it is the treatment of land contaminated with asbestos-containing materials (ACM), of which the majority is AC. This material has come from illegal dumping, inappropriate demolition of AC structures in the past and inadequate remediation of contaminated land sites.

Strategies for remediation and validation of ACM-contaminated soils must incorporate risk-management approaches rather than ad hoc procedures.

Many public, environmental and government stakeholders have been unduly concerned that disease can arise from casual and brief contact with a small amount of non-friable AC in soil. Together with substantial and often ill-informed press coverage, the sight of workers in suits and respirators performing trivial tasks associated with minor amounts of AC in good condition has created a public perception that these tasks are dangerous to workers and communities.

Further, there may be an expectation that buried AC is required to be considered as friable asbestos material, and a friable licensed asbestos-removal contractor is needed for its removal. It should be noted the common definition of friable asbestos material is any material that contains asbestos and is in the form of a powder or can be crumbled, pulverised or reduced to powder by hand pressure when dry. The current Safe Work Australia publications on asbestos (Safe Work Australia, 2016a, 2016b) indicate that a risk-assessment approach should be made to determine control and remediation strategies. It is important that a risk assessment considers appropriate techniques whenever any disturbance of asbestos-contaminated soil is required.

There is a widespread perception that AC becomes friable if it is being processed by earth-moving equipment. While some pieces may fracture or become slightly abraded, essentially none becomes 'friable' by the broad definition of that term (Safe Work

Australia, 2013). Further, laboratory analysis of thousands of soil samples taken from AC-contaminated sites, often in the immediate vicinity of the AC itself, indicate no respirable asbestos fibres are released into the soil (Pickford et al., 2004).

The public image of asbestos as inherently dangerous has impelled a push for 'zero' tolerance in soils that is impossible to achieve, either scientifically or practically. Consequently, soil that is contaminated with AC is removed to an approved waste-management depot. Small amounts of AC contamination are often removed by hand, and the remaining soil is inspected and sometimes retested for the presence of asbestos before being certified satisfactory for the intended use. Other forms of remediation, including full-scale screening, are possible for a limited number of situations. All of these options are expensive, and the disposal of soil exhausts scarce landfill sites. It is therefore important for regulators to develop and adopt formal and practical guidelines for the management of asbestos-contaminated soil.

7.8.5.2 Naturally occurring asbestos

Naturally occurring asbestos (NOA) can be found in certain workplace locations where mining, construction and road building occur. There may be veins within the Earth and certain rock formations which, when disturbed during excavation or drilling, may release asbestos fibres into the air. It is important for workers that this potential risk be considered by H&S practitioners in locations where geological information indicates that NOA may be present. Where NOA is to be disturbed, specialised work practices and control strategies must be employed to protect workers and the community.

7.8.5.3 Asbestos removal

The asbestos-removal industry has developed over the last 30 years, with special isolation and sealing requirements to prevent the spread of asbestos fibres during a removal procedure. The merits of removing ACM that is in good condition before the end of its normal service life is not examined here. In most cases, no health-related need has been identified for such removal, since fibre monitoring invariably has revealed results of <0.01 fibres/mL. Sometimes employee demands, or a proactive approach, may drive removal works to occur. As a general rule, the planned removal of ACM (particularly friable ACM) is a good principle and can avoid accidental exposure in the future. Generally, removal is necessary or prudent before building alterations or refurbishment.

When an asbestos-lagged installation requires maintenance, or where an insulating product fails in service, asbestos removal is generally necessary. An example of a commonly failing product is sprayed-on decorative/acoustic surface finishes that lose adhesion as they age and can fall down in large slabs. Water penetration or continuous external damage may also damage these types of products. Fire-rated structural beams that have lost sections of insulation must also be targeted for remediation involving asbestos removal and reinstatement with an asbestos-free insulation. Where sprayed asbestos exists as insulation on the inside surfaces of roof spaces or air-conditioning systems, removal is a high priority.

When a building or structure is to be demolished or extensively refurbished, all asbestos present should be removed beforehand. For the H&S practitioner, the greatest priority is to ensure the removal is conducted safely and in accordance with local legislation. Both the asbestos removal worker and the bystander require suitable exposure protection. Poor work practices in the lucrative business of removing asbestos insulation from commercial, high-rise and domestic dwellings in the 1980s and 1990s resulted in poorly conducted removals with remnants of asbestos insulation left in place. There may be legacy issues in some buildings due to these previous practices.

7.8.5.4 The Safe Work Australia codes of practice for asbestos removal and management

The detailed guidelines for asbestos are set out in these Australian documents:

- *Code of Practice: How to Manage and Control Asbestos in the Workplace* (Safe Work Australia, 2016a). This applies to persons conducting a business or undertaking, and specifies:
  - managing risks, including formulating an asbestos management plan
  - the need for proper identification of asbestos hazards
  - requirements for an asbestos register
  - assessment of risk
  - roles and responsibilities
  - how to control hazards
  - how to choose a removalist
  - proper demolition and disposal
  - safe practices for ACM.
- *Code of Practice: How to Safely Remove Asbestos* (Safe Work Australia, 2016b) applies to those involved in the task of safely removing asbestos. Important topics covered include:
  - construction of isolation barriers
  - respiratory protective equipment for asbestos removal work
  - purpose and use of decontamination units (construction and operation)
  - negative air pressure units
  - inspection techniques for air leaks, smoke testing
  - the purpose and value of air monitoring
  - different kinds of ACM removal procedures
  - clean-up procedures and vacuum-cleaner types
  - correct asbestos disposal
  - sealing systems for remaining fibres
  - value of monitoring airborne asbestos fibre during final inspections.

  Of particular interest in the removal document is the section on choosing respiratory protection. It includes estimates of the expected asbestos fibre concentrations in air for different activities and the appropriate type of respirator to be worn using guidance from Australian/New Zealand Standard AS/NZS 1715: 2009 Selection, Use and Maintenance of Respiratory Protective Equipment (Standards Australia, 2009a). Some activities (e.g. dry stripping) involve fibre concentrations of several hundred fibres/mL or greater, justifying the need to check the code for respiratory protection requirements.
- *Guidance Note on the Membrane Filter Method for Estimating Airborne Asbestos Fibres* (NOHSC, 2005). This document deals with the sampling and analysis of airborne asbestos fibres, discussed in section 7.8.7.

7.8.6 *THE RISKS OF ASBESTOS EXPOSURES*

7.8.6.1 Asbestos dose

All three asbestos-related diseases discussed in section 7.8.3 are dose related. The larger the inhaled dose of asbestos fibre, the greater the risk of developing disease. Dose (otherwise known as cumulative exposure) is a function of the amount of asbestos fibres in the air and the duration of exposure. This is usually expressed in terms of fibre/mL years—that is, the *time-weighted average* airborne-fibre concentration inhaled by the worker in fibres/mL of air multiplied by the length of time the worker is exposed. In this instance, *time-weighted average* airborne-fibre concentration must be estimated for the entire period of exposure, not just over a single day. Periods of non-exposure and variable exposure must therefore be taken into account:

Dose = average airborne concentration of inhaled fibre × years of work   (Equation 7.1)

Using historical industrial exposures and response data in attempting to extrapolate today's risk (with its low exposures) presents considerable problems. It cannot be assumed that any of the asbestos diseases follow a linear dose–response model (i.e. there may be a threshold level below which asbestos exposure has no effect). Research on the fibre contents of the lungs of people without asbestos disease has quashed the fallacy of the 'one-fibre' theory (i.e. that a single asbestos fibre in a person's lungs can cause an asbestos-related disease). It has been shown that non-exposed people who die in old age from non-asbestos related causes can have significant quantities of asbestos fibres present in their lungs, presumably from environmental exposure (Berry, 2002; Rogers et al., 1994).

The risk of developing asbestosis in any Australian manufacturing industry is remote, as previous poor industrial conditions have been eradicated and asbestos is now a prohibited substance. While the Safe Work Australia exposure standard for asbestos in air is set at a level that will preclude any occurrence of asbestosis, it is intended primarily to prevent the rarer lung cancer and mesothelioma, which can occur after considerably smaller exposures.

Regarding the ES, it is important to know that the current ES of 0.1 fibres/mL is a very low level compared with fibre concentrations before awareness of the asbestos health issues, where an asbestos worker could be exposed to tens or hundreds of fibres/mL each working day. In addition, as for any ES, the recommended level is not intended to separate safe and unsafe conditions. These standards have been arrived at after extrapolating from historical occupational hygiene surveys and epidemiological data. Further, the improved sensitivity of today's measurement techniques (at least ten times more sensitive than 50 years ago) means that conditions of the 1950s and earlier may have been much worse than the few measurements available from that time suggest. In other words, workers may actually have had exposures of hundreds to thousands of fibres/mL in terms of modern measurement methods.

The risk of lung cancer associated with today's levels of airborne asbestos fibres also appears to be negligible, since the high levels of exposure that led to lung cancer no longer exist. Epidemiologists have been able—albeit with difficulty—to distinguish lung cancer attributable solely to asbestos exposure from lung cancer that might actually have been caused by cigarette smoking (Hodgson and Darnton, 2000). Since it is clear that smoking far exceeds asbestos as a cause of lung cancer, H&S practitioners will be far more effective in the task of promoting health in the workplace if they focus on altering lifestyle habits (e.g. promoting quit-smoking programs) in conjunction with controlling asbestos exposure.

Today, the risk of developing mesothelioma from the current low levels of exposure is also unlikely. The ES has been set to cater for the two fibre types strongly implicated in mesothelioma—crocidolite and amosite—and also for chrysotile. Previous occupational exposure data related to industry workers who had high fibre exposures. The exposure data for those who developed mesothelioma outside the industry or in situations not directly associated with asbestos work have not been recorded. The methods of extrapolating backwards linearly from high to very low doses are unreliable.

7.8.6.2 Practical assessment of the risk

The risk posed by asbestos arises from inhalation of respirable asbestos fibres. A visual assessment of the workplace is therefore the first important step. If a workplace looks dusty, no control procedures are in place and no respiratory protection is being used, workers could be exposed excessively to airborne fibres. Accurate fibre identification and fibre counting are the minimum requirements for assessing the risk where ACM are being handled or disturbed.

The following steps indicate the basic procedures for risk assessments for asbestos. H&S practitioners will not be able to conduct the identification or counting processes without extensive training; however, they must be able to assess the information about the types of asbestos fibre present, airborne asbestos fibre concentrations and risk factors.

1 *Ascertain that asbestos is present.* Asbestos is an established constituent of many products, such as brake linings and AC building products. For insulation materials in particular, asbestos is only one of many materials used; others include:
   - synthetic mineral fibre—fibrous glass, rock wool, refractory ceramic fibre
   - vermiculite
   - shredded cellulose (using newsprint).
   Optical microscopy is most commonly and appropriately used to distinguish these other materials from asbestos fibres. Although many are obviously not asbestos fibres (e.g. they are bright pink or yellow), others are difficult to detect with the naked eye. If reliable information is not available, collecting and analysing a sample of the material may be necessary to confirm or eliminate the presence of asbestos.

2 *Collect an appropriate sample.* A representative sample is needed for laboratory analysis. There may be more than one kind of fibre in the sample, and it may include different kinds of asbestos. Sampling for ACM is a destructive process, and it is important to minimise disturbance during sampling and ensure the area is suitably cleaned or sealed after the sample is removed. While collecting samples for analysis, it is important to protect the sampler by using precautions such as dampening the material and using an appropriate respirator (Safe Work Australia, 2016a).

   Enough material should be collected to include all types of material present and ensure that the sample is representative of the material. For example, in sprayed applications or furnaces, different types of material may have been applied at different times, therefore, the sample to be collected should extend from the surface through all layers to the bottom. Usually a sample of 10–50 g will be sufficient and should be packed carefully in a labelled and sealed container so it will maintain its form and neither disintegrate nor cause contamination during transport or in the testing laboratory.

To differentiate old AC building products from the newer asbestos-free materials, a piece approximately 5 centimetres square should be submitted. Vinyl-asbestos floor tiles are difficult to analyse, so a minimum 10 cm square is needed, while materials such as gaskets and friction blocks can be submitted whole.

3　*Have the asbestos fibre type identified positively.* Crocidolite and amosite are more hazardous and generate higher airborne asbestos concentrations than chrysotile, so it is important to know which type of fibre is present, especially for risk-assessment purposes. Analysis should only be carried out by a specialist laboratory accredited in this field by NATA, and to AS 4964 (Standards Australia, 2004). Capital costs of the equipment used in identification are significant, and a high degree of skill is essential.

Analysis requires the observation of a number of optical properties using complex diagnostic criteria to distinguish and identify different kinds of fibres.

The analytical techniques used include low- and high-power stereomicrosopy, polarised light microscopy (PLM) and dispersion staining microscopy. Sometimes, for confirmatory purposes, other additional techniques can be used, such as:
- infrared spectroscopy
- x-ray diffractometry
- electron microscopy—scanning (SEM) or transmission (TEM) electron micro-scopy incorporating X-ray analysis.

Examples of different types of fibres most commonly submitted for identification are shown in Figure 7.14. Note particularly the wavy shape of chrysotile fibres compared with straight amosite and crocidolite fibres. Synthetic mineral fibres are commonly very large in diameter compared with asbestos fibres, or show long filaments of uniform diameter.

PLM and dispersion staining microscopy are also able to discriminate between different asbestos fibres in mixtures, aggregates and soils, and in general have a detection limit around 0.1 g asbestos/kg of sample, depending upon the type and condition of asbestos and the complexity of the matrix. When a mixture of chrysotile and other asbestos fibres is detected, the risk associated with handling the material is usually assessed on the basis of the more hazardous type of fibre. X-ray and infrared spectro-photometric techniques are not able to differentiate between the non-fibrous form of asbestos minerals and the fibrous form. However, when combined with a technique to ensure that the fibrous form is present, x-ray and infrared spectrophotometric techniques with spectral subtraction facilities are able to differentiate between types of asbestos in mixtures, although they are not suitable for identifying low and 'trace' levels of fibres in a mixture. SEM or TEM, although specialised and expensive and requiring skilled interpretation, can provide useful information about the different types of fibres when unequivocal identification by the traditional methods is not possible.

4　*Does the source material contain much or only a little asbestos?* Experienced use of a stereomicroscope, PLM and dispersion staining methods can sometimes give very broad estimates of the amount of asbestos present in the sample. For instance, for

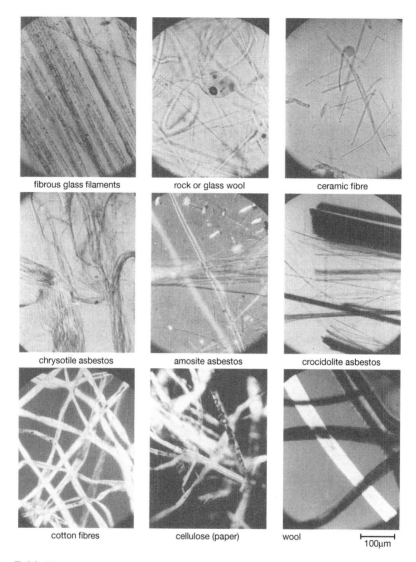

fibrous glass filaments   rock or glass wool   ceramic fibre

chrysotile asbestos   amosite asbestos   crocidolite asbestos

cotton fibres   cellulose (paper)   wool

100μm

Figure 7.14 Photomicrographs of different types of fibre commonly found in workplaces

control and removal purposes, it may be useful to know that a sample contains, say, 80–100 per cent of amosite asbestos by weight, or perhaps less than 10 per cent.

5   *What is the type of asbestos-containing material?* Knowledge of how the fibres are contained or the form in which they are present in a material can assist in assessing risk. Is the material friable or non-friable (Safe Work Australia, 2016a)?

6   *What is the condition of the asbestos?* Exposure risk depends on the amounts of fibre released into the workplace. Ascertaining the state or condition of the asbestos is therefore crucial to its control. Some ACMs release almost no fibres to the workplace (e.g. AC in situ, sealed gaskets, vinyl-asbestos floor tiles, encapsulated fireproofing, and resin-bonded friction materials).

Other ACMs can release considerable amounts of fibre, particularly loose forms of insulation. If the ACM appears to be in poor condition, plans should be made for its removal because it is failing to do the job it was intended to do and is also more liable to release fibres as it ages because of poor adhesion.

7   *What is the procedure for handling the material in the workplace?* Using processes that minimise airborne fibres will reduce exposure risk. The use of wet dust-suppressing methods is important, especially for removal processes.

8   *Are there any control procedures in place?* Control procedures are crucial to the safe handling of asbestos. Workplace methods that prevent dust from being generated are the primary means of control. Suitable respiratory protection is also added when dust-control procedures cannot control the release of fibres. Refer also to section 7.8.5.4 on asbestos removal.

7.8.7   AIR SAMPLING FOR ASBESTOS

Air sampling for asbestos in the occupational environment in Australia is carried out according to the *Guidance Note on the Membrane Filter Method for Estimating Airborne Asbestos Fibres* (NOHSC, 2005). This filter method employs light microscopy, even though electron microscopy is sometimes used for special environmental investigations. It is important that the H&S practitioner understands the results the membrane filter method produces as well as the method's limitations. Laboratories accredited by NATA or by state regulatory authorities should always be sought to conduct this work. Asbestos-removal regulations in most jurisdictions require monitoring to be conducted, and it must conform to the requirements of the *Guidance Note* (NOHSC, 2005).

Airborne dust is collected on a membrane filter with a pore size of 0.8 μm, usually housed in a three-piece conductive cowl as shown in Figure 7.15.

After sampling, the filter is rendered transparent and mounted on a microscope slide together with a cover slip. A phase-contrast optical microscope is used to count the number of fibres, geometrically defined as those that are at least 5 μm in length, less than 3 μm in diameter, and with a length-to-diameter aspect ratio of more than 3:1. One disadvantage of the optical counting method is that it cannot distinguish between true asbestos fibres and other fibres, such as fine cellulose or ceramic fibres, which leads to conservative estimates (i.e. overly high concentrations) for environments containing mainly non-asbestos fibres. False positives will almost certainly be obtained even if the air sample derives from an asbestos-free environment, because of the presence of fibres from plant matter, carpets, fabric and clothes.

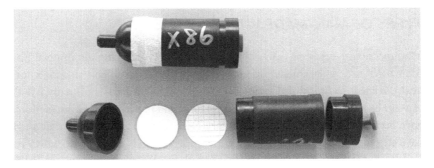

Figure 7.15 Asbestos and SMF sampling cowl, assembled and disassembled

7.8.7.1 Exposure monitoring

The membrane filter method was developed initially for measuring fibre levels in workplaces using asbestos. In any such work environment there was usually a mixture of dusts, of which the largest component was asbestos fibres.

In Australia today, exposure monitoring is unlikely to be needed because of the ban on asbestos use. If it is required, then advice from an occupational hygienist or specialist laboratory with significant experience in the methodology should be sought.

7.8.7.2 Control monitoring

Control monitoring employs the membrane filter method, and can be used in three situations:

- around the outside of an isolation enclosure to ensure maintenance of seal
- to check on continued optimum performance of the filters on a negative-pressure air fan attached to an isolating enclosure
- to provide final clearance monitoring after asbestos stripping is complete.

A level of 0.01 fibres/mL has been recommended by Safe Work Australia for control monitoring outside asbestos-removal work areas. Should the results from control monitoring exceed 0.01 fibres/mL, then various actions are required (e.g. investigate cause, implement additional controls, cease work until situation resolved) to ensure the integrity of the removal process (Safe Work Australia, 2016b).

7.8.7.3 Environmental sampling

For special investigations for situations with very low fibre concentrations, electron microscopy (SEM or TEM) can be used. These methods have the added advantage that fibres can be positively identified by energy dispersive x-ray analysis. They also permit the differentiation of other fibres from asbestos, and detection of fibres that are too small to be seen by standard optical phase-contrast microscopy. For these techniques, special sampling and preparation techniques are needed.

SEM and TEM are not used routinely in Australia for exposure or control monitoring because of the high cost of both equipment and analysis. There are no formal standards for environmental asbestos in air.

7.9   FIBROUS DUSTS: MAN-MADE VITREOUS FIBRES

7.9.1   TYPES OF MAN-MADE VITREOUS FIBRES

Man-made vitreous fibre (MMVF), formerly known as synthetic mineral fibre (SMF), has been used as a replacement for most asbestos-based insulation materials for a mixture of health-related, technical and economic reasons. The properties of many MMVF mean they can also be used in applications where asbestos could not (e.g. glass-fibre reinforcing). MMVFs have been around for more than 80 years. Some initial concern that they might turn out to have effects similar to those of asbestos has proved largely unfounded. MMVFs present some problems in the workplace because they have the ability to cause contact irritation, mainly due to the coarse fibres. The major commercial types of MMVF are listed in Table 7.3.

Table 7.3 Major commercial types of man-made vitreous fibres

Type	Raw material	End-product examples
Filaments	Borosilicate glass	Continuous-filament reinforcing fibres, woven cloth, electrical insulators
Wools	Basalt + fluxes Borosilicate glass	Insulation and acoustic batts, tiles and preformed sections
Ceramic fibre	Alumina + silica	Blanket, boards, modules, plasters, textiles, etc.

Most MMVFs have large fibre diameters compared with asbestos fibres. Continuous-filament glass fibres are used in textiles, and to reinforce plastics and concretes in typical applications such as swimming pools, boats, surfboards and plumbing materials. Typical diameters range from 5 to 30 µm, depending upon the product, with very few or no respirable fibres present. There is generally a narrow range of fibre diameters in any single product.

Glass fibre or glass wool mainly takes the form of insulation mats or blankets, with a significant percentage of fibres of respirable size (less than 3 µm diameter), even though many of the fibres are in the range of 5 to 15 µm diameter. Rock wools (or slag wools) contain fibres in a range of sizes similar to that of glass wool; however, they have a larger percentage of respirable-sized fibres.

Refractory ceramic fibres are aluminosilicates, and are found mainly in the form of high-temperature insulation blankets. Common trade names are Kaowool and Fiberfrax. They range in diameter from sub-micrometre to around 6 µm, with a large proportion of respirable fibres. Refractory ceramic fibres are required for applications involving temperatures >900°C.

Airborne fibre concentrations for MMVFs are usually low unless hygiene practice is poor. Personal dust and respirable fibre exposures for workers installing building insulation (e.g. mineral wool or fibreglass batts) in attic spaces and walls are generally low. However, exposures can be much higher during installation of loose-fill materials. Removal of ceramic fibres can result in significantly higher exposures.

7.9.2   *HEALTH EFFECTS OF MAN-MADE VITREOUS FIBRES*

The major acute health effect from exposure to MMVFs is irritation of the skin, eyes and upper respiratory tract. Garments with close-fitting collars and cuffs are useful for reducing skin contact and hence skin irritation. Exposure to high airborne dust concentrations can be prevented by using P1 particulate respiratory protection.

With regard to lung diseases, factors such as fibre dose (cumulative exposure), dimension (diameter and length) and durability (biopersistence) are important. Other factors to consider include chemical composition of the fibre type (filament, wool or ceramic fibre) and whether the MMVF is new or old. Older-style MMVFs have health effects different from those of newer MMVFs, which are specifically designed to have low biopersistence in the lungs (AIOH, 2016).

MMVFs do not remain in the lungs very long because they dissolve there. The new fibre types generally dissolve in the lungs even more quickly (i.e. biosoluble fibres). While this development is an attempt to further distance MMVFs from asbestos fibres in terms of physical properties and health effects, the clearance rates of any type of MMVF are very rapid compared with those of asbestos.

The current workplace ES for all forms of MMVF is 0.5 fibres/mL; in addition to this, MMVFs can be dusty, so a dual ES of 2 mg/m³ is applicable (Safe Work Australia, 2013).

If a workplace is involved in activities that use MMVF, refer to the NOHSC *Code of Practice for the Safe Use of Synthetic Mineral Fibres* (NOHSC, 1990). If sampling and analysis of respirable airborne MMVF is required, the NOHSC's *Technical Report and Guidance Note* (1989) should be consulted. This publication provides guidance on the membrane filter method, which has many similarities to the asbestos-monitoring method (NOHSC, 2005).

In 2002, the IARC Working Group on the Evaluation of Carcinogenic Risks to Humans (2002) reviewed available epidemiological data on MMVFs and determined that:

- special-purpose glass fibres such as E-glass and '475' glass fibres are possibly carcinogenic to humans (Group 2B)
- refractory ceramic fibres are possibly carcinogenic to humans (Group 2B)
- insulation glass wool, continuous filament, rock (stone) wool and slag wool are not classifiable regarding their carcinogenicity to humans (Group 3).

Further information regarding MMVFs can be found in the AIOH position paper on synthetic mineral fibres (AIOH, 2016).

7.10   NANOTECHNOLOGY

Nanotechnology involves the precision-engineering of materials at the nanoscale (10^{-9} to 10^{-7} metres), at which point unique and enhanced properties can be utilised. These properties have led to the development of new products, procedures and processes. Many valuable uses of enabling nanotechnologies have already been identified. Nanotechnology in Australia includes work in the areas of materials, nanobiotechnology and medical devices, energy and environment, electronics and photonics, quantum technology and instrumentation, and software (Department of Industry, Innovation, Science, Research and Tertiary Education, 2011).

Despite these advancements, the properties associated with engineered and manufactured nanomaterials arising from, for example, high surface area per unit mass may give rise to health and safety concerns, and nanomaterials generally are more toxic than the corresponding macrosize substance (Toxikos and Safe Work Australia, 2013). There is considerable knowledge about the work health impacts of fine and ultrafine particulate air pollution that can be applied when considering the potential health effects of manufactured or engineered nanomaterials. While there is potential for both ingestion of and dermal exposure to nanomaterials, the main concern in the workplace is potential inhalation exposure.

7.10.1   KEY DEFINITIONS

The International Organization for Standardization (ISO) has published a number of definitions relevant to nanotechnologies in ISO/TS 80004-1:2015, including:

- Nanoscale—size range from approximately 1 nm to 100 nm.
- Nanomaterial—material with any external dimension in the nanoscale or having internal structure or surface structure in the nanoscale.
- Nanotechnology—application of scientific knowledge to manipulate and control matter in the nanoscale in order to make use of size and structure-dependent properties and phenomena, as distinct from those associated with individual atoms or molecules or with bulk materials. (ISO, 2015)

7.10.2   NANOTECHNOLOGY AND THE WHS REGULATORY FRAMEWORK

The Australian model work health and safety (WHS) laws include both general care duties and more specific obligations for managing the risks of hazardous chemicals. These provisions apply to nanotechnologies and nanomaterials; however, there are nanomaterial-specific issues that impact the application of these provisions, notably in the areas of:

- uncertainty about the hazardous properties of engineered nanomaterials, and
- capability in measuring the emissions and exposures of nanomaterials in workplaces, including personal exposure assessment.

In Australia and globally, significant work is being undertaken to address issues in this area and further information can be found on the Safe Work Australia website. Where there is limited understanding of hazards, Safe Work Australia advocates taking a precautionary approach to prevent or minimise workplace exposures.

7.10.3   SAFETY DATA SHEETS AND LABELLING

To comply with hazardous chemical regulations, manufacturers and importers must classify these chemicals correctly and provide appropriate safety data sheets (SDSs) and workplace labels based on that classification.

Information is provided regarding nanomaterials in the model codes of practice for SDS and workplace labelling (Safe Work Australia, 2010 and 2012). The codes note that SDS and labels should be provided for engineered nanomaterials unless there is evidence that they are not hazardous.

Extra parameters have been added to Section 9 ('Physical and Chemical Properties') of the model code of practice for SDS that are particularly relevant for nanomaterials, and also relevant for chemicals more generally. The additional parameters are particle size (average and range), size distribution, aggregation and/or agglomeration state, shape and aspect ratio, crystallinity, specific surface area, dispersibility and dustiness.

The ISO Technical Report on the preparation of SDS for manufactured nanomaterials (ISO/TR 13329) contains advice that can help manufacturers and importers provide accurate and relevant information in SDS for nanomaterials.

7.10.4 WORKPLACE EXPOSURE STANDARDS AND LIMITS

There are currently only a limited number of exposure standards for nanoscale materials. Exposure limits vary according to factors such as type of nanomaterial (composition), the size of particle and crystallinity, all of which can influence toxicity.

In order to increase the guidance available for decision-making on control measures, other values such as NIOSH's recommended exposure limits (RELs) may be used as guidance (QUT, 2012). As an example, the NIOSH proposed REL for carbon nanotubes (CNT) is based on a measurement detection limit. NIOSH recommends that efforts should be made to reduce airborne concentrations of CNT as low as possible below the REL (NIOSH, 2013).

7.10.5 HAZARDOUS PROPERTIES OF NANOMATERIALS

7.10.5.1 Health hazards

There has been a significant amount of research conducted on nanomaterial toxicity and potential health hazards. Consistent with the research findings, there are many factors that impact on toxicity, which can lead to a range of hazard severities. Generally, nanomaterials are more toxic than larger particles.

As an example, CNT can exist in both fibre-like and non-fibre-like structures, and both are potentially hazardous. The Australian National Industrial Chemicals Notification and Assessment Scheme (NICNAS) recommends that these are classified as hazardous. Based on a number of studies (CSIRO, 2011; Poland et al., 2008), there is potential for CNT to cause mesothelioma.

As this area of science is constantly changing, it is important for the H&S practitioner to remain up to date with the latest health-related information on nanoparticles from organisations such as Safe Work Australia.

7.10.5.2 Physicochemical hazards

The physicochemical properties of nanoscale materials, such as high surface area per unit mass, make them widely useful for many applications—for example, as catalysts. There are also a number of potential safety hazards arising directly from the physicochemical properties, such as risk of fire or explosion, or unexpected catalytic properties. Published data related to physicochemical hazards of engineered nanomaterials are associated primarily with combustibility and explosivity of dusts (Toxikos, 2013), and data indicate the following:

- Similar to dusts, there remains the risk of explosion if airborne nanomaterial concentrations are sufficiently high and the dusts can be ignited.
- Regarding the minimum concentration in air (minimum emission concentration) required for an explosion, the mass concentration of nanomaterials in a dust cloud

needed for an explosion is orders of magnitude higher than measured nanomaterial airborne concentrations arising from fugitive emissions in a well-managed workplace.

- The minimum ignition energy varies considerably with material type. Nanoscale metal powders are sensitive to ignition (low minimum ignition energy), and carbon nano-materials are not (high minimum ignition energy).

7.10.6 *MEASURING NANOMATERIALS*

A number of challenges are associated with the measurement of nanomaterials in air. First, there are many different types of varying shapes and sizes. Second, nanoparticles have a high tendency to agglomerate, aggregate or to stick to other particles and surfaces (Seipenbusch, Binder and Kasper, 2008). Finally, there are significant amounts of background nano-particles in air from natural or incidental sources—for example, combustion products such as diesel exhaust emissions. This background level can vary significantly, which can make it difficult to detect and quantify emissions of nanoparticles from processes.

In relation to hazards, a number of parameters provide relevant information, includ-ing mass concentration, number concentration, size distribution, shape and chemistry, and surface area. There are a number of different instruments that potentially can be used to measure these parameters; however, there are many issues to consider. In practice, many of these instruments are only used for detailed research work—examples include the scanning mobility particle sizer, electrical low pressure impactor and fast mobility particle sizer.

Work has been undertaken with a focus on practical emissions and exposure meas-urement in the workplace. In 2009, the Organization for Economic Cooperation and Development (OECD) Working Party on Manufactured Nanomaterials (WPMN) published *Emission Assessment for the Identification of Sources and Release of Airborne Manufactured Nanomaterials in the Workplace: Compilation of Existing Guidance* (OECD WPMN, 2009). This is based on the US NIOSH document *Nanomaterials Emissions Assessment Technique (NEAT)* (Methner et al., 2010). The approaches to measurement recommended in these documents were validated in research undertaken by Queensland University of Technology and Workplace Health and Safety Queensland, investigating the operations of six nanotechnology processes with a number of different engineered nanomaterials. The research confirms that a three-tiered approach is effective in assessing worker exposure to emissions (QUT, 2012):

- *Tier 1.* The tier 1 assessment involves a standard occupational hygiene survey of the process area, plus measurement of aerosols, to identify likely points of particle emission.
- *Tier 2.* Tier 2 assessment involves measuring particle number and mass concentration to evaluate emission sources, worker breathing zone exposures and the effectiveness of workplace controls. A combination of instruments such as a portable condensation particle counter, optical particle counter and photometer can be used effectively.
- *Tier 3.* If further information is required, a tier 3 assessment can be undertaken. This involves repeating tier-2 measurements together with simultaneous collection of particles for offline analysis of particle size, shape and structure, mass and fibre

concentration and chemical composition. Offline particle analysis can be compared with real-time measurement results, and with exposure standards or other limit values.

It may not be necessary to undertake all three tiers of assessment. The findings of tier 1 and/or tier 2 may be sufficient to identify that controls are effective, or that work needs to be done to improve controls and prevent exposure.

In practical terms, in order to assess whether workplace controls are effective, the parameters that need to be measured are number concentration and mass concentration. This can be achieved using a combination of handheld instruments, such as a condensation particle counter, optical particle counter and photometer (e.g. DustTrak), in conjunction with conventional sampling techniques.

7.10.7   *ELIMINATING OR MINIMISING EXPOSURE TO NANOMATERIALS*

As is the case for substances with larger particles, nanomaterial exposure levels will be process and material dependent, with the highest exposures likely when handling 'free' or uncontained nanomaterials. As for hazards and hazardous chemicals generally, the hierarchy of control should be applied for nanomaterials as discussed in Chapter 4.

A range of substitution and modification options have been identified that can be used to make nanomaterials less hazardous—for example, making the nanomaterials more hydrophilic, more soluble or less biopersistent. As with any substitution control strategy, it is important that product properties can be maintained.

When appropriately designed and maintained, conventional engineering controls such as local exhaust ventilation can effectively reduce exposures to nanomaterials. Nanoparticles can move through filter media by diffusion, and there is a probability that they will impact on the filter fibres and be captured. This means that air-purifying respirators with P2 and P3 filters may reduce exposure to nanomaterials. Air supply respirators with higher protection factors are likely to be more effective. Respirator manufacturers may be able to assist in determining which products are most effective to capture the size range (and shape) of the particles of interest.

7.10.8   *RISK MANAGEMENT FOR NANOPARTICLES*

Safe Work Australia has published *Safe Handling and Use of Carbon Nanotubes* (Safe Work Australia, 2012), which provides two different CNT risk management approaches—risk management with detailed hazard analysis and exposure assessment, and risk management by control banding—either or both of which may be used.

Safe Work Australia (2010) has also developed an assessment tool for handling engineered nanomaterials. This is a useful tool for identifying hazards and developing relevant policies and procedures for nanomaterials in the workplace.

Further information regarding managing risks for manufactured nanomaterials can also be found in the ISO/TR 13121: 2011 document: *Technical Reports: Nanotechnologies—Nanomaterial Risk Evaluation* (ISO, 2011). This document includes health and safety information for identification, evaluation, decision-making and communicating risks for the public, consumers, workers and the environment.

7.11 REFERENCES

Alamango, K., Whitelaw, J. and Apthorpe, L. 2015, 'Have you got your head in the sand? Respirable crystalline exposures of restorative stonemasons', *Proceedings of the Australian Institute of Occupational Hygienists Annual Conference*, Perth, 5–9 December.

American Conference of Governmental Industrial Hygienists 2018, *Threshold Limit Values and Biological Exposure Indices*, ACGIH®, Cincinnati, OH.

Amman, C.A. and Siegla, D.C. 1981, 'Diesel particulates: What are they and why?', *Aerosol Science and Technology*, vol. 1, pp. 73–101.

Australian Institute of Occupational Hygienists (AIOH) 2016, *Position Paper: Synthetic Mineral Fibres (SMF) and Occupational Health Issues*, AIOH, Melbourne.

—— 2017, *Position Paper: Diesel Particulate Matter and Occupational Health Issues*, AIOH, Melbourne.

—— 2018, *Position Paper: Respirable Coal Dust and Its Potential for Occupational Health Issues*, AIOH, Melbourne.

Berry, G. 2002, 'Asbestos lung fibre analysis in the United Kingdom, 1976–96', *Annals of Occupational Hygiene*, vol. 46, no. 6, pp. 523–6.

Burton, K.A., Whitelaw, J.L., Jones, A.L. and Davies, B. 2016, 'Efficiency of respirator filter media against diesel particulate matter: A comparison study using two diesel particulate sources', *Annals of Occupational Hygiene*, vol. 60, no. 6, pp. 771–9.

CSIRO 2011, *Durability of Carbon Nanotubes and Their Potential to Cause Inflammation*, Safe Work Australia, Canberra, <www.safeworkaustralia.gov.au/sites/swa/about/Publications/Documents/582/DurabilityOfCarbonNanotubesAndTheirPotentailToCauseInflammation.pdf> [accessed 30 July 2018]

Department of Industry, Innovation, Science, Research and Tertiary Education 2011, *Nanotechnology: Australian Capability Report,* 4th ed., <www.innovation.gov.au/Industry/Nanotechnology/NationalEnablingTechnologiesStrategy/Documents/NanotechnologyCapabilityReport2011.pdf>.

Hines, J., Davies, B., Gopaldasani, V. and Badenhorst, C. 2017, 'Linking emissions based maintenance of diesel engines with worker exposure—pros and cons', in *Proceedings of the Australian Institute of Occupational Hygienists Annual Conference*, Canberra, 4–6 December.

Hodgson, J.T. and Darnton, A. 2000, 'The quantitative risks of mesothelioma and lung cancer in relation to asbestos exposure', *Annals of Occupational Hygiene*, vol. 44, no. 8, pp. 565–601.

IARC Working Group on the Evaluation of Carcinogenic Risks to Humans 2002, *Man-made Vitreous Fibres*, IARC, Lyon.

—— 2012, *Diesel and Gasoline Engine Exhausts and Some Nitroarenes*, IARC, Lyon.

International Organization for Standardization (ISO) 1995, *Air Quality: Particle Size Fraction Definitions for Health-related Sampling*, ISO 7708: 1995, ISO, Lyon.

—— 2011, *Technical Report: Nanotechnologies—Nanomaterial Risk Evaluation*, ISO/TR 13121(en), ISO, Geneva.

—— 2012, *Technical Report: Nanomaterials—preparation of material safety data sheets (MSDS)*, ISO/TR 13329(en), ISO, Geneva.

—— 2015, *Technical Specification; Nanotechnologies—Vocabulary—Part 1: Core terms*, ISO/TS 80004-1, ISO, Geneva.

Methner, M.M., Hodson, L., Dames, A. and Geraci, C. 2010, 'Nanoparticle Emission Assessment Technique (NEAT) for the identification and measurement of potential inhalation exposure to engineered nanomaterials—Part B: Results from 12 Field Studies', *Journal of Occupational and Environmental Hygiene*, vol. 7, pp. 163–76.

National Health and Medical Research Council (NHMRC) 1984, *Methods for Measurement of Quartz in Respirable Airborne Dust by Infrared Spectroscopy and X-ray Diffractometry*, NHMRC, Canberra.

National Institute for Occupational Safety and Health (NIOSH) 2013, *Current Intelligence Bulletin: Occupational Exposure to Carbon Nanotubes and Nanofibers*, Department of Health and Human Services, Washington, DC, <www.cdc.gov/niosh/docs/2013-145/pdfs/2013-145.pdf> [accessed 30 July 2018]

—— 2016, 'Diesel particulate matter (as elemental carbon), analytical method 5040', in *NIOSH Manual of Analytical Methods (NMAM)*, 5th ed., NIOSH, Atlanta, GA, <www.cdc.gov/niosh/docs/2014-151/pdfs/methods/5040.pdf>.

National Occupational Health and Safety Commission (NOHSC) 1989, *Technical Report on Synthetic Mineral Fibres and Guidance Note on the Membrane Filter Method for Estimation of Airborne Synthetic Mineral Fibres*, AGPS, Canberra.

—— 1990, *National Code of Practice for the Safe Use of Synthetic Mineral Fibres*, Canberra, AGPS, Canberra, <www.safeworkaustralia.gov.au/doc/code-practice-safe-use-synthetic-mineral-fibres-nohsc-2006-1990-archived> [accessed 3 August 2018]

—— 2005, *Guidance Note on the Membrane Filter Method for Estimating Airborne Asbestos Fibres*, 2nd ed., NOHSC: 3003, AGPS, Canberra, <www.safeworkaustralia.gov.au/doc/guidance-note-membrane-filter-method-estimating-airborne-asbestos-fibres-2nd-edition> [accessed 3 August 2018]

NSW Minerals Council 1999, *Diesel Emissions in Underground Mines: Management and control*, NSW Minerals Council, Sydney.

Organization for Economic Cooperation and Development Working Party on Manufactured Nanomaterials (OECD WPMN) 2009, *Emission Assessment for Identification of Sources and Release of Airborne Manufactured Nanomaterials in the Workplace: Compilation of Existing Guidance*, OECD, Paris, <www.oecd.org/dataoecd/15/60/43289645.pdf> [accessed 30 July 2018]

Pickford, G., Apthorpe, L., Alamango, K., Conaty, G. and Rhyder, G. 2004, 'Remediation of asbestos in soils: A ground breaking study', *Proceedings of the Australian Institute of Occupational Hygienists Annual Conference*, Fremantle, 4–8 December.

Poland, C.A., Duffin, R., Kinloch, I., Maynard, A., Wallace, W.A., Seaton, A., Stone, V., Brown, S., Macnee, W. and Donaldson, K. 2008, 'Carbon nanotubes introduced into the abdominal cavity of mice show asbestos-like pathogenicity in a pilot study', *Nature Nanotechnology*, vol. 3, pp. 423–8.

Pratt, S., Granger, A., Todd, J., Meena, G.G., Rogers, A. and Davies, B. 1997, 'Evaluation and control of employee exposure to diesel exhaust particulates at several Australian coal mines', *Applied Occupational and Environmental Hygiene*, vol. 12, pp. 1032–7.

Queensland University of Technology (QUT) 2012, *Measurements of Particle Emissions from Nanotechnology Processes, with Assessment of Measuring Techniques and Workplace*

Controls, Safe Work Australia, Canberra, <www.safeworkaustralia.gov.au/system/files/documents/1702/measurements_particle_emissions_nanotechnology_processes.pdf> [accessed 30 July 2018]

Rogers, A. and Davies, B. 2001, 'Diesel particulate (soot) exposures and methods of control in some Australian underground metalliferous mines', in *Proceedings of the Queensland Mining Industry Health and Safety Conference*, Townsville, 26–29 August.

Rogers, A., Leigh J., Ferguson, D., Mulder, H., Ackad, M., and Morgan, G. 1994, 'Dose–response relationship between airborne and lung asbestos fibre type, length and concentration, and the relative risk of mesothelioma', *Annals of Occupational Hygiene*, vol. 38, suppl. 1, pp. 631–8.

Rogers, A. and Whelan, W. 1996, 'Elemental carbon as a means of measuring diesel particulate matter emitted from diesel engines in underground mines', in *Proceedings of the 15th Annual Conference of the Australian Institute of Occupational Hygienists*, Perth, 1–4 December.

Safe Work Australia (SWA) 2010, *Work Health and Safety Assessment Tool for Handling Engineered Nanomaterials*, Safe Work Australia, Canberra, <www.safeworkaustralia.gov.au/system/files/documents/1702/work_health_safety_tool_handling_engineered_nanomaterials.pdf> [accessed 30 July 2018]

—— 2012, *Safe Handling and Use of Carbon Nanotubes*, Safe Work Australia, Canberra, <www.safeworkaustralia.gov.au/system/files/documents/1702/safe_handling_and_use_of_carbon_nanotubes.pdf> [accessed 30 July 2018]

—— 2013, *Workplace Exposure Standards for Airborne Contaminants*, Safe Work Australia, Canberra, <www.safeworkaustralia.gov.au/system/files/documents/1705/guidance-interpretation-workplace-exposure-standards-airborne-contaminants-v2.pdf> [accessed 30 July 2018]

—— 2016a, *Code of Practice: How to Manage and Control Asbestos in the Workplace*, Safe Work Australia, Canberra, <www.safeworkaustralia.gov.au/system/files/documents/1705/mcop-how-to-manage-and-control-asbestos-in-the-workplace-v2.pdf> [accessed 8 July 2018]

—— 2016b, *Code of Practice: How to Safely Remove Asbestos*, Safe Work Australia, Canberra, <www.safeworkaustralia.gov.au/system/files/documents/1705/mcop-how-to-safely-remove-asbestos-v3.pdf> [accessed 8 July 2018]

Seipenbusch, M., Binder, A. and Kasper, G. 2008, 'Temporal evolution of nanoparticle aerosols in workplace exposure', *Annals of Occupational Hygiene*, vol. 52, no. 8, pp. 707–16.

Standards Australia 2004, Method for the Qualitative Identification of Asbestos in Bulk Samples, AS 4964: 2004, SAI Global, Sydney.

—— 2006a, Fume from Welding and Allied Processes: Guide to Methods for the Sampling and Analysis of Particulate Matter, AS 3853.1: 2006, SAI Global, Sydney.

—— 2006b, Fume from Welding and Allied Processes: Guide to Methods for the Sampling and Analysis of Gases, AS 3853.2: 2006 (R2016), SAI Global, Sydney.

—— 2009a, Selection, Use and Maintenance of Respiratory Protective Equipment, AS/NZS 1715: 2009, SAI Global, Sydney.

—— 2009b, Workplace Atmospheres: Method for Sampling and Gravimetric Determination of Respirable Dust, AS 2985: 2009, SAI Global, Sydney.

—— 2009c, Workplace Atmospheres: Method for Sampling and Gravimetric Determination of Inhalable Dusts, AS 3640: 2009, SAI Global, Sydney.

Toxikos and Safe Work Australia 2013, *Evaluation of Potential Safety (Physicochemical) Hazards Associated with the Use of Engineered Nanomaterials*, Safe Work Australia, Canberra, <www.safeworkaustralia.gov.au/doc/evaluation-potential-safety-hazards-associated-use-engineered-nanomaterials> [accessed 30 July 2018]

TRGS 554 2008, *German Federal Institute of Occupational Safety and Health* (Bundesanstalt für Arbeitsshutz und Arbeitsmedizin—BAUA), Technische Regeln für Gefahrstoffe 554 (TRGS 554).

World Health Organization (WHO) 1996, 'Diesel fuel and exhaust emissions', *Environmental Health Criteria 171*, WHO, Geneva, <www.inchem.org/documents/ehc/ehc/ehc171.htm> [accessed 3 August 2018]

8. Metals in the workplace

Ian Firth and Ron Capil

8.1   INTRODUCTION

The world's industrial and pre-industrial civilisations have depended in numerous ways on metal-ore extraction and metal fabrication. Coinage, precious metals, the implements of war and industry—they have all been linked with occupational health hazards since the Bronze and Iron Ages. During the Industrial Revolution, and more recently in the technological age, metals have been implicated in occupational disease in many industries.

The toxic nature of metals and metal salts has also long been recognised, with lead and arsenic compounds often favoured by poisoners. Most people today are aware of the possibility of lead poisoning in children, who may eat or chew the sweet-tasting flakes of lead paint in old houses. The Mad Hatter in Lewis Carroll's *Alice in Wonderland* may have been sent 'mad' by mercury poisoning; psychotic symptoms were common among workers in the fur and hat-making industries in the early nineteenth century, owing to excessive mercury exposure (see section 8.9).

The current challenge is to monitor and measure those metals that are increasingly being used in various nanomaterials. Research in this area may raise concerns, since some metals are more toxic in nanoparticle form; this will be discussed briefly in Chapter 17.

8.2   MAJOR METALS OF CONCERN IN THE WORKPLACE

This chapter examines the more toxic of the most commonly encountered metals, namely lead, cadmium, mercury, chromium, zinc and nickel, and the metalloid arsenic. (Metalloids have properties of both metals and non-metals.) The following aspects are covered for each of these materials:

* typical occurrence and use
* basic toxicology
* assessment in the workplace
* typically used control procedures.

A few less occupationally significant metals, such as aluminium, beryllium, cobalt, copper, manganese, selenium and thallium, and the metalloids antimony and boron, are examined in less detail. This is not an exhaustive list, and the H&S practitioner should seek more authoritative references for metals not covered here.

8.3   NATURE OF METAL CONTAMINANTS AND ROUTES OF EXPOSURE

The extraction, processing, refining, fabrication and widespread use of metals and their salts produce hundreds of situations in which hazardous exposures can occur. Because most metals and their salts are solids, the majority of exposure to metals and metal salts in the workplace occurs through inhalation of their particulate (or aerosol) forms—dust, fume or mist. However, the contribution of ingestion should not be overlooked, as it is possible to transfer significant amounts of metals into the mouth during smoking and eating when personal hygiene is poor.

Most metals are solid at room temperature, though mercury, a few metal hydrides (e.g. arsine, stibine) and some organometallic compounds are common exceptions to this rule. These are either gases or can exert enough vapour pressure at room temperature to be present in the vapour state. In such cases, the metal can be inhaled as a vapour.

Some significant exposures also occur via the skin. Mercury salts, thallium and organometallic liquids can penetrate skin, and metals and metallic salts can enter the body through damaged skin, cuts and abrasions. In some cases (e.g. nickel and other skin-sensitising metals), the skin is the target organ, and direct skin contact is a route of exposure.

Processes giving rise to metals in a form that can be absorbed are:

- metal-ore extraction (e.g. mining of iron ore, manganese, lead, zinc, copper, uranium, etc., and their subsequent processing prior to smelting)
- metal smelting (e.g. arsenic, cadmium and selenium are liberated in lead and zinc smelting, and mercury is liberated in gold refining and in alumina refining)
- metal founding (e.g. lead, brass)
- metal machining (e.g. beryllium drilling, grinding or polishing, and cobalt in dental and hip prostheses)
- hot metal processing (e.g. hot zinc galvanising, metal recycling)
- welding, soldering, brazing and thermal cutting of metals (producing potentially hazardous metal fumes of cadmium, chromium, copper, iron, lead, magnesium, manganese, mercury, molybdenum, nickel, titanium, vanadium and zinc)
- handling powders of metal salts (e.g. lead battery manufacture, zinc and copper oxide manufacture, lead stearate use in PVC pipe manufacture).

8.4   METAL TOXICITY

The forms in which the metals exist are important. They may exist as the native material (e.g. chromium metal) or as various salts (e.g. chromium oxide) and, depending on these forms, their ions may exert a range of toxic effects, from dermatitis through neurotoxic effects to cancer. Some exposure standards have different exposure limits, depending on the form of the metal, its chemical valency or whether it is in the form of an inorganic salt or organometallic compound (e.g. chromium, nickel).

Assessing the toxic dose of various metals is often more complicated than for other hazardous substances. Indeed, some metallic elements are essential to human life because of their role in cellular functioning, bone structure or blood and enzyme systems. Fourteen metals, including sodium, potassium, calcium and magnesium, are involved in the body's basic building blocks. Trace elements, including zinc, selenium, iron, cobalt, arsenic and copper, are all essential in the right amounts—they have a narrow 'window of life' range of concentrations, with higher and lower concentrations being detrimental. A number of metal-based compounds, themselves potentially toxic to humans, have found great service in pharmacologically active drugs, including the early anti-syphilitic drugs, Mercurochrome and platinum-containing cytotoxic (anti-cancer) drugs.

8.5   ASSESSING EXPOSURE TO METALS IN THE WORKPLACE

Most monitoring for metals in the workplace requires sampling for dusts and fumes. In the case of electroplating, some metals—such as chromium and nickel—become airborne as mists, which are monitored in much the same way as dusts and fumes containing metals. However, monitoring of some metals or their compounds (e.g. mercury vapour, arsine and stibine) requires special techniques.

Because similar sorts of air-monitoring processes are used for most metals and metallic compounds, a procedure is detailed here for only one metal, lead. As the toxic effects of metals often result from a combination of absorption from the lungs and ingestion after deposition in the nasopharyngeal (nose and throat) region, the inhalable fraction of the particulate is most often appropriately sampled (although in some cases that will be explained later, the respirable fraction should also be sampled). See Chapter 7 for other practical details or, for more complete procedures, AS 2985 (Standards Australia, 2009a) on gravimetric determination of respirable dusts, AS 3640 (Standards Australia, 2009b) on gravimetric determination of inhalable dusts and MDHS 14/4 (HSE, 2014a) on gravimetric analysis of respirable, thoracic and inhalable aerosols.

Air monitoring may indicate compliance with the relevant exposure standard (refer to Chapter 3). To assess the exposure from all routes (inhalation, ingestion and skin absorption), however, biological monitoring of exposure may be necessary in particular circumstances to evaluate the accumulated dose experienced by individual workers. The general principles of biological monitoring are discussed in Chapter 10. Throughout this chapter, reference is made to Safe Work Australia's (2018b, 2018c) workplace exposure standards (WESs) for airborne contaminants, available from the web-based Hazardous Chemical Information System (HCIS), the American Conference of Governmental Industrial Hygienists' (ACGIH®, 2019) threshold limit values (TLVs®), the US National Institute for Occupational Safety and Health (NIOSH) recommended exposure limits (RELs), the Occupational Health and Safety Administration's (OSHA, 2017) permissible exposure limits (PELs), and the UK Health and Safety Executive's (HSE, 2011) workplace exposure limits (WELs).

8.6 METHODS OF CONTROL

Although metal contaminants are often present as dust, fumes or mists, control procedures vary greatly depending on how the contaminant is generated. Further, some of the more toxic metals take greater effort to control (e.g. lead dusts require more stringent control procedures than iron dusts). The example of lead (section 8.7) provides the detail that may typically be required for an H&S practitioner involved in the control of hazardous metals in the workplace.

Workers in industries where toxic metals such as lead, cadmium, mercury, arsenic, chromium, nickel and zinc are handled must be fully informed of the routes of exposure, the nature of the health hazards and the measures required to prevent hazardous exposure, including respiratory protective equipment (RPE) and its use and maintenance. Appropriate RPE must be selected on the basis of the assessed exposure. Hand washing and separate eating facilities must be provided and their use enforced to prevent any possibility of accidental ingestion in the workplace. The need to prohibit smoking as a further guard against ingestion must be stressed.

8.7 LEAD

8.7.1 USE AND OCCURRENCE

The soft, bluish-white to dull-greyish coloured malleable metal lead is obtained by smelting ores containing lead sulfide (galena), lead sulfate or lead carbonate. Lead ores often contain zinc and other toxic metals such as cadmium in minor concentrations. Chile, Australia and the United States are the largest producers and exporters of lead; however, about half of all lead produced each year comes from recycled material (Bell, 2018). Lead's main industrial use is in lead-battery manufacture, but it also finds applications in automotive paints, solders, ceramic glazes, metal alloys (e.g. gun metal), bearings and lead shot.

Industrial workplaces where there is **potential exposure to** *inorganic lead* include:

- the lead mining and refining industry
- the battery industry, both manufacture and reclamation
- the radiator repair industry
- propeller grinding
- lead lighting
- non-ferrous metal foundries manufacturing gun metal or leaded bronzes (Figure 8.1)
- the spraying of lead-based paints
- the sanding or torch cutting of lead-painted metals (e.g. bridge painters, demolition workers)
- the assaying of gold and silver

Figure 8.1 Brass founding may result in some exposure to lead fume

- indoor shooting galleries and rifle ranges
- house painting, in the sanding of some houses painted prior to the 1950s
- ceramic glazing.

Organic lead **exposure** occurs in petroleum workers potentially exposed to tetramethyl or tetraethyl lead. These are being phased out of use and are no longer used in most countries, but they are an issue in some developing countries. Additionally, PVC pipe manufacturers use lead stearate as a stabiliser.

8.7.2   TOXICOLOGY

Lead is the metal most likely to harm both workers and members of the public. More is known about the toxic effects of lead than of any other metal. Toxicity depends mainly on particle solubility and size, since these determine how easily the metal is absorbed. The greatest hazard in the workplace has typically been inhaled lead, either as particulate (dust) or as very fine lead fume, but workers may also accidentally ingest lead if there is lead dust on their hands or face when they smoke or eat. Some inhaled dust and fume will also ultimately be swallowed after coughing or clearing the throat. Soluble lead salts are very toxic if swallowed. The smaller the particles, the more rapid their absorption, and the more acute and severe their toxic effect. Thermally generated lead fumes are often involved in lead poisoning: fumes inhaled into the lungs can pass easily through the alveolar walls directly into the bloodstream. These fumes contain the easily soluble lead suboxide, common in the grey fume that occurs in and around lead smelters and brass foundries. For inorganic lead, absorption through the skin is not a significant route of exposure.

Once in the body, lead is transported in the bloodstream to all tissues, and is stored predominantly in the bones, where it replaces calcium. Mobilisation of lead from bone to blood is slow and can lead to slight elevations of blood lead levels for many years after exposure ceases. Absorbed lead is excreted from the body primarily via the kidneys, in urine. Blood lead levels are expressed in micromoles per litre ($\mu mol/L$) or micrograms per 100 mL or decilitre ($\mu g/dL$), and are a good reflection of absorption of inorganic lead into the body.

Lead is a neurotoxin that can slow the transmission of impulses along the nerves. It has been implicated in affecting intellectual development in the young (exposed to lead during gestation and early childhood) and is also associated with kidney dysfunction, elevated blood pressure and sperm abnormalities. Other serious effects can accompany acute and chronic lead intoxication. Historically, the major toxic effect of lead has been on the haemopoietic (blood generation) system, resulting in anaemia. At very high levels, which are no longer typical, lead absorption can cause constipation, abdominal pain, blue lines along the gums, convulsions, hallucinations, coma, weakness, fatigue, tremors and wrist drop.

The International Agency for Research on Cancer (IARC) classifies inorganic lead compounds as being probably carcinogenic to humans, Group 2A (IARC, 2018).

8.7.3   STANDARDS AND MONITORING

Regulations require both:

- monitoring of air in the workplace for lead
- biological monitoring of the worker for lead in the blood.

Blood lead levels should be used as the primary indicator of both inhalation of airborne lead and ingestion via eating and smoking. Air monitoring in the workplace should also be considered as a complementary measure to evaluate the effectiveness of controls for airborne lead (AIOH Exposure Standards Committee, 2018).

8.7.3.1   Air monitoring

Adequate control of exposure to airborne lead should be employed to maintain workplaces within a time-weighted average (TWA) occupational exposure limit (OEL) of 0.05 mg/m^3. The ACGIH®, NIOSH and Safe Work Australia use this limit value for lead and its inorganic compounds, although some jurisdictions still use 0.15 mg/m^3 (e.g. UK HSE). The TLV®-TWA of 0.05 mg/m^3 is intended to maintain worker blood lead levels below 30 μg/dL (1.45 μmol/L). Monitoring provides information about:

- effectiveness of control measures
- reasons for some high blood-lead levels
- specific work factors that may be hazardous
- the correct level of intervention required for control (e.g. local exhaust ventilation (LEV) or RPE).

Personal air monitoring must be undertaken using a standard method (e.g. AS 3640; Standards Australia, 2009b). The basic principles are:

- Workplace air is sampled using an IOM sampler or equivalent (Figures 7.7a and 7.7b) fitted in the worker's breathing zone, using a portable monitoring pump running at 2 L/min attached to the worker's belt or a pump harness.
- The particulate or fume is trapped on an appropriate membrane filter.
- The sampled material is dissolved in nitric acid.
- The amount of lead is measured by atomic absorption spectrometry.
- The resultant concentration of lead in the air is calculated, taking into account the sample volume.

Analysis for lead requires laboratory facilities. Most H&S practitioners will probably need to limit their involvement to sampling and the subsequent calculations and reporting.

8.7.3.2   Biological monitoring

Air monitoring may indicate compliance with the OEL. Monitoring the worker's blood lead levels may still be necessary where:

- it is required by regulation
- there may be accidental uptake (e.g. via smoking or eating)
- control of the lead hazard has failed

- the worker's tasks (e.g. irregular duct cleaning) might still result in increased exposure to lead
- workers have a history of excessive exposure to lead
- primary control processes are not used and RPE is the only defence
- a health and safety inspector requests blood-lead monitoring.

Biological monitoring for lead represents more than just checking hygiene in the workplace. It is an active intervention necessary for maintaining the health of exposed workers and ensuring that lead is not absorbed in deleterious quantities. Wherever lead is used some lead exposure is inevitable, but symptoms do not appear in most workers if the blood lead level can be kept below the action level chosen (e.g. 0.97 μmol/L or 20 μg/dL of blood). Specific action levels may not apply to all workers. Lead is toxic to the foetus, and lower levels of exposure will be necessary for most women. Safe Work Australia has model regulations (Part 7.2) relating to lead exposure for women of childbearing age (Safe Work Australia, 2018a). For females of reproductive capacity, a 'lead risk job' is one in which blood lead is likely to exceed 5 μg/dL (0.24 μmol/L).

Measurement of pre-existing blood lead levels may be required by regulation before a worker begins a job where lead is a potential hazard. There are standards for the sampling and analysis of blood for lead, such as AS 2636 (Standards Australia, 1994), AS 2411 (Standards Australia, 1993a) and AS 4090 (Standards Australia, 1993b). Some other test indicators of the biological effect of lead, such as zinc protoporphyrin (ZPP), can be used to supplement the basic blood-lead measurement, but they are not generally used in routine surveillance.

Regulations can have a complex regime of actions to control workers' blood-lead levels. For example, in the Safe Work Australia (2018a) amended model regulations, monitoring frequency can vary from every six months when less than 10 μg/dL, to every six weeks if greater than 20 μg/dL (0.97 μmol/L) for all but females of reproductive capacity, where 5 μg/dL (0.24 μmol/L) will trigger six-weekly monitoring. Removal from exposure is mandatory at greater than 30 μg/dL (1.45 μmol/L) for males, with return to work allowed only when blood lead returns to less than 20 μg/dL. Some companies use different blood-lead action levels than these; for male workers, examples include:

- an acceptable level is <15 μg/dL; repeat monitoring in eight months
- the level triggering counselling and review of control measures is 30 μg/dL; repeat blood test in one month and measure ZPP or use other effect test
- the transfer level is 40 μg/dL; repeat in one month and measure ZPP
- the return level is 30 μg/dL, with ZPP below 7 μg/g haemoglobin.

Testing of urine for lead level is not a good indicator of exposure or body burden in the case of inorganic lead because it is cleared from the blood and bone at quite different rates. However, urine testing is the method of choice for assessing exposure to organic lead compounds such as tetraethyl lead.

8.7.4 CONTROLS

Where levels of lead in air or blood indicate that exposure is deleterious, the reason for such exposure needs to be determined and appropriate controls instituted. Control of both the inhalation and ingestion routes will usually be required.

Lead is essential in many industries, but substitution with another metal or a change in form of the lead should always be considered. Though the possibility of elimination or substitution is limited, it has occurred in a number of cases, including the manufacture of fishing weights and some PVC stabilisers. In addition, changing the form of the material, such as using pelletised forms of lead stabilisers instead of fine powders in PVC pipe production, reduces the probability of lead becoming airborne. As lead fumes are generated at more than 450°C, reducing the temperature of the molten metal will also reduce airborne lead exposure.

The most common secondary control procedures, with examples, are:

- enclosing processes to control dust or fume:
  - lead alloying plants and lead oxide furnaces are completely enclosed to prevent any escape of lead dust (Figure 8.2a)
- using dust-minimising techniques:
  - keeping process materials wet
  - mixing spilled material with wet sawdust as a dust suppressant
  - using a vacuum cleaner (with a HEPA filter) in place of sweeping
- use of dust or fume extraction equipment:
  - fume extraction hoods or LEV over high-temperature furnaces (>500°C)
  - portable hand tools fitted with dust extractors for sanding/grinding
  - exhausted enclosures in plate stacking in battery manufacture
- administrative controls:
  - permitting blood levels to increase above a background level but limiting the increase to a specified level
  - removing lead-affected workers from lead work
  - ensuring that eating, drinking and smoking do not occur in the workplace
  - maintaining good housekeeping
  - ensuring that facilities for good personal hygiene are provided and used
- use of RPE, with the type chosen depending on the concentration of lead in the workplace atmosphere, how it is generated, the protection factor required and the physical demands of tasks being undertaken. For example:
  - thermally produced fumes will require medium efficiency particulate (P2) filtration for air concentrations up to 10 times the TWA OEL (Figure 8.2b)
  - higher concentrations will require the use of powered air purifying respirators with P2 filters.
  - airline respirators will be necessary for filling batteries in submarines.

Instruction, training and maintenance of monitoring and health-surveillance programs are all very necessary in workplaces where lead is handled. Workers must also know and understand the hazards of handling lead. In addition to its toxic qualities, lead is heavy, and lifting and moving it safely by hand can require ergonomic considerations.

8.7.5   LEAD REGULATIONS

Lead is subject to regulation in most countries. In Australia, as an example, regulations in each state are to be based on the *Model Work Health and Safety Regulations* (Safe Work Australia, 2018a). These embody the concepts of a lead process (certain conditions have

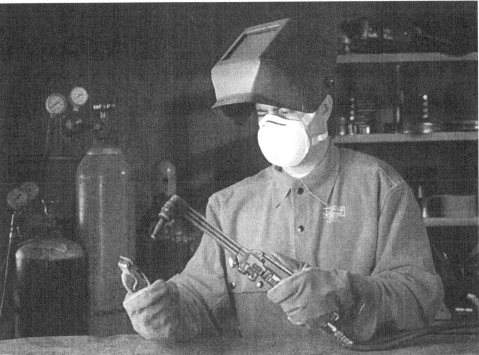

Figure 8.2 Examples of fume control and RPE against thermally generated lead fume
(a) Full extraction on a lead-to-lead oxide-converting furnace
(b) P2 particulate filtration respiratory protection

to be met before the task is considered a lead hazard) and a lead risk job (a job involving lead exposure that results in various blood-lead levels—hence categories—depending on individual circumstances). There are also special requirements with regard to equal-employment opportunities for women and men. The requirements emphasise biological and air monitoring, followed up with a strict control regime. H&S practitioners involved in the assessment or management of lead risk jobs will need to know:

- how the operations produce lead dust or fume contamination
- the exposure to airborne lead in each job or similar exposure group
- the categories of job involving a risk of lead exposure and the health-surveillance requirements
- the blood-lead levels of individual operators (subject to medical confidentiality)
- the control processes and the relative effectiveness of each
- procedures to undertake when action levels in blood or air are exceeded.

For much of this work, the H&S practitioner will have to collaborate with a medical practitioner experienced in interpretation of lead exposure. The assessment of control measures may be facilitated by knowledge of the blood-lead levels of individual workers. In order to preserve medical confidentiality, it may be advisable to ask workers to sign a release form for their blood-lead results. The H&S practitioner must ensure that this information is used solely to help the workers control their lead absorption rates.

8.8   CADMIUM

8.8.1   USE AND OCCURRENCE

The white, ductile metal cadmium finds a number of industrial uses because of its low melting point, conductivity and resistance to corrosion. It is used in the manufacture of nickel-cadmium (NiCAD) batteries, in cadmium electroplating to apply a protective coating to steel, in welding rods, brazing solders, low-melting-point safety valves and metal alloys. Cadmium salts are also widely used in pigments, rubbers, paints, inks, plastic stabilisers, fireworks, rectifiers, solar cells and television phosphors.

Exposure occurs principally by inhalation, usually from processes that involve handling the material or its salts as powders, or where thermally generated fumes occur in the workplace. Recovery of cadmium from NiCAD scrap batteries and welding of cadmium-plated metals are also potential sources of exposure. Accidental ingestion is rare.

8.8.2   TOXICOLOGY

Cadmium shows both acute and chronic toxic effects. Cough, headache, eye irritation, chill and fever, with chest pain, may follow acute inhalation of a cadmium fume, with possible delayed lung damage (pulmonary oedema, pneumonitis). A metal worker over-exposed to cadmium fume may develop a typical metal-fume fever in the evening (or even days later), and not relate it to work carried out the previous day.

Chronic effects of exposure include damage to the kidneys, sometimes with the formation of kidney stones, as well as to the respiratory system (fibrosis). IARC classifies cadmium and cadmium compounds as being carcinogenic to humans, Group 1 (IARC, 2018).

8.8.3   STANDARDS AND MONITORING

The Safe Work Australia WES-TWA for airborne cadmium oxide, fume or metal is 0.01 mg/m³ (Safe Work Australia, 2018b), a level set to prevent kidney disease in long-term cadmium workers. The US NIOSH and OSHA (2017) have a lower limit of 0.005 mg/m³, while the UK HSE has a limit of 0.025 mg/m³ (HSE, 2011). ACGIH® (2019) recommends the 0.01 mg/m³ limit but has also adopted a respirable-particulate fraction exposure standard of 0.002 mg/m³, set to minimise the potential for cadmium accumulation in the lower respiratory tract, which could induce lung cancer.

8.8.3.1   Air monitoring

Workplace monitoring requires sampling the personal breathing zone with an Institute of Occupational Medicine (IOM) inhalable dust sampler or equivalent and a respirable cyclone dust sampler. Other practical details are similar to those outlined for lead sampling. Laboratory analysis is required and is described in methods MDHS 10/2 (HSE, 1994) and MDHS 91/2 (HSE, 2015) on detection and measurement of inorganic cadmium and its compounds in air.

8.8.3.2   Biological monitoring

Cadmium is a scheduled hazardous substance for which health surveillance must be provided if the risk of exposure is found to be significant. The surveillance includes testing of respiratory function, questionnaires and urinary and blood testing.

For workplaces that use cadmium regularly, blood testing for cadmium may be necessary. The Biological Exposure Index (BEI®) for blood cadmium is 5 µg/L (ACGIH, 2019). Alternatively, urinary cadmium, for which the BEI® is 5 µg/g creatinine, can be measured. Urinary excretion of cadmium is related to body burden, recent exposure and renal damage, so the interpretation of urinary cadmium levels is not simple.

8.8.4   CONTROLS

The highly toxic nature of cadmium requires that its use in the workplace be extremely well controlled. Elimination and substitution are rarely feasible, so prevention of airborne dust and fume production is mandatory. This is achieved by measures such as minimising temperatures in welding and soldering and using mechanical cutting instead of thermal cutting of cadmium-coated products.

Where LEV is employed, it will need to be of a high standard to control the dust or fume hazard. The filtration and recovery of cadmium or its salts from air discharged from the extraction system also require consideration.

RPE for use with cadmium or its salts may be required in certain work operations where higher-level controls (i.e. engineering controls, ventilation) are impractical or cannot adequately control the hazard. Respirators with medium particulate filtration efficiency (P2) may be used for airborne concentrations up to ten times the TWA OEL. However, high-temperature soldering/brazing or thermal cutting with cadmium-containing materials can generate concentrations up to 50 mg/m³ (cadmium evaporates significantly at its melting point). Mistaken reliance on filtration RPE could have disastrous consequences. For thermally generated fumes, one of the following is needed:

- primary fume/dust control and medium-efficiency particulate filtration (P2)
- an air-supplied system with backup high-efficiency particulate (P3) protection, while ensuring there is no subsequent exposure of bystanders.

All workers involved with cadmium in the workplace require thorough instruction in the hazards and routes of exposure, and methods of safe handling and use. Particular attention must be paid to the use and maintenance of RPE whenever it is required, as well as to the need for biological monitoring.

8.9   MERCURY

8.9.1   USE AND OCCURRENCE

Metallic mercury is a heavy, silvery liquid obtained from roasting cinnabar ore. Use of mercury-containing compounds has a long history, dating at least from Roman times. In the seventeenth century, the use of mercury nitrate in the hat trade for carroting of fur (raising the scales on fur shafts) was widespread. Mercury has found more recent use in submarine ballast, mercury-fulminate explosive detonators, barometers, thermometers, pressure pumps, electric lamps and mercury rectifiers, chlor-alkali cell electrodes, dry-cell batteries, electrical switches, chemical catalysis, dental amalgams, pesticides and the extraction of gold from ores. Cyanide leaching of low-grade gold and silver ores also collects any mercury contained in the ore, which is volatilised off by heating during further processing for the gold. Other uses are as a mould and fungus inhibitor for wood, paper and grain, in some medicinal preparations, and in paints for inhibiting marine growth on ships' hulls. It can still be found providing a frictionless 'float' bearing for the lens assembly in a few lighthouses. There have been calls to severely limit the use of mercury, owing mainly to its tendency to accumulate in brain and foetal tissues and in breast milk, with possible adverse consequences for foetal and child growth.

Mercury may occur as a contaminant in ore (e.g. bauxite) and coal, to be concentrated to environmentally significant levels by processing.

8.9.2   TOXICOLOGY

Liquid mercury vaporises readily at room temperature, so inhalation of mercury vapour is the primary route of entry, although both the metal and its compounds can be absorbed through the skin.

Cases of gross poisoning with skin ulceration and gastrointestinal symptoms are now a thing of the past. Accidental poisonings of children with mercury and its compounds still occur, but they are rare.

Mercury accumulates mainly in the brain and kidneys. The main target organ is the central nervous system (including the brain), although kidney damage may also occur with some mercury salts. The toxic action of mercury compounds occurs by precipitation of protein and inhibition of sulfydryl enzymes. These damage the central nervous system, causing headaches, tremors, weakness and psychotic disorders that may present as shyness, irritability and excitability. The classical mad hatter's disease, with its spidery writing and withdrawn behaviour, is typical of mercury poisoning but rarely seen these days. Some people may develop sensitivity to mercury whereby their skin reacts even to the vapour, causing contact dermatitis. Ingestion of organomercury compounds has resulted in a particular type of central nervous system debilitation, which is often irreversible. This has occurred in Japanese people who have eaten fish containing high levels of methylmercury (so-called Minamata disease) and in people from the Middle East who have consumed seed wheat treated with mercury.

Recent data indicate that mercury can affect both male and female reproduction, and lead to children with abnormal cognitive and physical functioning. The IARC (2018) considers that mercury and inorganic mercury compounds are not classifiable as to their carcinogenicity to humans, and gives these compounds a Group 3 classification.

8.9.3 STANDARDS AND MONITORING

The Safe Work Australia WES for mercury vapour and inorganic mercury compounds is 0.025 mg/m³, expressed as a TWA (Safe Work Australia, 2018b). Note that mercury and its compounds can have a notation for skin absorption. The same exposure standard is used by the ACGIH® (2019), while NIOSH (OSHA, 2017) recommends a REL-TWA of 0.05 mg/m³ and the HSE (2011) has a WEL-TWA of 0.02 mg/m³. The latter was set to minimise the potential for damage to the central nervous system and kidneys, and to provide some assurance that the cognitive and physical abilities of children would not be adversely affected.

8.9.3.1 Air monitoring

Several methods are available for monitoring mercury in the workplace, all of which rely on the fact that mercury can be trapped and measured as a vapour. They are:

- indicator stain tubes (see Chapter 9, Figure 9.2); this method is generally useful for exposures greater than the TWA OEL
- the direct monitor (Figure 8.3) using ultraviolet detection or a gold film amalgam; this method is useful both above and below the TWA OEL
- impingement into an acid permanganate solution (Figure 8.3) followed by cold-vapour atomic absorption spectrometry—the most sensitive method
- passive or active sampling (see Chapter 9, section 9.15) onto a solid sorbent device (using Hydrar or hopcalite as the sorbent), followed by cold-vapour atomic absorption spectrometry. This method is also useful both above and below the TWA OEL and is the most commonly employed. Refer to method MDHS 16/2 (HSE, 2002) on determination of mercury and its inorganic divalent compounds in air.

8.9.3.2 Biological monitoring

Inorganic mercury is a scheduled hazardous substance. Health surveillance may be required if the risk from exposure to mercury is found to be significant, and any affected workers

Figure 8.3 Mercury monitoring methods
(left) portable gold film meter; (right) permanganate impinger method

should immediately be removed from further exposure. The preferred indicator of elemental mercury absorption is its concentration in urine. ACGIH® recommends a BEI® of 20 µg/g creatinine, from a sample taken pre-shift.

Some mercury is normally contributed from the diet (e.g. from eating fish). Those exposed occupationally to mercury show elevated levels of mercury in urine; background mercury levels of unexposed persons are generally less than 5 µg/g creatinine, while levels above 35 µg/g creatinine are considered significant. Urinary levels of about 50 µg/g creatinine are seen after occupational exposure to about 0.04 mg/m³ mercury in air. Blood levels of mercury can be monitored, but sample collection is much more complicated than simple collection of urine samples and interpreting the significance of the results can be difficult.

8.9.4   CONTROLS

The Minamata Convention on Mercury (2018), a United Nations multilateral environmental agreement, addresses the adverse effects of mercury through practical actions to protect human health and the environment from emissions and releases of mercury and mercury compounds. The primary methods used to prevent mercury uptake in the workplace are as follows:

- *Replacement of mercury-containing equipment*, such as thermometers with non-mercury-containing alternatives such as alcohol thermometers or thermocouples. Mercury from mercury-in-glass thermometers that break in hot ovens will rapidly evaporate.
- *Substitution of mercury-containing processes* wherever possible (e.g. use of composite restoration material in place of mercury dental amalgams; use of organic chemicals

for fur carroting; use of carbon-in-pulp leaching of gold rather than the older-style mercury retorting; replacement of neon/fluorescent tubes that contain mercury).

- *Prevention of vapour generation.* Mercury evaporates readily, so it should be stored in closed, water-sealed containers. Where mercury metal is heated, the hazard is enormously increased, so operations such as welding or brazing on mercury-in-metal thermometers must be strictly controlled.
- *Enclosure of processes where mercury vapours may be generated* to prevent escape of vapours. Where not possible, LEV to a high standard is required. All mercury-contaminated air exhausted from a process should be routed through a mercury scrubbing device using a sulphur- or iodine-impregnated carbon pack, or bubbled into a tank that contains a mercury-complexing agent in conjunction with a demister. Mercury tends to condense in ventilation ducts. Having smooth-walled ducts that slope towards a gravity collection trap should control this condensation.
- *Designing areas for handling mercury to cope adequately with spills,* with impervious flooring (e.g. use epoxy, polyurethane or vinyl sheeting; wood or carpeting should be avoided) and bunded (built-up) edges to prevent dispersion of the small mercury droplets. Proprietary products (HgX®) or a paste of slaked lime and flowers of sulphur (1:1) can be used to assist in removing the metal from inaccessible and otherwise hard-to-clean cracks and crevices, equipment and the like. Contaminated materials such as carpet are best discarded, but their disposal must comply with local hazardous waste requirements.
- *Preventing mercury from spraying into small droplets* that are difficult to coalesce and collect. A given volume of mercury has a much larger surface area when broken into small droplets and evaporates much more rapidly. Mercury should not be collected with a vacuum cleaner unless it is specifically designed for mercury (e.g. Tiger-Vac® or Nilfisk®). Any vacuum pick-up system must have an adequate mercury-vapour trap.
- *Using personal protective equipment for mercury vapour* when significant mercury spills have to be cleaned up. Respirators impregnated with a mercury-vapour absorbent can be used with relatively small spills.
- *Using airline breathing apparatus for major spills,* particularly if the ambient temperature is elevated. The use of overalls and gloves will be necessary if any skin contact is likely.
- *Instructing workers handling mercury in the hazards of the material,* as well as in adequate methods of containment and control, and in decontamination procedures should a major spill occur.

8.10 ARSENIC

8.10.1 *USE AND OCCURRENCE*

Arsenic, particularly its trioxide, finds a curious place in history as a poison favoured by murderers. Metallic arsenic has a few industrial uses in alloys with lead for bearings, cable sheaths and battery grids. In contrast, compounds of arsenic (oxides and complex salts with other metals) are widely used in the manufacture of weed killers, insecticides, wood preservatives, anti-fouling paints and fungicides. Tobacco crops were once widely sprayed with arsenicals (arsenic compounds). A well-known copper-chrome-arsenic (CCA) preparation has been widely used in the production of logs and timbers for outdoor and garden use, although in Australia and other countries CCA use has been restricted in residential situations. CCA should not be used on high-contact timber structures, including garden furniture, picnic tables, exterior seating, children's play equipment, patio and domestic decking and hand rails.

Arsenic compounds are no longer used as paint pigments for obvious public health reasons, but they are still used in some fireworks. Arsenic has found new applications in the manufacture of semiconductors and radiation detectors. Workers in these industries potentially risk exposure to arsenic or its compounds. Smelting of ores containing arsenic impurities (e.g. copper, lead and zinc ores) is also an important source of potential exposure.

Users of arsenic-containing agricultural chemicals are also at risk. End-users of CCA-treated timber should not be at undue risk, provided that timber offcuts are not burned (e.g. in barbecues). Burning can convert the arsenic bound into timber to the volatile and hazardous arsenic trioxide. The health implications of frequent contact with CCA-treated timber structures, particularly for children, are uncertain (Australian Pesticides and Veterinary Medicines Authority, 2005).

8.10.2 TOXICOLOGY

The toxicity and action of arsenic and its compounds vary, depending on the metal's chemical form. Arsenic and its compounds can be both ingested and inhaled as dusts, and arsenicals can cause corrosive or ulcerative effects in the skin and mucous membranes. The major routes of entry in the workplace are by inhalation. Acute effects include haemorrhagic gastritis, muscular cramps, facial oedema, peripheral neuropathy, corrosive actions and skin lesions. Chronic effects include irritation to the nasal mucosa, with penetration of the nasal septum in some workers exposed over long periods to low levels of arsenic dusts. Dermatitis may be observed, with heavy skin pigmentation and peripheral vascular disorders in some people who ingest arsenic-contaminated water.

The IARC (2018) considers that arsenic and arsenic compounds are carcinogenic to humans and gives arsenic a Group 1 classification. Arsenic causes cancers of the lung (e.g. in smelter populations) and respiratory tract (e.g. in workers making arsenical pesticides). There is evidence that ingestion of inorganic arsenic is associated with skin and perhaps liver cancer. Paradoxically, arsenic is also a therapeutic agent which is still used in the treatment of cancer and trypanosomiasis. Some skin cancers and hyperkeratosis are attributable to medicinal arsenic exposure.

Arsenic is one of a group of metals that form a volatile hydride by reacting with nascent hydrogen (e.g. from contact with acids). Arsenical contamination can generate this hydride—arsine—accidentally. A gas with a garlic-like odour, it presents as an extremely acute inhalation hazard. For example, in a Queensland case, several people—including children—were affected when ground contaminated with old arsenical cattle dip came into contact with an acid source, probably superphosphate, and arsine was liberated. Onset of symptoms occurs within a few hours of exposure, with headache, giddiness, abdominal pain and vomiting. The urine is stained by haemolysed, excreted blood cells. Anaemia and jaundice follow, which may result in kidney failure.

8.10.3 STANDARDS AND MONITORING

The Safe Work Australia WES-TWA is 0.05 mg/m³ for inorganic arsenic salts, the metal and its soluble salts, and 0.16 mg/m³ (or the equivalent 0.05 ppm) for the gaseous arsine (Safe Work Australia, 2018b). ACGIH® (2019) has an exposure standard for arsine that is ten times lower (0.005 ppm) and has a TLV®-TWA of 0.01 mg/m³ for elemental and inorganic arsenic compounds, to minimise the potential for adverse effects, including cancer, on the skin, liver, peripheral vasculature, upper respiratory tract and lungs. The UK HSE (2011) WEL-TWA for arsenic and its compounds (except arsine) is 0.1 mg/m³, with arsine being 0.05 ppm.

8.10.3.1 Air monitoring

Monitoring methods depend on the form of the arsenic. Those most widely used in the workplace involve personal air sampling, with a sampling pump connected to filter(s) or a solid sorbent tube.

Arsenic trioxide vapours (a major source of exposure in smelting) must be collected on filter papers treated with sodium hydroxide or sodium carbonate. Any sampling head suitable for use with treated filters will suffice. HSE method MDHS 41/2 (HSE, 1995) contains practical details.

Arsenic particulates are collected on mixed cellulose ester filters using the IOM sampler or equivalent. Refer to AS 3640 (Standards Australia 2009b) or MDHS 41/2 (HSE, 1995) and MDHS 14/4 (HSE, 2014a) for sampling details.

Arsine is collected using a charcoal sorbent tube followed by analysis using atomic absorption spectrophotometry with graphite furnace, as detailed in NIOSH method 6001 (NIOSH, 2016).

Most H&S practitioners should be able to conduct the sampling with appropriate training, but all the methods require laboratory analysis.

Simple direct-reading indicator stain tubes (see Chapter 9, section 9.11) are available for both arsenic trioxide vapour and arsine. These indicator tubes may require a very large number of pump strokes to be readable at concentrations much lower than the TWA OEL. The gaseous-arsine indicator stain tubes are more manageable, requiring only 20 strokes, with a measuring range of 0.05–3 ppm.

8.10.3.2 Biological monitoring

Inorganic arsenic is a scheduled hazardous substance and may require health surveillance if exposure is significant. Although air monitoring remains the preferred means of workplace surveillance, biomonitoring of urine in workers exposed to arsenicals is recommended so recent exposures can be reviewed. The preferred biological indicator of the absorption of elemental arsenic and soluble inorganic-arsenic compounds is the concentration of inorganic and its methylated metabolites in urine. This is an analysis which separates dietary arsenic from inorganic arsenic and the ACGIH® (2019) recommends a BEI® of 35 µg As/L, taken at the end of the working week.

If an analysis is performed by a laboratory that does not separate dietary arsenic from inorganic arsenic then high background levels can be found, giving a false positive indication. Unexposed individuals have background dietary arsenic in urine levels below

10 µg/L in European countries, slightly higher in the United States, but around 50 µg/L in Japan. Seafood can be a prime source of dietary arsenic (mainly as organic compounds). A laboratory experienced in trace-level arsenic determinations can advise on the sampling procedure. When the laboratory performs the speciation analysis of inorganic arsenic and its methylated metabolites then background dietary arsenic levels do not interfere with the analysis.

Exposure to 0.01 mg/m^3 of arsenic in the air for eight hours will most likely result in a urinary concentration of about 35 µg As/L.

8.10.4 CONTROLS

A number of control procedures can be required when working with arsenic-containing compounds. Substitution of arsenic preparations in agriculture by organochlorine and organophosphate pesticides has markedly reduced the potential for occupational exposure. All processes involving the handling of powders should be totally enclosed. When this is impractical, a high standard of LEV must be employed to remove dust from the workplace.

Processes utilising arsine must be fully enclosed, with extraction systems and reliable leak detectors. Accidental arsine generation in smelters and metal shops must be prevented by keeping dross dry and forbidding all hazardous reaction ingredients (acids, alkalis, zinc and aluminium).

Skin protection with impervious gloves is mandatory for workers handling liquids (e.g. in the CCA wood treatment industry—see Figure 8.4), where splashes to the hands

Figure 8.4 Gauntlet hand-arm protection is required when handling arsenic-treated timber

and forearms are common. Face shields may also be required when working with solutions of arsenic salts.

Where RPE is required (i.e. where engineering controls have not reduced air concentrations to required levels), careful attention is needed in its selection. The H&S practitioner should take account of:

- the nature of the arsenic hazard (dust, vapour or gas)
- the required minimum protection factor.

For example:

- Arsenic-containing dusts require particulate-filter respirators with filtration efficiency (P1, P2 or P3) according to the concentration of dust in the workplace.
- Arsenic trioxide vapour requires a filter treated with the appropriate absorbent, soda lime, which can be used up to 1000 ppm only.
- Arsine requires supplied air with a full-face mask. Normally, this would be for emergency use only, in the event that arsine is generated accidentally.

In addition to being fully instructed with regard to health effects and control techniques, workers should regard health surveillance, which includes medical assessments, biological arsenic monitoring and skin examination, as central to their wellbeing when working with arsenic or any of its compounds.

8.11 CHROMIUM

8.11.1 USE AND OCCURRENCE

The hard, grey metal chromium, obtained from chromite ore, and many chromium salts find uses in a variety of industrial applications. Chromium metal is extensively used as an alloy in stainless steel, special tooling metals, welding rods and electrical resistance wires. Chromium exists in several valency states, as II (chromous, basic), III (chromic, trivalent chromium or amphoteric) and VI (hexavalent chromium or chromate, acidic). The chromium II salts are relatively unstable and not widely used. The chromium III and VI salts find wide use, the chromium VI salts because of their strong acid and oxidative properties. Chromium III salts are used most widely, including in the production of pure chromium metal and chromium VI compounds.

Typical workplace processes where chromium exposure can occur include:

- chromium electroplating
- manual metal arc (MMA) and flux-cored arc (FCA) welding of stainless steels
- aluminium anodising
- chromium-based timber treatments (CCA)
- tanning of leather hides
- manufacture and use of spray paints containing chromium salts as pigments or zinc chromate as a rust inhibitor
- chromium bichromates of ammonia, sodium and potassium used as mordants in dyeing

- photography and photo-engraving
- manufacture and use of high-temperature chromium-containing cements for aggressive environments (e.g. refractory products).

8.11.2   TOXICOLOGY

While the metal chromium is inert, its salts are irritating and destructive to human tissue. Uptake of chromium in the workplace occurs mainly by inhalation. Chromium VI salts in particular are an irritant, and may cause dermatitis and skin ulcers, and in extreme cases ulceration and perforation of the nasal septum. Chrome 'holes' around fingernails, finger joints, eyelids or sometimes on the forearms may occur, though they are not proliferative and may be caused more by the strong oxidative power of these materials than by the chromium itself.

Studies of workers producing chromates, bichromates and chromic acid, however, have established that prolonged inhalation of chromium VI dust causes lung cancer. Exposure to roasted chromite ore may likewise cause cancer. The IARC (2018) considers that chromium VI compounds are carcinogenic to humans and gives chromium VI a Group 1 classification. For this reason, chromium VI salts are considered environmentally hazardous; many local government authorities require them to be converted to a reduced form prior to disposal.

Chromium metal, raw chromite ore and chromium III salts may cause respiratory tract and skin irritation and dermatitis, but are not considered carcinogenic. The IARC (2018) considers that metallic chromium and chromium III compounds are not classifiable as to their carcinogenicity to humans and gives these compounds a Group 3 classification. In fact, some chromium III salts are essential nutrients.

8.11.3   STANDARDS AND MONITORING

Because the toxic effects of chromium salts depend on oxidation state, different exposure standards are proposed for different salts, as shown in Table 8.1. Note that the ACGIH® (2019) has recommended lower inhalable standards for both chromium III and VI soluble compounds, and also a STEL (0.0005 mg/m³) for chromium VI soluble compounds. The OEL-TWA of 0.5 mg/m³ for the metal and the salts in lower oxidation states is believed

Table 8.1 Exposure standards for chromium and its salts

Chromium compound	SWA WES-TWA	ACGIH® TLV®-TWA
Chromium metal	0.5 mg/m³	0.5 mg/m³
Chromium II compounds	0.5 mg/m³	–
Chromium III compounds	0.5 mg/m³	0.003 mg/m³
Chromium VI compounds	0.05 mg/m³	0.0002 mg/m³

Sources: Safe Work Australia (2018); ACGIH® (2019).

to be sufficient to minimise the potential for respiratory tract and skin irritation and dermatitis. For chromium VI salts, the lower TWA OEL in most cases is primarily to protect long-term workers against increased risk of cancer in addition to respiratory tract and skin irritation and dermatitis, although the lower ACGIH values are to also minimise respiratory sensitisation and the likelihood of asthmatic responses in already sensitised individuals. It should be noted that NIOSH (OSHA, 2017) has also recommended a REL-TWA of 0.0002 mg/m^3 for chromium VI compounds.

8.11.3.1 Air monitoring

Air monitoring for chromium particulates requires personal sampling with an IOM inhalable dust sampler or equivalent. Refer to AS 3640 (Standards Australia, 2009b) or MDHS 14/4 (HSE, 2014a) for details. Other technical details are fully covered in methods such as MDHS 91/2, *Metals and Metalloids in Air by X-ray Fluorescence Spectrometry* (HSE, 2015), MDHS 52/4, *Hexavalent Chromium in Chromium Plating Mists* (HSE, 2014b) and *NIOSH Methods* 7600, 7605, 7302 and 7304 (NIOSH, 2016).

Laboratory advice should be sought before attempting to monitor chromium, as a number of technical problems are present. Sampling chromic acid mists requires a non-metallic open-faced filter holder, and a filter material (e.g. PVC) that will withstand the action of acid and not affect the stability of the chromium. Sampling of chromium in welding fume is time critical; the chromium oxidation state will depend on the welding technology (inert-gas shielded or manual metal arc, etc.) and the time after thermal generation of the fume.

8.11.3.2 Biological monitoring

As chromium is a scheduled hazardous substance, health surveillance for chromium workers may be required, depending on the significance of risk. Urine testing is required for those working with water-soluble chromium VI salts. The ACGIH® (2019) recommends a BEI® of 25 µg/L for total chromium in urine, taken at the end of the last shift of the working week. A BEI® of 10 µg/L for the increase of total chromium in urine during a shift is also recommended. An annual physical examination with emphasis on the respiratory system and skin and weekly inspection of hands and forearms are advisable if there has been significant exposure (e.g. greater than half the OEL). Air monitoring provides a better indication of the level of exposure than blood or urinary monitoring.

8.11.4 CONTROLS

Control of chromium dusts in the workplace is of paramount importance in preventing exposure by inhalation. Enclosed systems and high-quality LEV are priority control procedures.

In chromium electroplating, the chromium or chromic acid mists generated over air-agitated tanks require several levels of control:

- surface active additives to prevent formation of stable bubbles
- floating surface balls to prevent a large surface-to-air interface
- push–pull ventilation to capture any escaped mists (see Chapter 5, Figure 5.16).

Skin protection is necessary where splashes of liquid may occur, or where dried salts may be picked up by sweat on the skin (e.g. in the leather-tanning industry). A paraffin and lanolin barrier cream should be used as an added protection. A 10 per cent CaNa EDTA ointment should be applied to any cuts or abrasions contaminated by chromium salts (this converts all chromium VI to chromium III, which can be safely bound or chelated). Full impermeable protective equipment against accidental exposure should be used where appropriate in electroplating.

Because of chromium's carcinogenic potential, RPE should be a last resort, to be used only when all other methods of controlling exposure have been exhausted (e.g. when stripping chromium cements inside a warm furnace flue). The choice of RPE depends on the following considerations:

* valency state of chromium (III or VI)
* airborne concentration of chromium
* protection factor required
* physical demands of tasks being undertaken.

Workers involved in handling chromium salts, chromium electroplating or welding of stainless steels should be fully instructed in the health effects and specific hazards of their particular task. Attention should be paid to safety measures and controls, and their correct implementation.

8.12   ZINC

8.12.1   USE AND OCCURRENCE

Metallic zinc is obtained by treatment (froth flotation and roasting) of zinc sulfide ore. Smelting produces an impure grade of zinc suitable for galvanising, and electrolytically refined, higher-grade zinc is suitable for metal die-casting and moulding. Zinc reacts rapidly with oxygen to form a protective layer of zinc oxide that prevents any further reaction. Zinc finds use in the protective galvanising of steel and roofing iron, in light-weight metal castings for toys and fancy goods, in zinc oxide for paints and vehicle tyre fillers, in zinc dry-battery cases and in alloys such as brass.

Workers most likely to be affected by zinc-fume fever (see section 8.12.2) are welders working on galvanised iron in poorly ventilated or unventilated conditions, hot-dip galvanisers and workers in zinc smelters, and in zinc oxide and granulated zinc production plants. Zinc powder also presents a combustion and explosion hazard.

8.12.2   TOXICOLOGY

Because zinc is an essential metal, occupationally related poisonings are not numerous. While some poisonings do occur with zinc metal, most arise from zinc compounds. The major route of intake in the workplace is inhalation, with possible subsequent ingestion after clearance from the lung. Zinc and its compounds are readily absorbed owing to their solubility. The major health consequence of zinc inhalation is metal-fume fever, sometimes known as 'brass founder's ague' or just 'zinc chills'. This fever occurs only as a result of exposure to freshly generated fumes of either the metal or zinc oxide. The particles must be sufficiently small to enter into the lung space. Fever usually begins some hours after exposure and is accompanied by increased leukocyte count. A continuously elevated leukocyte count can confer some resistance to a recurrence of fever. There is no indication of chronic health effects from zinc exposure, but other substances in the fume (e.g. lead, arsenic, cadmium or oxidant gases from welding) may cause chronic illness.

Zinc chloride fumes (from smoke bombs or soldering fluxes) are known to be toxic, causing inflammation and corrosion of lung tissue and subsequent rapid lung fibrosis. Zinc chloride solutions have a caustic action on the skin. Zinc phosphide used in rat poison is toxic when inhaled or ingested.

8.12.3 STANDARDS AND MONITORING

The Safe Work Australia (2018b) WES-TWA and the NIOSH (OSHA, 2017) REL-TWA for zinc oxide fume (or respirable zinc oxide) is 5 mg/m^3, with a STEL of 10 mg/m^3. However, the ACGIH® (2018) recommends a TWA limit of 2 mg/m^3 for zinc oxide as respirable dust, with a STEL of 10 mg/m^3, to prevent metal-fume fever. The Safe Work Australia (2019) WES-TWA for inhalable zinc dust is 10 mg/m^3, while for zinc chloride the TWA value is 1 mg/m^3, with a STEL of 2 mg/m^3, as also set by the ACGIH® and NIOSH.

It is therefore necessary to know whether an exposure was to zinc fume or dust. In cases where the zinc is thermally generated (e.g. welding, burning of zinc in a candle furnace, smelting or hot-dip galvanising), it is appropriate to assume that it will be finely divided fume. Concentrations up to ten times the TWA OEL may be reached during unventilated welding of galvanised steel.

There are no simple, direct methods for measuring zinc fume in the field. Monitoring for zinc fume or zinc oxide particulate requires personal sampling with a respirable dust sampler. Refer to AS 2985 (Standards Australia, 2009a) or MDHS 14/4 (HSE, 2014a) for details. Monitoring for zinc dust and zinc chloride requires personal sampling with the IOM inhalable dust sampler or equivalent. Consult AS 3640 (Standards Australia, 2009b) for details. Measurement requires laboratory assessment, usually by atomic absorption spectroscopy. Sampling in hot-dip galvanising works requires special efforts to protect the sampling head from direct splashes of hot metal, which will burn holes in the filter. In this case the use of a seven-hole inhalable dust sampler may be more beneficial.

8.12.4 CONTROLS

Processes using zinc should be controlled by enclosure, adequate LEV or other engineering measures. The metal and oxide fumes are not life threatening, and RPE with a protection factor suitable for metal fumes can often be considered satisfactory when other controls are impractical. For welders working regularly on galvanised steels, airline RPE rather than filter RPE should be considered if work practices consistently place them in the plume of metal fume originating from the welding task. Special RPE products for welding are available from a number of manufacturers. The Welding Technology Institute of Australia (Weld Australia) (2015) supplies a good publication, titled *Fume Minimisation Guidelines*.

Those who work with zinc and zinc products should be fully instructed in the nature of the associated health hazards and the control procedures required to prevent exposure.

8.13   NICKEL

8.13.1   USE AND OCCURRENCE

Nickel is obtained from a mixture of ores including sulfides, silicates and laterites. One industrial process (Mond) converts the sulfide to nickel oxide by roasting, followed by conversion to nickel carbonyl by reaction with hydrogen and carbon monoxide, and subsequent reduction to pure nickel. Nickel carbonyl may also occur in the chemical, glass and metal-plating industries, where it is used as a catalyst in various chemical reactions, in glass plating and in the forming of nickel films and coatings.

Nickel finds use mainly in stainless steels and in a huge number of alloys, in which nickel confers high resistance to corrosion and to extremes of temperature. It is used in nickel-cadmium batteries, nickel plating, Monel alloy, nickel chemical-reaction catalysts, magnetic tapes, coins, jewellery and electronic, electrical and engine parts. Workers are exposed to nickel-containing dusts and fumes during plating and grinding, mining and nickel refining, and in steel plants, foundries and other metal industries. Exposure can also arise from welding of metals containing nickel.

8.13.2   TOXICOLOGY

Nickel exerts two major workplace-related health effects. The first is nickel itch or nickel eczema on the arms of those whose skin is exposed to nickel (e.g. platers). The itch may spread to the face and other parts of the body, constituting an allergic response. Workers so sensitised may not be able to return to this kind of work. Nickel-plated watches, jewellery and earrings may likewise cause an allergic reaction on the skin, produced when sweat comes in contact with the metal. Nickel can sensitise the skin by direct contact, and there is good evidence that it can also do so after being ingested.

Inhalation of nickel carbonyl can result in acute symptoms of headache, dizziness and vomiting, followed by chest pain, dry cough, shortness of breath and extreme weakness, depending on the degree of exposure. Nickel carbonyl causes haemorrhagic pneumonia. Most inhaled nickel ends up in the brain and lung. The process of nickel roasting produces nickel fumes that are considered to be carcinogenic. The IARC (2018) considers that nickel compounds are carcinogenic to humans and gives these compounds a Group 1 classification. Nickel metals and alloys are classified as Group 2B by the IARC (possibly carcinogenic to humans). The compounds principally implicated in causing respiratory cancer are sulfidic nickel, particularly nickel sub-sulfide and oxidic nickel, which includes a range of insoluble nickel compounds. There is debate about whether soluble nickel compounds are carcinogenic (AIOH, 2016).

8.13.3   STANDARDS AND MONITORING

There are a number of Safe Work Australia WES-TWAs (Safe Work Australia, 2018b) for nickel compounds, depending on the form of the nickel. The ACGIH® (2019)

recommends some different values, as shown in Table 8.2. Nickel carbonate, nickel sulfide and nickel oxide are insoluble compounds of nickel, whereas nickel chloride, nickel sulfate and nickel nitrate are soluble compounds. The AIOH (2016) recommends an OEL-TWA of 0.1 mg/m³ for all forms of nickel, as the inhalable fraction.

Table 8.2 Exposure standards for nickel and its compounds

Nickel compound	SWA WES-TWA	ACGIH® TLV®-TWA
Nickel metal	1.0 mg/m³	1.5 mg/m³
Soluble compounds of nickel	0.1 mg/m³	0.1 mg/m³
Nickel carbonyl	0.05 ppm or 0.12 mg/m³	0.05 ppm
Nickel sub-sulfide	1.0 mg/m³ (Category 1 carcinogen)	–
Insoluble compounds as Ni	–	0.2 mg/m³ (A1 carcinogen)

Source: ACGIH® (2019); Safe Work Australia (2018b).

Indicator stain tubes are available for direct measurement of nickel carbonyl in air, although the method is not particularly sensitive. Better sensitivity and precision in the assessment of inorganic nickel in workplace air are obtained by methods such as MDHS 42/2 (HSE, 1996b), MDHS 91/2 (HSE, 2015) or NIOSH method 7302 (NIOSH, 2016). As with other metals, the dust or fumes are collected on a filter, dissolved by acid and measured by atomic absorption spectroscopy, inductively coupled argon plasma (ICP) or x-ray fluorescence (XRF).

8.13.4 CONTROLS

Because of the irritation and carcinogenic effects of some nickel dusts and aerosols, control of processes involving nickel by enclosure or LEV is mandatory. Workers in nickel electro-plating should have skin protection against nickel salts and access to adequate washing facilities in case of skin splashes.

Refining processes involving nickel carbonyl must be completely enclosed, preventing any exposure. Only air-supplied RPE is suitable for entry into any area of a plant where nickel carbonyl is processed.

Workers handling nickel metal, its salts or nickel-containing compounds should be instructed in the hazards—particularly the need to be aware of skin conditions. Workers also require instruction in the use of controls appropriate to the potential exposure.

8.14 MINOR METALS OF WORKPLACE CONCERN

The preceding sections have dealt with the more common hazardous metals that the H&S practitioner is likely to encounter; however, there are some others, classed as metals of minor significance because of their relatively rare use and occurrence, not because of their toxicology, which may be both complex and fascinating. Their classification as 'minor' also stems from the fact that they are seldom the subjects of extensive workplace health or hygiene investigations.

For example, boron is a toxic metalloid, but its use as a high-energy neutron shield in nuclear reactors is of little consequence to the average H&S practitioner, as there are few reactors in most countries. On the other hand, compounds of boron find wide use in ceramic glazes, fireproofing materials, low-grade insecticides, organic chemical synthesis, printing and painting. So there may well be good reasons to consider these materials as other than 'minor'. H&S practitioners seeking more extensive information on the metals of minor significance should consult some of the references at the end of this chapter.

Details on sampling procedures for these metals are not presented here, but the *NIOSH Manual of Analytical Methods* (2016) covers most of them. The H&S practitioner should contact a specialist hygiene laboratory if measurement of these metals is required. There are no simple techniques available to the H&S practitioner for assessing any of them in the field.

8.14.1   ALUMINIUM

Aluminium is a relative newcomer to the family of commercial metals. In its mineral form, bauxite, aluminium is the most abundant metal in the Earth's crust. Mined bauxite is refined into alumina (aluminium oxide) using a chemical refining process (Bayer), whereby finely ground bauxite is digested in a hot caustic soda solution, then clarified, precipitated and calcined. The alumina is then smelted into aluminium using electrochemical reduction, whereby the alumina is dissolved in molten cryolite in cells through which a direct electrical current is passed via carbon cathodes and anodes. The molten metal formed in each cell is drawn off at regular intervals for casting into various shapes. Aluminium is often alloyed with small amounts of copper, magnesium, silicon, manganese and other elements to impart a variety of useful properties.

Aluminium oxide is used as an abrasive, in refractory material and for electronic applications. The uses of aluminium are numerous: in automobile engines, transmissions, bodies and suspension components, shipbuilding, aircraft manufacture, home products such as doors, window frames, roofing and insulation, electrical wire and transmission cable, packaging material such as aluminium foil, drink cans and wrap, aluminium coatings for telescope mirrors, and decorative paper, packages and toys. Aluminium compounds and materials also have a wide variety of uses in the production of glass, ceramics, rubber, wood preservatives and pharmaceuticals, and in waterproofing textiles. Salts of aluminium include alum, which is used in water treatment, and natural aluminium minerals (e.g. bentonite and zeolite), which are used in water purification, sugar refining, brewing and the paper industry.

Aluminium metal readily oxidises to aluminium oxide, forming a protective layer against further ordinary corrosion. Powder and flake aluminium are flammable and can form explosive mixtures in air, especially when treated to reduce surface oxidation (e.g. pyro powders). The toxicity of aluminium is still controversial. Safe Work Australia (2018b) exposure standards are set to minimise the potential for irritation of the respiratory tract, with air exposure standards of 10 mg/m³ for aluminium metal dust and oxide, 5 mg/m³ for aluminium welding fumes, and 2 mg/m³ for soluble salts. The ACGIH® (2019) has set a TLV®-TWA of 1 mg/m³ (as respirable fraction) for aluminium metal and insoluble compounds to protect from lower respiratory tract irritation, pneumoconiosis and neurotoxicity.

There appears to be no serious adverse effect on respiratory health associated with exposure to bauxite in open-cut bauxite mines. There is a belief among the general public that aluminium plays some role in Alzheimer's disease, although research findings so far do not support this notion. Of more importance for those working in the aluminium electrolytic smelting industry is the potential for a condition known as 'pot-room asthma', characterised by cough with dyspnoea, wheezing and chest tightness, usually occurring within the first year of exposure to fumes from smelting cells or pots. While peak concentrations of fluoride in air have been implicated as a causative agent, the actual cause of this asthma is yet to be clearly determined. For those working in aluminium smelters, the production of carbon cathodes and anodes can cause exposure to polycyclic aromatic hydrocarbons (PAHs) from coal-tar pitch volatiles (CTPVs). PAHs are carcinogenic, and thus of primary concern.

General control procedures for smelting aluminium should principally be aimed at preventing pot-room fumes and CTPVs from entering the workplace atmosphere. Expert opinion and advice are needed to accurately assess exposures in an aluminium smelter.

8.14.2 ANTIMONY

Metallic antimony finds use in alloys with lead (e.g. battery grids), in cable sheaths, pewter, ammunitions and some solders. Salts of antimony are used in paints, rubber, glass and ceramics. One regular application of antimony trioxide is as a fire retardant in weather-proofing/insulation membranes for housing construction. The toxic effects of antimony resemble those of arsenic, and include irritation of mucous membranes, gastrointestinal symptoms, sores in the mouth and skin lesions. The hydride of antimony, stibine, is an extremely toxic haemolytic agent. Antimony trisulfide is also very toxic and is reported to cause heart failure. This effect on cardiac muscle may be shared by other antimony compounds as well.

The use of antimony in lead-acid battery grids leads to a small possibility of stibine production during battery charging. Good ventilation procedures should be followed, particularly in large battery facilities.

Generally, however, antimony does not find significant use in industry. The OEL-TWA is commonly 0.5 mg/m³ for all forms, but it should be noted that the production process for antimony trioxide is suspected of being carcinogenic.

General control procedures for working with antimony should be aimed at preventing antimony compounds from entering the workplace atmosphere. Expert opinion and advice should be sought if antimony exposure is a possibility.

8.14.3 BERYLLIUM

Beryllium finds its principal use as an alloying agent in steel, copper, magnesium and aluminium. Beryllium oxide is also a component of ceramics. Its use as a phosphor in fluorescent tubes has long since been abandoned. Its major application today is in the nuclear industry, aerospace products and high-stress, low-strain metals and alloys.

Beryllium is cited by the ACGIH® (2019) as a confirmed human carcinogen, causing lung cancer. The IARC (2018) considers that beryllium and beryllium compounds are carcinogenic to humans and gives these compounds a Group 1 classification. Mining of the beryl ore is not associated with lung disease. Beryllium poisoning is related to the very toxic effects of beryllium dust, which affects a number of organs, and delayed effects from material deposited in the body can occur. Sensitisation or development of a beryllium-specific, cell-mediated immune response arises in between 2 and 19 per cent of exposed individuals. Sensitisation usually precedes the development of chronic beryllium disease, or berylliosis, characterised by a Type IV, delayed-hypersensitivity, cell-mediated immunity. Relatively small amounts of inhaled material can induce acute effects. Expert diagnostic opinion, with particular emphasis on complete occupational history, is extremely important in establishing beryllium-induced disease. Medical control in beryllium-using industries rests mostly on regular health surveillance, including chest x-rays and beryllium lymphocyte proliferation testing (BeLPT).

The Safe Work Australia (2018b) WES-TWA for beryllium is 0.002 mg/m³, while NIOSH (OSHA, 2017) recommends an REL of 0.0005 mg/m³ as a ceiling value, and the ACGIH® (2019) has a very stringent TLV®-TWA of 0.00005 mg/m³ (as inhalable fraction), set to prevent beryllium sensitisation and chronic beryllium disease (berylliosis). Thus any grinding, polishing or machining process on beryllium-containing metals or ceramics should be considered very carefully. Good occupational hygiene practices will be required to keep the risks of working with beryllium to acceptable levels. Industrial or research operations using beryllium should be assessed carefully. See MDHS 29/2 (HSE, 1996a) for air monitoring.

8.14.4   BORON

Boron is an essential element for plants and animals, including humans. Boron compounds are widely used in applications from household cleaning products (e.g. detergents and bleaches) to boron-fibre technology. Borax and boric acid are used in smelting, glazes for ceramic ware and production of glass and glass-related products (e.g. insulation and textile fibreglass, Pyrex), and borax is still used in fireproofing of pulped-cellulose insulation, low-activity insecticide mixtures and leather tanning. These compounds do not pose significant hazards in normal use, and consequently the OEL-TWA ranges from 1 to 10 mg/m³, depending on the type of boron compound. H&S practitioners need to ensure that neither inhalation nor ingestion of dusts can occur.

Several other boron compounds, including boron trifluoride and boron trichloride, find use in chemical synthesis and organic chemical reactions.

There are several boron hydrides (e.g. diborane, pentaborane, decaborane and others) which are highly toxic. They are used mainly in high-energy rocket 'zip' fuels and in pharmaceutical and rubber vulcanising. They are very irritating to skin and mucous membranes, and induce headache, chills, dizziness and weakness. These compounds also present extreme fire and explosion risks. Good occupational-hygiene practice (engineering control, personal protective equipment) is required to provide adequate control of these substances.

8.14.5 COBALT

Cobalt, being a relatively rare element, is also relatively rare in the workplace. The metal finds use in Al-Ni-Co permanent magnets, as a binder in hard-metal cutting tools, in some electrical alloys and in some dental metal alloys. Some cobalt salts, such as cobalt blue and cobaltous chloride, have been used in glass and ceramic enamels for colouring. Cobalt is an essential element in the formation of vitamin B12.

Cobalt metal workers have developed a pulmonary disease often known as 'hard metal disease', though exposure to other materials might also be implicated. The effects of cobalt poisoning include diarrhoea, loss of appetite, hypothermia and possible death. Inhalation of cobalt dust can cause pulmonary oedema in animals, so should be avoided. An allergic dermatitis resulting from exposure to minute amounts has been seen in some workers in the building, metalwork, pottery, leather and textile industries.

Exposure to cobalt by inhalation should be limited, preferably by isolation, LEV or other engineering controls. This is particularly important in grinding operations that produce fine dust. Control of airborne dusts should also reduce the need for skin protection. Additional skin protection, including barrier creams, may prove necessary in situations where such dusts are difficult to control. RPE must be of a high standard for mechanically generated dusts, preferably an airline respirator or high-efficiency filter respirator, depending on the protection factor required.

The Safe Work Australia WES-TWA for cobalt metal and compounds is 0.05 mg/m³ (Safe Work Australia, 2018b). The ACGIH® (2019) recommends a lower limit value of 0.02 mg/m³.

8.14.6 COPPER

Another essential element, copper finds common use in electrical wiring, alloys of brass, bronze or Monel, plumbing services and cookware. Copper is also used in insecticides and as an algaecide and a bactericide, and in electroplating. Industrial exposure to copper fume is not common—copper melts at about 2350°C—although the fume can give rise to a metal-fume fever similar to that caused by zinc. Copper salts, copper carbonate and copper sulfate are all relatively safe to use, provided proper precautions are taken to prevent inhalation or accidental ingestion of dusts. Ingestion of copper has occurred accidentally when acidic foodstuffs have been in contact with copper vessels (e.g. fruit juice, moonshine liquor from copper stills). Copper exposure presents no significant problem in industry, other than for workers with Wilson's disease, a rare genetic inability to excrete excess copper.

The OEL-TWAs for copper fume and copper dusts are typically 0.2 mg/m³ and 1 mg/m³ respectively.

8.14.7 MANGANESE

Manganese finds its major use in steel making, but it is also used in the production of aluminium alloys and cast iron, in alkaline manganese dry batteries, as manganese dioxide for brick colouring, as an oxidising agent in dyes and chemicals, as a chemical catalyst, as manganates and permanganates for disinfecting and bleaching and as a fungicide. Mining and smelting of manganese ore (pyrolusite) create the greatest exposure hazards. Subsequent environmental contamination around smelters can also occur. Other small exposures occur in welding with manganese-containing rods and in (dark) paint pigments. The organic manganese compound methylcyclopentadienyl manganese tricarbonyl (MMT) has been used as an additive to increase octane rating in petrol.

Manganese is an essential trace element, but in excessive amounts it can cause neuro-logical disorders involving the central nervous system. Symptoms of manganism range from apathy, anorexia and mental excitement to speech disturbance, clumsiness and a stone-faced appearance. In the established phase of the disease, staggering gait, muscular rigidity (e.g. finger or hand deformation), spasmodic laughter or tremors may occur. Manganese-poisoning victims may be cripplingly debilitated but otherwise well. There was concern by some that MMT might contribute to neurological effects similar to those caused by tetraethyl lead in petrol, but this has not eventuated.

Metal-fume fever can follow inhalation of fine manganese fume, and a manganese-induced pneumonia can occur. Manganese poisoning has also been associated with a decline in lung function and reduced fertility of male workers.

Prevention of manganese-related disease depends mostly on preventing inhalation of manganese-containing dusts and fumes. Manganese can also be accidentally ingested. LEV and other engineering controls should be adequate to maintain exposures within the Safe Work Australia (2018b) WES-TWA of 1 mg/m^3 for manganese fume, dust and compounds. The ACGIH® (2019) recommendation is 0.1 mg/m^3 (as inhalable fraction) and 0.02 mg/m^3 (as respirable fraction), levels set to protect against central nervous system effects. Use of RPE, subject to achievement of appropriate protection factors, could be appropriate, provided that conditions permit wearing it.

Medical supervision of those exposed occupationally to manganese dust is advisable. Manganism exhibits three distinct phases, so detection of any slight neurological abnor-mality may allow workers to be removed from exposure before symptoms worsen.

8.14.8   SELENIUM

Selenium is a non-metallic element that can occur in a metal-like form. This is obtained as a by-product of refining copper (and sometimes zinc) ore, so exposure to selenium fume can occur during the smelting and refining of these ores. Some of its forms are volatile, so more difficult to contain. Occupational exposure occurs primarily via inhalation, although in some cases it can be by direct skin contact. Selenium is used in the manufacture of glass, pigments, ceramics and semiconductors; in photocopiers and cameras; in photoelectric cells, steel, rectifiers in electronic equipment; as catalysts; and in the vulcanising of rubber. Selenium compounds are also used in the treatment of a number of animal and human diseases—for example, it is a common ingredient in anti-dandruff shampoos. It is an essential trace element for animals and humans, but can concentrate to toxic levels in the body, causing selenosis.

The acute toxicities of inorganic selenium compounds vary greatly. Hydrogen selenide, selenium oxychloride, selenium dioxide and selenium hexafluoride are highly poisonous, while the element and sulfides are much less toxic. Selenium dust, selenium dioxide and hydrogen selenide are more likely to be encountered in the workplace. The chronic effects of nearly all forms of selenium in humans appear to be similar: depression, languor, nervousness, dermatitis, upset stomach, giddiness, a garlic odour of breath and sweat, excessive dental caries, and in extreme cases loss of fingernails and hair. Most knowledge of selenium's toxic effects is derived from clinical toxicology and the incidence of selenium poisoning as a result of ingestion of seleniferous grains and other food; there have been no reports of disabling chronic disease or death from industrial exposures. The Safe Work Australia (2018b) WES-TWA standard is 0.1 mg/m³ for selenium compounds (excluding hydrogen selenide), which is half that of the ACGIH® (2019) and OSHA (2017) 0.2 mg/m³ limit, set to minimise irritation of the eyes and upper respiratory tract as well as systemic effects such as headache, garlic breath, skin rashes and the like. The OEL-TWA standard for hydrogen selenide is typically 0.05 ppm or 0.16 mg/m³.

General control procedures for working with selenium should be aimed at preventing selenium compounds from entering the workplace atmosphere. Expert opinion and advice should be sought if selenium is encountered in a workplace.

8.14.9 THALLIUM

This metal was once widely used in the preparation of salts for rodenticides because it is undetectable by taste or smell, but it is no longer readily available in developed countries. Thallium is also used in special optical lenses for scientific equipment (infrared cells), glass tinting, semiconductors and photoelectric cells. It is also alloyed with lead, zinc, silver and antimony to enhance their resistance to corrosion. A rare application of thallium is in Clerici's solution, thallium malonate, which has a high specific gravity and is used in heavy-mineral test separations in the mineral sands industry. Thallium may also be encountered in the production of cement and in the handling of pyrites and flue dusts.

Most deaths due to thallium are from poisoning—either accidental (when poisoned grain is mixed in with grain to be milled) or deliberate (suicidal or homicidal). Thallium is also absorbed through the skin. Symptoms of severe poisoning include swelling, joint pain, vomiting, mental confusion and loss of hair.

The effects of exposure during the preparation of thallium salts have been milder, thanks mostly to the care taken in handling the salts, but they are usually more gradual than acute poisonings. A worker's occupational history is extremely important in diagnosis.

The WES-TWA (Safe Work Australia, 2018b) and the PEL-TWA (OSHA, 2017) standard is 0.1 mg/m³, while the ACGIH® (2019) TLV®-TWA is 0.02 mg/m³ (as inhalable fraction). Monitoring of urine may be a useful measure for those regularly involved in the preparation of thallium compounds.

If thallium must be used, processes must be isolated and strict precautions taken to prevent dispersal of airborne dusts in the workplace (LEV) and skin contact (personal protective equipment). Hand-washing facilities and separate eating facilities are mandatory, and eating and smoking in the workplace and the wearing of work clothes to and from home should be forbidden.

8.15   REFERENCES

American Conference of Governmental Industrial Hygienists (ACGIH) 2018, *2018 Threshold Limit Values (TLVs®) for Chemical Substances and Physical Agents and Biological Exposure Indices (BEIs®) with 7th Edition Documentation CD-ROM*, ACGIH®, Cincinnati, OH.

—— 2019, *2019 TLVs® and BEIs®*, ACGIH®, Cincinnati, OH.

Australian Institute of Occupational Hygienists (AIOH) Exposure Standards Committee, 2016, *AIOH Position Paper: Nickel and Its Compounds and Their Potential for Occupational Health Issues*, AIOH, Melbourne, <www.aioh.org.au/resources/aioh-library> [accessed 1 September 2018]

—— 2018, *AIOH Position Paper: Inorganic Lead and Occupational Health Issues,* AIOH, Melbourne, <www.aioh.org.au/resources/aioh-library> [accessed 7 August 2019]

Australian Pesticides and Veterinary Medicines Authority (APVMA), 2005, *The Reconsideration of Registrations of Arsenic Timber Treatment Products (CCA and Arsenic Trioxide) and Their Associated Labels: Report of Review Findings and Regulatory Outcomes Summary Report*, APVMA, Canberra, <https://apvma.gov.au/sites/default/files/publication/14316-arsenic-summary.pdf> [accessed 27 April 2018]

Bell, T. 2018, 'Metal profile of lead', *The Balance*, 1 October, <www.thebalance.com/metal-profile-lead-2340140> [accessed 28 April 2018]

Health and Safety Executive (HSE) 1994, *Cadmium and Inorganic Compounds of Cadmium in Air*, MDHS 10/2, HSE, London, <www.hsl.gov.uk/resources/publications/mdhs/mdhs-revisions> [accessed 27 April 2018]

—— 1995, *Arsenic and Inorganic Compounds of Arsenic (except Arsine) in Air*, MDHS 41/2, HSE, London, <www.hsl.gov.uk/resources/publications/mdhs/mdhs-revisions> [accessed 27 April 2018]

—— 1996a, *Beryllium and Beryllium Compounds in Air*, MDHS 29/2 HSE, London, <www.hsl.gov.uk/resources/publications/mdhs/mdhs-revisions> [accessed 27 April 2018]

—— 1996b, *Nickel and Inorganic Compounds of Nickel in Air (except Nickel Carbonyl): Atomic Absorption Spectrometry*, MDHS 42/2, HSE, London, <www.hsl.gov.uk/resources/publications/mdhs/mdhs-revisions> [accessed 27 April 2018]

—— 2002, *Mercury and Its Inorganic Divalent Compounds in Air*, MDHS 16/2 HSE, London, <www.hsl.gov.uk/resources/publications/mdhs/mdhs-revisions> [accessed 27 April 2018]

—— 2011, *Workplace Exposure Limits*, EH40/2005, 2nd ed., HSE, London, <www.hse.gov.uk/coshh/basics/exposurelimits.htm> [accessed 27 April 2018]

—— 2014a, *General Methods for Sampling and Gravimetric Analysis of Respirable, Thoracic and Inhalable Aerosols*, MDHS 14/4, HSE, London, <www.hse.gov.uk/pubns/mdhs/pdfs/mdhs14-4.pdf> [accessed 27 April 2018]

—— 2014b, *Hexavalent Chromium in Chromium Plating Mists,* MDHS 52/4, HSE, London, <www.hse.gov.uk/pubns/mdhs/pdfs/mdhs52-4.pdf> [accessed 27 April 2018]

—— 2015, *Metals and Metalloids in Air by X-ray Fluorescence Spectrometry*, MDHS 91/2 HSE, London, <www.hse.gov.uk/pubns/mdhs/pdfs/mdhs91-2.pdf> [accessed 27 April 2018]

International Agency for Research on Cancer (IARC) 2018, *Agents Classified by the IARC Monographs, Volumes 1–122*, <https://monographs.iarc.fr/ENG/Classification/index.php>.

Minamata Convention on Mercury 2018, <www.mercuryconvention.org> [accessed 30 August 2018]

National Institute for Occupational Safety and Health (NIOSH) 2016, *NIOSH Manual of Analytical Methods (NMAM®)*, 5th ed., Centers for Disease Control and Prevention, Atlanta, GA, <www.cdc.gov/niosh/nmam/default.html> [accessed 27 April 2018]

Occupational Safety and Health Administration (OSHA) 2017, *Permissible Exposure Limits—Annotated Tables*, OSHA, Washington, DC, <www.osha.gov/dsg/annotated-pels/index.html> [accessed 27 April 2018]

Safe Work Australia (SWA) 2018a, *Model Work Health and Safety (Blood Lead Removal Levels) Amendment Regulations 2018–Model Provisions*, SWA, Canberra, <www.safeworkaustralia.gov.au/doc/model-work-health-and-safety-blood-lead-removal-levels-amendment-regulations-2018-model> [accessed 27 July 2018]

—— 2018b, *Workplace Exposure Standards for Airborne Contaminants*, SWA, Canberra, <www.safeworkaustralia.gov.au/doc/workplace-exposure-standards-airborne-contaminants> [27 April 2018]

—— 2018c, *Hazardous Chemicals Information System (HCIS)*, SWA, Canberra, <http://hcis.safeworkaustralia.gov.au> [accessed 27 July 2018]

Standards Australia 1993a, Venous Blood—Determination of Lead Content—Flame Atomic Absorption Spectrometric Method, AS 2411: 1993, SAI Global, Sydney.

—— 1993b, Whole Blood—Determination of Lead Content—Graphite Furnace Atomic Absorption Spectrometric Method, AS 4090: 1993, SAI Global, Sydney.

—— 1994, Sampling of Venous and Capillary Blood for the Determination of Lead or Cadmium Concentration, AS 2636: 1994, SAI Global, Sydney.

—— 2009a, Workplace Atmospheres—Method for Sampling and Gravimetric Determination of Respirable Dust, AS 2985: 2009, SAI Global, Sydney.

—— 2009b, Workplace Atmospheres—Method for Sampling and Gravimetric Determination of Inhalable Dust, AS 3640: 2009, SAI Global, Sydney.

Welding Technology Institute of Australia 2015, *Fume Minimisation Guidelines—Welding, Cutting, Brazing and Soldering*, WTIA, Sydney, <https://wtia.com.au/resources/fume-minimisation-guidelines> [accessed 27 April 2018]

9. Gases and vapours

Aleks Todorovic

9.1   INTRODUCTION

The occupational work environment may contain vast amounts of atmospheric contaminants, many of which can be categorised into gases and vapours. Although many gases are essential to life, air that is contaminated with hazardous gases and vapours risks workers' health and wellbeing.

This chapter will concentrate on a few of the most common gases and vapours that are seen in industry and examine their implications for health and hygiene in the workplace. The toxicological review is limited to the gases noted, and some mechanisms to control those compounds in order to prevent them from posing a health risk are discussed. It must be noted that an enormous number of gases and vapours are harmful to health, so the reader is advised to consult other advanced texts and resources for expanded information. It is also recommended that practitioners regularly consult other resources, such as websites with regulatory information, for recent developments.

Today's H&S practitioner has a choice of a wide array of passive, active and real-time detection devices for measurement of the workplace environment. The reader will be guided into a detailed review of the various ways in which gases and vapours can be monitored, the variation between the different types of monitoring that are employed and the limitations of such monitoring techniques.

The nature of the potential hazards that exist within confined spaces is explained and, finally, the ways in which the H&S practitioner can employ the latest emerging technologies to monitor gases and vapours and to analyse data will be discussed.

9.2   GASES

Gases range from simple gases such as helium to complex hydrocarbon compounds found in the petrochemical industry. Gases may be heavier, lighter or have the same density as air, and may be colourless and odourless.

Many gases can be stored as pressurised liquids until released for use. As gases will fill any available volume, if not controlled they may enter and contaminate the workplace environment. Gases used in workplace processes include:

- chlorine in water treatment plants
- nitrous oxide (laughing gas) in anaesthesia
- ammonia in refrigeration plants
- oxygen from liquid oxygen sources

- phosphine/methyl bromide in grain fumigation
- ethylene oxide in sterilisation machines in hospitals.

Other gaseous contaminants may arise as by-products of industrial processes. Typically, this occurs where some sort of chemical reaction has taken place, or where the gas is produced in the breakdown of a complex chemical—for example:

- carbon monoxide from incomplete burning of natural gas in ovens and kilns
- oxides of nitrogen from diesel exhausts
- ozone from photocopying or some electric arc welding machines
- formaldehyde off-gassing from particle board
- hydrogen sulfide from sewers and wastewater treatment plants.

9.3   VAPOURS

Vapours are the gaseous state of any material that would under normal temperature and pressure exist as a solid or liquid. In the workplace, thousands of organic chemicals used as solvents, adhesives, paints, chemical reactants, catalysts, sterilants or processing aids produce vapours through natural evaporation, heating or spraying.

Organic chemicals in the form of solvents are by far the largest source of vapours in workplaces. These organic compounds are often chosen as solvents because of two special abilities: they will dissolve a material and they will readily evaporate into the atmosphere. This last property leads to many of the problems of exposure. Some typical examples of organic solvents are acetone, chloroform, ethanol, chlorofluorocarbons (CFCs), hexane, perchlorethylene, benzene, toluene and xylene. Organic solvent-based chemicals are found in the factory, the farm, the office and the home.

Elements such as mercury can produce hazardous vapours. In industries such as the petroleum industry, other hazardous vapours can be produced, such as benzene, toluene and alkyl lead vapours.

A few materials actually sublime—that is, convert from the solid directly to the vapour. Examples include naphthalene (mothballs) and paradichlorobenzene (deodorant).

9.4   WARNING SIGNS AND INDICATORS

9.4.1   ODOUR

Gases and vapours are mostly invisible. Many have strong and characteristic odours which give warning of their presence in the workplace, but others have no warning odour (e.g. carbon monoxide) and with others, harmful effects may be caused at concentrations well below the odour threshold. With some substances, warning odorants are added to prevent inadvertent exposure and/or to permit detection of the gas's presence. For example, the odorant ethyl mercaptan is used in liquefied petroleum gas (LPG) and natural gas supplies. Lastly, individuals vary greatly in their ability to detect odours, and even in the same person this ability may vary from day to day. For example, the range of odour detection of methyl ethyl ketone (MEK) lies between 0.07 and 339 ppm. In other words, while odour can be useful in detection, there are a number of limitations to its use.

Odour is often a good warning indicator for bacterial decay; however, it cannot easily indicate the degree of exposure. For some gases (e.g. hydrogen sulfide), higher concentrations cause olfactory fatigue and the contaminant can no longer be smelled even when the concentration of the gas may be at lethal levels. Although odour is often mistakenly associated with harm, it is an unreliable indicator of harmful conditions, and is not a measure of toxicity. Some gases are deadly yet do not smell, while others are odorous at levels far below those that will cause harm—they just smell unpleasant.

For many volatile materials, the odour threshold may be one-tenth to one-thousandth of the workplace exposure limit (OEL). Table 9.1 shows the relationship between approximate odour thresholds and OEL for a number of common workplace gases and organic vapours.

Table 9.1 Comparison of odour thresholds and OEL for some common workplace gases and vapours

Substance	Odour threshold (ppm)	Australian WES TWA (ppm)
Acetone	0.4–11 745	500
Ammonia	0.043–60.3	25
Benzene	0.47–313	1
Ethylene oxide (oxirane)	0.82–690	1
Hydrogen sulfide	0.00004–1.4	10
MEK (methyl ethyl ketone)	0.07–339	150
Ozone	0.0031–0.25	0.1 Peak
Styrene (monomer)	0.0028–61	50
Toluene	0.021–157	50
1,1,1 Trichloroethane	0.97–715	100
Xylene	0.012–316	80

Sources: Murnane, Lehochy and Owens (2013); Safe Work Australia (2018).

Although odour cannot be relied upon as a means of measuring hazardous gases in the workplace, it is still a very useful property. Workers are equipped with a very sensitive odour detector: the nose. In fact, the nose is often much more sensitive to certain chemicals than analytical devices. Odours are easily masked, however, and the nose is very poor at distinguishing mixtures of materials.

9.4.2   IRRITATION

Some gases and vapours may reveal their presence by various irritating effects:

*   respiratory irritation, coughing, asthma
*   lachrymatory action on the eye (tearing)
*   cloudy vision (at high concentrations) or other visual disturbances
*   acidic taste
*   metallic taste (organometallic compounds).

Acidic gases, including chlorine and sulphur dioxide, may be evident because of respiratory irritation rather than smell. In some workers, formaldehyde may cause eye irritation before it can be detected by smell. Investigation of the workplace may reveal a range of processes that produce gases or vapours, and the H&S practitioner will need to distinguish between those that have reliable warning properties and those that do not.

9.5   OCCUPATIONAL HEALTH ASPECTS OF GASES AND VAPOURS

For health and safety purposes, gases and vapours can be classified into three groups:

*   irritants
*   asphyxiants
*   those with miscellaneous effects.

Both irritants and asphyxiants tend to give rise to relatively acute responses, leading to rapidly observed effects. The miscellaneous effects may be acute or chronic.

9.5.1   IRRITANT GASES AND VAPOURS

Irritants cause inflammation of the tissues exposed to them. Symptoms of exposure can range from mild irritation of the mucous membranes (eyes, nose and throat) to severe damage to the lung (e.g. by ammonia, chlorine, phosgene, formaldehyde and nitrogen dioxide).

9.5.2   ASPHYXIANTS

Asphyxiant gases fall into two groups:

*   *simple asphyxiants*, a vapour or a gas that can cause unconsciousness or death by suffocation (lack of oxygen). Most simple asphyxiants are harmful to the body only when they become so concentrated that they reduce oxygen in the air (normally 21 per cent) to dangerous levels (18 per cent and under). Asphyxiation is one of the principle potential hazards of working in confined spaces.
*   *chemical asphyxiants*, which interfere with the body's ability to take up and transport oxygen; they include exposure to low concentrations of carbon monoxide, hydrogen sulfide or hydrogen cyanide.

9.5.3   *GASES AND VAPOURS WITH MISCELLANEOUS EFFECTS*

Many gases and nearly all vapours from solvents fall into this category. Effects include acute effects—for example, on the central nervous system (CNS), and chronic toxic effects on many different organs of the body. Table 9.2 presents some typical examples of miscellaneous effects caused by gases and vapours that may occur in industrial workplaces.

Table 9.2 Gas or vapour hazards demonstrating various health effects

Gas or vapour	Health effect
Benzene	Leukaemia
Carbon disulfide	Cardiac disease
Coal tar pitch volatiles	Skin sensitisation, lung cancer
Ethyl glycol monoethyl ether	Foetotoxic effects
Fluorocarbons	Cardiac arrhythmias
Helium	Vocal changes
Methyl bromide	Cardiac effects
Mercury	Brain damage, kidney damage, gingivitis, tremors, erethism
n-hexane	Peripheral-nerve neuropathy
Nitroglycerine	Vasodilation, decreasing blood pressure
Nitrous oxide	Analgesia
Oxygen (deficiency)	Brain disturbance, CNS effects
Oxygen (excess)	Pulmonary inflammation, lung oedema
Toluene	Headache, confusion, loss of memory
Trichloroethylene	Psychoactive effects
Vinyl chloride	Angiosarcoma of the liver

9.6   UNDERSTANDING EXPOSURE MEASUREMENT IN THE WORKPLACE

The H&S practitioner needs to understand what exposure is occurring in the workplace. While the exposure standards are listed as a single value for an eight-hour average and/or a fifteen-minute average, or a peak limitation, in reality the exposure changes continually throughout a work shift. Exposures to a particular contaminant are averaged over the period of the shift to give the daily exposure of the worker as a single value, which is then compared with the exposure standard for that contaminant. Figure 9.1 shows the output

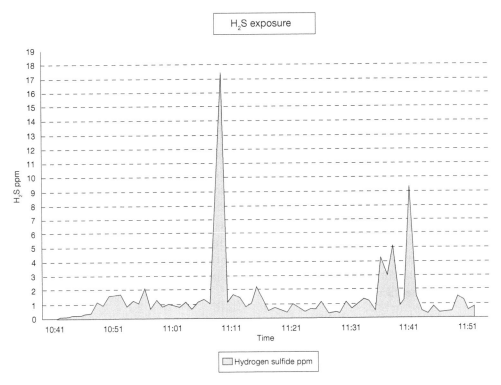

Figure 9.1 Hydrogen sulfide levels at a bitumen tank hatch

from a data logger measuring at one-minute intervals the hydrogen sulfide (H_2S) exposure of a worker near the hatch of a tank. This highlights the changing concentrations that can occur over a work shift. Any instantaneous reading from a 'grab sample' will be only one result at one time in a changing environment. (A grab sample is an air sample collected over a short period, usually between one and five minutes.)

In the above case, if no further exposure occurs in the eight-hour shift, then the hydrogen sulfide TWA is only 0.2 ppm—well below the OEL of 10 ppm. Yet clearly, during one short period, the instantaneous readings were well in excess of this standard.

Further information on the variation of exposure with work routine can be compiled using video exposure monitoring (VEM), which combines video of a worker with continuous measurements, which are presented in graphic form on screen.

9.7   MONITORING OF GASES AND VAPOURS

Many gases and organic vapours occur in various workplaces, and a variety of different techniques are required to assess them.

Collection of gas and vapour samples for monitoring is relatively simple compared with dust sampling. The H&S practitioner merely measures the total concentration in the atmosphere. Generation of mists or aerosols changes the characteristics from a gas phase to a finely dispersed liquid, however, so care must be taken to understand the nature of the chemical and its form when selecting the sampling technique.

While some gases and vapours may be dissolved in the moisture of the upper airways, others can reach the alveolar regions of the lung, from where they may pass into the blood-stream. While a considerable proportion of inhaled gas or vapour will be exhaled, exposure measurement is based on the total amount available for inhalation.

There are two key factors involved in the monitoring of gases and vapours:

- identification—which gas (or gases) or vapours—which contaminants—are present?
- quantity— how much of the gaseous contaminant is present?

The answers to these questions will decide how the worker's exposure is measured.

Determining *what* is in the air (identification) can be done by collecting an air sample and analysing it in the laboratory, or by the use of direct-reading instruments such as portable gas chromatographs, or instruments for detecting specific contaminants such as hydrogen sulfide or carbon monoxide. Determining *how much* contaminant is present in the worker's environment requires monitoring the specific contaminant over a certain time period. This requires either a direct-reading instrument or the collection of the contamin-ant during the work period and subsequent laboratory analysis.

Other factors include:

- the measurement task
- selectivity to the target gas or vapour and sensitivity to interfering gases and vapours
- susceptibility of the sampler to environmental factors
- fitness for purpose, e.g. weight, size, durability
- training requirements for the reliable operation, maintenance and calibration
- the total cost of purchase and operation, including calibration and maintenance
- compliance with the performance requirements of appropriate national or local governmental regulations.

9.8   OPTIONS FOR MONITORING THE WORKPLACE AIR

Several techniques are available for gas and vapour monitoring:

- conventional monitoring of the air—more than 98 per cent of all gases and vapours are investigated in this fashion
- monitoring the worker for uptake of a gas or vapour by examining a biological index for the particular substance. This may be done by measuring excretion of the substance or measuring a metabolite of the inhaled substance. The results are then compared with biological exposure indices (BEI®). This is covered in Chapter 10.
- analysing the air exhaled by the worker at some time after the end of a work shift (e.g. measuring the carbon monoxide in exhaled air).

9.9   MONITORING THE WORKPLACE AIR

For the present, emphasis will be placed on air monitoring because of its simplicity and practical utility. The H&S practitioner will find that quite a large number of regularly occurring gases and vapours can be measured conveniently with modest equipment.

As discussed in Chapters 4 and 6, air monitoring is a requirement when respiratory protection programs are introduced. Two approaches are widely used:

- direct reading instruments for use in the field
- sample collection with subsequent laboratory analysis.

9.10 DIRECT-READING INSTRUMENTS

The ability to examine exposure profiles in real time using simple hand-held devices enables the identification of specific tasks or procedures that require exposure control without the need for laboratory analysis, which may take considerable time. The use of direct-reading instruments improves the effectiveness of the H&S practitioner and the safety of workers. The benefits of employing direct-reading equipment are both qualitative and quantitative. Other reasons for using direct-reading instruments in the field are:

- the ability to obtain a result for a known chemical immediately
- the capacity to identify high short-term exposures that may lead to acute effects, and thus facilitate entry into confined spaces
- the ability to check on extraction processes and other control procedures, identifying leaks and major breakdowns
- the potential for the instruments to serve as alarms for evacuation or remedial action
- the ability to provide estimates of longer-term exposures if used with correct sampling procedures.

Hundreds of common gas or vapour contaminants require monitoring in the workplace, and modern chemical and electronic technologies have provided a range of instruments for measuring many of them. A good number are well within the operational capability of an appropriately trained H&S practitioner.

Here we examine only the instrument types that find regular use in workplace surveillance. Their use may range from occasional to constant, depending on the nature of the hazard, how often processes change and the effectiveness of the control procedures.

9.11 DIRECT READING: COLOURIMETRIC DETECTOR TUBES

For many gas and vapour monitoring tasks, the most convenient testing instrument is a colourimetric detector tube or detector tubes. Figure 9.2 shows the Dräger Accuro. Detector tubes operate on the principle of a reaction between the test gas (contaminant being measured) and a chemical reagent bound to an Inert matrix.

A hand-operated bellows or a piston pump draws a calibrated volume of contaminated air through a tube containing a reagent chemical. Contaminants in the air will react with the reagent in the tube to produce a progressive colour change along the tube as the air passes through. The length of the stain in the tube is proportional to the concentration of contaminant present and is read directly from the graduations marked on the tube.

Detector tubes are inexpensive, and tubes are available for a wide range of applications. No special expertise is required to operate them and they are excellent for monitoring

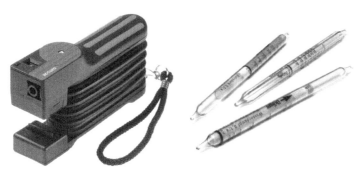

Figure 9.2 Dräger Accuro, with a selection of detector tubes

single known contaminants. Disadvantages include cross-sensitivities to other unknown contaminants, which can lead to false readings. In addition, the tubes often have a limited shelf life and are not always available in the range of concentrations required.

Considerable expertise is required to interpret results. Samples collected by this means are grab samples, representing concentrations or a 'snapshot' at particular points in time. Potential problems arise when concentrations are potentially subject to rapid change. The user must consider the work task being monitored to account for measurements of contaminants that may be discontinuously present and how the results should be interpreted in terms of the relevant exposure standard.

H&S practitioners responsible for work areas where gases and vapours constitute a significant proportion of air contaminants should consider obtaining these useful gas-detector tubes. (The kit includes a hand-operated sampling pump and accessories; the appropriate detector tubes need to be specified for the gas/vapour of interest and are purchased separately.) Several brands are available.

9.11.1   LIMITATIONS OF COLOURIMETRIC TUBE SYSTEMS

Detector tube systems should not be used to establish whether or not a contaminant is present, only for measuring already known contaminants. The following are some common problems:

- Chemical mixtures may negatively influence the results of detector tubes. Cross-sensitivity from the presence of interferents other than the target analyte may lead to false positive or false negative results.
- False negatives can arise if the incorrect measuring range is chosen.
- False negatives can arise if the identity of the contaminant is not known and an incorrect tube is selected.
- Humidity and temperature may affect detector tube performance, leading to possible false readings.
- Readings may need to be adjusted to account for variations in atmospheric pressure (e.g. at high altitudes).
- There is an inability to provide dynamic results for concentrations that vary with time.

9.11.2 PRACTICAL INSTRUCTIONS

H&S practitioners intending to purchase and use one of these devices should consult the various manufacturers. Dräger, MSA, Gastec, Kitagawa and Honeywell produce different designs, including hand bellows, thumb pump and piston types. All manufacturers offer a wide range of tubes, but a measuring range will need to be chosen to suit the particular workplace and its contaminants. Pumps and tubes from different systems cannot be interchanged, for there is no guarantee that the pumped sample volumes will be identical for each system unless they are confirmed to have the same sampling volume and flow rate. Manufacturers or suppliers usually offer training in the use of this equipment, together with appropriate selection of measuring tubes. As with all monitoring equipment, the manufacturer's instructions should always be read carefully before use.

9.11.3 PASSIVE DOSIMETER TUBES

Passive dosimeter tubes closely resemble standard detector tubes, which are placed in the worker's breathing zone. Rather than a piston or a bellows pump drawing the sample into the tube, contaminants passively diffuse into the tube. The length of the stain is proportional to the concentration of the atmospheric contaminant being sampled. These tubes provide a time-weighted average reading rather than instantaneous peak measurement.

9.12 DIRECT-READING BADGES

Several specific direct-reading badges are now available (e.g. for formaldehyde, isocyanates, ethylene oxide). When exposed to contaminating vapours of a specific kind the badge colour changes, and as the analyte concentration and exposure time increase the colour change becomes more intense. To use these badges successfully, the H&S practitioner should be certain of the identity of the contaminant to be measured.

9.13 SPECIFIC DIRECT-READING MONITORS

As technology has advanced, so too has the range and sophistication of specific direct reading monitors and detectors. Instruments for gases such as carbon monoxide, sulphur dioxide, phosphine, carbon dioxide, hydrogen cyanide, chlorine, nitrogen dioxide, nitric oxide, ammonia, flammable gases, hydrogen sulfide, hydrogen, formaldehyde and chlorofluorocarbons, as well as for oxygen, are widely used. These instruments work on various principles of electrochemistry, infrared spectrometry, photoionisation or direct-reading paper tape colourimetry. Figure 9.3 shows a carbon monoxide monitor being used to sample a work area.

Instrument monitors are made for fixed locations as well as for portable use. Fixed-location instruments typically are found in processing plants where fugitive gases and vapours can pose potential hazards to unprotected workers. In oil refineries and petrochemical facilities, fixed-location detectors are deployed to detect leaks or emissions from flanges and seals in processing areas, bulk storage areas and plant modifications. These fixed detectors are normally part of an integrated monitoring system whereby readings

Figure 9.3 Carbon monoxide monitor in use

are relayed to a central monitoring point normally located in a control room or on a centralised status panel. Instruments for personal use are often fitted with visible, audible and vibration alarms set at or just below the appropriate OEL.

Modern instruments continuously monitor the surrounding atmosphere and have onboard data-logging capabilities that allow for vast amounts of data to be collected. Data logging permits calculation of TWA concentrations, identification of any peak exposure, and identification of any exposure pattern in the workplace (as illustrated in Figure 9.1, above). These instruments are easily able to be connected to computers to allow for the data to be downloaded and interrogated further.

Figure 9.3 shows one of many instruments that are useful for direct field measurements. Modern direct-reading monitors are easy to use, but their application needs to be properly considered. Some designs are primarily used for personal monitoring whereas others are used as survey monitors.

Direct-reading monitors can range in price from several hundred dollars to thousands of dollars, so frequent use may be needed to justify their purchase. Many models can detect up to six gases in a single instrument package. For example, the oxygen/explosive gas/hydrogen sulfide and carbon monoxide type is indispensable for confined-space entry (e.g. sewers, silos, silage pits and mines) and particularly useful for other confined-space applications (discussed in Section 9.21). Figure 9.4 shows two multi-sensor types.

When considering purchase of one of these instruments, it is advisable to bear in mind:

- the need for regular calibration checks
- servicing requirements (electrochemical cells have limited lives, from six months up to ten years, depending on sensor type)
- the fact that use is strictly limited to particular contaminants
- the possibility of cross-contamination or sensor poisoning
- the high initial cost, although some instruments can be hired.

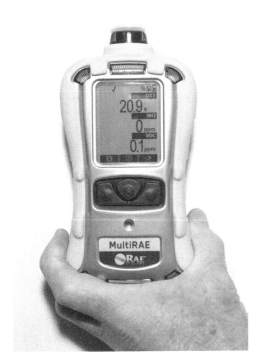

Figure 9.4a MultiRAE 6 gas monitor

Figure 9.4b Dräger X-AM 5600 6 gas monitor

Catalytic explosive sensors can be poisoned by sulfide, silicone compounds (e.g. insect repellants), phosphates and alkyl leads.

9.14 GENERAL PURPOSE DIRECT-READING ANALYSERS

Several devices are now available for general-purpose gas and vapour detection:

- portable gas chromatographs
- opto-acoustic gas analysers
- infrared analysers.

These very versatile instruments provide occupational hygienists with great sensitivity and selectivity in monitoring workplace organic vapours, but are not generally recommended for use by H&S practitioners. In a workplace that encounters some dozens of organic vapours (e.g. a chemical manufacturing plant, solvent reprocessing plant, large manufacturing plant), general-purpose devices can provide an efficient and rapid means of surveillance during various processes.

Another kind of general testing instrument, the photoionisation detector (Figure 9.5), finds use with a wide range of organic vapours. This device cannot distinguish between the different contaminants it is able to detect; rather, it provides a single cumulative reading for all the detectable substances present at any moment. However, this instrument can be calibrated for known substances and is useful for single-contaminant situations.

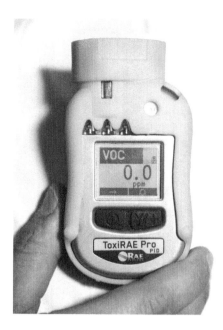

Figure 9.5 ToxiRAE Pro PID (photoionisation detector)

It has excellent sensitivity to many compounds encountered in the workplace, particularly solvents such as acetone, MEK and toluene.

Recent advances in photoionisation detection instruments mean they can now be used as personal monitors—for example, the ToxiRAE Pro PID shown in Figure 9.5.

9.15   SAMPLE COLLECTION TECHNIQUES

Conventional sample collecting techniques may need to be applied where:

- time weighted average concentrations are required, or
- no direct reading instruments are available.

Chapter 2, on occupational health, indicated that much of our knowledge of the long-term hazards from exposure to hazardous gases and vapours is based on average long-term exposures in the workplace. While a few short-term high or peak exposures may lead to irritation, narcotic effects and feelings of nausea, giddiness and so on, it is often the long-term lower exposures that lead to sensitivity, organ damage, neurotoxic effects and cancer. For this reason, we may need to estimate exposure over longer periods.

Measuring exposure to chemicals implies there is a OEL that can be used to determine compliance or otherwise. While OELs vary slightly from country to country, the method of measurement does not. Reference should be made to the regulated standards in the country in question. If none exists for a particular chemical, then most professionals use either the American Conference of Governmental Industrial Hygienists (ACGIH®) Threshold Limit Values (TLVs®) (ACGIH, 2018) or the British HSE Workplace Exposure Limits as guidelines.

To comply with these standards, time-weighted average exposure measurements generally are required, principally as personal samples.

Five different techniques are regularly used. In order of practical importance, they are:

- collection of gas or vapour by pumping the contaminated air through a small sampling tube filled with a suitable absorbent
- passive adsorption of the contaminating gas or vapour onto a badge or tube dosimeter worn on the lapel of the worker
- collection of the contaminating gas or vapour by bubbling it through a suitable liquid in a liquid impinger (this is less common)
- passing the gas or vapour through a filter impregnated with a material which reacts with the contaminant
- collection of samples in bags or pumped cylinders.

9.15.1   TUBE SAMPLING

This is the most commonly employed method of reliable vapour monitoring in the workplace. Sampling techniques for monitoring and interpretation can be found in AS2986.1 Workplace Air Quality—Sampling and Analysis of Volatile Organic Compounds by Solvent Desorption/Gas Chromatography—Pumped Sampling Method (Standards Australia, 2003a). Other methods are the US NIOSH Analytical Methods, the US OSHA Methods or the UK HSE Methods of Determination of Hazardous Substances (MDHS).

The media inside the tubes are selected to optimise absorbance or adsorbance for a specific contaminant, and normally are activated charcoal, silica gel or a number of special chemicals. A small calibrated sampling pump is set at a constant flow, anywhere between 20 and 250 ml/min, and draws the contaminated air through the adsorbent. Sampled tubes are usually stable for periods of days or weeks if properly capped and stored in a refrigerator.

Sampled tubes have to be assessed in a chemical laboratory. The contaminant can be stripped from the adsorbent by:

- the action of a powerful solvent such as carbon disulfide
- thermal desorption in a hot furnace (this requires special tubes).

The concentration of contaminant in the air is measured by either gas chromatography or high-performance liquid chromatography. These measurement facilities generally are available only in commercial, research or government laboratories. The sampling, however, can be done by any H&S practitioner who has the necessary low-flow sampling pumps, and who can apply the appropriate Australian Standard, NIOSH, OSHA or HSE method and is appropriately trained.

Where a government inspector is auditing risk assessment procedures under the Australian Work Health and Safety Regulations (or their equivalent), the employer or the H&S practitioner may be requested to provide time-weighted average (TWA) contaminant concentrations for various workers. Use of a pumped sample tube that is analysed by a competent laboratory is generally an acceptable method of meeting this need.

There are some limitations to this method:

- Most permanent gases cannot easily be determined by such methods. (For hydrocarbons, only butane and heavier gases can be measured by this method.) The limitations of the analysis should be discussed with the testing laboratory.

- Breakthrough of a sorbent tube is when the tube becomes too full and has no further capacity to retain the contaminant on the media. It normally occurs when the sample volumes are too large or the flow rates are too great. The result of breakthrough is an under-estimation of the worker's exposure.

Breakthrough can be determined by using a tube with two segments: a normal trapping segment and a back-up segment. Should contaminant be present in the back-up layer, breakthrough can be assumed to have occurred and the sample should be voided. Total volume of sampled air should be kept low enough that breakthrough is prevented, or at least minimised. These sampling tubes again come with a wide variety of adsorbents and with different weights of adsorbent packing. The typical tube is the standard NIOSH tube, which has a 100 mg packing with a 50 mg back-up section. For higher volumes, the 1000 mg 'jumbo' sampling tubes, again with a back-up section, can be used. These sampling tubes are designed for analysis by solvent desorption. Breakthrough cannot be ascertained in tubes which are thermally desorbed, as these samples allow only 'one shot' analysis—that is, repeat testing of the sample is not possible because all the contaminants are destroyed in the analysis.

Figure 9.6 shows the arrangement of a portable low-flow sampling pump and an organic-vapour sampling tube and holder. The sampling tube is placed near the breathing zone of the operator, within about 15–30 centimetres of the nose.

In petrochemical industries, which handle highly flammable materials, any equipment taken on site, even a small sample pump, must be 'intrinsically safe'—that is, it must not be a source of ignition in these combustible atmospheres. Requirements for intrinsic safety vary from company to company, and approval must be obtained to use such equipment on site before a monitoring survey is undertaken (usually through a daily safe-work permit system). As a guideline, in the United States monitoring equipment rated intrinsically safe by Underwriters Laboratories Inc. (UL) is accepted; in Europe, equipment meeting the code ATEX II 2 G EEx ib IIC T4 is accepted; and in Australia, instruments with ANZEx or IECEx certification for intrinsic safety are accepted. These ratings are usually displayed on the equipment. Refer to Figures 9.7a, b and c.

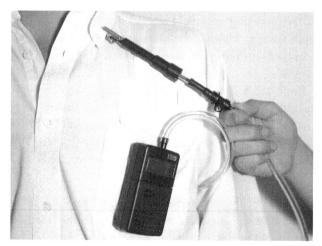

Figure 9.6 Worker being fitted with a low-flow sample pump and adsorbent charcoal tube

Caution: Take careful note of sample pump and monitoring equipment requirements for petrochemical industries.

Figure 9.7a SKC® Airchek 5000

Figure 9.7b SKC® Pocket pump

Figure 9.7c SKC®Airchek 3000

9.15.2   *PASSIVE ADSORPTION SAMPLING*

Passive samplers or tubes are a simple and convenient method by which to sample for organic vapour contaminants at a workplace. Figure 9.8 illustrates a typical organic-vapour sampling badge, which would be placed in the breathing zone of a worker. Figure 9.9 shows a specific sampling badge for formaldehyde. In these passive approaches, sampling pumps are not required.

These samplers contain layer/s of material capable of absorbing the contaminant of interest. They also contain other sorbent materials to remove possible interferents. Passive samplers operate by allowing the contaminant molecules to diffuse through the sampler and bond to the sorbent material inside. Techniques for using this method are to be found in AS2986.2 Workplace Air Quality—Sampling and Analysis of Volatile Organic Compounds by Solvent Desorption/Gas Chromatography—Diffusive Sampling Method (Standards Australia, 2003b), US NIOSH methods, OHSA methods or UK HSE methods.

For large surveys of organic vapours (e.g. in commercial printing, boat building, spray painting, motor body repairs and the petrochemical industry), passive sampling is a very cost-effective way for the H&S practitioner to undertake vapour sampling. However, before any sampling for organic vapours begins, it is important to contact both the

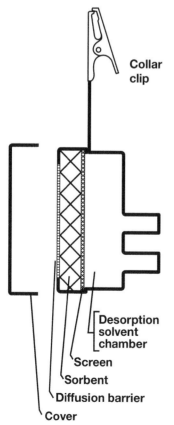

Collar clip

Desorption solvent chamber

Screen

Sorbent

Diffusion barrier

Cover

Figure 9.8 SKC® badge, a typical organic-vapour sampling badge

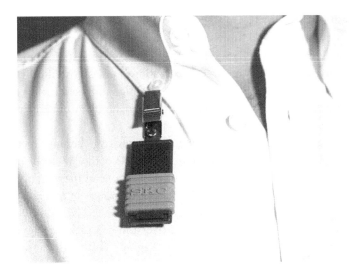

Figure 9.9 UMEX®
badge

sampling device manufacturer/representative and the analytical laboratory for guidance on the correct sampler for the task, and allow the laboratory to arrange proper calibration procedures.

This is expensive laboratory time for which the H&S practitioner will have to pay. There are several passive adsorption badges available for specific contaminants, but the badges or tubes must be analysed by a competent laboratory.

Once collected and properly capped, the samples are sent to an appropriate laboratory for analysis. As with tube sampling above, the contaminants are desorbed with a specific solvent or by heat. Figure 9.10 shows a laboratory gas chromatograph fitted with an automated thermal desorber suitable for unattended analysis of organic-vapour samples.

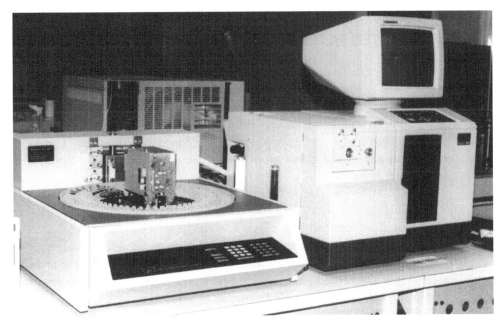

Figure 9.10 Laboratory gas chromatograph fitted with an automated thermal desorber

9.15.2.1 Limitations of passive sampling

As passive samples are designed to work in moving air, it is important to ensure that the worker is moving during sampling. They are generally not designed for static applications to obtain 'workplace environmental' vapour concentrations, unless specifically designated as suitable for this, as stagnant air can 'starve' the sampler and yield useless results. Typically, a passive sampler must be worn for more than four hours to obtain sufficient material for the laboratory to analyse.

If the workplace is being monitored for solvent mixtures (e.g. hexane, methylene chloride and xylene), the H&S practitioner will have to provide complete detailed information to the processing laboratory or, if possible, provide a sample of the solvents.

On no account should exposed sample tubes or badges and containers of solvent be transported or sent in a single enclosure.

Passive badge or tube samplers are extremely sensitive, and can pick up vapour that may leak or diffuse through seals and inappropriate packaging. This will ruin sampling efforts completely.

9.15.3   IMPINGEMENT INTO LIQUID SAMPLES

A number of gases or vapours contaminating workplaces cannot successfully be trapped on tubes or badges, but they can be trapped by bubbling them through water or some other solvent. These bubblers are called liquid impingers. Contaminated air is drawn by a small pump through the absorbing liquid, at around 1 litre per minute. For safety reasons, it is often better to collect area samples rather than personal samples if the absorbing solution contains corrosive substances (acids or alkalis) or hazardous solvents, unless spill-proof impingers can be used. In many cases, though, this sampling method has been superseded by more modern techniques.

9.15.4   IMPREGNATED FILTERS

Some vapours can be trapped successfully on filters chemically impregnated with a substance that is reactive to the contaminant in question. The method is similar to the vapour tube technique, but more versatile in that it can simultaneously trap particulates. For instance, an aluminium smelter will produce both gaseous fluoride and particulate fluoride, both of which can be trapped on a filter doped with citric acid. This method can likewise trap aerosol and gaseous forms of certain paints, which exist simultaneously in spraying operations.

Other examples are (a) glass-fibre filters impregnated with 2,4-dinitrophenylhydrazine for sampling of aldehydes and (b) glass-fibre filters impregnated with methoxy-phenylpiperazine for sampling isocyanates.

All these sample-collection techniques provide time-weighted average exposure readings because they are collected and integrated over a long sampling period. Where information is also needed on short-term or any peak exposure, the H&S practitioner will have to use one of the grab sample techniques as well.

9.15.5 *COLLECTION OF SAMPLES IN BAGS OR PUMPED CYLINDERS*

In some cases, the above absorption methods are unsuitable to identify the contaminants in the workplace and other techniques need to be used.

Some samples can be collected successfully in impervious bags (Figures 9.11a and b) or in evacuated pressurised cylinders or canisters (Figures 9.11c and d) and returned to a laboratory for analysis. This technique is required for many of the permanent gases (e.g. oxygen, carbon monoxide, methane, hydrogen), which cannot be trapped by any other means. However, the vessel must retain the sample adequately, without loss. Some plastics will adsorb organic vapours, and small molecules (particularly hydrogen) will diffuse through plastics. The H&S practitioner must ensure that the plastics are specified as suitable for their intended purpose. As an example, some compounds are photosensitive, so the sample bag chosen must be able to preserve the integrity of the sample.

To obtain a representative sample, bags are simply inflated and emptied, and refilled several times with the gas that is to be tested. Small sampling pumps with a direct outlet can be used for this purpose. A hand-operated pressure pump can also be used to collect gas analysis samples in bags. Pressurised cylinders are under negative pressure, and the air sample is collected upon opening the valve. Flow restrictors maintain flow stability over the desired sample period from less than one minute to twelve hours. The large cylinder can be fitted with a vacuum gauge to monitor the flow.

This is essentially a grab sample, and will provide information only on the contaminants present at the time it was collected; it cannot be used to determine the time-weighted average. It can only be used as a time-weighted sample if the sample has been collected uniformly over the whole shift—that is, if the bag or cylinder was filled at the same flow rate over the shift. It is also indicative only of the general environment where the sample was taken, and not the personal exposure of the worker. It is, however, useful to evaluate whether engineering controls are operating correctly.

9.16 TESTING COMPRESSED BREATHING AIR

Compressed breathing air finds a number of applications in workplaces, including in underwater breathing apparatus, compressor-fed airline breathing apparatus for abrasive blasting, spray painting in booths and breathing sets for firefighting and rescue work.

In all these situations, the air produced by a compressor must be perfectly suitable for breathing under all operating conditions. Compressors can introduce contaminants to the air through incorrect siting of air intakes or through operation of the compressor itself. Oil vapour arises from compressor oil, carbon dioxide arises from contaminated intake air and carbon monoxide can arise from partial breakdown of the compressor oil as well as from intake air sources. Water vapour will be more concentrated in the compressed air unless it is adequately removed by driers.

Special tests need to be conducted on compressed breathing air to ensure its quality. The instrument most widely used for this purpose is the Dräger Aerotest. Cylinder air or air directly from a compressor receiver (after passage through a conditioner) can be tested. Air compressor systems require regular testing, and should be checked thoroughly after initial installation or any servicing. This is generally best left to outside experts.

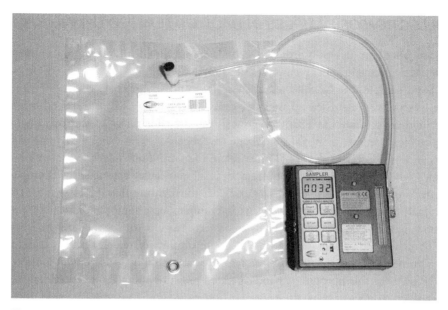

Figure 9.11a Positive-pressure sample bag

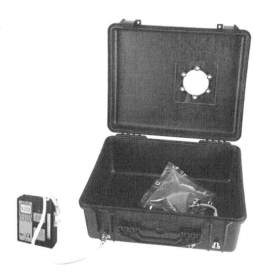

Figure 9.11b Negative-pressure sample bag

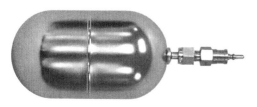

Figure 9.11c Small pressurised canister

Figure 9.11d Large pressurised canister

9.17   TIPS FOR PERSONAL MONITORING

Every personal monitoring sample has a 'story' attached to it, and this should be recorded by answering the following questions:

* Why was there potential for exposure?
* What chemicals and processes were used?
* What did the worker do during the sampling period?
* Did they come into contact with these chemicals?
* What was the physical form of the contaminant?
* How was the chemical used—as a liquid, as an aerosol, sprayed?
* Were there any other chemicals used nearby to which this person could inadvertently have been exposed (fugitive emissions)?
* Was respiratory protection used? Was it worn? Was it suitable? Did it work correctly?
* Was other personal protection used? Gloves, apron? (Often, we measure air contaminants, but greater exposure may occur through unprotected skin contact with liquids and vapours.)
* Were engineering controls used—ventilation/extraction systems?
* Were there any spills, process upsets, known fugitive emissions?

9.18   MONITORING THE WORKER: BIOLOGICAL MONITORING

Monitoring the worker's health directly is another way of checking on the quality of the work environment. This is referred to as biological monitoring, and the worker is the 'sampling device', accounting for all exposure routes. Biological tests can provide information on the uptake of a substance by the worker, whereas air testing cannot. For example, blood cyanide tests on workers in the cyanide manufacturing industry can indicate whether they need to be removed from further cyanide exposure.

Few industries make regular use of biological monitoring, however—partly because it may involve medical tests and partly because it cannot be done without the complete cooperation of the worker. Confidentiality of medical test results may present another problem. Most H&S practitioners will not become involved in such kinds of sampling. Further detail is found in Chapter 10.

9.19   COMMON GAS OR VAPOUR HAZARDS

This section deals with just a few gases and vapours among the many that occur in workplaces—explosive gases, carbon monoxide, hydrogen sulfide, ammonia, chlorine, nitrous oxides (NO_x), benzene and mercury vapour. These substances give rise to typical and recurrent complaints that require investigation by occupational hygienists. Making use of the H&S practitioner's basic recognition skills, basic toxicological information, assessment methods and control procedures, this section examines:

* typical occurrence
* brief toxicology of the materials

- OELs and how the risks are assessed
- possible control procedures that can be recommended.

The following profiles are very brief and are presented only as basic guidelines. The H&S practitioner who encounters any hazardous gases and vapours in the workplace needs to research each one to be fully informed on safety, health and medical factors, and control processes.

9.19.1   FLAMMABLE/EXPLOSIVE GASES

Many gases encountered in the workplace can be flammable or, in the right mixture with air and with an ignition source, explosive.

9.19.1.1 Lower and upper explosive limits for flammable gases and vapours (LEL/UEL)

Before a fire or explosion can occur, three conditions must be met simultaneously. A fuel (combustible gas) and oxygen (air) must exist in certain proportions, along with an ignition source such as a spark or flame. The ratio of fuel to oxygen required varies with each combustible gas or vapour. The minimum concentration of a particular combustible gas or vapour necessary to support its combustion in air is defined as the gas or vapour's lower explosive limit (LEL). Below this level, the mixture is too 'lean' to burn. The maximum concentration of a gas or vapour that will burn in air is defined as the upper explosive limit (UEL). Above this level, the mixture is too 'rich' to burn. The range between the LEL and UEL is known as the *flammable range* for that gas or vapour. Table 9.3 shows the wide flammable range of many gases, particularly hydrogen and acetylene.

The values shown in this table are valid only for the conditions under which they were determined (usually room temperature and atmospheric pressure).

Some of these gases will be asphyxiants, while others will be toxic or irritants. Each gas should be assessed for its health hazards as well as its flammability.

When assessing the flammability/explosivity of the atmosphere in a workplace, a particular direct-reading instrument is required. Many modern direct-reading monitors include an 'explosimeter' to measure the percentage LEL in the work environment. The explosimeter *does not* identify the gas; it only determines where the work environment falls on the continuum of explosivity. It is essential to measure flammability for any confined space entry (see section 9.21). These monitors must be intrinsically safe, or they themselves could spark a fire or explosion.

An explosimeter is a device used to measure the amounts of combustible gases in a sample. When a certain percentage of the LEL of an atmosphere is exceeded, an alarm signal on the instrument is activated. The device, also called a combustible gas detector, uses catalytic bead sensors and operates on the principle of resistance being proportional to heat. Other explosive sensors use non-dispersive infrared (NDIR) technology to measure explosive gas concentrations.

Note that the detection readings of an explosimeter are accurate only if the gas being sampled has the same characteristics and response as the calibration gas. Most explosimeters are calibrated to methane or pentane.

Table 9.3 Selected chemicals and flammable range

Gas or vapour	%LEL	%UEL	Gas or vapour	%LEL	%UEL
Acetone	2.6	13.0	Hydrogen	4.0	75.0
Acetylene	2.5	100.0	Hydrogen sulfide	4.0	44.0
Ammonia	15.0	28.0	Isobutane	1.8	8.4
Benzene	1.3	7.9	Isobutylene	1.8	9.6
1,3-Butadiene	2.0	12.0	Methane	5.0	15.0
Butane	1.8	8.4	Methanol	6.7	36.0
n-Butanol	1.7	12.0	Methylacetylene	1.7	11.7
Carbon monoxide	12.5	74.0	Methyl bromide	10.0	15.0
Carbonyl sulfide	12.0	29.0	Methyl Cellosolve®	2.5	20.0
Chlorotrifluoroethylene	8.4	38.7	Methyl chloride	7.0	17.4
Cyclopropane	2.4	10.4	Methyl ethyl ketone	1.9	10.0
Dimethyl ether	3.4	27.0	Methyl mercaptan	3.9	21.8
Ethane	3.0	12.4	Pentane	1.4	7.8
Ethanol	3.3	19.0	Propane	2.1	9.5
Ethyl acetate	2.2	11.0	Propylene	2.4	11.0
Ethyl chloride	3.8	15.4	Propylene oxide	2.8	37.0
Ethylene	2.7	36.0	Toluene	1.2	7.1
Ethylene oxide	3.6	100.0	Trichloroethylene	12.0	40.0
Gasoline	1.2	7.1	Xylene	1.1	6.6

Many combustible-gas monitors also measure oxygen levels. It should be noted that catalytic explosive sensors require at least 10 per cent oxygen to give an accurate reading. They should therefore not be used for this application if the oxygen level is below 10 per cent, as a false low reading may be obtained, causing under-estimation of the risk. In applications where the oxygen concentration is below 10 per cent, such as in gas pipeline maintenance, NDIR sensor fitted gas monitors are widely used.

9.19.2   CARBON MONOXIDE

9.19.2.1 Use and occurrence

Carbon monoxide is a colourless and odourless toxic gas that burns with a pale-blue flame. Prior to the widespread introduction of natural gas as a domestic fuel, town gas supplies contained carbon monoxide. Town gas is still used as an oven gas in coke and steel works.

Carbon monoxide is used to render some reactive gases and other materials inert, but concerns about it in the workplace arise principally from incomplete combustion of fuels in ovens and furnaces, internal combustion engines (e.g. forklifts used in cold stores) and coal mines. Smouldering materials (e.g. poured metal moulds in foundries and building fires) produce considerable quantities of carbon monoxide. Although workers and vehicle drivers generally are familiar with the dangers, accidental carbon monoxide poisoning is relatively common. Escape from a dangerous situation is often impossible, because the first indication of gross exposure may be collapse.

9.19.2.2 Toxicology

Carbon monoxide (CO) falls into the class of chemical asphyxiants, which block the transport of oxygen in the blood. The affinity of CO to the haemoglobin in blood is over 300 times greater than that of oxygen. Inhaled carbon monoxide is rapidly and extensively absorbed into blood and distributes throughout the body. Short exposures of 2000 ppm result in asphyxia and death, whilst at 50 ppm discomforting effects are noticeable. When carrying carbon monoxide, arterial blood does not become the bluish colour of venous blood, as in cases of normal asphyxia, so the blood and the lips remain a bright cherry red. The tissues of the central nervous system and the brain are the most susceptible. Methylene chloride, a commonly used industrial solvent, metabolises to carbon monoxide, and exposure to high concentrations may produce chemical asphyxia.

Symptoms of carbon monoxide poisoning include, progressively, headache, nausea, drowsiness, fatigue, collapse, unconsciousness and death. These states correspond to increasing percentages of blood carboxyhaemoglobin (COHb), from 10 per cent to 40 per cent. Smokers' blood may contain between 5 and 10 per cent COHb.

9.19.2.3 Standards and monitoring

SafeWork Australia lists a TWA-OES for CO of 30 ppm for an eight-hour day, which has a bioequivalent of 5 per cent COHb. The Exposure Standard Working Group believes this value should protect against adverse behavioural effects arising from carbon monoxide exposure as well as minimising the risk to those persons with sub-clinical cardio-vascular disease. The ACGIH BEI (ACGIH, 2018) is set at 3.5 per cent COHb as being bioequivalent to the TWA of 25 ppm, not to be exceeded at any time during the work shift to prevent adverse neurobehavioural changes and to maintain cardiovascular exercise capacity. There are some variations possible for higher short-term exposures for reduced periods, as shown in Table 9.4.

Because of the insidious nature of carbon monoxide, monitoring for it should be carried out in work areas *wherever* there is a possibility that the gas is present. Detector tubes are useful for taking grab samples of potential exposure and provide an instantaneous reading of airborne concentrations. Specific direct-reading monitors are readily available for carbon monoxide from a variety of manufacturers (MSA, Dräger, RAE Systems); many are available as pocket-sized personal gas monitors, with visible and audible alarms that can be set at predetermined levels to alert the user to a potentially hazardous atmosphere.

Portable CO monitors allow for the measurement of a TWA and STEL. Furthermore, the use of portable gas monitors allows the user to track how exposure varies with time by

Table 9.4 Australian WESs for carbon monoxide

Time	TWA (ppm)	STEL (ppm)
8 hrs	30	no value
1 hr	No value	60
30 min	No value	100
15 min	No value	200
<15 min	No value	400

Source: Safe Work Australia (2013).

graphing exposure vs time. This helps identify where the main exposure is occurring and can be useful for targeting controls.

9.19.2.4 Controls

Engineering controls are mandatory to prevent build-up of carbon monoxide and subsequent exposure of workers. Underground car parks and tunnels must be fitted with extraction systems to remove vehicle exhaust. Ovens, gas furnaces and burners must be correctly vented to carry away both burnt and incompletely burnt combustion products.

Respiratory protective equipment is not recommended for long-term protection against carbon monoxide, since it is more appropriate to use proper ventilation. Emergency escape equipment consists of either self-contained breathing apparatus or a self-rescuing gas converter that turns carbon monoxide into carbon dioxide, with the accompanying production of much heat. Firefighting and rescue equipment usually consists of self-contained breathing apparatus.

9.19.2.5 Training and education

All workers exposed to carbon monoxide (e.g. from coke oven gases) must be instructed and trained in safe work procedures and in the appropriate use of self-contained breathing apparatus.

9.19.2.6 Medical requirements

Heavy smokers are at greater risk of ill-effects from carbon monoxide exposure because a proportion of their haemoglobin is already incapacitated. Workers with emphysema may also be at greater risk. Because of the possibility of severe health problems in workers with cardiovascular disease, exposure to carbon monoxide should be restricted to ensure that the carboxyhaemoglobin content of all workers' blood is maintained below 5 per cent.

9.19.3 HYDROGEN SULFIDE

9.19.3.1 Use and occurrence

Hydrogen sulfide is a colourless gas with a rotten-egg odour that some people are able to smell at levels of 0.2 ppb. It is produced by anaerobic sulphur fixing bacteria, especially associated with raw sewage. It is also found in crude oil, marine sediments, tanneries, pulp and paper industry. As it is heavier than air, it collects in pits, within protective berms or in other low-lying areas. It is a standard recommended gas to test for before entering a confined space.

9.19.3.2 Toxicology

Hydrogen sulfide is a systemic poison that prevents utilisation of oxygen during cellular respiration, shutting down power source for many cellular processes and leading to respiratory paralysis and asphyxia at high concentrations. It also binds to hemoglobin in red blood cells, interfering with oxygen transport. Exposure to hydrogen sulfide occurs primarily by inhalation, but can also occur by ingestion (contaminated food) and skin (water and air). Once taken into the body, it is rapidly distributed to various organs, including the central nervous system, lungs, liver, muscle, as well as other organs which may cause muscle cramps, low blood pressure and death at high levels.

Eye, throat and lung irritation, nausea, breathing difficulties, headache and chest pain have been reported in cases of continuous exposure to concentrations above 10 ppm. Although easily smelled in very low concentrations, in high concentrations hydrogen sulfide may cause a temporary loss of smell (see section 9.4.1).

9.19.3.3 Standards and monitoring

The Safe Work Australia WES for hydrogen sulfide is 10 ppm TWA and 15 ppm STEL, while the ACGIH lists a TWA at 1 ppm and a STEL to 5 ppm (ACGIH, 2018).

Although indicator stain tubes can be used to measure hydrogen sulfide, the main means of continuous detection is direct-reading electrochemical instruments. Detectors for hydrogen sulfide as a single gas or in combination with other gases are now quite cost effective for long-term monitoring and are standard on most sites where hydrogen sulfide is a common risk.

9.19.3.4 Controls

Attention must be given to the design of industrial plants to reduce the possible production and build-up of hydrogen sulfide. Where it is not possible to significantly reduce or eliminate the gas, fixed-installation detection equipment and personal detection monitors should be used throughout the plant.

Where engineering controls are not feasible, then personal protective equipment of respirators, filters, breathing apparatus and protective clothing must be used. If a gas alarm is sounded, escape filters or breathing apparatus must be worn.

9.19.4   AMMONIA

9.19.4.1 Use and occurrence

Ammonia is a colourless gas with a strong, pungent smell. It is commonly used in refrigeration, ice-making, as a cleaner, as a fertiliser and in the production of other pharmaceuticals, resins and paper products. Ammonia is usually compressed into liquid form for transport or storage.

Ammonia has a lower flammability limit of 16 per cent by volume. Although it is difficult to ignite in the open air, there have been several explosions and fires caused by the release of ammonia in a confined space.

9.19.4.2 Toxicology

Ammonia is an upper airway irritant. Intense pain in the eyes, mouth and throat normally results from single intense exposures. This is due to the formation of ammonium hydroxide when it comes into contact with water. Longer single exposures will produce ulceration of the mouth and nose, cyanosis and pulmonary oedema.

Exposure for 30 minutes at levels higher than 2500 ppm is usually fatal.

9.19.4.3 Standards and monitoring

The Safe Work Australia WES for ammonia is 25 ppm TWA and 35 ppm STEL. Ammonia is detectable by the nose at concentrations of 1 to 5 ppm, but workers who are regularly exposed to ammonia become desensitised and *will lose their ability to smell ammonia* at levels below 50 ppm.

Although detector tubes can be used to measure ammonia, the main means of continuous detection is direct-reading electrochemical instruments, both as fixed installations in plants and as portable personal monitors. PID sensors can also be used to measure the explosive limits of ammonia.

9.19.4.4 Controls

Engineering controls should be implemented in the design of industrial plant to reduce the risk of ammonia leaks or spills. A closed handling system is the optimum solution. Where this is not possible, use of the gas should be restricted to small amounts in well-ventilated areas.

For storage areas or large-scale operations, detectors should be in place in case of leaks and protective equipment of suitable respirators and protective clothing should be donned. Escape breathing apparatus should be available in these working areas.

9.19.5   *CHLORINE*

9.19.5.1 Use and occurrence

Chlorine is supplied in cylinders or in tanks as a liquid under pressure. It is a highly reactive heavy gas which is greenish-yellow in appearance with a characteristic pungent odour. Chlorine solution in water is a strong acid and is corrosive. The substance is a strong oxidant that reacts violently with gases, combustible substances and reducing agents. It reacts with most organic and inorganic compounds, causing fire and explosion hazard, and attacks metals, some forms of plastic, rubber and various coatings.

Its principal applications are in the production of a wide range of industrial and consumer products, such as plastics, solvents for dry cleaning and metal degreasing, textiles, agrochemicals and pharmaceuticals, insecticides, dyestuffs, and household cleaning products. The most significant of organic compounds in terms of production volume are 1,2-dichloroethane and vinyl chloride, intermediates in the production of PVC. Other particularly important organochlorines are methyl chloride, methylene chloride, chloroform, vinylidene chloride, trichloroethylene, perchloroethylene, allyl chloride, epichlorohydrin, chlorobenzene, dichlorobenzenes, and trichlorobenzenes. The major inorganic compounds include hydrochloric acid, dichlorene oxide, hypochlorous acid, sodium chlorate and chlorinated isocyanurates.

9.19.5.2 Toxicology

Chlorine is corrosive to the eyes, skin and respiratory tract. Rapid evaporation of the liquid may cause frostbite. Inhalation may cause asthma-like reactions and pneumonitis. It may also cause lung oedema, but only after initial corrosive effects on eyes and/or airways have become manifest. The symptoms of lung oedema are often not apparent for several hours, and are aggravated by physical effort. Rest and medical observation are therefore essential. Exposure to chlorine in sufficient concentration or over a long period can be lethal.

Chlorine's effects on the respiratory tract and lungs may result in chronic inflammation and impaired function. It may also damage the teeth.

It is highly inadvisable to let chlorine or its compounds enter the workplace environment.

9.19.5.3 Standards and monitoring

Chlorine is detectable by the nose at 0.08 ppm as a suffocating/sharp/bleach odour. The Safe Work Australia WES for chlorine is 1 ppm peak limitation. By contrast, the ACGIH recommends lower occupational exposure limits—TLV: TLV: 0.1 ppm as TWA; 0.4 ppm as STEL; A4 (not classifiable as a human carcinogen) (ACGIH, 2018). Chlorine is immediately dangerous to life and health (IDLH) at 10 ppm.

Although indicator stain tubes can be used to measure chlorine, the main means used for continuous detection are direct-reading devices, as these provide much more information on the work environment.

9.19.5.4 Controls

Engineering controls should be included in the design of new industrial plant to reduce the risk of leaks or spills. A closed handling system is the optimum solution. Where this is not possible, the use of chlorine should be restricted to small amounts in a well-ventilated area.

For storage areas or large-scale operations, detectors should be put in place in case of leaks and protective equipment of suitable respirators and clothing should be worn. Escape breathing apparatus should be available in these working areas.

9.19.6   OXIDES OF NITROGEN (NO$_x$)

9.19.6.1 Use and occurrence

Oxides of nitrogen that the H&S practitioner needs to consider are mainly nitrogen dioxide (NO$_2$) and nitric oxide (NO), commonly referred to as NO$_x$. Nitric oxide (NO) is a colourless, odourless gas that is only slightly soluble in water. The main sources of NO are combustion processes. Nitric oxide is then partly oxidised to form nitrogen dioxide.

NO$_2$ is a heavy red brown gas with a pungent odour that is produced from diesel-powered machinery. Although its acrid, biting odour is detectable by the nose at low concentration, it can also anaesthetise the nose at continuous low concentrations (4 ppm).

9.19.6.2 Toxicology

Although there is limited evidence that nitric oxide alone is toxic, there is some suggestion that it can react with haemoglobin in the blood to form nitrosyl haemoglobin and methaemoglobin, reducing the blood's ability to transport oxygen. Exposure to nitric oxide along with carbon monoxide (the other main component of diesel emissions) further increases oxygen starvation of the tissues.

The affinity of NO for haemoglobin is 1400 times that of CO. Inhaled nitric oxide can rapidly react with oxygen in the lungs to form nitrogen dioxide, which is a potent pulmonary irritant. NO_2 is a respiratory irritant and occupational exposure is via inhalation, with common symptoms of exposure being irregular breathing, reduced lung function and wheezing.

9.19.6.3 Standards and monitoring

SafeWork Australia lists an eight-hour shift TWA-OES for NO_2 of 3 ppm and a STEL of 5 ppm, which is derived from supporting documentation from ACGIH®. However, in 2012 the ACGIH adopted a TWA for NO_2 of 0.2 ppm, which was based on controlled human exposure studies to NO_2 where resultant respiratory systems effects were observed. Safe Work Australia has adopted an eight-hour shift TWA-OES of 25 ppm for nitric oxide. This recommended exposure limit is designed to reduce the potential for respiratory tract irritation.

Portable instruments and colorimetric tubes are widely used to measure for NO and NO_2. The same factors and issues need to be accounted for as in the measurement of CO when using these devices, such as cross-sensitivity to other compounds. Care needs to be taken in using NO_2 portable gas meters when considering the adoption of a 0.2 ppm TWA. Although the resolution on the sensor is 0.1 ppm, drift common in many gas monitors may lead to the instrument recording an excursion over the prescribed alarm point.

9.19.6.4 Controls

Exhaust ventilation is the main method used to control acid gases and aerosols within mine tunnels and above acid pickling and electroplating baths. Regular testing of diesel vehicles' exhaust emissions is also a reliable control method. For example, in underground coal mines vehicles are tested once a month directly at the exhaust pipe. Results must be within certain levels for the vehicle to remain in use underground. Personal protective equipment must be used to protect against splashes to skin and eyes, and specialised respirator filters are available for use against all acidic gases.

9.19.7 BENZENE

9.19.7.1 Use and occurrence

Benzene is a colourless liquid with a sweet (aromatic) odour. It evaporates very quickly and dissolves to some extent in water. It is highly flammable and vapour/air mixtures are potentially explosive. Benzene's LEL is 1.35 per cent and its UEL is 6.65 per cent.

Benzene is a major industrial chemical, derived mostly from petroleum. It is widely used as an intermediate in the manufacture of products from plastics to pesticides and pharmaceuticals. Key industries involved in the production or use of benzene include:

- petroleum refining
- coke and coal production
- rubber tyre manufacturing
- storage sites (tank farms)
- transportation services (ships, tanker trucks)
- laboratories
- manufacture of plastics, synthetic rubber, glues, paints, furniture wax, lubricants, dyes, detergents, pesticides and some pharmaceuticals.

Natural sources of benzene include volcanoes and forest fires; it is also present in cigarette smoke.

Benzene is highly mobile in soils. Leaking petrol storage tanks and pipelines have resulted in soil and groundwater contamination, which is of particular concern where town water is drawn from underground sources.

9.19.7.2 Toxicology

Benzene is irritating to the eyes, skin and respiratory tract. If it is swallowed, aspiration into the lungs may result in chemical pneumonitis. It also affects the central nervous system, and exposure to levels far above the OEL may lead to unconsciousness and death.

Prolonged exposure dries out and defats the skin, and may damage the bone marrow, reducing blood cell production and impairing the immune system. In sufficient doses, it is carcinogenic.

Benzene can be measured in breath, blood or urine, but only for about 24 hours after exposure owing to its relatively rapid removal by exhalation or biotransformation. Most people in developed countries have measurable baseline levels of benzene and other aromatic petroleum hydrocarbons in their blood. In the body, benzene is converted by enzyme action to a series of oxidation products, including muconic acid, phenylmer-apturic acid, phenol, catechol, hydroquinone and 1,2,4-trihydroxybenzene. Most of these metabolites have some value as biomarkers of human exposure, since they accumulate in the urine in proportion to the extent and duration of exposure, and they may still be present for some days after exposure has ceased. The current ACGIH® biological exposure limits for occupational exposure are 500 μg/g creatinine for muconic acid and 25 μg/g creatinine for phenylmercapturic acid in an end-of-shift urine specimen.

9.19.7.3 Standards and monitoring

NIOSH (2017a, 2017b) puts the IDLH concentration of benzene in air at 500 ppm. Safe Work Australia sets a WES of 1 ppm TWA and classes benzene as a Category 1 carcinogen. By contrast, the ACGIH® recommends a lower occupational exposure limit—TLV: 0.5 ppm as TEL of 2.5 ppm with a skin notation and lists as A1 (confirmed human carcinogen) (ACGIH, 2018).

Monitoring can be done by detector tubes and direct-reading instruments; however, for comparison with the WES, a time-weighted sample must be obtained, either by charcoal tube/pump, diffusion badge or direct-reading instruments (specifically modified for benzene) with TWA capabilities.

A more extensive explanation of monitoring approaches can be found in RAE System Application Note AP–236.

9.19.7.4 Controls

A closed-loop system should be implemented to prevent release of products containing benzene. Other measures include ventilation extraction systems, explosion-proof electrical equipment and lighting. Open flames, sparks and cigarettes should NEVER be used where benzene may be present. Compressed air should be used for filling, discharging or handling products containing benzene. Hand tools should be non-sparking and the build-up of electrostatic charges prevented (e.g. by grounding or earth-bonding).

Respiratory protection is required when the exposure is likely to exceed the action level.

9.19.8   MERCURY VAPOUR

In recent years, a new hazard has emerged in the oil and gas industry. Mercury vapour is contained in certain crude oils, and can deposit out in various process units in a petroleum refinery. The hazards of mercury are covered in Chapter 8. This section will cover only monitoring equipment and its applications in the oil industry; the toxicology is covered in section 8.9.2.

9.19.8.1 Occurrence in the oil industry

Mercury can be found in trace amounts in some crude oil, and because it can form a vapour it may be transferred through the production process into the refinery. It can amalgamate with aluminium components in heat exchangers and may cause catastrophic failures in cryogenic units. The most likely locations for mercury deposits are in exchangers, towers, knockout pots and accumulators; however, the air around a refinery does not usually contain mercury. Most mercury is removed following steaming/cleaning.

When refinery equipment is opened during a turnaround or shutdown, it is important to assess whether mercury is present. Refinery exchangers and vessels need to be steamed out before entry. Mercury-vapour monitors can be used to 'sniff' for mercury in and around steam plumes bled off from vessels and equipment. In most cases, the steaming is done into a closed process and the steam can be 'bled' briefly to check for mercury in the air.

Mercury can condense to free liquid and become trapped in the walls of steel vessels and piping. Even after cleaning, some mercury is likely to remain in these walls and may leach out during hot work, because the heat helps liberate it from the steel. This is a health hazard during shutdowns and turnarounds.

Mercury vapour should be measured before any entry into confined spaces suspected of containing it.

9.19.8.2 Standards and monitoring

A mercury-vapour analyser is used to detect the presence of elemental mercury vapour and measure its concentration in the air. If readings exceed the Australian WES of 0.025 mg/m^3 TWA, then respiratory protection, disposable coveralls, nitrile gloves and rubber/PVC boots are required.

Air monitors may be used in 'sniffer' mode to locate sources of contamination, and should be used after clean-ups to ensure that removal of mercury was complete.

Several direct-reading instruments are available, including:

- the Jerome range of mercury vapour analysers
- the Nippon Instruments EMP-1A mercury gas monitor
- the Ohio-Lumex RA-915+ portable mercury vapor analyser.

There are also colourimetric detector tubes available, but these give only a spot reading and are less accurate than direct-reading devices. A Dräger tube gives readings within ±19 per cent of the actual concentration, while a Kitagawa tube is accurate to ±15 per cent. Tubes are used if mercury meters cannot be used for testing in a hazardous area (since they are not intrinsically safe), or if readings are above 1.0 mg/m³ (1000 µg/m³), which exceeds the direct-reading device's range.

The Jerome J405 mercury vapour monitor and 431-X analyser, both hand held, provide a far more accurate means of measuring mercury vapour than a Dräger or Kitagawa tube. They have a measuring range from 1 to 1000 µg/m³.

Since none of the mercury meters are intrinsically safe, a hot work permit will be required to operate the meter in process areas where flammable vapours may be present.

Measuring devices detect only elemental mercury vapour. To guard against skin contamination by mercury or its compounds, high standards of personal hygiene must always be maintained. If clothing is contaminated it must be removed, and if skin is accidentally contaminated the worker must shower, scrubbing thoroughly.

9.19.8.3 Controls

A confined space entry permit requires that the atmosphere inside such a space be tested; results should be entered under 'other tests' or as an additional comment. If the mercury vapour concentration is less than 0.012 mg/m³, no respiratory protection is required. If it exceeds 0.012 mg/m³ but is less than 0.6 mg/m³, a full-face respirator with a type Hg filter should be worn.

If the concentration is greater than or equal to 0.6 mg/m³, then the vessel or other space should be further cleaned and/or ventilated, then retested until the levels are acceptable before entry. Unless a vessel is very well ventilated, mercury vapour concentrations inside it can be expected to correlate with temperature, since mercury will be liberated from the walls of a steel vessel more rapidly as temperature increases (see Figure 9.12).

In the event that hot work is necessary in a confined space, the H&S practitioner should either perform a job safety analysis (JSA) or review it. The results will indicate what PPE is necessary.

Care must be taken when heating steel or other metallic vessels or enclosures, even after they have been decontaminated of mercury, since mercury vapour may still be emitted in high concentrations from below the immediate metal surface of the vessel. Mercury is absorbed into, as well as absorbed onto, metal surfaces when it is cooled, and mercury vapour may be released when the surface is reheated. Mercury inside vessels and exchangers can be seen best at night using a torch: the inside walls of the vessel will 'sparkle'. In some instances, hygienists have noted a foul odour in the air (not H2S/RSH) when significant levels of mercury are present.

Figure 9.12 Free mercury leaking from a partly opened exchanger flange
Source: Petroch Services.

Figure 9.13 illustrates the dramatic change in vapour pressure with temperature, and hence the increasing concentration of mercury vapour in air. This is why conditions can change during hot work such as welding and grinding.

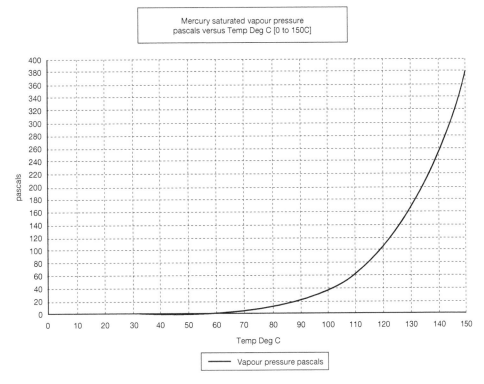

Figure 9.13 Change in mercury vapour pressure with temperature
Source: Petroch Services.

9.20   ATMOSPHERIC HAZARDS ASSOCIATED WITH CONFINED SPACES

Confined spaces present a number of special hazards in the workplace: risk of explosion, engulfment, oxygen deprivation and being overcome by toxic gases and vapours. The potential for the development of these hazardous conditions is affected by:

- the work being performed
- external sources of contamination
- products used or produced in conjunction with the space
- natural processes that occur in the space (such as fermentation, decomposition, etc.)
- the physical nature of the space,

Adequate oxygen levels are vital, and the risk posed by atmospheric contaminants increases dramatically. In fact, more occupational health-related deaths occur in confined spaces than anywhere else. A common scenario involves a worker who drops a tool in a pit and climbs in to retrieve it. He collapses; his workmate assumes he has had a heart attack and enters the pit to rescue him, then both workers die from lack of oxygen, inhalation of toxic gases or both.

No worker should enter a confined space until the atmosphere has been tested and found to be safe.

Confined space refers to an enclosed or partially enclosed space (not a mine shaft or mining tunnel) that:

- has a limited or restricted means of entry/exit
- is large enough for a worker to enter and perform certain tasks
- is not designed for constant worker occupancy
- is likely to pose a risk to health and safety from:
  - an atmosphere that does not have a safe oxygen level
  - contaminants, including airborne gases, vapours and dusts, that may cause injury from fire or explosion
  - harmful concentrations of any airborne contaminants
  - engulfment.

According to the Australian Model Work Health and Safety Regulation 2015, a confined space can be inside a vat, tank, pit, pipe, duct flue, oven, chimney, silo, reaction vessel, container, receptacle, underground sewer, shaft, well, trench, tunnel or similar enclosed or partially enclosed structure. The risks associated with confined spaces include:

- loss of consciousness, injury or death owing to the immediate effects of contaminants
- fire or explosion from the ignition of flammable contaminants
- asphyxiation resulting from oxygen deficiency when the oxygen level falls below 19.5 per cent
- enhanced combustibility and spontaneous combustion when the oxygen level exceeds 23.5 per cent
- asphyxiation resulting from engulfment by 'stored' material—for example, grain, sand, flour, fertiliser.

Working in confined spaces may also greatly increase the risks of injury from the following:

- *mechanical hazards* associated with plant and equipment, which may result in entanglement, crushing, cutting, piercing or severing of limbs or digits. Sources of mechanical hazards include augers, agitators, blenders, mixers, stirrers and conveyors.
- *ignition hazards* associated with plant and equipment in or near the confined space. The presence of ignition sources in a flammable atmosphere may result in fire or explosion, and the death or injury of workers. Examples of ignition sources include open flames, sources of heat, static or friction, non-intrinsically safe equipment, welding and oxy-cutting, hot riveting, hot forging, electronic equipment such as cameras, pagers and mobile phones, and activities that generate sparks, such as grinding, chipping and sandblasting.
- *electrical hazards*, which may result in electrocution, electric shock or burns
- *uncontrolled substances* such as steam, water or other materials, whose presence or entry may result in drowning, being overcome by fumes, engulfment by earth or rock, and so on
- *noise*, the levels of which can be greatly increased in confined spaces
- *manual handling hazards* arising from work in cramped, confined areas. These may be increased by the use of personal protective equipment such as an airline or harness which restricts movement, grip and mobility.
- *radiation hazards* from lasers, welding flash, radio frequency (RF) waves and microwaves, radioactive sources, isotopes and x-rays
- *environmental hazards*, including heat or cold stress, which may arise from work or process conditions, wet or damp environments, slips, trips and falls arising from slippery surfaces
- *biological hazards*, including infection by microbes and contact with organisms such as fungi, which may cause skin disease or result in respiratory illness, and mites in grain, which may result in dermatitis. Viruses and bacteria may also present a hazard. Exposures to *Leptospira* species and *Escherichia coli* are of particular concern for people who work in sewers. Insects, snakes and vermin are also potential hazards.
- *traffic hazards*, which may arise where confined-space entry or exit points are located on walkways or roads. Workers entering or exiting the space risk being struck and injured by vehicles such as cars or forklift trucks. The potential may also exist for others to fall into the space.

A number of situations may arise during work in confined spaces in which the oxygen level will decrease and/or other gases be generated. Some examples are:

- removal of oxygen by activated carbon, or by some soils undergoing microbiological activity
- decrease in oxygen level by the oxidation (rusting) of freshly grit-blasted metal surfaces
- partial oxidation of hydrocarbons, in the presence of catalysts such as alumina, to produce carbon monoxide and decrease oxygen levels, thus generating two hazards
- decrease in oxygen and the generation of toxic gases from the reaction of pyrophoric materials (pyrophoric iron will self-ignite in the presence of oxygen)

- the presence of hydrocarbons under a 'nitrogen blanket'. Some vessels are purged with nitrogen gas to minimise the risk of flammability, creating a nitrogen-rich atmosphere and all but eliminating oxygen. As catalytic explosive sensors require at least 10 per cent oxygen to give an accurate reading, they should not be used in such conditions, as a false low reading can be obtained, leading to under-estimation of the risk. The best practice is to use an infrared explosive sensor or a thermal conductivity sensor, neither of which needs oxygen to give a measurement.

In summary, confined space work requires special procedures, training and testing to ensure safe operations and protect the worker. More information can be found in the Safe Work Australia (2015) confined spaces code of practice and Australian Standard AS 2865–2009 Confined Spaces (Standards Australia, 2009).

9.21   EMERGING MONITORING TECHNOLOGY

Technological advances in producing direct reading instruments over the past decade have brought about new and reliable approaches to monitoring toxic vapours and gases. Combining this with the development and growth of wireless communications and technologies may change the way H&S practitioners monitor for gases and vapours in the workplace. Cloud-based software programs and storage devices as well as mobile smartphone-capable applications are rapidly becoming important resources for the H&S practitioner.

9.21.1   WIRELESS TECHNOLOGY AND REMOTE MONITORING

Technology is now available for continuous personal and area monitoring from a central location (Figure 9.14).

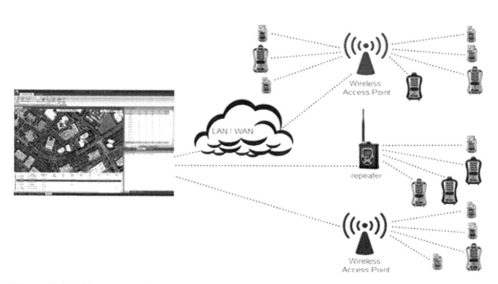

Figure 9.14 Remote monitoring layout

By gathering information from multiple points or workers at once, this allows the H&S practitioners to view data remotely in real time, and provides a dynamic approach to hazard control. Real-time personal monitoring equipment with integrated GPS (global positioning system) is available for monitoring of volatile organic compounds, benzene, and other toxic and hazardous gases. Broadcasting real-time peak level alarms as well as TWA and STEL alarms to a central host computer offers a cost-effective way of increasing safety and minimising exposure risks to employees or contractors. Real-time communications between wirelessly enabled instruments allows the H&S practitioner to *detect change* in a workplace, which inherently reduces the risk of exposure to gas and vapour hazards that adversely affect the health and wellbeing of workers and surrounding communities. Examples are wireless area-monitoring systems that may be configured to warn of potential hazard exposure well before a plume or release enters a working area. Being battery powered, these systems are rapidly deployable from one part of a site to another, and are especially useful in areas where access to electricity may be limited.

As the gathered data are stored in the cloud via a central computer, the H&S practitioners may review them at any time from any computer connected to the internet. This is especially useful as a complementary monitoring technique—for example, in the offshore petroleum industry—as it enables the monitoring of unique, uncommon or unplanned maintenance tasks where exposure levels would otherwise be very difficult to capture.

Interoperability of wireless detection and monitoring systems also allows the sharing of data with first responders such as fire or ambulance services, which helps to increase safety during emergencies.

Portable wireless-enabled gas monitors such as those used in confined space entry have the option to employ 'man-down' technology. Should a worker be incapacitated through a fall, health incident or a sudden catastrophic release of a toxic gas, a remote alarm will be sounded to co-workers either outside the space or close by. This technology reduces the response time of emergency service workers to the incident.

9.21.2   *DATA COLLECTION AND EVALUATION*

Transmission of occupational exposure data in real time from a monitoring device allows for integration into proprietary software systems. These software solutions have been developed to provide a cost-effective and time-saving mechanism for the analysis of the data for the occupational hygienist. Calculations and analysis that could otherwise take hours or days to complete can be performed in minutes.

9.21.3   *NEW AND EMERGING RESOURCES*

There are many online databases that can be referred to for exposure limits and detection options, as well as software programs in which entering a chemical name or CAS number instantly yields basic information on exposure limits, sampling equipment and media, and so on. There are also many instructional videos on YouTube showing how to detect gases and how to use various types of equipment. Most suppliers are using this medium to supply training in the use of their equipment.

9.22   REFERENCES

American Conference of Governmental Industrial Hygienists (ACGIH) 2018, *Documentation of the Threshold Limit Values for: Chemical Substances, Physical Agents and Biological Exposure*, ACGIH®, Cincinnati, OH.

Murnane, S.S., Lehochy, A.H. & Owens, P.D. (eds) 2013, *Odor Thresholds for Chemicals with Established Occupational Health Standards*, 2nd ed., AIHA, Akron, OH.

National Institute of Occupational Safety and Health (NIOSH) 2017a, *NIOSH Manual of Analytical Methods* (NMAM®), 5th ed., Centers for Disease Control and Prevention, Atlanta, GA, <www.cdc.gov/niosh/nmam/default.html> [accessed 28 October 2018]

—— 2017b, *Documentation for Immediately Dangerous to Life or Health Concentrations (IDLH): NIOSH Chemical Listing and Documentation of Revised IDLH Values*, <www.cdc.gov/niosh/idlh/intridl4.html> [accessed 28 October 2018]

Safe Work Australia (SWA) 2013, *Guidance on the Interpretation of Workplace Exposure Standards for Airborne Contaminants*, SWA, Canberra.

—— 2015, *Work Health and Safety (Confined Spaces) Code of Practice*, SWA, Canberra.

—— 2018, *Workplace Exposure Standards for Airborne Contaminants*, SWA, Canberra, <www.safeworkaustralia.gov.au/system/files/documents/1804/workplace-exposure-standards-airborne-contaminants-2018_0.pdf> [accessed 28 October 2018]

Standards Australia 2003a, Workplace Air Quality—Sampling and Analysis of Volatile Organic Compounds by Solvent Desorption/Gas Chromatography—Pumped Sampling Method, AS2986.1: 2003, Standards Association of Australia, Sydney.

—— 2003b, Workplace Air Quality—Sampling and Analysis of Volatile Organic Compounds by Solvent Desorption/Gas Chromatography—Diffusive Sampling Method, AS2986.2: 2003, Standards Association of Australia, Sydney.

—— 2009, Confined Spaces, AS/NZS2865: 2009, Standards Association of Australia, Sydney and Standards New Zealand, Wellington.

10. Biological monitoring of chemical exposure

Gregory E. O'Donnell and Martin Mazereeuw

10.1   INTRODUCTION

The role of the occupational hygienist is to assess exposures to hazards and, when found to be excessive, to suggest ways to reduce exposures to an acceptable level. Occupational hygienists can assess exposure by performing air monitoring; however, air monitoring may not always be the best option to assess the exposure for several reasons, including the nature of the chemical hazard, its chemical properties, how it is being used and the work processes and scenarios involved. It therefore may not always give an accurate measure of the real short- or long-term dose.

Biological monitoring is defined as the assessment of exposure to a chemical by the measurement of the chemical or its metabolites in blood, urine or exhaled breath. This measurement should be indicative of an internal dose due to the exposure. It should be used as an early indicator of exposure before health effects or harm occurs. Biological monitoring is a measure of the internal dose of a chemical via all routes of exposure, including inhalation, ingestion and dermal absorption. It therefore measures the actual total exposure dose of the individual to the chemical rather than a predicted dose by measuring the chemical in the external breathing zone of the worker. Biological monitoring thus integrates exposures from multiple pathways and sources; however, it cannot identify the source or pathway, or the relative contributions from the sources. Advances in analytical chemistry and the acknowledgement of the uncertainties involved in external air measurements have led to biological monitoring fast becoming the preferred method to estimate chemical exposure. It is an evolving area and more analytical methods are increasingly becoming available.

The theoretical pathway from *occupational exposure* to a chemical to the development of disease can happen by exposure via all absorption routes. Inhalation usually plays a major contribution to the internal dose by inhalation of gases, vapours, fumes, dusts or mists. *Biological monitoring* can be effective in estimating the internal dose; however, other effects can occur in the body which may be a response to the exposure but could also relate to other influences. These parameters can be measured by further *biochemical monitoring*, including general health-based markers such as liver function tests and other general pathology tests. A step further to the acquirement of disease is the measurement of biological effects to the exposure of the chemical. These tests can include the measurement of the activity of the cholinesterase enzyme in whole blood for the exposure to organophosphate insecticides, the level of zinc protoporphyrin for exposure to lead, chromosomal aberrations, sister chromatid exchange and other protein, haemoglobin and DNA adducts. These types of measurements are usually called *biological effect monitoring*. The final group of tests are used to determine the onset of *disease*. This process and available monitoring strategies are shown in the schematic illustration in Figure 10.1.

Figure 10.1 Schematic illustration of the pathway of occupational exposure to chemicals and the monitoring strategies employed to assess the exposure

10.2   WHEN TO USE BIOLOGICAL MONITORING

While the prime objective of biological monitoring is to determine whether individuals are at an increased risk of experiencing adverse health effects, its use can fulfil other functions, such as determining whether a particular chemical is above a mandated regulatory limit. Biological monitoring can be very useful if the exposure could be occurring via exposure routes other than inhalation—for example, where skin absorption or inadvertent ingestion may be significant exposure routes. These routes of exposure may not always be obvious to the observer. Biological monitoring can reveal unanticipated exposures, including exposures that have occurred outside the workplace.

When controls have been put in place, biological monitoring can be used to monitor whether these controls are working. Personal protective equipment (PPE), such as protective clothing, gloves and respirators, is only effective when used in the correct manner, properly maintained and free from any previous contamination. Biological monitoring can verify the effectiveness of PPE. Air monitoring cannot confirm whether PPE is effective because it is very difficult or even impossible to take an air sample in the breathing zone when a respirator is worn. Biological monitoring can be used to confirm assessments of exposure estimated by air monitoring.

Biological monitoring is also a useful technique to determine whether an exposure has taken place after an incident has occurred. It is usually too late to take an air monitoring sample; however, a biological monitoring sample can provide some indication of the extent of exposure.

10.3   HEALTH MONITORING

Health monitoring is the assessment of the workers' health status by clinical, biochemical, imaging or instrumental testing to detect any clinically relevant, occupationally dependent change of a single worker's health (Manno et al., 2010). Health monitoring can only be performed by or under the supervision of a registered medical practitioner with experience in health monitoring. Biological monitoring can be used as part of the health monitoring scheme and can be used to help interpret clinical findings.

10.4   SELECTION OF SAMPLE MATRIX

Biological monitoring can be performed in a variety of matrixes, including (in order of decreasing significance) urine, blood, exhaled breath, oral fluid and hair. The main matrixes used for biological monitoring are considered below.

10.4.1   URINE

By far the most frequently used matrix in occupational exposure assessment is urine. It is easy to collect and much less invasive than blood. It can be collected easily before a shift and at the end of a shift to estimate exposure throughout a shift. It can be repeatedly collected throughout the work week to show any accumulation and excretion patterns.

10.4.2   BLOOD

Most compounds can be measured in blood; however, the sampling of blood is invasive and requires trained medical staff. It is usually performed in a clinical setting. If blood sampling is performed, any individuals handling the samples need to be trained in the risks of infection and should be offered appropriate immunisations to minimise any risk.

10.4.3   EXHALED AIR

Exhaled air can be used for some volatile organic compounds; however, if the chemical is quickly metabolised in the blood then exhaled air is not suitable and urine or blood may be a more appropriate sample.

10.4.4   ORAL FLUID

Oral fluid has come to prominence in the area of monitoring for illicit drugs of abuse. Spot test kits using immunoassay analysis are commercially available to perform this analysis and a confirmatory test is performed later in a laboratory. Oral fluid has not yet been extensively employed in other areas; however, other test kits may become available in the future.

10.4.5   HAIR

Hair is not often used in the occupational biological monitoring area. It is increasingly being used in the forensic science area, as it is easy to collect, relatively stable, difficult to adulterate or substitute and can provide a long chronic estimate of the exposure of the previous months or years. However, it is prone to external contamination and pre-analysis washing regimes are yet to be standardised.

10.5 ACCOUNTING FOR DAILY URINE VOLUME IN BIOLOGICAL MONITORING

When two workers are exposed to the same amount of chemical during a work shift, the worker who has consumed more liquid (water, tea or coffee etc.) in that shift will excrete more urine. The amount of chemical excreted will consequently appear to be diluted; hence reporting the concentration of the chemical per litre of urine will give a false impression that the hydrated worker has been exposed to a much reduced amount of the chemical. There are a number of ways to account for the differences in urine volume excreted, and these are listed below. The ideal urine reference parameter should provide little inter-individual variation, and should show a constant day-to-day excretion. The parameter should have little influence on other biological factors such as diet or physical activity and should be easy to analyse (Greim and Lehnert, 1998). For the most accurate normalisation, the chemical should have come to a steady state concentration in blood; the timespan between the last urination is an important variance and a better correction would be a regression model using an analyte specific correction (Hertel et al., 2018). The reference parameters that can be used for biological monitoring are listed below.

10.5.1 VOLUME

It is cheap and easy to collect a 24-hour urine sample; however, the collection is time-consuming and therefore not always carried out, making it unreliable. It also presents a danger of sample contamination.

10.5.2 OSMOLALITY

Osmolality is the depression of the freezing point and is independent of diuresis. It is, however, time-consuming and influenced by pH and ethanol, and has poor international acceptance.

10.5.3 SPECIFIC GRAVITY

Specific gravity usually is performed by reporting the test results to a standard specific gravity of 1.020 g/mL. It is cheap and easy, and is relatively independent of the duration of the collection period; however, it is influenced by proteinuria and glucosuria as well as numerous medicines.

10.5.4 NORMALISATION TO CREATININE

The analysis of creatinine in urine is the internationally accepted procedure to account for different urine volumes excreted by individual workers each day. Creatinine is a breakdown product of the muscles and a constant amount is excreted each day by each individual. Expressing the test result of the analysis to the amount of creatinine instead of per litre of urine nullifies the effect of different volumes of urine. The advantage of creatinine as the reference parameter is that it reduces the influence of diuresis-related fluctuations and has a good correlation with 24-hours urine collection. It is unaffected by proteinuria and glucosuria or by salts; however, it is influenced slightly by age, sex, meat consumption, certain medicines, heavy physical work and long periods of fasting. It is also unsuitable for substances that are not mainly filtered in the glomeruli of the kidneys.

The examples in Table 10.1 illustrate the process of creatinine adjustment. In the case of exposure to benzene, the measurement of the metabolite S-phenylmercapturic acid (S-PMA) is performed and an example of exposure to mercury is shown.

Table 10.1 Examples of creatinine-adjusted calculations for S-PMA and mercury

Analyte	Worker 1	Worker 2
S-PMA	10.0 µg/L	10 µg/L
Creatinine	0.5 g/L	2.0 g/L
Creatinine adjusted S-PMA	20 µg S-PMA/g creatinine	5 µg S-PMA/g creatinine
Analyte	**Worker 3**	**Worker 4**
Mercury (Hg)	0.05 µmol/L	0.10 µmol/L
Creatinine	0.0020 mol/L	0.0250 mol/L
Creatinine adjusted mercury	25.0 µmol Hg/mol creatinine	4.0 µmol Hg/mol creatinine

As shown in Table 10.1, Worker 1 has the same concentration of S-PMA in the urine as Worker 2. However, Worker 1 has one-quarter the concentration of creatinine in the urine, showing that Worker 1 is well hydrated and the urine S-PMA concentration is diluted compared with Worker 2. After normalising to the amount of creatinine, Worker 1 shows a higher concentration of S-PMA relative to creatinine and therefore a higher dose. In the second example in Table 10.1, Worker 3 has half the concentration of mercury in urine 0.05 µmol/L as Worker 4, who has 0.10 µmol/L. After adjusting for the concentration of creatinine, it is shown that Worker 3 has more than six times the estimated dose compared with Worker 4.

10.6   CHEMICAL ELIMINATION FROM THE BODY

Generally, the pharmacokinetics of exposure of a worker to a chemical goes through four main stages: (1) absorption, (2) distribution, (3) metabolism and (4) excretion. The chemical is taken up initially by absorption through the lungs or the skin or by the digestive tract. It is then distributed to different organs and tissues in the body by the biological fluids. As the chemical goes through this process, it is absorbed and excreted by various organs. Each of these organs can theoretically be described as a compartment with its own excretion rate before the chemical is reabsorbed by the next compartment, which again has its own absorption and excretion rates, influenced by the preceding compartment. The true pharmacokinetic model can quickly become very complex, and for practical reasons is usually treated as one or two compartments showing a one or two phase elimination curve. This is illustrated in Figure 10.2. The time that it takes to excrete half of the concentration of the chemical from the blood (or urine) is known as the elimination half-life, $t_{1/2}$. It should not be confused with the stability of the chemical in the biological media after sampling.

When the time it takes to eliminate the chemical from the body is longer than the time between exposures, an accumulation can result. With repeated day-to-day exposures over the course of the working week, this can result in the chemical not being totally eliminated from the body until exposure is discontinued. This scenario is illustrated in Figure 10.3. The black line shows when elimination is completed before the next exposure. The grey line shows when the elimination is longer than the next exposure, showing an accumulation over the working week.

The elimination half-life is important when estimating the internal dose, and is relevant to determining when the most appropriate time is to take a biological monitoring sample. Generally speaking, most polar compounds (parent compounds or metabolites) are excreted fairly quickly within hours and certainly less than one day. These compounds are best sampled at the end of the shift or the following morning. These compounds include most solvents, organophosphate insecticides, herbicides, isocyanates, 4,4'-methylene-bis-chloroaniline, fluoride, methyl bromide, cyanide and cytotoxic drugs. Compounds with elimination half-lives greater than one day accumulate over the working week and hence are best sampled at the end of the last shift of the work week. These compounds include most metals, chlorinated solvents such as trichloroethylene and perchloroethylene, and polycyclic aromatic hydrocarbons. Compounds with longer elimination half-lives include

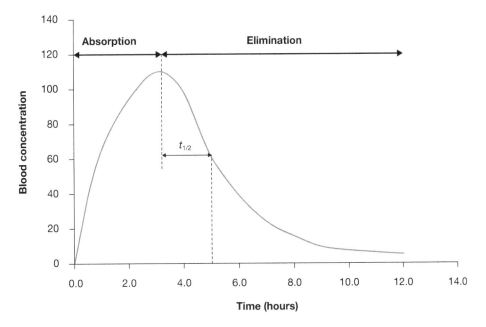

Figure 10.2 The absorption and elimination of the concentration of a chemical in blood following first-order kinetics and the corresponding elimination half-life, $t_{1/2}$

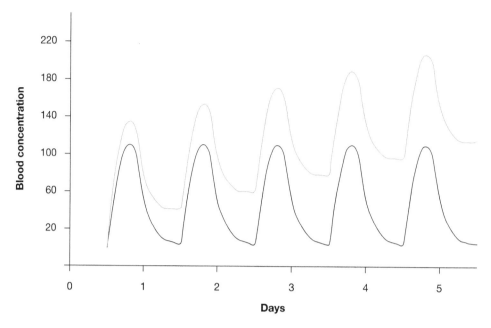

Figure 10.3 The absorption and elimination of the chemical over a working week. The dark line (black) shows elimination before the next shift and the lighter line (grey) shows the accumulation over the working week.

some heavy metals such as mercury, cobalt, cadmium, lead and manganese, and endocrine disruptor chemicals such as polychlorinated biphenyls and organochlorine insecticides. These chemicals are absorbed into body organs and lipophilic tissue; they are slowly released into the blood and reach a steady-state concentration that allows sampling at any time.

10.7   EXPOSURE STANDARDS

Biological monitoring occupational exposure standards have been developed over the years by a number of prominent bodies from countries such as the United States, Germany, the United Kingdom and the European Union. The values proposed by these bodies have largely been adopted internationally. These bodies evaluate the current research, including background population studies, workplace exposure scenarios, chemical incidents, epidemiological studies and health risk assessments. Typically, health risk thresholds are estimated and a limit is set considerably lower after applying an uncertainty factor. The general approach in research is to determine biological equivalents using the forward dosimetry model where an amount of the chemical is measured in the biological media resulting from a known concentration in air (Hays et al., 2007). This is, of course, only applicable to exposures where the main route of exposure is inhalation. When dermal exposure is predominant, other studies—such as patch testing—are more applicable. A pragmatic approach has been introduced in the United Kingdom in recent times, which equates the 90th percentile of the resulting concentration of the chemical in biological fluid when good occupational hygiene controls are in place. This is particularly useful when limited research data are available and also for carcinogenic chemicals where no threshold limits can be established. The foremost biological occupational exposure standards and their standard setting bodies are listed below:

- the Biological Exposure Index (BEI®), set by the American Conference of Governmental Industrial Hygienists (ACGIH®) in the United States (ACGIH, 2018)
- biological occupational tolerance value (Biologischer Arbeitsstoff Toleranz-Wert or BAT), set by the German research foundation, Deutsche Forschungsgemeinschaft (DFG) (DFG, 2018)
- biological limit values (BLV) from the European Union's Scientific Committee on Occupational Exposure Limits (SCOEL). Information on the occupational exposure values from various EU countries can be found in SCOEL publications (SCOEL, 2018).
- biological monitoring guidance value (BMGV) from the Health & Safety Executive (UK) (HSE, 2018)
- SafeWork NSW's biological occupational exposure limits (BOEL), which have largely been adopted from the above organisations and are published with the associated guidance notes by Safe Work Australia (2013).

Currently, most biological exposure standards, except for lead in blood, are not called up in regulations in Australia. However, health monitoring guidance published by Safe Work Australia (2013) references biological exposure levels at which actions should be taken to improve controls.

Exposure limit values are not values where health effects occur, but rather values below which health effects should *not* occur. They are warning levels that should trigger an investigation into the work environment, processes and exposure controls. Some organisations set their own action limits, which are usually below an official exposure limit. Care must be taken when assuming safe levels of chemical exposure, as the impact of exposures will vary among individuals, depending on their genetics and other environmental factors. A regular exceedance of the exposure limit can possibly affect the long-term health of the individual; however, lower exposures and/or small but extended exposures are not necessarily harmless and should be reduced as far as reasonably practicable.

10.8 BIOLOGICAL MONITORING PROGRAM

The regulations that apply to health monitoring can be found in the Model Work Health and Safety Regulations, which are available through the Safe Work Australia website (Safe Work Australia, 2018). These model regulations have been adopted by most Australian work health and safety jurisdictions; however, duty holders should ensure they are familiar with the specific regulations applying to their particular state, territory or jurisdiction.

Biological monitoring is essentially a workplace exposure measurement analogous to air or swab samples, and therefore can be undertaken by any competent professional. When biological monitoring is undertaken as part of health monitoring, however, it must be conducted or supervised by a medical professional with experience in health monitoring. This could be a general practitioner; however, a medical professional with more specialised workplace exposure skills such as an occupational physician is preferred. It is worthwhile to discuss monitoring strategies with a medical professional in the planning phase of the assessment (AIHA, 2004).

The use of urine samples for biological monitoring is considered here, and is the usual preferred option. The available tests are usually focused on this media for reasons noted earlier. If, however, the collection of blood is necessary, then a qualified phlebotomist must be involved.

To develop a biological monitoring program, the following aspects should be addressed.

10.8.1   PROGRAM MANAGER

A program manager should be appointed; they should be a qualified professional with relevant knowledge on biological monitoring and interpretation of the data, as well as the work processes used in the workplace.

10.8.2   MONITORING PROCESS

The following aspects are important when setting up the monitoring process.

10.8.2.1 What, where and who to measure

Before the commencement of a program, a thorough understanding of the work processes, the use of the chemicals and the workers involved is important for a meaningful review of the occupational exposures. At this stage an assessment should be made to determine if biological monitoring is the most appropriate method to assess the particular exposure.

10.8.2.2 Analytics

The hygienist should determine if there is a laboratory available that is capable of performing the required analysis. Analytical chemistry has progressed recently with the introduction of very sensitive instruments, and is key to the recent increase in use of biological monitoring. Care must be taken to select the target molecule and the associated analytical method. The target molecule can be the original chemical or its metabolite. When testing for the original chemical, contamination from clothes, hands and hair could occur and invalidate the results, hence correct sampling protocols should be sought from the laboratory.

The following aspects are important to discuss with the laboratory:

- *Specificity.* Ideally a target molecule is specific for the hazardous chemical and has no or minimal interference from non-occupational influences, such as diet or general metabolism within the body. A typical example is the use of phenol and S-phenylmercapturic acid (S-PMA) as target molecules for benzene exposure. Although both are metabolites of benzene, phenol is a relatively common molecule and testing will not be free from non-occupational influences. Phenol is therefore only valuable as a marker for high exposures to benzene where the phenol levels found in urine are beyond typical background levels. S-PMA, on the other hand, is highly specific and there are no known interferences that could influence the reliability of the measurement (Lauwerys and Hoet, 2001).

- *Selectivity.* The selectivity of a method is the ability to separate and quantitatively identify the target chemical from the sample matrix. Urine is a complex and variable matrix, and highly selective analytical methods typically are necessary. With the advent of mass spectrometric detection systems coupled to gas chromatography instruments, inductively coupled plasma instruments and more recently to liquid chromatography instruments has allowed the analytical power required for selective biological monitoring analysis.
- *Sensitivity.* The analysis should have sufficient response to clearly show a change in the chemical concentration.
- *Limit of quantitation.* The analysis should be able to measure well below the exposure standard or reference value, and ideally right down to the background environmental levels.
- *Quality assurance.* A laboratory ideally should be accredited to a recognised quality standard, such as ISO 17025 or ISO 13485, and successfully participate in a relevant interlaboratory proficiency testing scheme.

10.8.2.3 The location, timing and frequency of the sample collection

The collection location should be fit for purpose. The timing of the sample collection should be less than three times the elimination half-life of the chemical. The use of pre-shift and post-shift samples will allow you to exclude non-occupational exposures and might reduce the impact of dietary background influences. When the elimination half-life is longer than the interval between exposures, pre-shift and post-shift samples can show the accumulation of the chemical over the working week. Increasing the sampling frequency will create a more detailed overview of the investigated exposure. The run time for the entire study should be determined and in line with the aim of the study.

10.8.2.4 Reference level

The found levels should be compared against a known exposure limit or known environmental background levels (see above). Where no exposure limit is available, biological monitoring may still be very useful, provided the testing will give a quantitative result that assists in the management of the exposure and assessing the controls. In this instance, it is important to ensure that the right metrics are available to establish internal reference values. The use of a comparison of pre-shift and post-shift samples is often useful to determine if an exposure has occurred.

10.8.2.5 Statistics

Statistics can help to interpret large and complex data sets; they can also help with determining the minimum required sample numbers while maintaining a statistically sound study. Advice from a statistician should be considered in the design phase of the monitoring program.

10.8.2.6 Additional data

Additional information on diet, smoking, start and end times, chemicals and application processes used, episodes of skin contact, safety data sheet information, non-work exposure sources, controls and PPE used and timing of an exposure incident (as applicable) will be necessary for the interpretation of the results, and should be collected at the time of sampling.

10.8.2.7 Sample volume and storage

The specimen must be collected in amounts that are sufficient for analysis. This is particularly important for multiple analyses on a single sample; the laboratory can advise on this. It is best practice to store samples as fast as is practical after collection in a separate portable esky, fridge or freezer with a biological hazard label. The storage unit should not be used for food and drinks. In rare cases, biological activity or chemical degradation might affect the analyte levels in the specimen, so a lower storage temperature will prolong the quality of the sample. There are no universally applicable rules for all chemicals, but fridge storage will typically extend the shelf-life of samples to several weeks and when frozen a further extension, to six to twelve months, can be expected for many chemicals.

10.8.2.8 Sample transport

Transport to the laboratory should be done by road courier or by air transport companies. Air transportation of biological monitoring samples should comply with the International Air Transport Association (IATA) Dangerous Goods Regulations. Transportation should be done overnight or, if feasible, using a same-day delivery to the laboratory. To maintain sample integrity, it is strongly recommended that transportation over the weekend or during public holiday periods be avoided. Furthermore, while in transit, the samples should be kept under cold conditions until they are received by the laboratory. This is usually achieved by transporting the samples in an esky loaded with frozen esky bricks or ice packs. Wet ice should never be used. Dry ice is not necessary for short transport periods, but for longer journeys and international transport it might be worth considering. Specialised transport companies provide transportation services around the world, while maintaining a constant specimen temperature. Please note that the very low temperature of dry ice (<–65°C) can damage marker pen writing and strongly affects the glue layer of labels, making them drop off the containers. Special labels are available for dry ice transport.

10.9   CONSULTATION

All workers and/or their representatives should be consulted in the planning phase of a biological monitoring program to discuss the following.

10.9.1   *THE WHS REGULATIONS*

The Model WHS Regulations on health monitoring, links to specific jurisdictional regulations and health monitoring guidance material can be found on the Safe Work Australia website (Safe Work Australia, 2018) and should be explained to the staff involved.

10.9.2   *CONSENT PROCESS AND PRIVACY*

Prior to collecting samples, *informed consent* must be obtained from the individual worker. This involves providing a clear overview of the purpose of the monitoring program, outlining what samples are taken and any risk involved, what analyses are to be performed and who will have access to the test results, and how the data will be used. Many workplaces choose to seek consent to biological monitoring on a pre-employment basis, as biological monitoring data can be a key piece of information in managing exposure risks.

There might be concerns over the use of the sample for other analyses, like illicit drugs, medication or alcohol. Strict controls must be put in place to ensure that the samples are solely used for workplace exposure monitoring. Testing for drugs and alcohol at work has its own place and should not be entangled with biological monitoring to evaluate exposure to hazardous chemicals.

10.9.3 REFUSAL

A worker has the choice to refuse to participate in a biological monitoring program. The benefits of biological monitoring should be fully explained and assurances of strict privacy controls should be clarified for participants.

10.9.4 FEEDBACK AND ACTIONS IN CASE OF EXPOSURE

Prior to the start of the biological monitoring program, participants should be counselled on the aims of the program and how a found level of exposure will be handled, the impact on their work and how access to a medical professional can be obtained.

10.9.5 PRIVACY AND CONFIDENTIALITY

In the United States, OSHA stipulates that a test report from a biological monitoring program that is undertaken to monitor exposure only is not a medical record; however, a biological monitoring test report that is used in conjunction with health monitoring is a medical record. Nevertheless, a medical practitioner is obliged by law to give the employee the results of the health monitoring and to give the employer the general outcome of the monitoring. Employers who conduct exposure monitoring can remove personal identifiers when privacy and confidentially is of concern.

Employers are required under WHS regulations to provide health monitoring results to the work health and safety regulator if the results indicate that an illness has developed or that controls should be improved. In some circumstances, there is also a requirement for the medical practitioner to notify the work health and safety regulator and/or the health department of the results.

10.9.6 RESULTS AND FEEDBACK

All biological monitoring analytical results should be passed on to the worker, as well as being interpreted and explained by a competent person. Where available, the exposure should be compared with an occupational exposure standard. Access to a medical professional should be made available when:

- it is required as part of a health monitoring program under WHS law
- a medical interpretation of the results is requested by the worker

- symptoms are present that could be related to the exposure, or
- exposure levels are above an occupational exposure standard or similar guidance value.

10.10   EVALUATION

Most biological monitoring reports contain guidance on how to interpret the test result. Direct comparison to an exposure standard is the usual approach to determine whether excessive exposure has occurred on an individual basis. In the absence of an available exposure standard, information on the population background levels can be found from population studies that have been undertaken. A particularly useful source of information is the National Health and Nutrition Examination Survey (NHANES) that has been undertaken by the Centers for Disease Control and Prevention in the United States for many years (CDC, 2018).

To evaluate whether exposure has occurred on a group basis in a particular workplace, pre-shift and post-shift samples are useful. The mean of the difference between the pre-shift and post-shift samples can be assessed by using Student's paired t-test (DePoy and Gitlin, 2015). This can show that an exposure has occurred during that shift but does not compare it to a particular occupational limit.

The accepted procedure to compare the group exposures to an occupational limit is to first take the natural logarithm of the data, as exposure data is usually lognormally distributed, transforming it to a normal distribution. A one-sided comparison of the mean to the occupational limit is then performed by the Student's t-test.

Based on a reliable data set, improvements in the workplace should be made and the effectiveness of the implemented changes evaluated using the same monitoring. The frequency of the testing should be considered and would typically be reduced for work processes that are under control, while in the opposite situation a continuation or increase of the frequency might be desirable. Upon completion of a monitoring program, surveillance samples should be planned for ongoing low-frequency monitoring, if required on a risk basis.

The biological monitoring process and results should be documented with a sufficient level of detail for reporting and interpretation by a person who was not involved in the process. Company policies and regulatory requirements around retaining documentation on biological monitoring should be considered.

10.11   REFERENCES

American Conference of Governmental Industrial Hygienists (ACGIH) 2018, *TLVs and BEIs Threshold Limit Values for Chemical Substances and Physical Agents & Biological Exposure Indices*, ACGIH®, Cincinnati, OH.

American Industrial Hygiene Association (AIHA) 2004, *Biological Monitoring: A Practical Field Manual*, AIHA, Fairfax, VA.

Centers for Disease Control and Prevention (CDC) 2018, *National Health and Nutrition Examination Survey (NHANES)*, <www.cdc.gov/nchs/nhanes/index.htm>.

DePoy, E. and Gitlin, L.N. 2015, 'Statistical analysis for experimental-type designs', in E. DePoy and L.N. Gitlin, *Introduction to Research: Understanding and Applying Multiple Strategies*, Elsevier, New York, pp. 282–310.

Deutsche Forschungsgemeinschaft (DFG) 2018, *List of MAK and BAT Values,* Wiley-VCH, Bonn.

Greim, H. and Lehnert, G. 1998, 'Creatinine as a parameter for the concentration of substances in urine', in H. Greim and G. Lehnert, *Biological Exposure Values for Occupational Toxicants and Carcinogens Vol 3*, Wiley-VCH, Weinheim, pp. 35–44.

Hays, S.M., Becker, R.A., Leung, H.W., Aylward, L.L. and Pyatt, D.W. 2007, 'Biomonitoring equivalents: A screening approach for interpreting biomonitoring results from a public health risk perspective', *Regulatory Toxicology and Pharmacology*, vol. 47, pp. 96–109.

Health and Safety Executive (HSE) 2018, *Biological Monitoring Guidance Values*, <www.hsl.gov.uk/online-ordering/analytical-services-and-assays/biological-monitoring/bm-guidance-values> [accessed 14 September 2018]

Hertel, J. et al. 2018, 'Dilution correction for dynamically influenced urinary analyte data', *Analytica Chimica Acta*, vol. 1032, pp. 18–31.

Lauwerys, R.R. and Hoet, P. 2001, *Industrial Chemical Exposure: Guidelines for Biological Monitoring*, 3rd ed., Lewis, Boca Raton, FL.

Manno, M. et al. 2010, 'Biomonitoring for occupational health risk assessment (BOHRA)'. *Toxicology Letters*, vol. 192, pp. 3–16.

Safe Work Australia 2013, *Health Monitoring for Exposures to Hazardous Chemicals,* <www.safeworkaustralia.gov.au/system/files/documents/1702/guide-pcbu-health-monitoring-exposure-hazardous-chemicals.pdf> [accessed 14 September 2018]

—— 2018, website, <www.safeworkaustralia.gov.au> [accessed 14 September 2018]

Scientific Committee on Occupational Exposure Limits (SCOEL) 2018, *Scientific Biological Limit Values*, SCOEL, <http://ec.europa.eu/social/main.jsp?catId=148&intPageId=684&langId=en> [accessed 14 September 2018]

11. Indoor air quality

Sue Reed and Michael Shepherd

11.1   INTRODUCTION

Poor indoor air quality (IAQ) can be a significant health, environmental and economic problem, and has thus become a public health and liability issue for employers, as well as for building managers and owners. The definition of IAQ is often interpreted differently, depending on the discipline describing it. This chapter uses the broad definition of IAQ, which may also be referred to as indoor environment quality (IEQ), as specified in the Australian Building Code Board's (ABCB) *Indoor Air Quality Handbook* (ABCB, 2018). This chapter will only review indoor air quality, as the other issues—such as thermal, light and noise—are covered in chapters 12, 14 and 15. IAQ also has a direct impact on the personnel costs of any business, as it has been shown that there is an association between IAQ and occupant productivity.

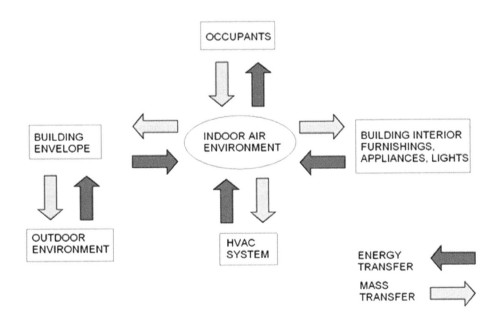

Figure 11.1 Indoor air quality model
Source: Spengler, Samet and McCarthy (2001, p. 572).

Spengler, Samet and McCarthy (2001) describe the relationship between the building environment and IAQ as shown in Figure 11.1.

If the equilibrium between two or more of these parameters is unbalanced, a range of building-related illnesses may result, many of which have identifiable causes (e.g. Legionnaires' disease, caused by bacterial growth in cooling towers). A sub-set of building-related illnesses—termed 'sick building syndrome' (SBS) and described by the World Health Organization (WHO, 2010)—includes a range of subjective symptoms such as mild irritation of eyes, nose and throat, headaches and lethargy. SBS symptoms are believed to arise from multiple causes that, while not clearly understood, are associated mainly (although not exclusively) with air-conditioned office buildings. Australian studies indicate that Australian buildings have similar IAQ issues to buildings in other developed countries with regard to building-related illnesses and SBS-like symptoms.

People differ markedly in their sensitivity to airborne pollutants. Infants, young children, the elderly and those with suppressed immunity are more vulnerable to respiratory illnesses associated with environmental tobacco smoke, house dust mites and gas combustion products such as nitrogen dioxide. People with asthma are sensitive to a variety of pollutants, which act as triggers for the condition. The issue of multiple chemical sensitivity (see section 11.2.3) and the possible role of indoor air pollutants are still under debate. The protection of such sensitive population groups is an issue in the selection of appropriate IAQ indicators for residential, health, aged care and educational buildings, but will not be discussed in depth in this chapter. Indicators for other types of buildings need to take into account the likely entry of sensitive individuals (e.g. in an office building that is accessible to the public). This chapter deals primarily with commercial, non-residential buildings, but H&S practitioners need an understanding of the impact of IAQ in residential buildings (e.g. for workers who telecommute or work from home). In Australia, few indoor air pollutants have been documented sufficiently in the published literature to establish typical indoor levels (Reed, 2009).

Indoor air pollutants (pollutants of indoor origin) typically occur at higher concentrations than air pollutants found outdoors, and also differ from them. Outdoor air pollutants (e.g. fine particles, ozone, nitrogen dioxide) do intrude indoors through windows or ventilation systems, but generally at much lower concentrations (apart from gases such as carbon monoxide and carbon dioxide), owing to deposition or losses on building surfaces. Pollutants generated from indoor sources can increase to elevated concentrations compared with outdoor pollutants because of restricted air dispersion, even where ventilation complies with AS 1668.2 and/or the Building Code of Australia (BCA). Since most people spend around 90 per cent of their time indoors, they are exposed to much higher concentrations of indoor air pollutants, and for longer periods, than they are to outdoor pollutants.

In Australia, the most significant indoor air pollutants are now considered to be:

- in *commercial buildings*, particulate matter ($PM_{2.5}$), micro-organisms, carbon monoxide and nitrogen oxides (based on location of air-conditioning fresh air intakes in relation to car parks and roads), volatile organic compounds (VOCs) and formaldehyde (based on their potential for high indoor levels under specific conditions)
- in *residential buildings*, nitrogen dioxide, environmental tobacco smoke, moulds and house dust mites (all based on the observation of high indoor levels).

High levels of environmental tobacco smoke (ETS) were traditionally found in hospitality buildings such as hotels and restaurants, but this is no longer the case in Australia since smoking in such venues has been banned (Davidson et al., 2007). In most commercial buildings, tobacco smoke enters only where occupants smoke close to doorways or air intake vents, or just before re-entering the building. House dust mite allergen levels are very high in homes, some child-care facilities and schools in coastal regions, where a majority of the Australian population live. Nitrogen dioxide concentrations have been found to be high in many residences and schools with unflued gas heaters, and also in some large buildings (e.g. shops, warehouses), especially near major roadways.

The ventilation rate and the thermal comfort of a building space must be considered key factors affecting IAQ. Low ventilation rates will cause odours from occupants and pollutants from indoor sources to accumulate, reducing air quality. This problem should be limited at the construction stage by the BCA, but once a building is complete, further assessment is uncommon. The National Australian Built Environment Rating Scheme (NABERS) and Green Star programs have been developed to improve this situation. They assess buildings against a range of environmental parameters, particularly power and water usage, waste recycling, thermal comfort, air quality, acoustic comfort and lighting (NABERS, 2011). While the BCA's requirements have improved over time, they apply only to new buildings and not retrospectively to existing buildings. It is advisable to assess any building's ventilation rate relative to the current BCA, and AS 1668.2 is the first step in any IAQ investigation.

Occupants' acceptance of indoor spaces depends largely on thermal comfort. While guidelines for temperature control exist in a number of countries, this is not regulated in the BCA, though it is mentioned in relation to the sustainability of buildings and many large employers provide workers with guidelines or fact sheets on the subject.

11.2 WHAT IS IAQ AND WHO IS RESPONSIBLE?

11.2.1 DEFINITION OF IAQ

It is important to be clear about what is meant when discussing IAQ. To an engineer, it may mean the building's ventilation rate; to an architect, the occupants' perceptions of the building; to a microbiologist, the levels of bacteria, fungi and plant spores in the air. For an occupational hygienist or toxicologist, IAQ may refer to all these issues as well as exposures to specific pollutants. IAQ is one component of IEQ, which also includes thermal comfort and lighting.

People often assume that indoor air is very similar to the air just outside a given building. Variations in outdoor air quality (e.g. from traffic pollution) may indeed be reflected in the indoor air, depending on the ventilation filtration system and on the intake location and amount of make-up air. However, a wide range of pollutants are also generated within indoor spaces from the activities of the occupants and from the materials and appliances they use. These add to any pollutants introduced from outdoor air. Most urban dwellers spend about 91 per cent of their time indoors, 5 per cent in travel or in vehicles and only 4 per cent outdoors (Newton et al., 2001).

In Australia, the National Health and Medical Research Council (NHMRC, 1993) defines 'indoor air' as the air inside spaces where people may spend more than one hour a day. These spaces include offices, homes, shopping centres, hospitals, schools and recreational facilities. Indoor air quality can also be defined as the totality of attributes of indoor air that affect a person's health and wellbeing (Department of Environment and Energy, 2018).

The definition of indoor air normally excludes buildings that relate directly to industrial, construction and mining activities, as they normally are covered by exposure standards (ES) set by Safe Work Australia and/or state OHS/WHS regulations, but no direct guidance on general IAQ in the workplace is provided. The distinct difference between the industrial scenario and other workplaces is the well-known 'healthy worker effect', whereas good IAQ is aimed at protecting the general population, which includes sensitive people with existing illnesses, the young and the elderly. An industrial indoor workplace may contain only a few pollutants with a limited period of exposure (typically 40–48 hours per week), whereas indoor air in other buildings may contain a diverse range of pollutants. Although these are usually present at very low concentration, they are experienced for much longer periods, which in some cases can be nearly 24 hours a day, seven days a week. This causes a dilemma for the H&S practitioner, who may be able to apply a WES in an industrial setting but not the corresponding office building. One approach to this problem is to apply a WES in an office that members of the general public cannot enter and an IAQ goal for one that is accessible to the general public (WHO, 2010). In general, however, ESs are set at levels much higher than would be tolerated by many office workers (especially the more sensitive groups), so this approach has limited scope. Clearly, the protection of sensitive individuals below and above working age must be a prime concern when selecting IAQ indicators for health-care, educational and aged-care buildings. It should be noted that the Department of Environment and Energy (DEE) definition of IAQ refers not merely to health but also to wellbeing. This increases the challenge of assessing IAQ, since wellbeing is to a large extent a matter of individual feelings and is therefore not easy to measure.

11.2.2   RESPONSIBILITY FOR REGULATING IAQ

In many developed countries, responsibility for the regulation of IAQ falls to environmental or health agencies, and sometimes to building authorities. For example, the US Environmental Protection Agency (EPA) carries out extensive research and industry/community activities. In 2000 the WHO established air quality goals for Europe, stating clearly that these were relevant to both indoor and outdoor environments. More recently, the WHO has published guidelines for a limited number of pollutants in indoor air (WHO, 2010). In Australia, state environmental agencies have had limited involvement with IAQ measurement (Ferrari 1991; Newton et al. 2001). In recent years, Environment Australia and the Department of Health and Ageing (Department of Health and Ageing, 2002) have undertaken research and community education activities related to IAQ, though no organisation has taken clear regulatory responsibility.

The Australian Building Codes Board (ABCB) regulates building practice in Australia through the BCA, which—with small variations—has been adopted by most states (ABCB, 2018). The BCA is performance based rather than prescriptive. Since compliance with the BCA is a legal requirement for the design and construction of all buildings in Australia, a reference to an Australian Standard in the BCA's Housing Performance Provisions makes observing that standard a legal requirement. Australian Standards relating to IAQ are AS 1668.2 The Use of Ventilation and Airconditioning in Buildings (Standards Australia, 2012) and AS/NZS 3666.1 Air-handling and Water Systems of Buildings—Microbial Control. Part 1: Design, Installation and Commissioning (Standards Australia, 2011b). The BCA itself contains a specific performance provision for IAQ, but only in relation to ventilation with outdoor air (see section 11.6).

In addition, WHS legislation in Australia puts the onus on the person with the control of a business, employer and/or building owner to ensure that the indoor air is of good quality and will not adversely affect workers or visitors. A person conducting a business or undertaking has the primary duty under the act to ensure, so far as is reasonably practicable, that workers and other persons are not exposed to health and safety risks arising from the business or undertaking.

The regulations place more specific obligations on a person conducting a business or undertaking in relation to the work environment and facilities for workers, including requirements to ensure, so far as is reasonably practicable, that the layout of the workplace, lighting and ventilation enable workers to carry out their work without risks to health and safety. To ensure that indoor air does not contain pollutants that may be hazardous to health, it is important that such air be monitored and evaluated against the appropriate guidelines.

In the twentieth century, IAQ in Australia was defined as acceptable if it met NHMRC goals (NHMRC, 1996), which were only recommendations and were rescinded in 2002; nonetheless, they are still being used by a number of organisations. In 2018, the ABCB published guidelines in its handbook to meet modern building requirements and the 2016 National Construction Code (ABCB, 2016). Guidelines are also published by organisations such as NABERS to meet their assessment schemes (NABERS, 2015).

11.2.3 BUILDING-RELATED ILLNESSES, SICK BUILDING SYNDROME AND MULTIPLE CHEMICAL SENSITIVITIES

Health effects associated with indoor air range from severe (asthma, allergic responses, cancer) to subjective, generally non-specific symptoms that sufferers attribute to poor-quality air or the environment. Collectively, all such effects are termed 'building-related illnesses'. Many arise from identifiable causes, such as specific pollutants (e.g. formaldehyde, dust mite allergen, *Legionella* species), poor ventilation, humidifiers, poor thermal comfort, poor lighting and psychosocial factors. However, a range of subjective symptoms, termed 'sick building syndrome' (SBS) by the World Health Organization (WHO, 2010), may occur in 30 per cent or more of occupants of specific buildings. Wolkoff (2013) summarises the symptoms of SBS as follows:

- irritated, dry or watering eyes (sometimes described as itching, tiredness, smarting, redness, burning and difficulty in wearing contact lenses)
- irritated, runny or blocked nose (sometimes described as congestion, nose bleeds, itchy or stuffy nose)
- dry or sore throat (sometimes described as irritation, oropharyngeal symptoms, upper-airway irritation, difficulty swallowing)

- dryness, itching or irritation of the skin, occasionally with rash (also described in clinical terms such as erythema, rosacea, urticaria, pruritis, xerodermia)
- headache, lethargy, irritability and poor concentration.

Extensive investigation of SBS has failed to pin down any specific cause of the symptoms. They appear to arise from the combination of many factors, which are often vague or not clearly linked to symptoms. The following are some potential causes.

- *Fresh air intake.* A lack of fresh air intake into either ventilated or unventilated rooms raises carbon dioxide levels and increases relative humidity. This can cause a range of issues, which in some cases can be alleviated by increasing fresh air intake.
- *Ventilation systems.* Air-conditioning systems have been associated with a range of symptoms, which may be caused by poor air distribution, poor maintenance, lack of sufficient fresh air make-up for the use of the building, or the development of an environment conducive to the growth of micro-organisms and dust mites.
- *Airborne chemical pollution.* Many items within a building may emit pollutants, including the building structure, wall and surface coatings, floor coverings, fabrics, office chemicals, furniture, building activities and the occupants themselves.
- *Micro-organisms (mites, fungi, bacteria) and particulates.* A mixture of organic and non-organic dust from poorly maintained air-conditioning systems, poor maintenance, excess moisture, environmental damage such as floods and furnishings may allow micro-organisms to multiply and particulates to build up.
- *Temperature.* Ideal indoor temperatures vary according to the time of year, type of building (naturally or mechanically ventilated) and location of the building. Ranges outside 20°C to 25°C have been shown to increase symptoms, but possibly only where relative humidity is low or under particular conditions of (low) air movement.
- *Relative humidity.* This is the ratio of the amount of water vapour in the air to the amount of water vapour air can hold at that temperature. The optimum relative humidity range should be kept between 35 and 65 per cent, in accordance with ISO 7730. Relative humidity below 40 per cent may be associated with symptoms such as dry eyes and dry skin. High relative humidity may lead to fungal growth.
- *Lighting.* Certain symptoms may be promoted by poor lighting, the absence of windows or flicker from fluorescent tubes.
- *Personal and organisational factors.* Symptoms attributed to SBS are more frequent among women, workers in routine jobs, those with a history of allergy, workers using computers and those who perceive that they have little control over their indoor environment.

Assessment of building-related illnesses in Australia has previously been very limited, and generally has been restricted to studies of office environments, but the available data indicate that dissatisfaction with office IAQ may be common. Williams (1992) reported that occupants of 62 per cent of 228 low-rise office buildings in suburban Melbourne experienced unacceptably 'stuffy' and 'drowsy' conditions. Eighty-two per cent of the buildings failed to meet existing ventilation requirements, largely owing to changes in layouts and uses since the buildings were constructed. A multi-state survey of 511 government office workers in 1990 found that 91 per cent experienced discomfort or illness associated with poor ventilation and temperature control (McKenna, 1990). Complaints included: too hot (72 per cent), too stuffy (72 per cent), drowsiness (48 per cent), headaches (48 per cent) and sore throat (55 per cent). Rowe and Wilke (1994) found that more than 45 per cent of occupants of eight office buildings in the Sydney CBD and suburbs were dissatisfied with either thermal comfort or indoor air quality. Dingle and Olden (1992) reported that occupants of a new four-level office building in Perth complained of dry eyes (65 per cent), tired and strained eyes (54 per cent), reflection/glare (41 per cent), fatigue (57 per cent), sore throat (28 per cent) and migraine (36 per cent). Rabone and colleagues (1994) found that work-related symptoms in 401 occupants of a 'sick' office building in Sydney were strongly associated with the stuffiness of the air. Rowe (2003) found differences in occupants' perceptions of air quality, thermal comfort and self-reported SBS symptoms in Sydney offices, with conditions worse in mechanically ventilated/air-conditioned offices than in naturally ventilated offices. When the occupants were supplied with personally controlled cooling/heating units, their perceptions improved.

A NSW Standing Committee on Public Works (2001) inquiry into SBS in Australia concluded that it significantly affected the health and productivity of building occupants. The committee called for regulations and improved standards and codes relating to ventilation systems and their maintenance, low-emission building and fit-out materials, commissioning of new buildings and indoor pollutant criteria for existing buildings. Ten years on, these have yet to be implemented in New South Wales or any other state, but many states have implemented NABERs or Green Star schemes, which have similar requirements.

Multiple chemical sensitivity (MCS) refers to a chronic medical condition characterised by a diverse array of symptoms, which are reportedly triggered in a small number of individuals by exposure to chemicals at levels far below those that cause illness in the rest of the population. The aetiology of MCS is not fully understood, but it has been suggested that the worldwide prevalence of medically diagnosed MCS is 0.2–4 per cent (National Industrial Chemicals Notification and Assessment Scheme and the Office of Chemical Safety and Environmental Health, 2010). Poor IAQ, in addition to cosmetics, food additives, household products and other items, has been blamed as a major cause of MCS, but this has not been substantiated by rigorous studies.

11.2.4 *RELATIONSHIP OF IAQ TO PRODUCTIVITY*

Improvements in IAQ have the potential to reduce:

- the costs of health care
- sick leave
- diminished productivity caused by workers' discomfort and/or adverse health symptoms
- the costs of investigating complaints by workers and building visitors.

Thermal comfort and lighting also influence worker performance and perception of IAQ without affecting health. Staff salaries account for the bulk of operating costs in many organisations, and even very small increases in productivity may justify the cost of building improvements. Based on a review of the literature, Fisk and Rosenfeld (1997) estimated that in the United States, the potential annual benefits—including improvements in productivity—were:

- a 10–30 per cent reduction in acute respiratory infections, allergy and asthma symptoms
- a 20–50 per cent reduction in acute non-specific health symptoms
- a 0.5–5 per cent increase in the performance of office work
- annual cost savings and productivity gains exceeding US$30 billion.

A study by Bell and colleagues (2003) also found that improvements in IAQ had a positive effect on indoor workers' health and productivity.

11.3   INDOOR AIR QUALITY GUIDELINES

Regulation of the quality of indoor air in Australia is minimal, especially compared with that of outdoor air and industrial workplace air. This was the case in most countries until the early 2000s, when Finland, Germany, Hong Kong, Japan, the Netherlands and the United States initiated IAQ regulations and/or guidelines. In the past, such guidance as existed in Australia was provided by the NHMRC and the National Occupational Health and Safety Commission (NOHSC), now Safe Work Australia. There remains a need for a more structured approach to the evaluation and control of IAQ. Implementing such an approach is made more difficult by the lack of a single governmental authority with national responsibility for IAQ. The Department of Sustainability, Environment, Water, Population and Communities (DSEWPC) has begun to provide guidance on the importance of good IAQ and has funded limited research in the field. Occupational standards and environmental guidelines need to be developed to cover a variety of indoor environments and harmonised nationwide. Development of improved ventilation codes, voluntary reduction of pollutant emissions from manufactured products and improved public education all have a role to play in improving IAQ.

11.4   MAJOR INDOOR AIR POLLUTANTS AND SOURCES

The types of pollutants commonly found in Australian buildings (Table 11.1) include:

- formaldehyde and VOCs from new building materials in new or renovated buildings (less than six months old)
- formaldehyde in caravans/transportable buildings with high loadings of reconstituted wood-based panels
- house dust mite allergens in carpets and bedding (except well inland and in Central Australia)
- combustion products (nitrogen dioxide and carbon monoxide) from unflued gas heaters and stoves
- vehicle exhausts (benzene, 1,3-butadiene, particulate matter ($PM_{2.5}$), carbon monoxide, DPM) from adjacent garages or enclosed car parks

Table 11.1 Typical indoor and outdoor air pollutant levels in Australia

Pollutant	Typical indoor air concentrations ($\mu g/m^3$)			Typical outdoor air ($\mu g/m^3$)
	New house/ office	Established house	Established office	
Formaldehyde	100–800	20–120	40–120	10–20
Total VOCs	5000–20000	200–300	100–300	20–100
Nitrogen dioxide				10–50
no unflued gas heater	—	10–35	—	(300 peak)
unflued gas heater	—	60–1500	—	
Fine particles (PM_{10})				5–30
smoking	—	>90	100–300	
non-smoking	40–60	5–40	10–40	
Dust mite allergens (per gram of house dust)	<0.1	10–60 coastal <1 inland	<2 (data limited)	<0.1

Sources: Brown (1996, 1998, 2000, 2002); Mannins (2001).

- asbestos from the disturbance of friable and non-friable building materials, especially insulation products (e.g. switchboard backing, boiler/pipe lagging, sprayed-on fire-retardant).

11.4.1 MICRO-ORGANISMS IN INDOOR AIR

There is significant concern about micro-organisms as indoor air pollutants. This category includes viable and non-viable organic matter such as viruses, bacteria, fungi, protozoa, house dust mites, insect faeces and pollens, which may cause infectious disease, toxic effects or allergic reactions. The occurrence and health effects of *Legionella* bacteria are well understood in Australia, but there have been few published investigations of other micro-organisms. Overseas research has established associations between indoor dampness and ill-health, but there is insufficient evidence to identify or rule out specific causes such as inhaled spores from fungi such as moulds (Levik et al., 2005).

In any building, a variety of micro-organisms will be present at different times and in different locations. For disease agents to be transported through the air, there must first be a source in which they can breed, some means for them to multiply and a mechanism for their release and dispersion (Gunderson and Bobenhausen, 2011). The major indoor reservoir is often stagnant water or moist interior surfaces. Micro-organisms that enter the building from outdoors can survive and multiply in these environs. Airborne dispersion is relatively easy for micro-organisms found in ventilation systems or contaminated carpet. It should be noted that most people are exposed to fungi every day, but in low concentrations. Gunderson and Bobenhausen's (2011) review of the literature points to three key causes of elevated fungal populations:

- poor construction techniques
- a failure to rapidly identify/repair water incursion
- a failure to correctly operate and maintain air-conditioning systems.

A number of guidelines, including International Organisation for Standardisation (ISO) standards, rely on an initial site inspection for identifying potential sources of micro-organisms. Air sampling is regarded as a secondary or last-resort measure, and findings on both species and concentrations should be compared with those from outdoor air.

In one mechanically ventilated Sydney office building in the 1990s (Seneviratne et al., 1994), symptoms of sick building syndrome were associated with micro-organisms, particularly fungi. The building had damp walls and 'high' fungal spore counts (primarily *Cladosporium*, *Aspergillus* and *Penicillium* spp.; 600–2500 CFU (colony-forming units)/m³).

Using a centrifugal air sampler (Biotest RCS, Japan) as a simple screening test (without speciation) for micro-organism contamination of buildings, Wakelam, Lingford and Caon (1995) and Brown (1998) investigated approximately 60 commercial buildings between 1991 and 1998. While neither study related the measurements to occupant illness, both authors concluded that levels of fungi exceeding 500 CFU/m³ and of bacteria exceeding 500CFU/m³ could be considered high and that such measurements could complement visual inspection for moisture and growths on building interior surfaces.

Bacteria are ubiquitous in air and in the general environment, and can have adverse effects on human health as well as the integrity of building materials when they proliferate in indoor environments (Ghosh, Lal and Srivastava, 2015). The effects of exposure to bacteria in indoor air will depend on the species and the route of exposure. Bacteria entering building air via windows or air vents can come from soil and vegetation, compost, municipal landfills and sewage sludge. They can also be expelled by occupants breathing, coughing and sneezing. They may colonise the ductwork of cooling systems, water-cooling towers (e.g. *Legionella* species) or materials such as wallboard, wallpaper and flooring (Bates, 2000). In indoor environments, bacteria usually grow in areas that contain standing water, such as water-spray and condensation areas of air-conditioning systems (Ghosh, Lal and Srivastava, 2015). Indoor air consultants in Australia have found cases of standing water in non-draining condensate trays of air-conditioning systems. While the prevalence of this situation is unknown, building codes now specifically require self-draining condensate trays.

The Department of Health and Ageing (2004) concludes that:

- fungi, bacteria and allergens from house dust mites, cat dander and cockroaches are important indoor biological pollutants
- asthma triggers are manifold, but include allergens—usually small glycoproteins that can provoke an immune response—from house dust mites, animal skin and saliva to pollen, moulds and, in rare cases, foods.

As with other environmental allergens, indoor exposure to mould has been shown to correlate with wheezing and peak-flow variability in lung-function tests. Fungal exposure occurs both indoors and outdoors. Exposure to moulds such as *Alternaria* increases the risk of asthma symptoms and airway reactivity in sensitised children and the risk of sudden respiratory arrest in sensitised young adults with asthma.

Many studies have shown that sensitisation to environmental allergens is strongly associated with childhood asthma; however, the role of allergen avoidance in prevention of asthma in adults is unproven.

11.4.2 *DUST MITES AND SURFACE MICRO-ORGANISMS*

11.4.2.1 Dust mites

Dust mites are ubiquitous in nature and have been recognised as a cause of allergies for many decades. Their secretions and faeces can cause allergic reactions in susceptible people. Dust mites are typically 0.25–0.5 mm long (too small to see easily with the naked eye), and their food sources include flakes of shed skin, animal dander and other organic materials. They can be found in most indoor environments, where they favour warm, moist and dark areas such as upholstered furniture, long-fibred or deep-pile carpets and mattresses.

While it is not possible to remove all dust mites from most office environments, allergic reactions are typically related to the quantity, or dose, of allergen. Reducing the dust mite population through strategies such as regular vacuuming with a vacuum cleaner fitted with a HEPA filter, washing, controlling moisture/humidity and replacement of soft fabrics in furnishings will assist in the reduction of allergies.

11.4.2.2 Surface micro-organisms

Levels of micro-organisms on building surfaces reflect hygiene and cleaning practices. Areas that are difficult to access accumulate dust and moisture. Given that dust is ubiquitous in most indoor environments, controlling moisture is crucial to preventing microbial growth.

Surface micro-organisms can proliferate in air-conditioning ductwork, cooling coils and drip trays. They may also be found on walls, ceilings and in carpets, provided they have sufficient access to moisture. Keeping these surfaces dry and the relative humidity below 75 per cent is essential to preventing microbial growth on such surfaces. Levels of micro-organisms in air can vary considerably over short periods and show poor correlation with levels on indoor surfaces. Even when levels are measured, interpreting their significance is difficult, owing to a lack of health-based exposure limits.

11.5   ASSESSMENT OF IAQ

The following indicators (and their critical sources, where applicable) are recommended for IAQ assessment and control:

- *comfort indicators:*
  - thermal comfort criteria
  - optimal humidity range 35–65 per cent relative humidity
  - occupant symptom questionnaire (Wolkoff, 2013).

- *ventilation indicators:*
  - concentration of carbon dioxide under steady-state conditions: residences <1000 ppm, commercial buildings <800 ppm
  - mechanical ventilation system to current Australian Standards requirements (AS1668.2); 7.5 L fresh air/person/second is required for office buildings.
- *source indicators:*
  - asbestos fibres: applicable codes and regulations for hazard assessment of products
  - particulate matter ($PM_{2.5}$): compare to national environmental protection measures (NEPM) for $PM_{2.5}$ and PM_{10} (particulate matter with sampling cut points of 2.5 μm and 10 μm, respectively; the cut point is the particle aero-dynamic diameter at which the efficiency of particle capture is 50 per cent). Currently the NEPM for PM_{10} is 50 μg/m³ and for $PM_{2.5}$ it is 25 μg/m³ over a 24 hour period.
- *Legionella species:* use current codes and regulations
- *house dust mite:* measure allergens in dust to determine if below tenth percentile level for particular geographical area
- *micro-organisms:* moist or damp surfaces, with or without visible growths present, are unacceptable; no confirmed pathogens or toxigenic fungi should be present in air or surface samples
- *formaldehyde:* measure in relation to air toxics (NEPC, 2004) goal, 0.04 ppm (24-hour averaging period)
- *volatile organic compounds (VOCs):* a total VOC concentration >500 μg/m³ indicates that significant sources are present; the NHMRC's 'Goals for maximum permissible levels of pollutants in ambient air' (rescinded in 2002) assess the health significance of concentrations of carcinogens and irritants if potential sources are present
- *pesticides:* measure concentrations if visible residues are found or if building has 'leaky' floor, especially for post-construction application of termiticides
- *nitrogen dioxide:* measure in relation to the NEPM (NEPC) goal 0.12 ppm (based on one-hour averaging period) in all buildings (but particularly dwellings, schools and hospitals) while unflued gas appliances are operating (such as gas heaters and cookers)
- *carbon monoxide:* measure in relation to NEPM (NEPC) goal 9.0 ppm (based on eight-hour averaging period) in all buildings (but particularly dwellings, schools and hospitals) while unflued gas appliances are operating (particularly heaters), and in relation to workplace exposure standards in enclosed parking sites
- *ozone:* measure in relation to NEPM (NEPC) goal 0.10 ppm (based on one-hour averaging period) in dedicated rooms with heavy use of photocopiers, laser printers and other sources, and at outlets from ozone-based air sterilisers.

11.5.1   SAMPLING PROTOCOLS

Sampling protocols for indoor air have received little attention in the published literature and vary considerably, although there are a number of Australian and international standards relating to sampling methodology. Since indoor air sampling is carried out for a variety of reasons (e.g. epidemiological studies, compliance with exposure guidelines, resolution of occupant complaints, baseline monitoring and identification of pollutant sources), it is important that uniform methods be adopted for easy comparison of results.

A critical aspect of a sampling protocol is the duration of sampling. This should correlate with the likely biological effect of the pollutant. Short-term (tens of minutes) samples are needed for assessment of irritants, sensitisers and allergenic agents. Long-term samples (over several hours to days) are irrelevant in the case of strong tissue irritants with immediate effect but highly relevant for chronic disease agents. Seifert and colleagues (1989) note that the period and location of sampling must depend on whether the pollutant source is long term (e.g. formaldehyde from particleboard) or short term (e.g. use of spray products, such as furniture polish). For long-term sources, the recommended sampling locations and periods are under building conditions that allow for the estimation of the average pollutant levels as well as low concentrations and high concentrations.

Generally, most published methods that include sampling strategies for IAQ measurements recommend that the following be considered as a minimum:

* *building sample selection*—'zoned' according to known sources of indoor air pollutants in order to focus on high-exposure locations
* *building operation*—in a manner which maximises pollution during the period of pollutant sampling
* *sampling period*—over a timescale relevant to the likely biological effect of the pollutant
* *number of buildings*—sufficient for a reliable estimate of exposure for 'at-risk' populations.

11.5.2 MEASUREMENT METHODS

Australian standards exist for only a limited number of indoor air pollutants, and have not been reviewed since 2005, although in recent years more have been published as ISO standards. This section discusses current standard methods available for the specific measurement of indoor air pollutants, but the methods developed for assessing other occupational exposures may be used, provided they are sensitive enough. The H&S practitioner should also refer to chapters 7, 9 and 16 for guidance on specific instruments to use in air sampling.

11.5.2.1 Carbon dioxide

There is no relevant Australian standard, although carbon dioxide is commonly measured using instruments with real-time monitors with non-dispersive infrared (NDIR) sensors at ambient air concentrations from approximately 350 ppm to industrial levels of 5000 ppm.

11.5.2.2 Particulate matter including ultrafines

The NHMRC (1996) goal for particulate matter (PM2.5) has been superseded by the NEPM (see Table 9.3 in Chapter 9), based on PM_{10} (goal 50 µg/m^3 over a 24-hour period) and $PM_{2.5}$ (advisory reporting standard: 25 µg/m^3 over a 24-hour period; 8 µg/m^3 over a one-year period). These measures, AS/NZS 3580.9.6 Methods for Sampling and Analysis of Ambient Air—Determination of Suspended Particulate Matter—PM_{10} High Volume Sampler with Size-selective Inlet—Gravimetric Method (Standards Australia, 2003b) and AS/NZS 3580.9.10 Methods for Sampling and Analysis of Ambient Air—Determination of Suspended Particulate Matter—$PM_{2.5}$ High Volume Sampler with Size-selective

Inlet—Gravimetric Method (Standards Australia, 2006), are based on a size selection of particles with, respectively, a 50 per cent cut point at 10 μm and 2.5 μm equivalent aerodynamic diameters (EAD). Historically, workplace atmospheres have been evaluated for exposures to respirable dust, as in AS 2985 Workplace Atmospheres—Method for Sampling and Gravimetric Determination of Respirable Dust (Standards Australia, 2009). This describes a method for determination of respirable dust of a particle size that penetrates to the unciliated airways when inhaled; alternatively, it can be described by a cumulative lognormal distribution with a median EAD of 4.25 μm and a geometric standard deviation of 1.5. The practical detection limit for sampling periods longer than 60 minutes is greater than 10–100 μg/m^3, depending on microbalance sensitivity. This limit of sensitivity is acceptable for evaluating occupational exposure standards in the mg/m^3 range, but is problematic for evaluating exposures compared with the NEPM goals unless long sample times are used.

In recent years, direct-reading, portable laser light-scattering (LLS) instruments have been used for measuring particulate matter (PM$_{2.5}$). Generally, they have high sensitivity and allow changes over the work day to be evaluated. However, they are usually calibrated using standard particle sizes that may differ from the particle sizes in the environment monitored. For example, Kim and colleagues (2004) found good agreement between LLS instruments and gravimetric measurement of PM$_{2.5}$ for boilermakers exposed to welding fumes and ash, while Chung and colleagues (2001) found that an LLS instrument consistently over-estimated urban air PM$_{2.5}$ by a factor of 3 relative to gravimetric measurement. Heal and colleagues (2000) used an LLS instrument to monitor undisturbed indoor air and found that it over-estimated PM$_{10}$ by a factor of 2.0 and PM$_{2.5}$ by 2.2 relative to gravimetric measurement.

11.5.2.3 Formaldehyde

Formaldehyde is measured using AS 2365.6 Methods for the Sampling and Analysis of Indoor Air—Determination of Formaldehyde-impinger Sampling in Chromotropic Acid Method (Standards Australia, 1995b), in which fritted glass bubblers (two in series) containing a sodium hydrogen sulfite solution (approximately 15 ml) capture the formaldehyde from room air, generally as 30–60 L samples of air collected over a period of 30–60 minutes. The sample is subsequently analysed by UV spectroscopy. Alternative procedures are to sample with dinitrophenylhydrazine reagent and analyse using liquid chromatography or the use of thermal desorption molecular sieves in accordance with the methodology for sampling VOC levels at trace levels in accordance with US EPA Method TO-17.

11.5.2.4 Volatile organic compounds (VOCs)

VOCs in industrial workplaces are usually measured in accordance with AS 2986.1 Workplace Air Quality—Sampling and Analysis of Volatile Organic Compounds by Solvent Desorption/Gas Chromatography (Standards Australia, 2003a), equivalent to ISO Standard 16200 (ISO, 2000), in which air is sampled onto activated charcoal or other sorbents and subsequently solvent-desorbed for gas chromatography (GC) analysis. The method is sensitive typically down to 10 μg/m^3, depending on the VOC. Methods in which thermal desorption or cryo-focusing are used are generally 100 times more sensitive than solvent desorption, and are the preferred methods for measuring IAQ. For VOCs with boiling points above 60°C, these two basic approaches are specified in the US Environmental Protection Agency (USEPA, 1999a, 1999b) and ASTM (2001, 2003) standards.

Use of a sensitive photoionisation detector for measuring parts-per-billion (ppb) levels of VOCs has been suggested by manufacturers as an alternative to the above methods; however, this instrument does not identify the VOCs present, lacks high sensitivity, has cross-sensitivity to other gases and responds differently to different VOC mixtures. On the other hand, it does show fluctuations in total VOCs in real time, and can therefore be used to identify potential sources and chart differences based on location. When used in conjunction with standard sampling methods, it is a powerful tool.

11.5.2.5 Carbon monoxide

AS 2365.2 Methods for the Sampling and Analysis of Indoor Air—Determination of Carbon Monoxide—Direct-reading Portable Instrument Method (Standards Australia 1993) describes a method for measuring concentrations of 0–500 ppm.

11.5.2.6 Nitrogen dioxide

AS 2365.1.2 Methods for the Sampling and Analysis of Indoor Air—Determination of Nitrogen Dioxide—Spectrophotometric Method—Treated Filter/Passive Badge Sampling Procedures (Standards Australia, 1990) has a detection limit of 0.035 ppm for a two-hour sampling period and can measure down to the NEPM level of 0.12 ppm over one hour. The use of direct reading instruments with electrochemical sensors does not typically have the resolution for indoor air quality assessments.

11.5.2.7 Ozone

For ambient air, ozone is commonly measured using direct-reading UV absorption instruments according to AS 3580.6.1–2011 Methods for Sampling and Analysis of Ambient Air—Determination of Ozone—Direct-reading Instrumental Method with a Measurement Range to 0.5 ppm and Detection Limit Below 0.01 ppm (Standards Australia, 2011a). Such instruments are generally too large and expensive for typical indoor air assessments. Sampling tubes and portable instruments based on electrochemical cells may be used as alternative (non-standardised) methods, but may under-estimate levels owing to the high surface reactivity of ozone.

11.5.2.8 Pesticides

There are no relevant Australian standards for measuring pesticides in air. Testsafe Australia (n.d.) uses filter sampling with the detection limit dependent on the type of pesticide, which may not be sufficiently sensitive for many situations. A method published by the USEPA (1999c) has detection limits from 0.001 to 50 $\mu g/m^3$ for sampling periods of four to 24 hours.

11.5.2.9 Bioaerosols

Bioaerosol sampling involves collecting particles from the air on or in a preselected medium. Impaction, filtration and impingement are three common sampling techniques. Sampling standards for bioaerosols include the NIOSH method 0800 Bioaerosol Sampling (Indoor Air) and the four ISO 16000 Indoor Air standards, Part 16: Detection and Enumeration of Moulds—Sampling by Filtration; Part 17: Detection and Enumeration of Moulds—Culture-based Method; Part 18: Detection and Enumeration of Moulds—Sampling by Impaction; Part 19: Sampling Strategy for Moulds; and Part 21 Detection and Enumeration of Moulds—Sampling from Materials (ISO 2008a, 2008b, 2011, 2012, 2013)

11.5.2.10 House dust mite allergen

The indicator for house dust mite (HDM) exposure (CEC, 1993) should be the levels of allergen in accumulated dust. This dust is collected by vacuum sampling methods from mattresses, carpets and furniture, and analysed by immunochemical assay, preferably enzyme-linked immunosorbent assay for the main house dust mite allergen of concern (CEC, 1993). The indicator can be evaluated against hygienic threshold limits (HTL) proposed by Platts-Mills and de Weck (1988) or categories (based on measured values in residences) recommended by CEC (1993). These limits and categories (all based on allergen levels in collected dusts as µg/g dust) are:

* CEC very low <0.5 µg/g dust
* HTL for sensitisation 2 µg/g dust
* CEC low <5 µg/g dust
* HTL for acute attacks of asthma 10 µg/g dust
* CEC intermediate <15 µg/g dust
* CEC high <20 µg/g dust
* CEC very high >20 µg/g dust.

11.5.2.11 Radon

AS 2365.4 Methods for the Sampling and Analysis of Indoor Air—Determination of Radon (Standards Australia, 1995a) specifies three methods for radon measurement with detection limits of 3–40 Bq/m^3.

11.6   BUILDING VENTILATION

Published data on ventilation rates in Australian buildings are limited, but they indicate that ventilation rates in dwellings declined during the 1990s and 2000s as a result of modified construction practices for energy conservation (improved gap sealing, absence of fixed vents). Ventilation rates, and thus IAQ, declined in mechanically ventilated office buildings for similar reasons. In the 1990s, building codes were amended in an attempt to improve air quality. The design, operation and maintenance of mechanical ventilation systems can all contribute to poor IAQ, and all need to be addressed for IAQ control. Use of an alternative (improved) ventilation mode, such as displacement ventilation, can also improve IAQ. In displacement ventilation, 100 per cent outdoor air is introduced at floor level and rises by convective buoyancy to the ceiling, where it is extracted and exhausted (that is, there is no recycling of the building air). Such a ventilation mode must be incorporated into a building's design. However, its effectiveness may be impaired if occupants disturb the vertical movement of air.

AS 1668.2 Mechanical Ventilation for Acceptable Indoor Air Quality (Standards Australia, 2002) and AS/NZS 3666.1 Air-handling and Water Systems of Buildings—Microbial Control (Standards Australia, 2011b) are adopted through their reference in the Building Code of Australia (BCA) *Handbook on Indoor Air Quality* (ABCB, 2018). Section P.2.4.5 of the BCA has specific performance provisions for IAQ, but only in relation to ventilation with outdoor air, as follows:

(a) A space within a building used by occupants must be provided with means of ventilation with outdoor air which will maintain adequate air quality.

(b) A mechanical air-handling system in a building must control—

    (i)   the circulation of objectionable odours

    (ii) the accumulation of harmful contamination by micro-organisms, pathogens and toxins; and

    (iii) air temperature and humidity.

(c) Contaminated air must be disposed of in a manner which does not unduly create a nuisance or hazard to people in the building or other property.

Note that the BCA does not define 'adequate air quality'. This standard can be met in a variety of ways, but options that are deemed to satisfy the BCA are:

- the use of a mechanical ventilation system in accordance with AS 1668.2 (Standards Australia, 2002), or
- by direct ventilation (or ventilation borrowed from an adjoining room) where there are specified areas of permanent openings, windows, doors or other devices.

11.6.1   CHANGES IN VENTILATION OVER TIME

Generally, dwellings in Australia rely on natural air infiltration and operable windows for ventilation, while mechanical ventilation is the norm for large commercial buildings (humidity control generally is regarded as unnecessary in the largely temperate climate). Usually, AS 1668.2 (Standards Australia, 2002) has taken guidance from Standard 62 of the code of the American Society of Heating and Refrigerating Airconditioning Engineers (ASHRAE, 1973), so the history of this code should be reflected in changes to AS 1668.2. This history is relevant, since buildings were constructed to the standard of the day and were not modified as the standard changed.

The air-conditioning of buildings to provide occupant comfort came into wide use in American cinemas in the 1920s, and gradually spread to other building types. Ventilation rates from the 1940s onwards were influenced by the research of Yaglou and Witheridge (1937), who determined that a minimum outdoor-air ventilation rate of 5 L/sec per occupant was required to ensure that 80 per cent of entrants did not find body odours in the room objectionable. Fanger (1988) arrived at a similar ventilation requirement using different methods, as presented in Figure 11.2. This shows that the percentage of occupants dissatisfied with indoor environments increases markedly for ventilation rates below 5 L/sec/person.

One of the major changes evident in the history of both ASHRAE Standard 62 and AS 1668.2 was the need to maintain a minimum outdoor-air ventilation rate in

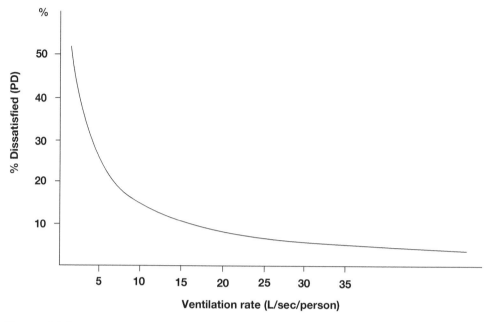

Figure 11.2 Perceived air quality (% of people dissatisfied) at different ventilation rates
Source: CEC (1993, p. 7).

mechanically ventilated buildings, which was increased from 2.5 L/sec/person in 1981 (ASHRAE, 1981) to 7.5 L/sec/person in 1991 and then 10 L/sec/person in 2012. The rationale for this increase was to keep levels of indoor carbon dioxide exhaled by occupants to below 1000 ppm, this being used as a surrogate for unacceptable body odours to 20 per cent of visitors entering an occupied space, as shown in Figure 11.3. Odours from other sources are expected to require much higher ventilation rates estimated at 20 L outdoor air/sec/person (Standards Australia, 2012).

The ventilation Standard AS 1668.2 Ventilation Design for Indoor Air Contaminant Control (Excluding Requirements for the Health Aspects of Tobacco Smoke Exposure) (Standards Australia, 2012) is structured somewhat differently from previous ventilation standards. It allows for either of two procedures to be used in ventilation design: a simple prescriptive procedure or a more complex engineered procedure. The former will be discussed briefly here, and the reader should consult the standard for a more thorough understanding. The prescriptive procedure introduces two factors, one for occupant-related contaminants and the other for contaminants related to building materials and other non-occupant sources. The minimum flow rate of outdoor air supplied by the air-handling system (L/sec) is determined for each contaminant. The required ventilation rate is the higher of the total for occupant-related contaminants and the total for building material/non-occupant-related contaminants.

AS 1668.2 2002 set ventilation rates for enclosures in which smoking was *not* prohibited based on the 'amenity effects' (odour) of environmental tobacco smoke (which accumulates and ages in indoor environments). The standard did not address the health aspects of environmental tobacco-smoke exposure, as smoking is now banned in most

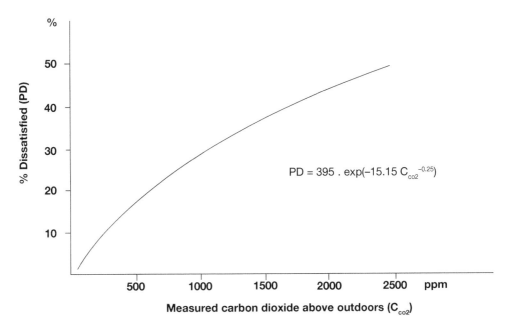

Figure 11.3 Perceived air quality (% dissatisfied) as a function of carbon dioxide concentration in buildings
Source: CEC (1993, p. 7).

indoor spaces. This has been addressed in appendices A and E of AS 1668.2 Supp 1—2002 (S2016), which looks at the potential cancers risks based on number of smokers in a building.

While no systematic investigation has been done, limited measurements have indicated that a large proportion of commercial buildings in Australia do not comply with AS 1668.2, which is not surprising in view of the increases in the ventilation requirements of AS 1668.2 relative to previous requirements. In the absence of an ongoing regulatory requirement to check that the systems continue to perform to design specifications, any IAQ investigation should first establish that the mechanical ventilation system does operate to current AS 1668.2 requirements before air sampling or IAQ assessment begins. If a building is not provided with sufficient ventilation to meet the minimum requirements of the code, further investigation by air sampling should be considered of no real use in determining how to achieve acceptable air quality—at least until the required ventilation conditions are achieved.

11.6.2 *PERFORMANCE AND MAINTENANCE OF VENTILATION SYSTEMS*

A typical heating/ventilation/air-conditioning (HVAC) system is shown in Figure 11.4. It consists of extensive ductwork, air filters, fans (supply air, make-up air, room exhaust), cooling coils (with a well-drained condensate tray beneath to take away condensed water before it pools), heating coils, dampers, room air diffusers, return air grille and so on. Note that the one component that is not normally in direct contact with air being supplied to the building is the cooling tower, unless it is close to the air intake for the building and air

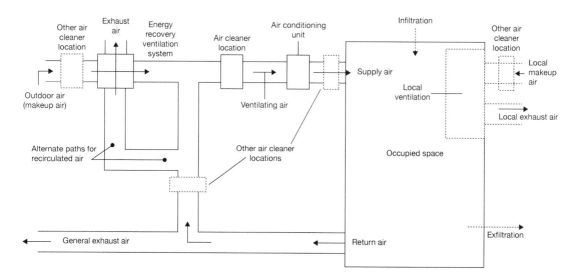

Figure 11.4 Schematic diagram of central mechanical ventilation/air-conditioning plant
Source: ANSI/ASHRAE Standard 62.1-2016 (ASHRAE, 2016, p. 4).

is contaminated with *Legionella* bacteria. Water-cooling towers must be located well away from building air inlets and meet rigorous regulations for chemical treatment, monitoring, cleaning and (in some states) registration—see AS/NZS 3666.1 Air-handling and Water Systems of Buildings—Microbial Control (Standards Australia, 2011b).

Refrigerative cooling systems do not have the same requirements but, except for small units, are much more expensive to purchase and use more energy. Also, large refrigerative cooling systems can weigh substantially more than their equivalent water systems, preventing retrofitting of the former based on a building's load-bearing capacity.

In a typical HVAC system, a high proportion of room air (of the order of 80–90 per cent) is usually recirculated by transport and treatment in the ventilation plant, with the remaining 10–20 per cent as 'make-up air' introduced from outside (i.e. the outdoor ventilation air referred to in AS1668.2), although it is becoming more common to recirculate 100 per cent of the air to reduce energy consumption. Quite aside from the required outdoor ventilation air, mechanical ventilation systems can lead to poor IAQ as a result of poor operation and maintenance of the above components, especially where water and condensation factors are involved that can lead to growth of micro-organisms. A written plan for regular maintenance should be developed and followed, and maintenance activities should be documented. Elements of periodic maintenance that are important to good IAQ include:

- changing of filters (besides blocking and restricting airflow, filters can become sources of odour and micro-organisms)
- checking correct installation of filters
- cleaning of condensate trays and cooling coils
- checking fan operation and operation of dampers that influence air-flow rates.

Measurement and balancing of the ventilation rate from an HVAC system are tasks for a specialist mechanical engineering consultant. The H&S practitioner can determine the need for this by measuring carbon dioxide levels in an occupied building as a surrogate for ventilation adequacy (see section 11.2). In general, ventilation measurement may be necessary:

- after significant changes in the building, the HVAC system or the occupancy and activity within the building
- after control settings are readjusted by maintenance personnel
- where accurate records of the system's performance are not available.

Another practical aspect of ventilation systems is duct cleaning, although the need for and benefits of this practice are open to question. Dust build-up occurs within ducts (usually made of sheet metal and covered with insulation) because particulates in the air supply naturally deposit over small areas in proportion to indoor surfaces. Until the deposition grows significantly enough to affect air-flow rates (by reducing the cross-section of the duct) or to be dislodged by air turbulence, or unless moisture transport to the duct interior leads to growth of micro-organisms, duct cleaning in most commercial buildings appears to be of little benefit. The common appearance of particles on ceiling surfaces adjacent to air inlet diffusers is likely to be merely an extension beyond the ductwork of the ordinary dust-deposition process. Duct cleaning may, however, be needed in mechanical ventilation systems for hospitals and food processing facilities where near-sterile surfaces are usually the goal.

AS/NZS 3666.1 Air-handling and Water Systems of Buildings—Microbial Control (Standards Australia, 2011b) is called up by the ABCB National Construction Code (NCC). Each state may also have legislation specifically directed at controlling the growth of *Legionella* species in cooling towers. AS/NZS 3666 prescribes a number of maintenance measures for HVAC systems, including:

- annual inspection of ductwork in the vicinity of moisture-producing equipment, and cleaning where necessary
- annual inspection of coils, trays, sumps, condensate lines, duct terminal units, air intakes and exhaust outlets, and cleaning where necessary
- annual inspection of air filters, where installed, and cleaning or replacement where necessary
- monthly inspection of humidifier components, and cleaning where necessary
- monthly inspection of cooling towers, and cleaning where necessary—with six-month maximum cleaning intervals unless otherwise approved.

11.6.3 *RELATIONSHIP BETWEEN INDOOR AIR AND OUTDOOR AIR*

As we have noted previously, town dwellers spend around 90 per cent of each day in a range of enclosed environments such as homes, workplaces, schools, shops, hospitality buildings and transit vehicles.

Figure 11.5 demonstrates the complex sources and distributions of air pollutants in a typical office building:

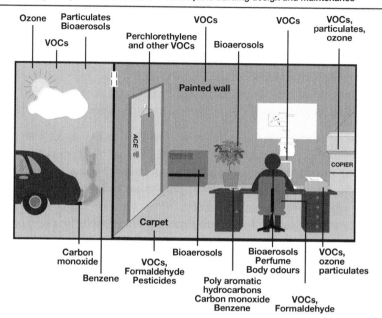

Primary sources of indoor air pollution
· Outdoor air · Building and construction materials and furnishings
· Building occupants and activities · Inadequate building design and maintenance

Figure 11.5 Primary sources of indoor air pollution
Source: http://nptel.ac.in/courses/105102089/air%20pollution%20(Civil)/Module-6/1.htm.

- outdoor air that is used to ventilate the office introduces outdoor air pollutants (e.g. carbon monoxide, VOCs such as benzene, particulates, micro-organisms)
- virtually all manufactured products used to construct, furnish and operate the office act as emission sources for a diverse range of air pollutants (e.g. VOCs, formaldehyde, and micro-organisms), some of which are also introduced with outdoor air
- the behaviour of occupants (e.g. cooking, use of appliances/equipment, pesticide use, body odours/perfumes) introduces specific pollutants (e.g. particulates, ozone, carbon monoxide, pesticides, VOCs)
- pollutants that are present in indoor air are limited in their dispersion by the enclosed nature of building environments, and their concentrations will rise if the ventilation rate is lowered.

It is important to understand that outdoor pollutants enter all buildings to some extent, with the level of air pollution indoors depending on the outdoor level, the extent and type of ventilation used and the nature of pollutant losses to indoor surfaces (e.g. ozone decays rapidly on contact with indoor surfaces, while carbon monoxide does not deplete at all). Air pollutants emitted from indoor materials and appliances will be present indoors *in addition to* pollution brought in from outdoors, to an extent that depends on the level of the product emissions, their persistence over time and the ventilation/indoor surface losses (Brown, 1996, 1998, 2000, 2002; Brown and Cheng, 2003; Brown, Mahoney and Cheng, 2004).

It is commonly found that pollutants originating indoors, where present, have a far greater impact on IAQ than air pollutants from outdoors. The important questions therefore relate to the number of sources of significant pollution in buildings, and whether sectors of the population are exposed to those pollutants at levels considered to pose risks to their health and wellbeing. Studies of the impact of outdoor air pollutants need to consider those urban environments where the highest pollutant levels occur—for example, near busy roads and in city centres.

11.6.4 THERMAL COMFORT FACTORS

Thermal comfort refers to people's satisfaction with the way they feel physically in a given environment. Thermal or physical comfort is affected by a number of factors, including air temperature, radiant temperature, relative humidity, air movement and an individual's clothing and metabolic activity. Differing perceptions of thermal comfort can also be influenced by ethnic background, age and gender. For all these reasons, it is not possible to satisfy thermal comfort for all individuals within a building. The most important factor in maintaining such levels is controlling indoor air temperatures and humidity levels. These should be kept within an acceptable range no matter what the weather is like outdoors.

Temperatures above 26°C are usually considered stuffy and are associated with fatigue and headaches. Temperatures below 18°C are usually considered cold and may be associated with complaints of chills and flu-like symptoms.

The optimal relative humidity range for indoor environments is 35–65 per cent. Eyes, noses and throats can become dry where humidity levels are low. Certain viruses such as influenza are immobilised in dry air. High relative humidity can be associated with fatigue, reports of stuffiness, headache and dizziness, particularly when coupled with high temperatures. Humid air can also encourage the growth of bacteria and moulds, and increase the rate at which formaldehyde and other organic compounds off-gas from building materials.

11.7 DYNAMICS OF INDOOR AIR POLLUTANTS

Indoor pollutant concentrations will vary substantially:

- from region to region—for example, levels of dust mite allergens and micro-organisms will be higher in tropical Queensland than in cold, temperate south-eastern Australia
- from building to building, depending on building age and type, the materials used, occupancy level and type and the ventilation system
- within a building over the course of a day as the activities of occupants fluctuate—for example, carbon dioxide levels will start to increase in the morning as people arrive and may not reach steady-state levels (as a measure of ventilation air delivery) until the afternoon
- within a building over the initial six to twelve months from construction, when pollutants such as formaldehyde and VOCs will be at their highest as they are emitted from new building materials, appliances and furniture
- within a building according to proximity to significant sources of pollutants, for example exhausts from copiers and printers.

The most significant factors influencing concentrations indoor pollutants are:

- the sources of the pollutants and their emission time profiles
- the design, maintenance and operation of the building ventilation system and other processes for pollutant removal (e.g. local air extraction, cleaning practices)
- the level of moisture control in the building.

A detailed discussion of the variation in indoor air pollution over time and building location is beyond the scope of this chapter, but information can be found via the references. The important considerations for the H&S practitioner are that IAQ is a dynamic variable and that this characteristic influences its measurement and control.

11.8   CONTROL OF IAQ

Ideally, IAQ control begins in a building's *design stage*, with:

- selection of low-emission building materials, appliances and furniture
- ensuring that ventilation meets or exceeds BCA requirements (e.g. the use, where feasible, of newer methods of mechanical ventilation such as displacement ventilation, hybrid ventilation (Spengler, Samet and McCarthy, 2001)
- ensuring that persistent condensation and water pooling are prevented on *all* interior surfaces and in all locations, especially in HVAC systems (the condensate tray is a key location where water can stagnate and emit fungal and bacterial spores into building air, so it must drain freely)
- establishing a plan for the operation and maintenance of building services
- documenting all the above actions for future review and assessment where needed.

It is more likely that the H&S practitioner will become involved in complaints about the IAQ in existing buildings where the cause is unknown, few details on the building's design and construction are available and the ability to control indoor conditions is limited. A step-by-step assessment plan should be used which addresses IAQ problems in the following logical sequence of steps:

1   Inspect the building to ensure that it functions as originally designed and to accepted general practice (e.g. use a checklist of building faults, building contents, cleaning practices).
2   Inspect operational components of the building for proper function (e.g. ventilation/heating/cooling appliances).
3   Apply a 'standard' indoor air environment questionnaire to occupants to determine personal wellbeing and environmental comfort. A starting point for such a questionnaire is the CDC Indoor Air Quality Questionnaire (CDC, 2008).
4   Take air samples to assess against IAQ goals where potential sources of pollutants are identifiable from inspection of the building (e.g. carbon monoxide and nitrogen dioxide should be sampled only where indoor combustion processes (heating, cooking, car parking) are active; radon should be sampled only in a habitable cellar). The logistics of sampling (where, when, how and for how long) must be planned to simulate the worst-case scenarios relative to known sources (e.g. sampling of nitrogen dioxide when

an unflued gas heater is operating and all external doors and windows are closed), or be based on sufficient repeat samples to estimate the full range of concentrations. Standard sampling and analysis procedures should be used when available.

5   Record the results of sampling and analysis for comparison with IAQ goals, with baseline results of similar buildings, and with future assessment of IAQ in the building.

During all steps, identify areas where action is necessary, take such action as appropriate and determine the effect this has on complaints before moving to the next step.

Generally, it is recommended that an H&S practitioner who needs to investigate suspected IAQ problems should consult an occupational hygienist who has IAQ experience and expertise.

11.9   RESOURCES/STANDARDS

In addition to standards that have already been covered in this chapter, there are a number of other relevant standards that have been published over the last ten years, which include:

*   ISO 16000-1: 2004 Indoor Air—Part 1: General Aspects of Sampling Strategy
*   ISO 16000-2: 2004 Indoor Air—Part 2: Sampling Strategy for Formaldehyde
*   ISO 16000-5: 2007 Indoor Air—Part 5: Sampling Strategy for Volatile Organic Compounds (VOCs)
*   ISO 16000-8: 2007 Indoor Air—Part 8: Determination of Local Mean Ages of Air in Buildings for Characterizing Ventilation Conditions
*   ISO 16000-15: 2008 Indoor Air—Part 15: Sampling Strategy for Nitrogen Dioxide (NO_2)
*   ISO 16000-19: 2012 Indoor Air—Part 19: Sampling Strategy for Moulds
*   ISO 16000-26: 2012 Indoor Air—Part 26: Sampling Strategy for Carbon Dioxide (CO_2)
*   ISO 16000-30: 2014 Indoor Air—Part 30: Sensory Testing of Indoor Air
*   ISO 16000-32: 2014 Indoor Air—Part 32: Investigation of Buildings for the Occurrence of Pollutants
*   ISO/16000-34 Indoor Air—Part 34: Strategies for the Measurement of Airborne particles
*   AS 2365.4-1995 (R2014) Methods for the Sampling and Analysis of Indoor Air—Determination of Radon

11.10   REFERENCES

American Society of Heating and Refrigerating Airconditioning Engineers (ASHRAE) 1973, *Standard for Natural and Mechanical Ventilation, Ventilation for Acceptable Indoor Air Quality*, Standard 62, ASHRAE, Atlanta, GA.

—— 1981, *Standard for Natural and Mechanical Ventilation, Ventilation for Acceptable Indoor Air Quality*, ASHRAE, Atlanta, GA.

—— 2016, *Ventilation for Acceptable Indoor Air Quality*, ANSI/ASHRAE Standard 62.1-2016, ASHRAE, Atlanta, GA.

ASTM International 2001, *Standard Test Method for Determination of Volatile Organic Chemicals in Atmospheres (Canister Sampling Methodology)*, Standard D5466–01, ASTM International, Philadelphia, PA.

—— 2003, *Standard Practice for Selection of Sorbents, Sampling, and Thermal*, Standard D6196–03, ASTM, Philadelphia, PA.

Australian Building Codes Board (ABCB) 2016, *National Construction Code* (Vols 1, 2 & 3), <www.abcb.gov.au/ncc-online/NCC> [accessed 3 July 2018]

—— 2018, *Indoor Air Quality Handbook*, ABCB, Canberra, <www.abcb.gov.au/Resources/Publications/Education-Training/Indoor-air-quality> [accessed 3 July 2018]

Bates, J. 2000, 'Microbial influence on indoor air quality in indoor air', in *Report on Health Impacts and Management Options*, Commonwealth Government, Canberra.

Bell, J., Newton, P., Gilbert, D., Hough, R., Morawska, L. and Demirbilek, N. 2003, *Indoor Environments: Design, Productivity and Health*, Report No. 2001–005-B, CRC for Construction Innovation, Brisbane, <http://eprints.qut.edu.au/26787/1/Final_Report.pdf> [accessed 3 July 2018]

Brown, S.K. 1996, 'Assessment and control of exposure to volatile organic compounds and house dust mites in Australian dwellings', in *Proceedings of the 7th International Conference on Indoor Air Quality and Climate, Nagoya, Japan*, vol. 2, pp. 97–102.

—— 1998, 'Case studies of poor indoor air quality in Australian buildings', in *Proceedings of the 14th International Clean Air and Environment Conference, Melbourne*, 18–22 October, pp. 205–10.

—— 2000, 'Air toxics in a new Victorian dwelling over an eight-month period', in *Proceedings of the 15th International Clean Air and Environment Conference, Sydney*, 26–30 November, pp. 458–63.

—— 2002, 'Volatile organic pollutant concentrations in new and established buildings from Melbourne, Australia', *Indoor Air*, vol. 12, no. 1, pp. 55–63.

Brown, S.K. and Cheng, M. 2003, 'Personal exposures of stevedores to VOCs, formaldehyde and isocyanates while loading and unloading new cars on shipping', Paper 306, 21st Annual Conference of the Australian Institute of Occupational Hygiene, Adelaide.

Brown, S.K., Mahoney, K.J. and Cheng, M. 2004, 'Pollutant emissions from low-emission unflued gas heaters in Australia', *Indoor Air*, vol. 14, suppl. 8, pp. 84–91.

Centers for Disease Control (CDC) 2008, *Indoor Air Quality Questionnaire*, <https://irp-cdn.multiscreensite.com/562d25c6/files/uploaded/CDC_Indoor%20Air%20Quality%20Questionnaire_2008.pdf> [accessed 4 July 2018]

Chung, A., Chang, D.P.Y., Kleeman, M.J., Perry, K.D., Cahill, T.A., Dutcher, D., McDougall, E.M. and Stroud, K. 2001, 'Comparison of real-time instruments used to monitor airborne particulate matter', *Journal of the Air and Waste Management Association*, vol. 51, no. 1, pp. 109–20.

Commission of European Communities (CEC) 1993, *European Collaborative Action: Indoor Air Quality and Its Impact on Man: Biological Particles in Indoor Environments*, CEC, Luxembourg, <www.inive.org/medias/ECA/ECA_Report12.pdf> [accessed 3 July 2018]

Davidson, M., Reed, S., Markham, J. and Langhorne, M. 2007, 'Do smoking bans in enclosed public places improve air quality?', in *AIOH 25th Annual Conference Proceedings*, AIOH, Melbourne.

Department of Environment and Energy (DEE) 2018, *Indoor Air*, DEE, Canberra, <www.environment.gov.au/protection/air-quality/indoor-air>.

Department of Health and Ageing (DHA) 2002, *Healthy Homes: A Guide to Indoor Air Quality in the Home for Buyers, Builders and Renovators*, DHA, Canberra, <www.health.gov.au/internet/main/publishing.nsf/Content/health-pubhlth-publicat-environ.htm> [accessed 3 July 2018]

—— 2004, *Asthma and Allergens: A Guide for Health Professionals*, DHA, Canberra, <www.nationalasthma.org.au/living-with-asthma/resources/health-professionals/information-paper/asthma-allergy>.

Dingle, P. and Olden, P. 1992, 'A temporary sick building?', *Australian Refrigeration, Air Conditioning and Heating*, vol. 46, no. 11, pp. 43–7.

Fanger, P.O. 1988, 'Introduction of the olf and decipol units to quantify air pollution perceived by humans indoors and outdoors', *Energy Buildings*, vol. 12, pp. 1–6.

Ferrari, L. 1991, 'Control of indoor air quality in domestic and public buildings', *Journal of Occupational Health and Safety Australia and New Zealand*, vol. 7, no. 2, pp. 163–7.

Fisk, W.J. and Rosenfeld, A.H. 1997, 'Estimates of improved productivity and health from better indoor environments', *Indoor Air*, vol. 7, no. 3, pp. 158–72.

Ghosh, B., Lal, H. and, Srivastava, A. 2015, 'Review of bioaerosols in indoor environment with special reference to sampling, analysis and control mechanisms', *Environment International*, vol. 85, pp. 254–72.

Gunderson, E.C. and Bobenhausen, C.C. 2011, 'Indoor air quality', in D.H. Anna (ed.), *The Occupational Environment: Its Evaluation, Control and Management*, 3rd ed., AIHA, Fairfax, VA, pp. 450–500.

Heal, M.R., Beverland, I.J., McCabe, M., Hepburn, W. and Agius, R.M. 2000, 'Inter-comparison of five PM10 monitoring devices and the implications for exposure measurement in epidemiological research', *Journal of Environmental Monitoring*, vol. 2, no. 5, pp. 455–61.

International Organization for Standardization (ISO) 2000, *Workplace Air Quality Sampling and Analysis of Volatile Organic Compounds by Solvent Desorption/Gas Chromatography*, ISO 16200, ISO, Geneva.

—— 2008a, *Indoor Air—Part 16: Detection and Enumeration of Moulds—Sampling by Filtration*, ISO 16000–16, ISO, Geneva.

—— 2008b, *Indoor Air—Part 17: Detection and Enumeration of Moulds—Culture-based Method*, ISO 16000–17, ISO, Geneva.

—— 2011, *Indoor Air—Part 18: Detection and Enumeration of Moulds—Sampling by Impaction*, ISO 16000–18, ISO, Geneva.

—— 2012, *Indoor Air—Part 19: Sampling Strategy for Moulds*, ISO 16000–19, ISO, Geneva.

—— 2013, *Indoor Air—Part 21: Detection and Enumeration of Moulds—Sampling from Materials*, ISO 16000–21, ISO, Geneva.

Kim, J.Y., Magari, S.R., Herrick, R.F., Smith, T.J. and Christiani, D.C. 2004, 'Comparison of fine particle measurements from a direct-reading instrument and a gravimetric sampling method', *Journal of Occupational and Environmental Hygiene*, vol. 1, no. 11, pp. 707–15.

Levik, M., Bakke, J.V., Carlsen, K-H., Jensen, J.A., Myhre, K.I., Nafstad, P., Omenaas, E. and Norderhaug, I.N. 2005, 'Indoor exposures and risk of asthma and allergy:

A systematic and critical review—preliminary report', in *Proceedings of the 10th International Conference on Indoor Air Quality and Climate*, Beijing, 4–9 September, vol. 5, pp. 3576–80.

Mannins, P. 2001, *Australia State of the Environment Report 2001 (Atmosphere)*, CSIRO Publishing, Canberra.

McKenna, D. 1990, untitled, unpublished paper for Community and Public Sector Union.

National Australian Built Environment Rating Scheme (NABERS) 2011, *NABERS Fact Sheet 1: About the NABERS Program*, NSW Department of Planning and Environment, Sydney, <https://nabers.gov.au/public/WebPages/DocumentHandler. ashx?docType=3&id=12&attId=0> [accessed 3 July 2018]

—— 2015, *NABERS Indoor Environment for Offices: Validation Protocol for Accredited Ratings*, NSW Office of Environment and Heritage, Sydney, <www.nabers.gov.au/ public/WebPages/DocumentHandler.ashx?docType=2&id=34&attId=0> [accessed 3 July 2018]

National Environment Protection Council (NEPC) 2004, *National Environment Protection (Air Toxics) Measure*, Secretariat for Standing Council on Environment and Water, Canberra, <www.environment.gov.au/atmosphere/airquality/standards.html#toxics> [accessed 10 December 2012]

National Health and Medical Research Council (NHMRC) 1993, *Volatile Organic Compounds in Indoor Air: Report of the 115th Session, June, Adelaide*, NHMRC, Canberra.

—— 1996, *Ambient Air Quality Goals and Interim National Indoor Air Quality Goals* (rescinded), NHMRC, Canberra, <www.nhmrc.gov.au/guidelines/publications/eh23> [accessed 3 July 2018]

National Industrial Chemicals Notification and Assessment Scheme (NICNAS) and the Office of Chemical Safety and Environmental Health (OCSEH) 2010, *A Scientific Review of Multiple Chemical Sensitivity: Identifying Key Research Needs*, NICNAS, Canberra, <www.nicnas.gov.au/Current_Issues/MCS/MCS_Final_Report_Nov_2010_ PDF.pdf>.

Newton, P.W., Vam, S., Bhatia, K. et al. 2001, *Australia State of the Environment Report 2001 (Human Settlements)*, CSIRO Publishing, Canberra, <www.environment.gov.au/ soe/2001/publications/theme-reports/settlements/index.html>.

NSW Standing Committee on Public Works 2001, *Sick Building Syndrome*, NSW Legislative Assembly, Sydney.

Platts-Mills, T.A.E. and de Weck, A.L. 1988, 'Dust mite allergens and asthma: A worldwide problem', *Bulletin of the World Health Organization*, vol. 66, no. 6, pp. 769–80.

Rabone, S., Phoon, W.O., Seneviratne, M., Gutirrez, L., Lynch, B. and Reddy, B. 1994, 'Associations between work related symptoms and recent mental distress, allowing for work variables and physical environment perceptions in a sick office building', in L. Morawska et al. (eds), *Indoor Air: An Integrated Approach*, Elsevier, Amsterdam, pp. 243–6.

Reed, S. 2009, 'What standards should we use for indoor air quality?', paper presented to 19th International Clean Air & Environment Conference, Perth, 6–9 September.

Rowe, D. 2003, 'A study of a mixed mode environment in 25 cellular offices at the University of Sydney', *International Journal of Ventilation*, vol. 1, pp. 53–64.

Rowe, D.M. and Wilke, S.E. 1994, 'Thermal comfort and air quality in eight office buildings: an interim report', paper presented to National Conference of the Australian Institute of Refrigeration, Air-Conditioning and Heating, 15 April, Surfers Paradise.

Seifert, B., Knoppel, H., Lanting, R.W., Person, A., Siskos, P. and Wolkoff 1989, *Report No. 6: Strategy for Sampling Chemicals Substances in Indoor Air*, Commission of the European Communities, <www.buildingecology.com/publications/ECA_Report6.pdf> [accessed 3 July 2018]

Seneviratne, M., Phoon, W.D., Rabone, S. and Jiamsakul, W. 1994, 'Health hazards from airborne microbes in workplaces: Can indoor air quality measurements tell us?', in *Proceedings of the 12th International Conference of the Clean Air Society of Australia and New Zealand*, Perth, 23–28 October, pp. 623–32.

Spengler, J.D., Samet, J.M. and McCarthy, J.F. 2001, *Indoor Air Quality Handbook*, McGraw-Hill, New York.

Standards Australia 1990, Methods for the Sampling and Analysis of Indoor Air—Determination of Nitrogen Dioxide—Spectrophotometric Method—Treated Filter/Passive Badge Sampling Procedures, AS 2365.1.2: 1990 (R2014), Standards Association of Australia, Sydney.

—— 1991, The Use of Mechanical Ventilation and Air-conditioning in Buildings—Mechanical Ventilation for Acceptable Indoor-air Quality, AS1668.2: 1991, Standards Association of Australia, Sydney.

—— 1993, Methods for the Sampling and Analysis of Indoor Air—Determination of Carbon Monoxide—Direct-reading Portable Instrument Method, AS 2365.2: 1993 (R2014), Standards Association of Australia, Sydney.

—— 1995a, Methods for the Sampling and Analysis of Indoor Air—Determination of Radon, AS 2365.4: 1995 (R2014), Standards Association of Australia, Sydney.

—— 1995b, Methods for the Sampling and Analysis of Indoor Air—Determination of Formaldehyde-impinger Sampling in Chromotropic Acid Method, AS 2365.6: 1995 (R2014), Standards Association of Australia, Sydney.

—— 2002, The Use of Mechanical Ventilation and Air-conditioning in Buildings—Mechanical Ventilation for Acceptable Indoor Air Quality, AS 1668.2: 2002, Standards Association of Australia, Sydney.

—— 2003a, Workplace Air Quality—Sampling and Analysis of Volatile Organic Compounds by Solvent Desorption/Gas Chromatography, AS 2986.1: 2003 (R2016), Standards Association of Australia, Sydney.

—— 2003b, Methods for Sampling and Analysis of Ambient Air—Determination of Suspended Particulate Matter—PM_{10} High Volume Sampler with Size-selective Inlet—Gravimetric Method, AS/NZS 3580.9.6: 2003, Standards Association of Australia, Sydney.

—— 2006, Methods for Sampling and Analysis of Ambient Air—Determination of Suspended Particulate Matter—$PM_{2.5}$ High Volume Sampler with Size-selective Inlet—Gravimetric Method, AS/NZS 3580.9.10, Standards Association of Australia, Sydney.

—— 2009, Workplace Atmospheres—Method for Sampling and Gravimetric Determination of Respirable Dust, AS 2985: 2009, Standards Association of Australia, Sydney.

—— 2011a, Methods for Sampling and Analysis of Ambient Air—Determination of Ozone—Direct-reading Instrumental Method with a Measurement Range to 0.5 ppm and Detection Limit Below 0.01 ppm, AS 3580.6.1: 2011, Standards Association of Australia, Sydney.

—— 2011b, Air-handling and Water Systems of Buildings—Microbial Control, AS/NZS 3666.1: 2011, Standards Association of Australia, Sydney.

—— 2012, The Use of Ventilation and Airconditioning in Buildings, AS 1668.2: 2012, Standards Australia, Sydney.

Testsafe Australia, (no date), *Chemical Analysis Branch Handbook*. 9th ed., WorkCover NSW, available online at: https://www.testsafe.com.au/__data/assets/pdf_file/0007/16387/ Chemical-Analysis-Branch-Handbook-9th-edition-TS033.pdf [accessed 15 March 2019]

US Environmental Protection Agency (USEPA) 1999a, *Compendium Method TO–17A Determination of Volatile Organic Compounds in Ambient Air Using Active Sampling onto Sorbent Tubes*, 2nd ed., USEPA, Cincinnati, OH, <www3.epa.gov/ttn/amtic/files/ ambient/airtox/to-17r.pdf> [accessed 3 July 2018]

—— 1999b, *Compendium Method TO–14A Determination of Volatile Organic Compounds (VOCs) in Ambient Air Using Specially Prepared Canisters with Subsequent Analysis by Gas Chromatography*, 2nd ed., USEPA, Cincinnati, OH, <www3.epa.gov/ttn/amtic/files/ ambient/airtox/to-14ar.pdf> [accessed 3 July 2018]

—— 1999c, *Compendium Method TO–10A Determination of Pesticides and Polychlorinated Biphenyls in Ambient Air Using Low Volume Polyurethane Foam (PUF) Sampling Followed by Gas Chromatographic/Multi-Detector (GC/MD) Detection*, 2nd ed., USEPA, Cincinnati, OH, <www3.epa.gov/ttn/amtic/files/ambient/airtox/to-10ar.pdf> [accessed 3 July 2018]

Wakelam, M., Lingford, A. and Caon, A. 1995, 'Measurement of indoor air quality indicators in Australian buildings', in L. Morawska et al. (eds), *Indoor Air: An Integrated Approach*, Elsevier, Amsterdam, pp. 91–4.

Williams, P. 1992, *Airconditioning System Faults Affecting Health and Comfort in Melbourne Office Buildings*, Department of Architecture and Building, University of Melbourne, Melbourne.

Wolkoff, P. 2013, 'Indoor air pollutants in office environments: Assessment of comfort, health, and performance', *International Journal of Hygiene and Environmental Health*, vol. 216, no. 4, pp. 371–94.

World Health Organization (WHO) 2010, *WHO Guidelines for Indoor Air Quality: Selected Pollutants*, WHO Regional Office for Europe, Copenhagen, <www.euro.who.int/__ data/assets/pdf_file/0009/128169/e94535.pdf> [accessed 3 July 2018]

Yaglou, C.P. and Witheridge, W.N. 1937, 'Ventilation requirements: Part 2', *ASHRAE Transactions*, vol. 43, pp. 423–36.

12. Noise and vibration

Beno Groothoff and Jane Whitelaw

12.1   INTRODUCTION

Noise is one of the most pervasive health hazards faced by the H&S practitioner. Industry is noisy, offices and places of entertainment can be noisy, traffic and transport vehicles of all kinds can be noisy. With the possible exception of libraries, few workplaces completely escape the intrusion of noise that is distracting, annoying or hazardous to hearing and health. Mechanisation and modern lifestyles have not alleviated the problem; in fact, the opposite is true, with widespread and sustained noise exposures occurring in an increasingly noisy world. Despite the virtual absence of long-range studies of city noise, many people have already noticed that road traffic, planes, trains and construction works, often combined with buildings with poor sound insulation, have led to increasingly noisy city environments and associated health problems.

Occupational exposure to noise, its physiological and psychological consequences and methods of control are all very large areas of study. Noise-induced hearing loss is a major occupational disease and accounts for a large percentage of workers' compensation claims.

Safe Work Australia's *Occupational Noise-induced Hearing Loss* report states (2010, p. xi) that, 'Occupational noise-induced hearing loss (ONIHL) is a significant health and economic problem in Australia.' Between July 2002 and June 2007 there were about 16,500 successful workers' compensation claims for industrial deafness involving permanent impairment due to noise. In its report *Listen Hear! The Economic Impact and Cost of Hearing Loss in Australia*, Access Economics (2006) estimates that 37 per cent of hearing loss is caused by noise; the cost of this amounts to about A\$4.3 billion annually: 'The economic burden of ONIHL is borne by workers and their families, business owners and managers, and the wider society' (Safe Work Australia, 2010, p. xi). The report further estimates that about 20 per cent of adult-onset hearing loss is caused by exposure to loud noises from all sources. In its *Occupational Disease Indicators* report, Safe Work Australia states (2014, p. 7) that, 'Among those who experience noise-induced hearing loss, 20% or more also suffer from tinnitus.'

The effects of certain industrial chemicals on hearing—either alone or in combination with noise—are now better understood. Because many workers are exposed to both noise and chemicals at work, the current *Code of Practice—Managing Noise and Prevention of Hearing Loss at Work* (Safe Work Australia, 2018) recommends that hearing conservation programs should take into account the effects of chemicals that are known to be ototoxic where workers are exposed to any of the ototoxic substances listed in Table 6 of CoP and noise level has $L_{Aeq,8h}$ greater than 80 dB(A) or $L_{C,peak}$ greater than 135 dB(C), or hand–arm vibration at any level and noise with $L_{Aeq,8h}$ greater than 80 dB(A) or $L_{C,peak}$ greater than 135 dB(C).

The effects of vibration on workers have also gained more attention as the effects of vibrating plant and equipment have come to be better understood. Legal precedents have also been set, with rather large compensation payouts in the United Kingdom for sufferers of vibration white finger or hand–arm vibration syndrome (HAVS) arising from long-term exposure to hand and arm vibration. Similarly, the effects of whole-body vibration on drivers of vehicles such as trucks and earth movers, as well as rigid-hull inflatable boats (RHIBs), are being studied more extensively to avoid or minimise adverse health effects.

In this chapter, the recognition, evaluation and control of noise and vibration are examined in general terms. The H&S practitioner is referred to the references for more detailed treatments. The chapter also introduces the phenomena of acoustic shock and oto-toxicity. Acoustic shock is a particular concern for workers in the call-centre industry and emergency services control rooms, and can have severe health consequences if no adequate protective measures are put in place.

12.2 REGULATORY NOISE LIMITS IN AUSTRALIA

To prevent noise-induced hearing loss (NIHL), Chapter 4, 'Hazardous Work', of the harmonised *Model Work Health and Safety Regulations 2019* (Safe Work Australia, 2019) sets as the exposure standard for occupational noise a limit of $L_{Aeq,8h}$ 85 dB(A), referenced to 20 μPa (an average of 85 dB(A) over an eight-hour exposure), or a C-weighted peak sound pressure level of $L_{C,peak}$ 140 dB(C), referenced to 20 μPa.

This does not mean that such noise levels should be tolerated in all occupations. For example, people in an open-plan office or a call centre could not work effectively with the distraction presented by noise levels of $L_{Aeq,8h}$ 85 dB(A) or even $L_{Aeq,8h}$ 75 dB(A). Effective communication and mental concentration are impaired by high levels of noise, and may in turn affect health and other aspects of safe work. The WHS legislation requires obligation holders to conduct risk assessments to prevent or minimise risk. This applies equally to low-level noise and high-level noise, both of which cause communication and other problems that can affect health. To minimise adverse health effects in these low-level noise work environments, the Code of Practice for Managing Noise and Prevention of Hearing Loss at Work recommends that noise levels be kept below:

- 50 dB(A) where work is being carried out that requires high concentration or effortless conversation
- 70 dB(A) where more routine work is being carried out that requires speed or attentiveness or where it is important to carry on conversations.

These levels include the noise from other work being carried out in the workplace.

12.3 PHYSICAL CHARACTERISTICS OF SOUND AND NOISE

Sound originates when a vibrating source causes variations in atmospheric pressure that are detected by the ear and interpreted by the brain. Sound may convey useful information. When sound is unintelligible or unwanted, or may cause damage to hearing, it is referred to as *noise*. A person perceives, evaluates and identifies the uniqueness of a sound by virtue of only three features: *combinations of frequencies*, their *relative intensities* and their rate of *onset and decay*. Differences between sounds (e.g. a bird call, a violin and a chainsaw) can all be ascribed to differences in these three features.

As sound is transmitted through the elastic medium of air, the air is compressed and rarefied to form a pressure wave (called a longitudinal wave), like the ripples that appear on a pond when a pebble is thrown into the water. The number of pressure variations per second is called the frequency of sound and is measured in Hertz (Hz) (see Figure 12.1).

The wavelength is the distance between two similar points on the sine (curved, peak-and-trough) wave. The velocity of sound (wavelength × frequency) depends on the mass

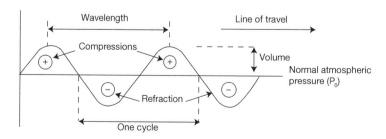

Figure 12.1 Sound as represented by an advancing pressure wave in air

and elasticity of the conducting medium. In air, sound propagates at about 344 m/s at 20°C. In water, sound propagates at about 1500 m/s, and through steel it propagates at about 6000 m/s. The spectrum of good human hearing ranges between about 20 and 20,000 Hz, and everyday sounds contain a wide mixture of frequencies. Speech communications rely on frequencies ranging between 100 and 5000 Hz. Audible sound pressure variations are superimposed on the atmospheric air pressure (about 100,000 Pa) and normally range between 20 µPa and 100 Pa.

12.4   SOUND PRESSURE, DECIBELS, LOUDNESS AND THE L$_{EQ}$

The quietest sound that can be detected by a young person with healthy ears is 20 millionths of a Pascal (20 µPa) at 1000 Hz. This has been standardised as the threshold of hearing (0 dB) for the purpose of sound-level measurements, and is used as a reference level (P_0). A sound pressure of 100 Pa (130 dB) is so loud it causes pain in the average person and is therefore called the threshold of pain.

Sound pressure variations of a pure tone fluctuate so that for half the duration of a complete cycle they are above (+) and the other half they are below (–) atmospheric pressure. The average pressure fluctuation is zero, since there are as many positive changes as negative ones. To overcome this, the average pressure is squared over a large number of cycles and the square root taken. The resulting value, the root mean square (rms), is used in sound level measurements, as it is proportional to the energy content of the sound wave. For pure tones, the rms is equal to 0.707 times the amplitude (volume) of the sine wave.

In terms of sound pressure detected by the ear, the loudest sounds heard at the threshold of pain are up to 10 million times greater than the softest ones. If calculations were made using a linear scale, the resulting numbers would be unwieldy. Additionally, the human ear does not respond linearly but rather logarithmically to sound stimuli. It is therefore more practicable to express acoustic parameters as a logarithmic ratio of the measured value to the reference value (20 µPa). This produces a more manageable basis for comparisons of sound pressures. The resulting unit, the Bel (after Alexander Graham Bell 1874–1922), is defined as the logarithm to the base 10 of the ratio between two acoustical powers or intensities. This unit is still large, so the decibel (one-tenth of a Bel) is generally used. The decibel is defined as:

$$L = 20 \log P/P_0 \ [dB] \qquad\qquad\qquad \text{(Equation 12.1)}$$

where:
L = root mean square (rms) sound pressure level, re 20 µPa
P = the rms sound pressure in Pascal, and
P_0 = reference sound pressure (20 µPa)

For a doubling of sound energy, the logarithmic scale increases by 3 dB, and a 100-fold increase in noise results in a 20 dB increase in scale. Some examples of indicative noise levels in decibels are shown in Table 12.1.

Table 12.1 Indicative noise levels of various sources

Sound	Sound pressure level dB re: (20 μPa)
Firearm	155–165
Jet engines	140–150
Threshold of pain	130
Jackhammer	110
Nightclub	100
Noisy factory	90–95
Passing heavy traffic	80–90
Office environment	45–70
Speech	60–65
Inside a home	40
Whisper	20
Threshold of hearing	0

Sound pressure levels represent only part of the picture. The apparent loudness of a sound depends very much on frequency, since the ear detects sound intensity with different sensitivities, depending on the frequency composition of the sound. The ear is most sensitive to frequencies between 1000 and 4000 Hz, with sensitivity falling off at low and very high frequencies. High-frequency sounds are heard much better than low-frequency ones of equal intensity. Further, an individual's interpretation and experience of loudness are subjective, depending on the harshness or intrusiveness of the noise (e.g. bagpipes compared with a flute). These factors are not reflected by the sound-level measuring instrument.

To relate measured sound levels to the response of the human ear, measuring instruments are fitted with sound-level weighting filters. The internationally adopted A-weighting is most commonly used, as it best corresponds to the response of the human ear. Sound pressure levels measured with the A-weighting filter are denoted as dB(A). This A-weighting response is illustrated in Figure 12.2, which shows clearly how the low-frequency response is weighted downwards—that is, a sound-level meter using the A-weighting reads lower than the actual (linear) response at frequencies below 1000 Hz.

The A-weighted scale has been generally adopted because it attenuates broadband frequencies in a way that reflects their association with noise-induced hearing loss. It is thought that the higher frequencies, considered to be above 1000 Hz, give rise to most hearing damage. The eight-hour noise-exposure limits of the exposure standard are based on A-weighted measurements (Safe Work Australia, 2011b).

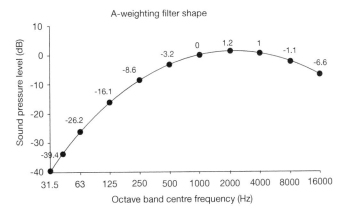

Figure 12.2 Frequency-response attenuation characteristics for the A-weighted network

Few workplaces have the constant noise levels of, for example, dust-extraction cyclones or electrical generators. Noise tends to be intermittent, like the clanking of machinery, the roar of passing traffic or the mechanical percussion of hammers, chippers or grinders. Thus noise levels may vary quite widely. WHS legislation accommodates this by requiring the average A-weighted noise level over an eight-hour work period ($L_{Aeq,8h}$) to be determined and compared with the legislated exposure limit. To obtain an 'average' value of the noise, an integrated value called the equivalent continuous sound (energy) level over a measurement period T, of say 60 seconds ($L_{Aeq,60}$), provides a measure of the acoustic energy of the fluctuating noise during the measurement period. The $L_{Aeq,60}$ is one parameter that finds convenient use in surveys of occupational-noise exposure levels.

12.5 EFFECTS ON HEARING

The principal health-related effect of noise exposure is hearing loss. Excessive noise can destroy the ear's ability to hear, and the damage is not reversible. The H&S practitioner should therefore put great emphasis on prevention. Damage to hearing depends on the loudness of the noise and the length of the exposure. Other health effects of noise exposure include cardiovascular effects such as increased blood pressure and heart rate, psychosocial stress, excretion of stress hormones and decreases in peripheral blood flow due to increase in vasoconstriction.

Noise also impairs communication, making it harder to do a job, and in the social sphere it spoils much of our enjoyment of life. Hearing normally deteriorates with age (presbycusis). Permanent hearing loss induced by noise can accelerate presbycusis and drastically affect quality of life.

The following are some hearing-related effects from noise exposure:

- *Temporary threshold shift* in hearing occurs during or immediately after exposure to significant loud sounds. Quiet sounds can no longer be heard, and the condition may last for periods ranging from minutes to hours and days. This occurs when the hair cells in the hearing organ (cochlea) become fatigued and reversibly desensitised.

- *Permanent threshold shift* usually results from long-term regular exposure to loud noise, but sometimes occurs from a single or a few extremely loud impact or impulsive noises. The ear canal acts as a resonating tube and actually amplifies sounds at between 3000 and 4000 Hz. This amplification increases the sensitivity of our hearing at these frequencies and also adds to the susceptibility of damage. Loud sounds, irrespective of their frequency composition, cause hearing damage, typically around the 4000 Hz range. NIHL is not reversible, as it arises from the destruction of the hair cells in the cochlea. The left-hand photograph in Figure 12.3 shows healthy hair cells—a single row of inner hair cells and three rows of outer hair cells. The right-hand photograph shows a large number of destroyed hair cells. The human ear does not have the ability to repair or regrow hair cells, and no medical treatment can restore them.
- *Acoustic trauma* normally results from high-intensity explosive or loud impact type impulsive noise, which can destroy the hair cells and other ear mechanisms after one or relatively few exposures.
- *Tinnitus* consists of the perception of sounds in the ears or head that do not originate from an external source (Jastreboff and Hazell, 2004). It ranges from being annoying to causing severe distress. It is permanent, and currently no cure is available. Tinnitus can have a devastating impact on sufferers and may even lead to suicide. Tinnitus sounds may be experienced as a tone, buzzing, hissing, whistling, ringing, cicadas, musical sounds, pulsating or a 'flickering' sound. The sounds are variable and are often complex.

There are basically three types of tinnitus:

- *Objective tinnitus* is usually uncommon and results from measurable bodily noise within the area of the ear, such as from blood flow through either normal vessels but with increased flow such as that caused by atherosclerosis, or abnormal vessels such as with tumours or malformed vessels, near the middle ear. There is no defect in hearing.

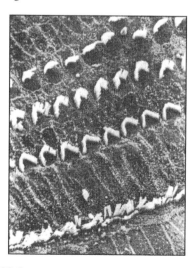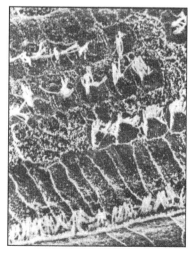

Figure 12.3 Destruction of hair cells as a result of noise-induced hearing loss
Source: US Dept of Labor.

- *Subjective tinnitus* is the most common cause of the tinnitus experience. The brain receives signals that are not caused by sound but interprets them as a perceived noise. It is considered to originate from abnormal neuronal activity in the auditory cortex due to damage in the auditory pathway chain, most commonly hair cell damage. There is a school of thought that considers the damage may cause a loss of suppression of intrinsic cortical activity and possibly the creation of new neural connections.
- *Phantom tinnitus* is where the vestibulocochlear nerves are damaged and hearing loss has occurred. The surviving nerve tissue compensates for the lack of information by spontaneously creating a phantom noise, which the brain interprets as sound.

Tinnitus is most likely experienced after exposure to loud noise that causes a temporary threshold shift—for example, the sound of a jackhammer or a firearm used without hearing protectors, or after a rock concert or drag racing. Loud noise through headphones is another important cause. Other causes of tinnitus include:

- conductive hearing loss, such as from wax in the ear
- Meniere's disease
- flu and head colds
- excessive alcohol consumption
- stress
- ototoxic drugs, such as those used in chemotherapy
- ototoxic chemical exposure in work situations
- head injury, including from head banging during rock concerts
- whiplash (from a car accident)
- tumour at vestibulocochlear nerves.

12.5.1 *TREATMENT OF TINNITUS*

There is currently no cure for tinnitus, and treatment must therefore consist of managing its presence. The management of tinnitus consists of two main strategies: relaxation therapy and sound therapy.

Relaxation therapy is probably the most difficult approach, as it depends on the commitment of the person to want to relax. In our busy world, most people do not know how to relax and must learn and practise to achieve this state. During relaxation, the heart rate and breathing slow down and brain activity decreases. This causes our bodies to feel mentally and physically refreshed afterwards, and may help to reduce the intensity of the tinnitus to manageable levels.

Sound therapy consists of using devices that introduce low-level sounds in various forms to mask the tinnitus. It aims to enable the brain to 'filter out' the tinnitus sounds. A second component of sound therapy consists of creating a positive understanding about tinnitus management and the emotional stress often associated with tinnitus. Sound therapy can be used as a self-help tool or as part of a management program delivered by qualified audiologists or specialty clinics. The sounds that are often used include environmental sounds, radio, smartphone apps providing broadband noise or gentle and relaxing music and wearable sound generators producing constant 'white noise'. These devices should be fitted by a tinnitus specialist as part of a tinnitus-management program. It is important that the generated sound is pleasant to listen to but basically blends into the background i.e. the sufferer is aware of the sound but it does not demand too much attention. It is important for those suffering tinnitus to avoid complete silence.

12.6   ASSESSMENT OF NOISE IN THE WORKPLACE

Two practical approaches to assessing noise in the workplace can be employed by the H&S practitioner. Both require sophisticated equipment and reasonable levels of competency for proper measurement and interpretation. Provided that appropriate equipment is available and the H&S practitioner is properly trained, the noise assessments could be carried out in house. In complex situations or where either equipment or competency cannot be secured, acoustic consultants should be sought to advise the H&S practitioner.

12.6.1   THE SOUND LEVEL METER

AS/NZS 1269.1 Occupational Noise Management, Part 1: Measurement and assessment of noise immission and exposure (Standards Australia, 2005b) states that an integrating sound level meter of at least Class 2 should be used. A Class 1 integrating sound level meter is the preferred instrument for measuring sound pressure levels. Such instruments, similar to that shown in Figure 12.4, provide simultaneous noise measurements in a range of different parameters, such as rms sound level $L_{Aeq,T}$, peak sound levels ($L_{C,peak}$), A-weighting and C-weighting, and linear (Z) dB measurements.

Figure 12.2 earlier in the chapter showed the A-weighting response curve (matching the human ear's sensitivity) of the sound level meter. This curve follows different values at different frequencies from the horizontal (0 dB) line. These values are used in noise control where the results of octave band analysis, measured linearly, have to be expressed as an A-weighted value. Another weighting often used in noise control is the C-weighting. This weighting follows a different curve which is essentially horizontal (flat) between

Figure 12.4 Sound level meter monitoring noise exposure of angle-grinder operator
Source: Environmental Directions Pty Ltd.

80 and 3150 Hz, and tapers off only outside this frequency range. The C-weighting is used predominantly when measuring peak sound pressure levels and determining the suitability of hearing protectors in a particular work environment.

Modern sound level meters can, depending on the software installed, provide 'octave' or 'one-third octave' (frequency) analysis at the same time as other parameters are measured. This enables measurement of the sound spectrum in a range of frequencies determined by the meter's capabilities. Frequency analysis is necessary for the proper analysis of noise in the design of noise-control measures, or provision of appropriate hearing protectors in a particular work environment. The use of frequency analysis is examined further in section 12.9.

The sound level meter is used in work locations principally for measuring noise emissions from machinery, equipment and processes, or the noise exposure of workers (noise immission).

A number of cheap and simple-to-operate sound level meters are available, which may be useful to the H&S practitioner who is inexperienced in conducting sound level measurements. There are also numerous sound level meter apps available for smartphones. If such an app is being used, the H&S practitioner should be careful to select one with a sound intensity exchange rate of 3 dB as many have a 5 dB exchange rate, based on US legislative requirements, and can therefore not be used. These cheap meters and phone apps do *not* meet the requirements of AS/NZS 1269 and should *never* be used to demonstrate compliance with WHS legislation. They may, however, be useful for indicative purposes—for example, in preliminary surveys or 'walk-through surveys' to determine whether or not the workplace or a work area is likely to contain excessive noise. Where this is found to be the case, more sophisticated instruments complying with the relevant standards should then be employed to conduct more detailed noise surveys using Class 1 instruments, and to identify options for noise control.

12.6.2 *PERSONAL NOISE DOSE METERS*

Many workers move in and out of noisy areas in their jobs, thus sustaining different noise exposures. Their total noise exposure is therefore the sum of many different partial exposures. The personal noise dose meter—also known as dosimeter (Figure 12.5)—provides an integrated measure of noise exposure over a known period, usually the work shift.

Modern noise dose meters provide data logging with computer-readable output in noise dose, $L_{Aeq,8h}$ and A-weighted noise exposure, $E_{A,T}$, expressed in Pascal squared hours (Pa^2h), and C-weighted peak readings; they also warn when daily exposure limits have been exceeded. The noise dose meter may be able to download its stored data into a computer (Figure 12.6, left) and provide a histogram of the worker's exposure (Figure 12.6, right).

In recent years noise dose badges have appeared and are now used frequently in favour of the larger traditional noise dose meters. These noise dose badges (Figure 12.7) perform similar functions as the traditional meters but are much smaller, with an in-built microphone, and thus do not have microphone cables attached to them which may get in the way of the worker performing tasks.

Feedback from workers wearing noise dose badges indicates that they are perceived to be less intrusive. This has had the positive effect that, in the course of conducting work,

Figure 12.5 Personal noise dose meter: correct microphone fitting relative to position of ear
Source: Environmental Directions Pty Ltd.

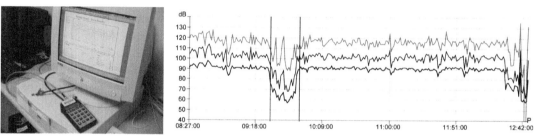

Figure 12.6 Downloading a histogram of a worker's exposure
Source: Environmental Directions Pty Ltd

Figure 12.7 Noise dose badge
Source: Air Met Scientific.

the worker basically forgets they are wearing it and are therefore less likely to interfere with the badge being worn, leading to more reliable measurement results.

All measurements must be made in accordance with measuring protocols recommended by the manufacturer and as required by the relevant Australian standards, that is, AS/NZS 1269 for occupational noise (Standards Australia, 2005b) and AS 1055 (Standards Australia, 2018) for environmental noise.

12.6.3 CALIBRATION

Both the sound level meter and the personal noise dose meter will require calibration with a field calibrator immediately before and after a noise survey, in accordance with the manufacturer's instructions. Most calibrators generate a pure tone of 1000 Hz and typically produce a sound pressure level of 94 dB, which equates to a sound pressure of 1 Pascal. These parameters are chosen so that the calibrator can be used with the different weightings, such as the A-weighting, C-weighting and linear, which all intersect at 1000 Hz.

If the calibrator uses a different frequency, such as 250Hz, the A-weighted response of the sound level meter must be adjusted. For instance, if a sound level meter set to A-weighting is calibrated at 250 Hz and 114 dB, the sound level meter indicator must read 105.4 dB(A) as the A-weighted response at 250 Hz is –8.6 dB relative to the reference level at 1000 Hz.

In addition to the abovementioned field calibration procedures AS/NZS 1269.1 (Standards Australia, 2005b) requires in clause 7.10 that complete measuring systems to be used for detailed or follow-up assessments must be calibrated in accordance with the relevant standards at regular intervals not exceeding two years. Such calibrations shall be performed by a suitably equipped and independently audited laboratory, with full traceability to national measurement standards (Standards Australia, 2005b). The two-yearly certification of sound-level measuring instruments would therefore normally be carried out by a NATA-registered laboratory. The certification of field calibrators should be carried annually by a NATA-registered laboratory to ensure their accuracy (Groothoff, 2015a, 2015b).

12.7 ADDING OF SOUND LEVELS

In cases where two noise sources operating side by side have been measured separately, their combined sound pressure level can be found using the addition method of Table 12.2. Because decibels are logarithmic values, the individual noise levels cannot simply be added.

For example, if two sources were measured at 90 dB and 93 dB, their combined sound pressure level could be found by taking the difference between 90 and 93, which is 3, looking up this value in the left-hand column of Table 12.2 and finding the corresponding correction value in the right-hand column, which in this case is 1.8.

Because the two sources operate side by side, their combined value must be higher than the highest individual value (93), and the outcome is therefore 93 + 1.8 = 94.8 dB, which can be rounded to 95 dB. Where the difference between two sources is 10 or more, it is generally considered that the lower of the two does not contribute to the higher value. For instance, one source at 85 dB(A) and one at 95 dB(A) will have a combined sound level

Table 12.2 Logarithmic addition of noise levels

Difference in dB value	Add to the higher dB value
0	3.0
1	2.6
2	2.1
3	1.8
4	1.4
5	1.2
6	1.0
7	0.8
8	0.6
9	0.5
10	0.4
11	0.3
12	0.2
13 or more	0

of about 95 dB(A). Where there are more than two sources operating with a difference of 10 dB or more, their combined level may be affected and needs to be determined by using Table 12.2. This may be the case with items of plant that start and stop intermittently, but that on occasion may all run at the same time and may cause exposure limits to be exceeded.

With a sound level meter, the actual sound pressure level can be measured and confirmed when the sources are generating noise simultaneously.

12.8   DAILY NOISE EXPOSURE

Because workers usually are exposed to different noise levels from different tasks, each lasting for a different period, it is helpful to have a means of integrating all these exposures into a single, useful noise exposure value. AS/NZS 1269.1 uses the 'eight-hour equivalent continuous A-weighted sound pressure level' ($L_{Aeq,8h}$) and the 'A-weighted noise exposure' ($E_{A,T}$) in Pa²h (Pascal squared hours) to express a worker's exposure to noise in a single numerical value (Standards Australia, 2005b). Modern WHS legislation limits a worker's eight-hour unprotected exposure to $L_{Aeq,8h}$ 85 dB(A). Traditionally, the eight-hour allowable exposure has been expressed as a daily noise dose (DND) with a value of 1.0—in other words, a DND of 1.0 is equal to $L_{Aeq,8h}$ 85 dB(A).

As we saw earlier, when sound propagates from its source, it causes small disturbances in atmospheric pressure. The static atmospheric pressure is about 10^5 Pa, and audible pressure variations are in the order of 20 µPa (20×10^{-6} Pa) to 100 Pa. As we saw in section 12.4, 20 µPa corresponds to the threshold of hearing and 100 Pa to the threshold of pain. The magnitude of the disturbance is measured with a sound level meter as a sound pressure level. The sound level meter measures the square of the relative value of the sound pressure in Pascals over a time period T. The result is expressed in Pa^2 and, if present for eight hours, is expressed as Pa^{2}h. A sound pressure level of 85 dB(A) corresponds to a sound pressure of 0.126 Pa. For an eight-hour work period, the Pa^{2}h value would be $8 \times 0.126 =$ 1.008 Pa^{2}h. For practical purposes the $E_{A,T}$ value of 1 Pa^{2}h corresponds to $L_{Aeq,8h}$ 85 dB(A). It follows that:

$$1 \text{ Pa}^2\text{h} = L_{Aeq,8h} \text{ 85 dB(A)} = \text{DND 1} \qquad \text{(Equation 12.2)}$$

As the noise exposure level increases, the permitted time of exposure for unprotected ears decreases. This is shown in Table 12.3 as eight hours of continuous exposure at 85 dB(A), or four hours at 88 dB(A), and so on. The 'doubling of sound intensity exchange rate' used by Australian authorities is 3 dB.

Both the daily noise exposure and Pascal squared hours methods provide for the integration of partial noise exposures over a given period, and enable normalisation of partial noise exposures to a five-day working week where the worker may work more or less than five days. Modern integrating noise dose meters will have built-in capabilities to calculate

Table 12.3 Period of maximum exposure for various equivalent continuous noise levels

Limiting dB(A)	Maximum exposure period allowed to stay within $L_{Aeq,8h}$ 85 dB(A)
82	12 hrs*
85	8 hrs
88	4 hrs
91	2 hrs
94	1 hr
97	30 min
100	15 min

* Extended shifts pose a degree of risk that is greater than from a normal eight-hour shift, due to additional damaging exposure once maximum temporary threshold shift is reached after ten hours and a shorter recuperation time between shifts. AS/NZS 1269.1 states that organisations should add an adjustment as listed in Table 2 of the standard to the normalised noise exposure level before comparing that level to the noise exposure criterion. In the case of a twelve-hour shift that adjustment would be 1 dB on top of the measurement result after having been normalised to an eight-hour exposure ($L_{Aeq,8h}$)

the daily noise dose (DND), $L_{Aeq,8h}dB(A)$, $E_{A,T}$ and Pa^2h (Pascal squared hours) and methods of integrating the noise exposure. AS/NZS 1269.1 (Standards Australia 2005b) provides tables to convert noise level measurements to Pascal squared values, and relatively simple tabular methods of converting the summed Pascal squared values to an equivalent $L_{Aeq,8h}$.

We must remember that exposures are determined for workers who are considered not to be wearing hearing protection devices. A worker wearing hearing protectors in a noisy workplace is in a situation of protected exposure, provided the hearing protectors are worn correctly and consistently. Of course, this cannot always be guaranteed. Note too that protected exposure does not mean *no* exposure, as the workplace noise is still present.

Shifts exceeding ten hours present a greater risk of hearing damage than eight-hour shifts. This is partly because of the additional damaging effect of continued exposure once the maximum temporary threshold shift is reached after about ten hours of exposure, and partly because of the reduced recovery time between shifts. Corrections must therefore be made to the noise exposure levels to accommodate the greater risk. AS/NZS 1269.1 incorporates in Clause 9.4 a correction table, reproduced in Table 12.4, which indicates the adjustments to be made to the normalised exposure level with extended work shifts.

Table 12.4 Correction table for extended shifts

Shift length, h	Adjustment to $L_{Aeq,8h}$ dB
<10	+ 0
>10 to <14	+ 1
>14 to <20	+ 2
>20 to 24	+ 3

Source: Standards Australia (2005b).

Suppose a worker works a fourteen-hour shift and is exposed during this shift to $L_{Aeq,14h}$ 89 dB(A). The normalised total daily noise exposure is therefore:

$$L_{Aeq,8h} = 89 + 10 \log_{10} (14/8) = 91.43 \text{ dB(A)} \qquad \text{(Equation 12.3)}$$

For a shift of fourteen hours, 1 dB must be added. The adjusted $L_{Aeq,8h}$ therefore becomes $91.43 + 1 = 92.43$ dB(A).

This value must be used in planning for noise management.

12.9   NOISE-CONTROL STRATEGIES

If the results of a noise survey identify the likelihood of exposure to excessive noise, further investigation is necessary to identify noise sources and prioritise control strategies to reduce the hazard. Workplace noise control strategies fall into three categories, ranged here from most to least effective:

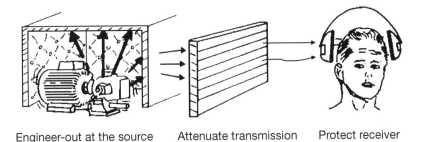

Engineer-out at the source   Attenuate transmission   Protect receiver

(a) **control at source:** elimination or modification of noise source or process
(b) **control of transmission path:** enclosures, barriers, sound-proofed control rooms
(c) **control of noise at the receiver:** enclosures, barriers, sound-proofed control room, training and the use of personal hearing protectors.

Fig 12.8 Essential elements to be addressed in noise-control strategies

• engineering out the noise hazard (*source* of noise)
• attenuating the noise hazard (*transmission* of noise)
• use of hearing-protection programs (*reception* of noise).

Hearing protectors should not be seen as noise-control devices, but rather as a temporary way of minimising noise exposure by reducing the noise that enters the ear canal.

Occasionally, the H&S practitioner may need to use two or even all three approaches to noise control. Attacking the source of the noise or preventing its transmission is always preferable to requiring workers to wear hearing protection. In practice, it is often difficult to enforce such requirements, so reliance on the use of hearing protectors by workers will, in the long run, cost more than implementing higher-order controls. Many (engineering) noise-reduction programs, although perhaps initially expensive, can be surprisingly cost-effective over the long term, and will remain consistently effective. Some excellent practical examples can be found in the references in section 12.15. Hearing protection should be used only if alternatives are not feasible, or reasonably practicable or economical, or as an intermediate phase until more permanent (higher order) noise-reduction measures are in place.

12.9.1   CONTROL AT SOURCE

Many noise hazards arise because little, if any, thought has been given to the acoustic qualities of the building—for example, modern concrete tilt-up buildings are popular for industrial areas as they are quick to erect but consist of acoustically highly reflect-ive concrete floors and walls and are often covered with a metal roofing system. This combination of hard reflective surfaces actually increases the noise production once the workplace is in operation. There is often also little or no thought given to the processes or the correct design or installation of equipment. High levels of continuous and percussive noise are often traded off for immediate ease of operation—for example, bending metal by hammering on it rather than using a machine or lever. Noisy motors, compressors, power saws and grinders can all be found in many workplaces, located near workstations with little regard to their impact on the worker.

The following are some ways to minimise noise or prevent its generation.

12.9.1.1 Substitution of processes

- Use welding instead of riveting.
- Use hot working of metals instead of cold forging.
- Use impact-absorbing materials (plastic, rubber, nylon) rather than metal.
- Use procedures such as lowering rather than dropping.

12.9.1.2 Minimising changes in force, pressure or speed that produce noise

- Eliminate impact noises—for example, use compression riveters rather than impact riveters.
- Replace hammering with slow application of force (Figure 12.9).
- Use hydraulic presses rather than mechanical-impact presses.

12.9.1.3 Reducing the speed

The higher the speed the higher the frequency, and so the louder (as perceived by the ear) and potentially more damaging the noise. To alleviate this:

- Use larger, slower machines rather than small, fast ones.
- Run machines at lower speeds, but with higher torque.
- Use air guns with air channels around the central channel, which reduce noise and pitch by not having all the air concentrated at high speed through the central air channel; rather, they spread the flow of air through several channels around the central orifice at lower speed while not reducing effectiveness. The air gun on the left in Figure 12.10 has air channels around the central orifice. This lowers the speed of the air and changes the pitch of the sound. This type of air gun can be up to 7 dB quieter than the 'traditional' air gun shown on the right.

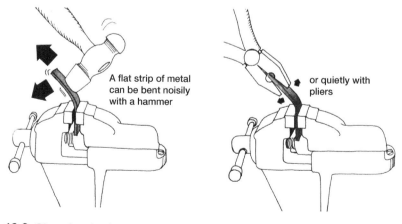

A flat strip of metal can be bent noisily with a hammer

or quietly with pliers

Figure 12.9 Changing the force, speed or pressure to reduce noise
Source: US Department of Labor.

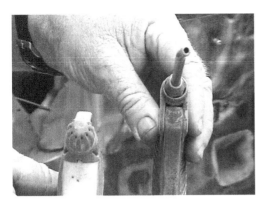

Figure 12.10 The airgun on the left spreads the airflow around the central orifice and is quieter than the traditional airgun on the right
Source: Environmental Directions Pty Ltd.

12.9.1.4 Preventing mechanical vibration from being converted into sound-generating sources

- Isolate the vibrating unit to prevent it from transmitting noise.
- Optimise rotational speed (usually decreasing speed) to prevent oscillating vibrations.
- Alter the size or mass to change resonant (i.e. natural) frequencies. Depending on their composition, size and shape, all objects tend to vibrate more freely at a particular frequency, called the resonant frequency. A machine transfers the maximum energy to an object when the machine vibrates at the object's resonant frequency.
- Dampen resonance. A ship's engine running at 125 rpm and connected directly to the propeller shaft is shown in Figure 12.11. This speed coincides with the resonant frequency of the propeller shaft and causes disturbing noise and vibration. When a reduction gear is added (Figure 12.11), the propeller shaft now runs at 75 rpm, shifting the vibration to a lower frequency and thus reducing noise and vibration.
- Use extra support or stiffeners to withstand vibrations. The circular-saw sharpening machine in Figure 12.12 causes an intense resonant (ringing) noise. Clamping the rubber sheet against the saw blade greatly reduces the resonance and thus the ringing noise.

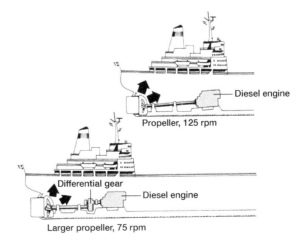

Propeller, 125 rpm

Diesel engine

Differential gear

Diesel engine

Larger propeller, 75 rpm

Figure 12.11 Dampening of resonant surfaces to reduce noise generation
Source: US Department of Labor.

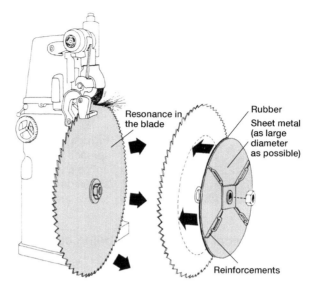

Figure 12.12 Stiffening of resonant surfaces to reduce noise generation
Source: US Department of Labor.

Figure 12.13 Various constructions for vibration dampening
Source: US Department of Labor.

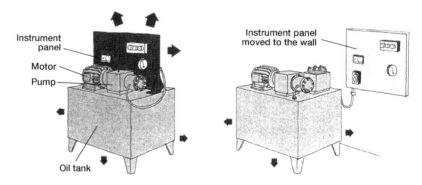

Figure 12.14 Reducing radiating area
Source: US Department of Labor.

- Use vibration-damping surfaces (Figure 12.13). Steel plates have very poor internal vibration damping. Adding coatings of intermediate layers with better damping properties can reduce vibration and thus noise.
- Reduce radiating area (Figure 12.14). The control panel on the left is mounted on the pump system and vibrates with the system. Isolating the control panel (as shown on the right) reduces the vibrating surface of the system and therefore the noise level.

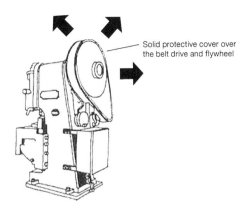
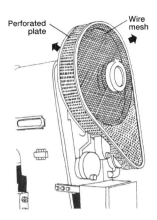

Figure 12.15 Use of non-resonant surfaces
Source: US Department of Labor.

- Use perforated non-resonant surfaces (Figure 12.15). On the left, the guard over the flywheel and belt consists of a solid metal cover. This causes the large surface area of the guard to vibrate and act as a sounding board for noise. Replacing the guard with a perforated one (as shown at right) reduces the vibration, and in turn the noise.
- Use active cancellation (artificial noise created 180 degrees out of phase) to negate the effect of the original source.

12.9.1.5 Reducing transmission possibilities

Fluid pumping systems can be very noisy because of the intense pressure shocks created in the liquid by the compressors driving the systems. If such systems are mounted rigidly to a building structure, noise can travel through the building and cause problems for the occupants. To avoid noise problems, compressors and pipe systems should be isolated from the building. As shown in Figure 12.16:

- use springs, dampers, flexible couplings and mountings
- ensure that ducts cannot carry sound.

12.9.1.6 Reduce the likelihood of noise being generated in air or fluid flow

- Use properly designed fans to reduce air turbulence. The air reaching a fan's rotor must be as unobstructed as possible to achieve quiet operation.
- Slowly reduce speed of air or fluid flow to avoid turbulence from sudden changes in volume and pressure drop.
- Use reduced pressures and velocities (Figure 12.17).
- Prevent rapid pressure change which produces 'cavitation' sound (left side of Figure 12.18) resulting from rapid pressure drops near control valves, propellers and pumps. To avoid cavitation noise, selected inserts can be placed in the fluid line so that the insert will not produce a greater pressure drop than required to prevent cavitation (see right side of Figure 12.18).
- Reduce turbulence on air exits (Figure 12.19). When the high-speed exhaust air escaping from the grinder's handle mixes with the relatively still outside air, the

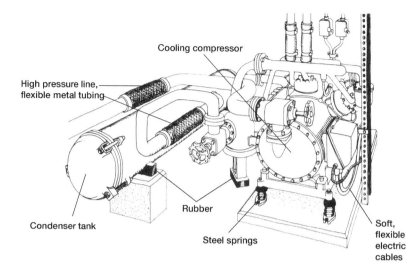

Figure 12.16 Use of a flexible connector to decouple noise source
Source: US Department of Labor.

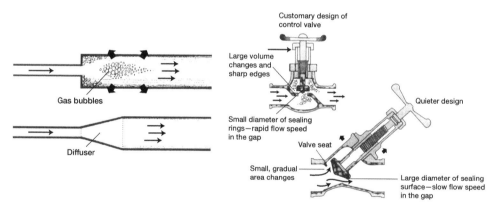

Figure 12.17 Slowly reducing pressure reduces noise transmission in liquids
Source: US Department of Labor.

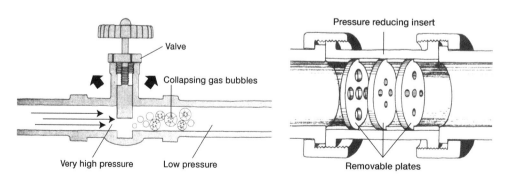

Figure 12.18 Use of pressure reducers to reduce 'cavitation' sound
Source: US Department of Labor.

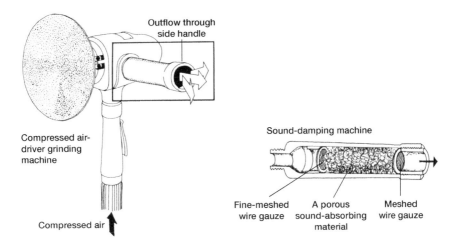

Figure 12.19 Minimising turbulence at air exits to reduce noise
Source: US Department of Labor.

resulting turbulence creates noise. Inserting a silencer made of porous sound-absorbing material tames the turbulence and reduces the exhaust noise.

12.9.1.7 Buying quiet

Noise-control techniques should be incorporated in the design and installation of equipment, since the cost of doing so is minimal compared to that of after-market design and installation. However, there is rarely an off-the-shelf solution for the suppression or control of noise, and it is necessary for the H&S practitioner to assess each workplace situation separately.

- While a piece of equipment may correctly claim to produce sound levels less than 85 dB(A), two or more operating side by side may still produce combined sound levels in excess of the regulatory noise level.
- Companies should be advised to include noise limits in purchasing specifications, bearing in mind the additive effects of more than one piece of equipment. The basic information to be requested from potential suppliers should include the A-weighted sound power level LWa. The sound power of a piece of equipment is independent of its environment, and sound power level can be converted to a sound pressure level. If enough H&S practitioners and employers insist on quiet machinery, manufacturers will begin to provide it.

A good example of 'buying quiet' is the German Blue Angel program, under which manufacturers of construction industry machines and equipment can have their products tested against environmental noise criteria. If their products meet the criteria, they are entitled to carry the Blue Angel logo. From the manufacturer's perspective this is a selling point, because German (and other European) building sites must comply with strict (local) government regulations, and quiet plant and equipment are often demanded in requests for tenders.

12.9.2 *CONTROL OF NOISE TRANSMISSION PATH*

The location of a machine or process is often a significant factor in the noise it transmits to the workplace. Several methods of noise attenuation focus on this point.

12.9.2.1 Isolating the noise source from the worker

If it is concluded that noise generation cannot be prevented, the following options can be tried:

- Locate the noise source at a distance from the workplace—for example, pumps, compressors, generators and the like can be installed outside the building (provided this does not lead to environmental noise complaints).
- Use enclosures around the noise source (see Figure 12.20).
- Confine sources to a noise-insulated room, using double walls with insulating material, double-glazed windows and solid core doors.
- Use an isolating enclosure around the worker.

12.9.2.2 Preventing much of the noise from reaching the worker

- Place noise sources away from natural reflectors—for example, in corners.
- Use sound absorbers on ceilings and walls (see Figure 12.21).
- Use noise baffles or deflectors to direct noise away from workers.

The efficacy of this method depends on frequency. Low-frequency noise tends to pass through openings and around objects (see Figure 12.21), while high-frequency noise is more easily deflected (see Figures 12.22 and 12.23).

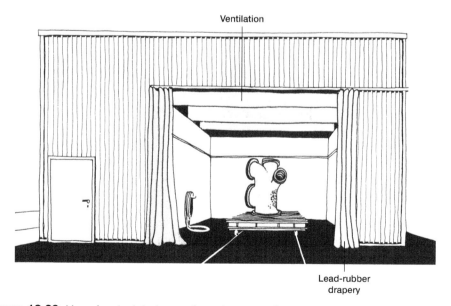

Figure 12.20 Use of an isolated room for noisy operations
Source: US Department of Labor.

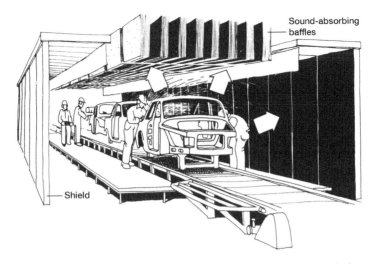

Figure 12.21 Sound-absorbing baffles can minimise sound movement indoors
Source: US Department of Labor.

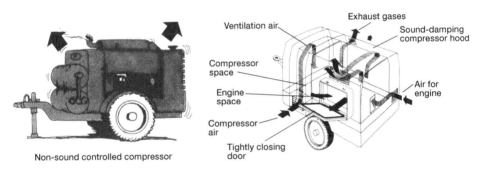

Figure 12.22 Use of sound absorbent to reduce sound transmission
Source: US Department of Labor.

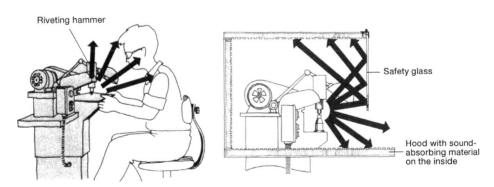

Figure 12.23 High-frequency noise can easily be deflected away from the worker
Source: US Department of Labor.

12.10   PERSONAL HEARING PROTECTION

Where it has been confirmed that workplace noise levels are in excess of any regulatory noise limits—for example, $L_{Aeq,8h}$ 85 dB(A)—and the noise is unable to be controlled at the source or along the transmission path, personal hearing protective devices (HPD) must be used until the noise exposure is reduced below the regulatory limits.

To have any chance of success, a hearing protection program requires the full involvement and cooperation of management. To secure such involvement, the H&S practitioner must demonstrate that a risk-management process has been carried out. The program must include, at a minimum:

- a noise-control policy
- a system for conducting noise level and noise level exposure surveys on a regular basis
- a program for the planning and implementation of engineering and administrative noise control, where possible
- a suitable hearing-protection training program, including regular refresher training, and records of training topics and attendance
- Frequent supervision and corrective actions by line management to ensure correct fitting and use of ear plugs to ensure attenuation is maximised
- selection and provision of personal hearing protectors and documented reasons why the selected hearing protectors were the most suitable
- provision for the use, maintenance, care and storage of the hearing protectors
- provision for audiological assessment of workers on commencement of employment (or the hearing protection program) and regularly thereafter
- ongoing monitoring and review of the effectiveness of the program.

AS/NZS 1269.3 (Standards Australia, 2005d) provides further guidance on the elements of and training requirements for the management of workplace noise through hearing protector programs.

Workers and management also need to be fully engaged in the program. They should be given instruction in:

- how the ear works, including how hearing loss occurs
- reasons why hearing protection is required—that is, importance of preservation of hearing and health; legislative requirements, limitations and other control options
- selection of personal hearing protectors
- fitting and use of hearing protectors, and the importance of good fit and comfort
- good and bad habits when wearing hearing protectors
- the requirement to wear the hearing protectors all the time when exposed to noise
- correct use and maintenance of hearing protection.

The H&S practitioner will find that implementing effective hearing protection programs is far more complicated than simply buying ear plugs or ear muffs for workers. No given ear muff or ear plug will fit or be comfortable for everyone. Different types of hearing protectors have different noise-attenuation capabilities, and may not necessarily be compatible with all work situations. The choice of hearing protectors depends very much on factors such as:

- the noise levels determined in the workplace
- the frequency spectrum of the noise
- the attenuation required to achieve compliance with regulations
- the worker's hearing ability
- the worker's acceptance of and the degree of good fit and comfort of the hearing protector. An ear plug should be inserted for at least three-quarters of its length to achieve the rated attenuation.
- compatibility with other personal protective equipment
- the need to communicate with others and to hear important signals and sounds
- cost.

Regardless of the type of hearing protector used, its effectiveness is drastically reduced when it is not worn correctly and consistently at all times in noisy environments. This is illustrated in Figure 12.24, which shows the reduction in effectiveness of hearing protectors with stated attenuation values of SLC_{80} 28dB(A) and 20dB(A) respectively.

If the protector with a stated reduction of 28dB(A) is removed for only five minutes during an eight-hour exposure time, the effective attenuation is reduced to only 20dB(A); if it is removed for ten minutes, the effective attenuation is reduced to about 16dB(A). The longer the hearing protector is not worn, the less effective it becomes.

To further illustrate the importance of the correct wearing of hearing protectors, there is documented evidence to suggest that 'real-world' attenuation may be up to 6 dB less than in laboratory tests for ear muffs and up to 9 dB for ear plugs (Berger, 2000; Eden and Piesse, 1991).

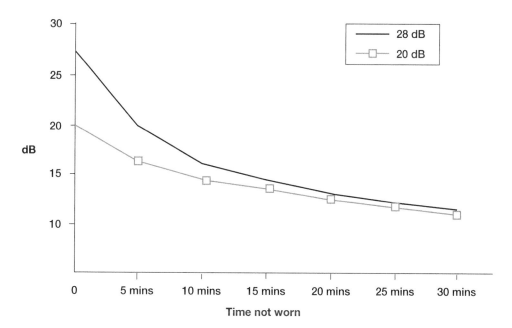

Figure 12.24 Reduction of effectiveness of a hearing protector with time not worn

12.10.1 TYPES OF HEARING PROTECTOR

There are four basic types of hearing protector, of which only two are in widespread use:

- *Acoustic helmets* cover a large part of the head and the outer ear and resemble those worn by helicopter and fighter pilots. These helmets are bulky, expensive and not suited for general industrial use. Noise attenuation may be as much as 50 dB at the lower frequencies and 35 dB at 250 Hz; they may also diminish noise conduction through bone.
- *Ear-canal caps* (Figure 12.25) are made from a light rubber or PVC type material and seal off the entrance to the ear canal without entering it as an ear plug does. Ear-canal caps are held in place by a spring headband. They are gaining popularity on construction sites because they can easily be placed over the ears when a noise—usually of short duration—starts suddenly and unexpectedly.
- *Ear plugs* (Figure 12.26) are widely used, both as single-use disposable types and reusable types. They are typically manufactured from foam rubber, plastic or silicone, and may come user-formable, pre-moulded or custom-moulded. All these types can provide good noise attenuation, but attention has to be paid to correct fitting. All ear plugs are designed to prevent noise reaching the inner ear by filling and blocking the ear canal. The formable types are designed to fit all ears; they are rolled between the thumb and finger into a cylindrical shape prior to insertion.

  Once the ear plug is rolled tightly with one hand, insertion in the ear is done by raising the other arm over the head, grabbing the pinna and pulling it up and towards the back of the head so as to straighten the ear canal and inserting the plug in the ear and holding it in place with a finger or thumb on it until the plug expands again and blocks the ear canal. The ear plug should be inserted sufficiently deeply so that when looking front-on, the plug should be hidden behind the tragus. When removing the

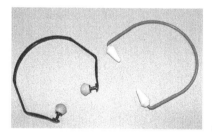

Figure 12.25 Types of ear-canal caps
Source: Environmental Directions Pty Ltd.

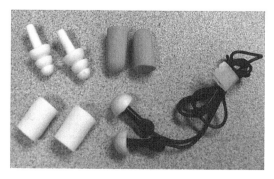

Figure 12.26 Selection of ear plugs
Source: Environmental Directions Pty Ltd.

ear plug, it would be good practice to grab the plug again between the finger and thumb and give it a quarter turn to break the seal and allow air to enter the ear canal before pulling the ear plug out completely. Reusable ear plugs (Figure 12.27) should be able to be inserted into the ear canal without having to be squashed to fit. Some pre-moulded ear plugs have an acoustic resonator chamber included to provide an almost flat attenuation across the frequency range. This type of ear plug may be beneficial for workers who have a higher need for communicating verbally amidst noise, or who have a mild hearing loss, as the attenuation at 2 and 4 kHz (4 kHz is where NIHL typically manifests itself) is not as great as with normal industrial-type ear plugs.

Individually moulded ear plugs are made from an acrylic or silicone mould of the wearer's ear canal. This means they can be worn only in the ear canal from which the mould was made. Individually moulded ear plugs can also be fitted with acoustic resonator chambers to provide an almost flat attenuation over the frequency range, and can therefore assist in communication by offering less distortion than ordinary individually moulded ear plugs. Individually moulded ear plugs must be expertly fitted, since performance and comfort may be poor if they are incorrectly shaped or sized. Flat attenuating or individually moulded ear plugs with built in acoustic resonators are sometimes also called musicians ear plugs, and may be beneficial to musicians as, compared with normal industrial ear plugs, the music is far less distorted, enabling their wearing during the making of music and preserving their hearing and thus their livelihood. The attenuation provided is affected by the expertise of the person doing the mould as well as the quality control of the process. They are not currently Standards Australia approved. An example of individually moulded ear plugs is shown in Figure 12.28.

- *Ear muffs* consist of two padded and internally insulated domes which cover the entire ear (see Figure 12.29). A spring-torsioned headband holds the padded cups to the sides of the head at a clamping force to provide the attenuation desired. When selecting ear

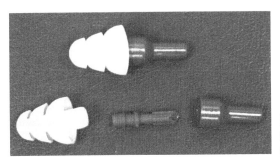

Figure 12.27 Example of a reusable flat attenuating ear plug
Source: Environmental Directions Pty Ltd.

Figure 12.28 Individually moulded ear plugs with resonator chambers fitted
Source: Environmental Directions Pty Ltd.

Figure 12.29 Ear muff hearing protectors
Source: Environmental Directions Pty Ltd.

muffs, ensure that the cup is just large enough to clear the ear lobes. It is important that the cushions attached to the cups are soft and that they are not cracked, as they are essential to provide a proper seal. They should be cleaned after use and their condition regularly checked. When hard or cracked, the cushions can easily be replaced.

In situations where communication is important, so-called noise-cancelling ear muffs can be used. These types of level-dependent ear muffs have electronically activated shut-off valves which enable the muffs to admit noise up to a certain level—usually 82 dB(A); above this level, the ear muff blocks the noise and acts as a normal ear muff.

Another type of ear muff has a built-in radio receiver. Although this may seem attractive for workers doing mundane or boring work, the volume of radio sound inside the cup combined with the level of workplace noise entering the cup may cause the eight-hour exposure limit to be exceeded. An employer providing this kind of ear muff to workers must ensure that the worker's in-ear noise level does not exceed the exposure limits imposed by WHS legislation.

12.10.2 FREQUENCY ATTENUATION

Sound can be transmitted to the inner ear via conduction through bone, leaks in the ear canal seal, oscillation of an ear plug, transmission of sound through an ear muff or vibration through a sealing cushion of an ear muff. For these reasons, attenuation of more than 50 dB is not possible. Some types of hearing protectors cannot attenuate low frequencies as well as others. Figure 12.30 compares the typical attenuations of an ear plug and an ear muff.

When selecting hearing protection, it is important to know which frequencies need to be attenuated. For example, if the noise is predominantly low frequency, a hearing protector that offers poor protection at low frequencies would not be an appropriate choice. On the other hand, a worker might need to hear some of the lower frequencies of speech. Overall, the aim is to ensure that the daily noise exposure of the hearing-protected worker

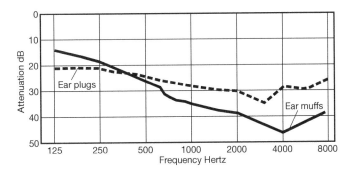

Figure 12.30 Noise attenuation of ear plugs and ear muffs

is still less than $L_{Aeq,8h}$ 85 dB(A) irrespective of the frequencies making up the noise in the workplace.

There are several possible approaches to selecting hearing protective devices (HPDs), ranging from a simple ad hoc assessment of the noise to a sophisticated assessment that takes account of all the frequencies involved. All approaches require some workplace noise measurements to be made. Failure to measure the workplace noise could result in either under- or over-attenuation, which can jeopardise a worker's hearing and safety respectively. There are several different methods of making the assessment, with each offering a different level of precision and useful for different purposes.

12.10.2.1 Classification method

The simplest way of selecting an HPD is to use the New Zealand classification method, as referenced in AS/NZS 1269.3 (Standards Australia 2005d). Selection depends on the extent to which the $L_{Aeq,8h}$ is above the critical level of $L_{Aeq,8h}$ 85 dB(A). For the classification method, a noise dose meter or sound level meter is used to measure the A-weighted average exposure in the workplace (if no noise dose meter is available, some equivalent measure such as $L_{Aeq,T}$ can be used, from which, based on a knowledge of workplace activities and work histories, an eight-hour exposure can be calculated). Selection of the appropriate class of hearing protector is then made from Table 12.5.

This method suits those H&S practitioners who can measure only an A-weighted workplace noise exposure. It does not take into account the actual frequency of the

Table 12.5 Class method for rating hearing protection

Class	$L_{Aeq,8h}$ dB(A)
1	less than 90
2	90 to less than 95
3	95 to less than 100
4	100 to less than 105
5	105 to less than 110

Source: Standards Australia (2005d).

workplace noise being measured, and can lead to inappropriate choices if the noise contains predominantly very high or very low frequencies.

12.10.2.2 Sound-level conversion method

The sound-level conversion (SLC_{80}) method was originally developed as a simplified version of the octave-band method. It uses the difference between the C-weighted L_{eq} and the A-weighted L_{eq} reaching the wearer's ear, and therefore accounts better for low-frequency components (<80 Hz) of the workplace noise as well as high-frequency ones (>3150 Hz). The SLC_{80} technique is a simple method for determining the HPD best suited for a hazardous noise area. Its name refers to a rating given to hearing protectors that provide the stated attenuation expected for 80 per cent of wearers.

The first step is to measure the workplace noise level in dB(C) at the worker's ear (i.e using C-weighting rather than A-weighting). The second step is to calculate the minimum SLC_{80} the HPD should have to provide adequate hearing protection for the wearer. This is determined by simply subtracting the 'desired' A-weighted in-ear noise level of the wearer from the C-weighted noise level measurement.

The 'desired' wearer's ear noise level is typically chosen to be between 75 and 80 dB(A), a level that presents minimal risk of NIHL. To enable communication in a noisy environment, the wearer must be able to hear and understand speech.

The HPD must therefore provide sufficient noise attenuation without blocking too much noise. This means the in-ear noise level should at least meet legal requirements, but preferably meet the target level of between 80 and 75 dB(A). The 'bigger is better' syndrome definitely does not apply here; over-protection should be avoided as much as under-protection. Over-protection leads to feelings of acoustic isolation and prompts workers to remove their hearing protector when communication is required.

All HPDs should, as a minimum, provide information of an SLC_{80} rating, a class rating and an octave band rating. Sometimes the manufacturer will provide an example of how the rating is applied. For example, if the work environment has a C-weighted noise level of L_{Ceq} 105 dB(C), wearing a hearing protective device with an SLC_{80} rating of 27 dB should provide an in-ear sound level of 105 − 27 = 78 dB(A). An HPD with an SLC_{80} of only 15 dB would permit in-ear noise levels of up to 90 dB(A), which is inadequate for use in that noise environment. Table 12.6 shows the relationship between class and SLC_{80} range for use in determining suitable hearing protection.

Table 12.6 Relationship between class and SLC_{80}

Class	SLC_{80} range
1	10 to 13
2	14 to 17
3	18 to 21
4	22 to 25
5	26 or greater

Source: Standards Australia (2005b).

12.10.2.3 Octave band method

The third method of measuring frequency attenuation is the more sophisticated octave band method. This method is the most suitable where special noise characteristics in the workplace need to be addressed (e.g. high-pitched noise, or a low-frequency component), and must also be applied where the $L_{Aeq,8h}$ exceeds 110 dB(A). It involves conducting a frequency analysis of the noise in the workplace where the hearing protector is to be used. The octave band attenuation of the hearing protector is then subtracted from the measured octave band results in the frequency range between 125 and 8000 Hz. Then the A-weighted sound level, as presented to the ear, is calculated for the resulting attenuated spectrum. Application of the octave band method requires a precision sound level meter designed for octave band analysis, because of the need to measure unweighted (i.e. linear) sound pressure levels.

To identify the degree of attenuation that the hearing protector needs to provide, two sets of data are required:

- the measured octave band analysis of the noise in the workplace
- the frequency attenuation of the hearing protector, provided by the manufacturer.

The following steps are required to calculate octave band attenuation:

- *Step 1:* Carry out a sound-level survey of the workplace, using octave band analysis at the frequencies shown in the following example. Use unweighted sound pressure levels, i.e. linear (never A- or C-weighted) levels. This is done by using a sound level meter that has the capacity to record the unweighted sound pressure levels of at least seven different frequency bands centred on the following frequencies: 125 Hz, 250 Hz, 500 Hz, 1 kHz, 2 kHz, 4 kHz and 8 kHz.

	Frequency (Hz)						
	125	250	500	1k	2k	4k	8k
Measured octave band level (dB re 20 µPa)	87	88	90	94	97	106	104

- *Step 2:* Obtain the frequency attenuation of the hearing protector. In each octave, subtract the mean-minus-standard-deviation octave band attenuation of the hearing protector from the measured octave band results.

	Frequency (Hz)						
	125	250	500	1k	2k	4k	8k
Attenuation of ear muff (mean-minus-standard-deviation) (dB*)	13	13	14	18	25	35	28

* AS/NZS 1270-2002 Table A2

- *Step 3:* Calculate attenuated sound levels.

	Frequency (Hz)						
	125	250	500	1k	2k	4k	8k
Unweighted sound pressure level* (from noise survey)	87	88	90	94	97	106	104
Mean-minus-standard-deviation (dB)	13	13	14	18	25	35	28
Attenuated sound level (dB)	74	75	76	76	72	71	76

- *Step 4:* Apply A-weighting corrections and calculate A-weighted attenuated sound levels. Make the A-weighting correction by adding the A-weighting corrections at the same frequencies. (The origin of the A-weighting correction values was explained in section 12.4.)

	Frequency (Hz)						
	125	250	500	1k	2k	4k	8k
Attenuated sound level (dB)	74	75	76	76	72	71	76
A-weighting correction (dB)	−16	−9	−3	0	+1	+1	−1
A-weighted attenuated levels	58	66	73	76	73	72	75

- *Step 5:* Rearrange the A-weighted attenuated sound levels from lowest to the highest—in this case: **58, 66, 72, 73, 73, 75, 76**. Now add the levels two at a time, using the procedure in Table 12.2, until all are used, as follows:
  - The difference between 58 and 66 is 8. Looking up the difference in Table 12.2 shows a correction factor of 0.6 to be added to the higher level. Combining 58 and 66 is 66.6.
  - The next level is 72. The difference between 66.6 and 72 is 5.4. Looking up Table 12.2, the closest correction factor is 1.2. Adding 1.2 to 72 is 73.2.
  - The next level is 73. The difference between 73 and 73.2 is 0.2. Looking up Table 12.2, the closest correction factor is 3.0. Adding 3.0 to 73.2 is 76.2.
  - The next level is 73. The difference between 76.2 and 73 is 3.2. Looking up Table 12.2, the closest correction factor is 1.8. Adding 1.8 to 76.2 is 78.
  - The next level is 75. The difference between 78 and 75 is 3. Looking up Table 12.2, the correction factor is 1.8. Adding 1.8 to 78 is 79.8.
  - The next level is 76. The difference between 79.8 and 76 is 3.8. Looking up Table 12.2, the closest correction factor is 1.4. Adding 1.4 to 79.8 is 81.2.

The attenuated level under the ear muff is therefore about 81.2 dB(A). This is well within the daily noise dose, $L_{Aeq,8h}$ 85 dB(A), but 1.2 dB over the target level of 80 dB(A), provided the worker always wears the hearing protector in this environment while exposed to noise.

To combine the attenuated A-weighted values of the hearing protector we can use this formula instead of the correction table:

$$L = 10_{\log 10} (10^{L1/10} + 10^{L2/10} + 10^{L3/10} + \ldots 10^{Ln/10}) \quad \text{(Equation 12.4)}$$

Applying the formula we should get:

$$L = 10_{\log 10} (10^{58/10} + {}^{1066/10} + {}^{1073/10} + {}^{1076/10} + {}^{1073/10} + {}^{1072/10} + 10^{75/10}) \; L = 81.2 \text{ dB(A)}$$

Reputable manufacturers of HPDs will provide full details on attenuation at the various frequencies on the packaging of their products as well as on their websites. Alternatively, when the HPD has been tested in accordance with AS 1270 (Standards Australia, 2002), the frequency attenuation can be found in the booklet *Attenuation and Use of Hearing Protectors* (National Acoustics Laboratories, 1998), available from the Australian Government Publishing Service or the National Acoustics Laboratories.

From the hearing protectors listed, select the one that provides adequate protection for the determined workplace noise, taking into consideration cost and wearer compatibility and acceptability.

H&S practitioners who believe that hearing protectors may be required in the workplace but who do not have sound level measuring equipment can ask a supplier or an acoustic consultant or occupational hygienist to conduct octave band noise analysis in each noisy location. The expected sound levels for the various hearing protectors can then be calculated from the manufacturer's published information and a decision made about which protector is suitable.

Other information on noise measurement can be found in:

- AS IEC 61672.1 Electroacoustics—Sound Level Meters, Part 1: Specifications (Standards Australia, 2004a)
- AS IEC 61672.2 Electroacoustics—Sound Level Meters, Part 2: Pattern Evaluation Tests (Standards Australia, 2004b)
- AS/NZS 1269 Parts 0–4—Occupational Noise Management (Standards Australia, 2005a, 2005b, 2005c, 2005d, 2014).

12.10.2.4 HML method

A fourth method, commonly used by European manufacturers and suppliers of hearing protectors, is known as HML (for high-, medium- and low-frequency noise reduction): H (high—2000 to 8000 Hz), M (medium—1000 to 2000 Hz) and L (low—63 to 1000 Hz). This is a simplified approach that claims the same accuracy as the octave band method. It requires measurement of both the C-weighted and A-weighted $L_{eq,T}$ measurements, and uses a nomogram approach to determine the predicted noise level reduction and the final A-weighted noise level that would reach the ear inside the hearing protector. This method can be used only if the manufacturer has specified HML values for the protector and it is possible to measure both the C-weighted and A-weighted sound levels simultaneously. It is a useful method for the H&S practitioner who wants to provide hearing protection that is better tailored to the noise profile of the workplace, and who is armed with a sophisticated sound level meter. The obtained data can be put into a free internet calculator, which is available at <www.noisemeters.co.uk/apps/naw/prot-hml.asp>.

12.10.2.5 NRR method

The NRR method is an American system of determining the expected attenuation provided by a hearing protector. This method cannot be used in Australia or compared with, say, SLC_{80} values, because of the different laboratory test method used in the United States and the different exchange rate for doubling of sound intensity. The United States uses a 5 dB exchange rate, whereas Australia and most other countries use the natural 3 dB rate.

12.11   OCCUPATIONAL AUDIOMETRY AND THE HEARING CONSERVATION PROGRAM

Occupational audiometry, the testing of workers' hearing acuity, may well identify hearing disabilities. Such tests may not necessarily indicate, however, that hearing loss is the result of current noise exposure.

Audiometry is not generally a task for the H&S practitioner. It requires well-trained, qualified, experienced personnel with properly calibrated test equipment and a specially constructed soundproof booth. The current standard for audiometric testing is AS/NZS 1269.4: 2014 (Standards Australia, 2014). The main changes from the 2005 edition of the standard are the replacement of the table of maximum allowable ambient noise levels for particular makes and models of earphones/enclosure combinations with a method to calculate the maximum permissible ambient noise level for any earphone/enclosure combination (Appendix C) (Standards Australia, 2014). It is important that, when conducting air conduction audiometry, the maximum permissible ambient noise levels stated in Table C1 of Appendix C are not exceeded. If these ambient noise levels are exceeded, then the audiometric test results are likely to be unreliable due to masking of the test signals. The H&S practitioner is therefore referred to Appendix C of the standard to ensure that the test environment does not exceed the required ambient noise levels before audiometric tests are carried out in that environment. To prove useful, audiometric testing must be part of an ongoing program in which hearing is tested regularly, preferably from the time the worker joins the workplace, to check whether there is any discernible deterioration above that attributable to ageing (presbycusis).

A health practitioner trained in conducting hearing tests (an audiometrist or audiologist) will test the worker's baseline auditory threshold for both ears. The frequencies used for both reference and monitoring audiometry are 500 Hz, 1000 Hz, 1500 Hz, 2000 Hz, 3000 Hz, 4000 Hz, 6000 Hz and 8000 Hz.

Reference audiometry must be conducted within three months after a worker starts employment, but ideally should be conducted before exposure to a noisy workplace occurs—for example, as part of a pre-employment assessment regime. Reference audiometric testing must be conducted immediately after a period of not less than sixteen hours of quiet—that is, noise exposures below 75 dB(A), which are unlikely to produce a temporary threshold shift.

The reference audiogram should be updated whenever a significant threshold shift has occurred, or at least every ten years, whichever occurs sooner. Future monitoring audiograms should then be compared with the most recent reference audiogram and records of previous reference audiograms shall be retained (Standards Australia, 2014).

Monitoring audiometry should be performed well into or at the end of the work shift and carried out within twelve months of initial reference audiometry and at least every two years on an ongoing basis, as prescribed by WHS legislation. The results should be compared with the reference audiometry test results to see whether the hearing threshold has changed. If it has, this will indicate inadequacies in the use of personal hearing protectors or changes in workplace noise exposure levels. For workers exposed to high exposure levels, >100 dB(A), more frequent audiometric testing—for example, every six months—may be required.

Where significant hearing impairment is detected at the initial reference audiometric test, the person should undergo a medical examination to see whether a repeat audiometric test, conducted on another day, confirms the original finding.

Where the monitoring audiometry results, when compared with the results of the reference audiometry, show:

- a shift in average threshold at 3000, 4000 and 6000 Hz greater than or equal to 5 dB, or
- a shift in mean threshold $\geq$ 10 dB at 3000 and 4000 Hz, or
- a change in mean threshold $\geq$ 15 dB at 6000 Hz, or
- a threshold shift $\geq$ 15 dB at 500, 1000, 1500 or 2000 Hz, or
- a threshold shift $\geq$ 20 dB at 8000 Hz,

the worker shall be requested to have a second confirmatory audiometric test on another day, again after sixteen hours in quiet conditions. If the threshold shift is confirmed during this second test, the person should be referred for a medical opinion. If the shift in threshold is confirmed as noise induced, the worker should be advised of this in writing. Figure 12.31 shows the various degrees of hearing loss and their relationship to everyday sounds via the scale on the right.

Audiometric testing can be of benefit to both employers and workers in excessively noisy industries (foundries, canneries, construction, metal industries, transport and storage), but only if it is an integral part of a rigorous noise-management program. Audiometric testing in isolation from other elements of a noise-management program serves only to record the deterioration in hearing. In combination with other elements, it can detect the early onset of NIHL and enable counter-measures to be put in place. The tests may also provide useful information for assisting in workers' compensation claims. For instance, an employer who knows the hearing ability of workers at the start of their employment and can demonstrate adherence to a robust noise-management program may well have a defence against future claims for NIHL compensation.

Noise-management programs are designed to prevent permanent hearing impairment, principally by maintaining noise exposure within the limits required by the legislation. Any of the technical control processes discussed above might be used in such a program. Certainly, the best are noise-reduction programs; however, if noise sources cannot be quietened sufficiently through higher-order noise-control measures, the worker's hearing

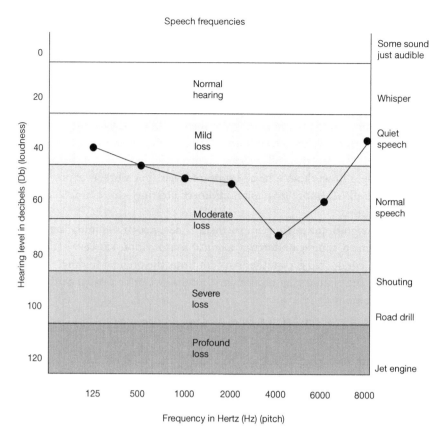

Figure 12.31 Various degrees of hearing loss
Source: Environmental Directions Pty Ltd.

must be protected by using an HPD. Management must then also provide training and instruction in, and supervision of, the efficient use of HPDs. Management's commitment to the use and benefits of HPDs must be matched by the rigid enforcement of the noise-management program.

12.12   ACOUSTIC SHOCK

12.12.1 INTRODUCTION

The Australian Communications Industry Forum (ACIF) G616 *Industry Guideline— Acoustic Safety for Telephone Equipment* (ACIF, 2013), defines an *acoustic incident* as:

> The receipt by a telephone user of any unexpected sound that has acoustic characteristics that may cause an adverse reaction in some telephone users. Depending on the characteristics of the sound and of the user, an acoustic shock may result from the incident.

An acoustic incident is typically a high-intensity, high-frequency monaural squawk, screech, shriek or howling tone that occurs without warning. This type of noise is typically produced by such things as misdirected fax calls, feedback between the microphone and speaker of the telephone or mobile phone signal interference.

Exposure to an acoustic incident usually does not last more than a few seconds, as the operator generally tears off the headset almost as soon as the signal is heard. If it is severe enough, however, an acoustic incident may lead to acoustic shock.

Acoustic shock is defined by the European Telecommunications Standards Institute (ETSI, 2000) and Australian Communications Industry Forum G616 (ACIF 2013) as:

> Any temporary or permanent disturbance of the functioning of the ear, or the nervous system, which may be caused to the user of a telephone earphone by a sudden sharp rise in the acoustic pressure produced by it.

Tests have shown that the signals in question are less than 120 dB sound pressure level at the eardrum because of the electronic limiters in the headset systems. Measurements by the Health and Safety Laboratory of the UK Health and Safety Executive (Health and Safety Laboratory, 2001) indicate that fax tones produce on average 83 dB(A), holding tones 88 dB(A) and carrier tones 95 dB(A) when measured at maximum volume.

12.12.2 HOW IS AN ACOUSTIC SHOCK EXPERIENCED?

When an acoustic incident occurs, telephone workers may initially feel startled or shocked. In extreme cases, they may quickly experience a varied combination of vertigo, nausea, vomiting, stabbing pain in the ear, feeling of fullness in the ear and/or facial numbness. Tingling, tenderness or soreness around the ear and neck and arms, and tinnitus, may also be experienced. There is seldom loss of hearing, and symptoms tend to disappear over time.

Secondary and tertiary symptoms may develop consistent with stress from trauma, including headache, fatigue, feelings of vulnerability, anger, hypervigilance and hyper-sensitivity to loud sounds, depression, substance abuse and anxiety, especially about returning to work.

12.12.3 WHY CALL CENTRES?

An explanation for the symptoms experienced by a call centre telephone operator after an acoustic incident must be sought in both the psychosomatic and physiological areas. The main psychosomatic factors are the stress of handling large volumes of telephone calls within certain time limits, the performance-monitoring systems applied and targets to be achieved, the company's management culture, the operator's perceived lack of control over the stressful work situation and the inability to anticipate an acoustic incident. Furthermore, the work environment, with often inadequate office equipment and acoustics, a lack of proper training, mistakes by incoming callers or clients such as sending a fax over a telephone line, interference from mobile phone signals and loud noises deliberately made by irate clients may all contribute. These issues combine to cause stronger physiological responses (e.g. a startle reflex) when an acoustic incident does take place than if the worker were exposed to a similar sound in an industry where the sound was anticipated.

The main physiological response to an acoustic shock occurs in the middle ear, which contains two muscles, the stapedius and the tensor tympani; these are attached to the ossicles (the middle ear bones, the malleus, incus and stapes). The tensor tympani muscle is attached to the malleus, and the stapedius muscle to the stapes. When loud sounds occur both muscles react, causing what is known as the 'aural reflex'. The stapedius muscle reacts to loud sounds, causing an increased pressure on the oval window membrane of the cochlea and the movements of the stapes. The tensor tympani muscle causes a startle reflex. If triggered, it restricts the movements of the malleus and incus and is capable of placing large forces on the alignment of the eardrum. The tensor tympani muscle's reflex threshold can be 'reprogrammed' to react at much lower sound levels.

With acoustic shock incidents, the loud noises occur with rise times of between 0 and 20 milliseconds. The time for the middle ear muscles to react by contracting is about 25 milliseconds. When they do contract, they do so with added force because of the combination of loud noise and the startle reflex. In extreme cases, this may lead to a tearing of the oval or round window membrane and subsequent leaking of fluid from the cochlea.

12.12.4 WHAT IS THE EVIDENCE FOR THE EXISTENCE OF ACOUSTIC SHOCK?

The available evidence indicates that acoustic shock and its effects constitute a real health issue because of the specific characteristics of call centres. In Australia alone, there have been several hundred reported cases of acoustic shock in call centre telephone operators. A minority of affected individuals are unable to continue working as telephone operators. Other countries report similar acoustic incidents with remarkably similar symptoms. The chance that all these cases are due to malingering is zero.

12.12.5 HOW CAN ACOUSTIC SHOCK BE MANAGED?

Management of acoustic incidents must focus on two main areas:

- prevention of sudden loud noises over the telephone through the incorporation of appropriate telephone equipment and electronic sound-limiting devices
- the reduction of stress levels in workers, through appropriate workplace design, management systems and worker education.

These must be implemented in combination, as either one will fail without the other.

12.12.6 PREVENTION OF LOUD NOISES OVER TELEPHONE LINES

Telephone systems should comply with AS/CA S004 Voice Performance Requirements for Customer Equipment (Standards Australia, 2013a), and are thus limited to 120 dB for handsets and 118 dB for headsets. However, this level does not prevent acoustic incidents. To prevent the loud noises from entering an operator's headset, a so-called volume-limiting amplifier must be incorporated between the telephone and the headset. Such a device searches for high-frequency acoustic tones and eliminates them within about 20 milliseconds. If the device is working correctly, the telephone operator is often not even aware that a loud tone has occurred.

12.12.7 REDUCTION OF STRESS LEVELS IN CALL CENTRE WORKERS

The two main areas for reducing negative stress are the work environment (including management systems) and the workplace design (including construction, layout and equipment).

WHS legislation is quite specific about the need to conduct risk assessments for different tasks and systems of work in order to ensure a safe and healthy work environment. It is less specific about workplace design. The acoustics in a given call centre will limit the number of operators because having too many close together leads to mutual speech interference. Proper acoustic design is therefore beneficial to owners because it optimises the number of telephone operators who can occupy a given space without speech interference. It is also beneficial to the telephone operators because, with better ability to communicate, their stress levels will decline to acceptable levels.

12.13 OTOTOXINS

12.13.1 INTRODUCTION

A wide variety of chemicals and medication may, alone or in concert with noise, result in hearing loss. These substances, known as ototoxins (*oto* = ear, *toxin* = poison), affect the hair cells and/or the auditory neurological pathways. Inhalation or absorption of certain chemicals through the skin may cause hearing loss independent of noise exposure, while other chemicals may have an additive or synergistic effect. Some chemical toxicants that do not cause permanent hearing loss themselves may, in combination with noise exposure, cause permanent hearing loss. Hearing damage is more likely to occur if the ear is exposed to a combination of substances or to the combination of a substance and noise. Because of the different mechanisms of interaction between noise and ototoxins, there are difficulties for both risk assessment and standard setting in the industrial environment. These are compounded by the use of different 'languages' by different professionals when measuring chemical agents and physical stressors, and the tendency in industry to attribute NIHL in noisy environments to noise alone.

Ototoxins can be divided into two general classes: workplace chemicals and medications. In this section, medications will be mentioned only briefly. Of the workplace chemicals, three major classes have been identified: solvents, heavy metals and chemical asphyxiants.

12.13.2 WORKPLACE CHEMICALS

In the vast majority of industrial cases, ototoxic hearing loss is caused by aromatic and aliphatic hydrocarbon solvents. These solvents are well recognised for their neurotoxicity to the central and peripheral nervous systems. The most common ototoxic solvents are alcohol, toluene, ethyl benzene, styrene, n-hexane, carbon tetrachloride, carbon disulfide, trichloroethylene, perchloroethylene and acrylonitrile. Other known ototoxic substances in the workplace include mercury, manganese, lead, arsenic and cobalt, and asphyxiants such as hydrogen cyanide and carbon monoxide. Carbon monoxide is also released as a metabolic by-product of the paint stripper methylene chloride.

The 2018 Safe Work Australia Work Health and Safety Code of Practice (CoP), *Managing Noise and Preventing Hearing Loss at Work* (Safe Work Australia, 2018), recommends that workers should have regular audiometric testing if they are exposed to any of the ototoxic substances listed in Table 12.7, where the airborne exposure (without regard to respiratory protection worn) is greater than 50 per cent of the national exposure standard for the substance regardless of noise level, or if they are exposed to ototoxic substances at any level *and* noise with $L_{Aeq,8h}$ greater than 80dB(A) or $L_{C,peak}$ greater than 135dB(C).

Table 12.7 Some common ototoxic substances

Type	Name	Skin absorption
Solvents	Butanol	✓
	Carbon disulfide	✓
	Ethanol	
	Ethyl benzene	
	n-heptane	
	n-hexane	
	Perchloroethylene	
	Solvent mixtures and fuels Stoddard solvent (white spirits)	✓
	Styrene	
	Toluene	✓
	Trichloroethylene	✓
	Xylenes	
Metals	Arsenic	
	Lead	
	Manganese	
	Mercury	✓
	Organic tin	✓
Others	Acrylonitrile	✓
	Carbon monoxide	
	Hydrogen cyanide	✓
	Organophosphates	✓
	Paraquat	

Occupations where noise and ototoxic substances are most commonly found together include:

- printing
- painting
- boat building
- furniture making
- petroleum product refinery/manufacture
- vehicle/aircraft fuelling
- firefighting
- weapons firing (armed forces, shooting clubs)
- rural/agriculture.

Research is ongoing to establish the effects of human exposure to workplace ototoxins. One remaining problem is to relate the results of animal studies to humans, as few studies have been conducted with workers to date. The occupational exposure studies conducted so far seem to confirm laboratory studies, and suggest that simultaneous exposure to noise and chemicals produces a significantly greater hearing loss than exposure to either agent alone. In the Code of Practice *Managing Noise and Preventing Hearing Loss at Work* (Safe Work Australia, 2018), the effects of known ototoxic chemicals and noise have been considered for the first time. The Code of Practice states that where workers are exposed to ototoxic substances at any level and noise with $L_{Aeq,8h}$ greater than 80 dB(A) or $L_{C,peak}$ greater than 135 dB(C), regular health monitoring is recommended. It is expected that in the near future it will be possible to predict the effect on hearing of chemical/noise combination. It is also expected that the individual effects of substances on the auditory system will be better understood and can be used in noise-management programs.

Agents that are considered synergistic with noise exposure (i.e. the interaction of the agents produces a threshold shift which is greater than the sum of the effects of the individual agents) include carbon disulfide, carbon tetrachloride, carbon monoxide, hydrogen cyanide, styrene, methyl ethyl ketone and methyl isobutyl ketone.

Agents that are considered to have either an additive or synergistic effect include toluene, ethyl benzene, styrene, carbon monoxide and hydrogen cyanide.

Agents that are known to cause auditory system impairment by themselves include:

- organic solvents such as toluene, styrene, xylene and trichloroethylene
- metals such as cobalt, mercury, lead and trimethyltin
- asphyxiants such as carbon monoxide and hydrogen cyanide.

As stated earlier, it is recommended that workers in any of the occupations listed above be included in audiometric testing programs. Reviewers of audiometric test results should be alert to the possible additive or synergistic effects between exposure to noise and ototoxins. Where necessary, they should suggest reducing exposure to one or both agents. Employers must therefore ensure that they know the airborne concentrations of the ototoxic substances to which their workers are exposed, and take the actions required to minimise these concentrations. Employers should also ensure that information on ototoxins and their effects is included in training sessions. It is further recommended that at least biannual audiograms be taken of employees whose airborne exposures to known ototoxic substances (without regard to respiratory protection worn) and the lower noise levels of the CoP (Safe Work Australia, 2018).

12.13.3 MEDICATION

Some medications have been identified as ototoxins, including antibiotics such as streptomycin, quinine, salicylates such as aspirin when used over long periods of time, anti-inflammatory, anti-thrombosis and anti-rheumatic agents and loop diuretics.

Employees who have any concerns about the ototoxic effects of medication should be encouraged to discuss their concerns with their doctor or pharmacist.

12.14   VIBRATION

Exposure to vibration is widespread in modern industries. Many tools, machines and vehicles such as chainsaws, jackhammers, chipping tools, tractors and earth-moving vehicles vibrate. Vibration occurs as a side-effect of industrial activities or may be deliberately introduced—for example, in concrete pours, where vibration is used to shake the wet concrete into place.

Prolonged exposure to vibration causes health effects, disorders and/or disease. According to Safe Work Australia (2015), during the period 2001–15 there were 5260 workers' compensation claims for injuries or illness attributed to exposure to vibration costing $134 million. The risk depends on the characteristics of the vibration, the part(s) of the body exposed and the duration of exposure.

12.14.1 HUMAN EXPOSURE TO VIBRATION

There are basically three kinds of vibration to which workers are exposed:

- vibrations transmitted to the whole body surface or substantial parts of it when the body is immersed in a vibrating medium such as air or water through which high-intensity sound is travelling
- vibrations transmitted to the body as a whole through the supporting surface—for example, in vehicles, on drill platforms or in the vicinity of working machinery, where vibration is transmitted through the feet, the buttocks or the supporting area of a reclining person
- vibrations applied to particular parts of the body, such as the head or limbs, by vibrating handles, pedals or head rests, or by hand-held power tools and appliances.

Vibrations of specific interest in the occupational environment are normally classified as either:

- whole-body vibrations, in the range of 1 to 80 Hz, or
- segmental vibrations, for example hand-arm vibrations, in the range of 8 Hz to 1 kHz.

Because the body acts as a mechanical system, its various parts resonate at various frequencies. When vibration occurs at or near any of the body's resonant frequencies, the effect on the body is greatly increased. The smaller the body part or limb, the faster it can vibrate and the higher its resonant frequency will be. For example, the head and shoulders resonate

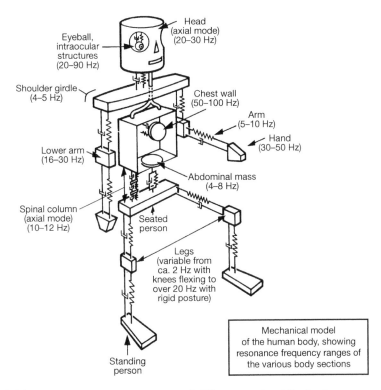

Figure 12.32 was described with labels: Head (axial mode) (20–30 Hz); Eyeball, intraocular structures (20–90 Hz); Shoulder girdle (4–5 Hz); Chest wall (50–100 Hz); Arm (5–10 Hz); Hand (30–50 Hz); Lower arm (16–30 Hz); Abdominal mass (4–8 Hz); Spinal column (axial mode) (10–12 Hz); Seated person; Legs (variable from ca. 2 Hz with knees flexing to over 20 Hz with rigid posture); Standing person; Mechanical model of the human body, showing resonance frequency ranges of the various body sections

Figure 12.32 Frequency response ranges of different parts of the body
Source: Brüel and Kjær (1989).

at a frequency of about 5 to 25 Hz, and the eyeball resonates in the range of about 30 to 80 Hz (Figure 12.32).

Exposure to vibration causes a number of physiological and psychological responses, which are outlined below.

12.14.1.1 Whole-body vibration

Research has shown that the human body is most sensitive to vertical vibrations in the range 4–10 Hz. Studies by Kroemer and Grandjean (1997) show that vibrations between 2.5 and 5 Hz generate strong resonance in the vertebrae of the neck and the lumbar region. Vibrations between 4 and 6 Hz cause resonances in the trunk, shoulders and neck, and vibrations between 20 and 30 Hz set up the strongest resonances between the head and shoulders of seated persons. Other health effects and some associated frequency ranges include:

- motion sickness in the range 0.2–0.7 Hz, with the greatest effect at 0.3 Hz
- faults in the vestibular system of the ear caused by disturbance to the inner ear's balancing system
- visual impairment affecting the efficiency of drivers of tractors and heavy vehicles and increasing the risk of accidents, in the range 10–30 Hz
- damage to bones and joints at frequencies below 40 Hz, especially in the lower spine—for example, ischaemic lumbago

- problems in the digestive system
- variations in blood pressure that may lead to cardiovascular problems
- disorders of menstruation, internal inflammation and abnormal childbirth in women exposed to 40–45 Hz vibrations
- increase in foetal heart rate when vibration (120 Hz) is applied to the mother's abdomen
- fatigue, loss of appetite, irritability, headache
- general reduced efficiency, which may lead to errors and/or accidents.

12.14.1.2 Hand–arm vibration

Segmental vibration affects a part of the body or an organ. The most prominent type of segmental vibration is hand–arm vibration, mainly experienced by operators of hand-held power tools such as jackhammers, pneumatically driven tools, chainsaws etc.

Vibrations caused by these types of tools are usually found in the higher frequencies, such as 40–300 Hz. In this range, vibrations may have ill effects on the blood vessels and nerve endings and blood circulation in the hands, causing:

- Raynaud's syndrome, also known as vibration white finger (VWF), or dead finger syndrome. Initially the whitening (blanching) of the fingers is localised on the tips of the fingers most exposed to the vibrating source, but with continued exposure it spreads to involve all fingers and the tips of the thumbs
- nerve and blood vessel degeneration, resulting in loss of the sense of touch, heat perception and grip strength
- endothelial damage to blood vessel wall elasticity, resulting in fibrosis and thus reducing the internal vessel diameter with subsequent restriction of blood flow
- pain and cold sensations between attacks of VWF
- muscle atrophy, tenosynovitis
- damage to joints and muscles in wrists and/or elbows
- decalcification of the carpal tunnel
- bone cysts in fingers and wrists.

The symptoms and effects of VWF are aggravated when the hands are exposed to cold and/ or the operator is a smoker.

The first symptoms are relatively mild (e.g. a tingling sensation in the fingertips) and tend to disappear when exposure ends. With continued exposure, however, symptoms become progressively more severe and eventually irreversible. Prevention is therefore a high priority.

12.14.2 EXPOSURE GUIDELINES

The Model Work Health and Safety Regulation, Part 4.2, 'Hazardous manual tasks', requires that risks to health and safety from manual tasks, including vibration, be controlled. The regulation does not give limits for vibration exposure. The Safe Work Australia (2018) *Code of Practice—Managing Noise and Preventing Hearing Loss at Work* does state a limit, but this is aimed primarily at minimising hearing loss by providing audiometric testing of workers exposed to hand–arm vibration at any level and to noise with $L_{Aeq,8h} > 80dB(A)$ or $L_{C,peak} > 135dB(C)$. Australian standards for the assessment and evaluation of whole-body and hand–arm vibration also do not specify limits for exposure, but they do provide information and guidelines regarding the risks of exposure level and duration. Although no vibration-specific exposure standard exists under WHS legislation, the general requirement to ensure health and safety at work also applies to vibration. Following the *Code of Practice—Hazardous Manual Tasks* (Safe Work Australia, 2011a) goes a long way towards demonstrating that proper diligence has been applied in a particular situation. Safe Work Australia (2015) has released a series of information material on both hand–arm and whole-body vibration. The series include guides to managing risks of exposure to vibration in workplaces as well as guides to measuring and assessing workplace exposure to hand-arm and whole-body vibration. Following the guides will further assist workplaces with demonstrating compliance with their duty of care. The guidance material can be accessed and downloaded from Safe Work Australia's website.

The current WHS legislation obliges manufacturers, suppliers and installers of plant and equipment to provide information that workers need to use the equipment without endangering their health and safety. Applying the control hierarchy is extremely important, as there are basically no effective options for minimising vibration exposure through PPE. An effective way of exercising proper diligence is to use international exposure guidelines, such as the European *Directive 2002/44/EC* on vibration, which provides 'daily exposure action' and 'daily exposure limit' values for both whole-body vibration and hand-arm vibration (European Union, 2002). In addition, H&S practitioners should prepare a documented health and safety program which puts it into practice and regularly reviews its effectiveness. Such a program should incorporate risk assessment and use of the relevant Australian standards. Those applicable to human vibration are:

- AS 2670.1 Evaluation of Human Exposure to Whole Body Vibration; Part 1: General Requirements (Standards Australia, 2001)
- AS ISO 5349.1: 2013 Mechanical Vibration—Measurement and Evaluation of Human Exposure to Hand-transmitted Vibration; Part 1: General Requirements (Standards Australia, 2013b)
- AS ISO 5349.2: 2013 Mechanical Vibration—Measurement and Evaluation of Human Exposure to Hand-transmitted Vibration; Part 2: Practical Guidance for Measurement at the Workplace (Standards Australia, 2013c).

In AS ISO 5349.1, vibration exposure evaluation is based on the measurement of vibration magnitude at the grip zones or tool handles and exposure times, and expressed as the vibration total value (VTV). Measurements of the VTV have values that are greater than measurements in a single axis of up to 1.7 times (typically between 1.2 and 1.5 times) the magnitude of the greatest component. The daily vibration exposure is based on the eight-hour energy equivalent acceleration value A8 (in AS 2763-1988—this was based on four-hour exposures). Conversion of four-hour exposures to eight-hour exposures can be done by multiplying the four-hour exposure by 0.7 to obtain the eight-hour exposure.

The standard provides useful information in a series of appendices, called annexes:

- Annex A gives definitions for frequency weighting W_h and band limiting filters applied with measurements.
- Annex B gives information on health effects.
- Annex C gives guidance to competent authorities for setting exposure limits or action values.
- Annex D gives information on human responses to vibration.
- Annex E gives guidance on preventative measures.
- Annex F gives guidance on uniform methods for measuring and reporting exposure of human beings to HAV.

Standards for human vibration in the workplace are expressed in terms of *acceleration* (m/s^2) and *duration of exposure*, and take the frequency of the vibration into account.

The European *Directive 2002/44/EC* (the 'Vibration Directive') was published on 6 July 2002 (European Union, 2002) and is the most used in Australian workplace situations for determining the acceptability or otherwise of human vibration exposure. In Safe Work Australia's guidance material, the Vibration Directive is referred to as containing the most widely used and accepted exposure action value and exposure limit value. The Vibration Directive requires employers to:

- assess the risk and exposure
- plan and implement control measures
- provide and maintain a suitable work environment
- provide training and information on vibration risks and their control to workers
- monitor and review the effectiveness of the risk-control program.

Where a risk has been identified, workers have a right to health surveillance. The 2015 Safe Work Australia guides for hand–arm vibration do not mention or refer to health surveillance. As a guide, the provisions in Appendix B of the now superseded Australian Standard 2763 of 1988 could be consulted (Standards Australia, 1988). Appendix B states:

> Segmental workers shall be medically examined.* The examination should be conducted by a medical practitioner or delegate (such as a qualified nurse who has received suitable instruction in the detection of vibration disease).
>
> * A segmental vibration worker is any person who is expected to encounter exposure of the upper limbs to vibratory motion within the frequency range of 5 Hz to 1500 Hz, at suspected weighted levels at or above 2.9 m/s^2 (Standards Australia 1988, p. 13)

The Vibration Directive provides for 'daily exposure action' (EAV) and 'daily exposure limit' (ELV) values, which are specified as eight-hour energy equivalent frequency-weighted acceleration values and expressed as A(8). For whole-body vibration, the directive also gives 'vibration dose value' (VDV) as an alternative. Exposure values are:

- for hand–arm vibration:
  - EAV: 2.5 m/s2 A(8)
  - ELV: 5 m/s2 A(8)

- for whole-body vibration:
  - EAV: 0.5 m/s^2 A(8) or VDV 9.1 m/s$^{1.75}$
  - ELV: 1.15 m/s^2 A(8) or VDV 21 m/s$^{1.75}$.

The above exposure values are derived from ISO 5349–1:2001 (ISO, 2001) for hand–arm vibration and ISO 2631–1:1997 (ISO, 1997) for whole-body vibration.

Because many exposures are shorter than eight hours, and because the VDV is more sensitive than the root mean square acceleration to shocks with high peak accelerations, the VDV options for the EAV and ELV will often be more protective than their A(8) counterparts (Nelson and Brereton 2005).

In industries such as mining, employees commonly work twelve-hour shifts. This increases the risk for workers exposed to vibration, and this increased risk needs to be accounted for. One way of doing this is by reducing the allowable exposure to vibration. For a twelve-hour shift, the allowable hand-arm action value should be reduced to 2.0 m/s^2 and the exposure limit to 4.0 m/s^2; however, it is highly unlikely that a worker's 'trigger time' would last the entire twelve-hour shift. For whole-body vibration, the allowable exposure action value should be reduced to 0.41 m/s^2 and the exposure limit to 0.94 m/s^2. Exposure to whole-body vibration may well persist for the whole of the twelve-hour shift in some industries.

The H&S practitioner should, however, use the accepted practice of working with normalised eight-hour shifts in order to demonstrate compliance or otherwise with legislative and Australian standard requirements.

European *Directive 92/85/EEC* provides guidance on limits for workers who are pregnant or have recently given birth, or are breastfeeding. With respect to vibration exposure, the directive states that work shall be organised in such a way that pregnant workers and those who have recently given birth are not exposed to work entailing any risk arising from unpleasant vibration of the entire body, particularly at low frequencies; micro-traumas, shaking and shocks, or situations where jolts or blows are delivered to the lower body (European Union, 2000, p. 19).

12.14.3 VIBRATION MEASUREMENT

Measuring whole-body or segmental vibration is rarely a task for the H&S practitioner, and the services of an expert with appropriate equipment will usually be required. However, with the rapid expansion of apps, the H&S practitioner can do some preliminary assessments with an appropriate smartphone and a whole-body vibration measuring app (Wolfgang and Burgess-Limerick, 2014).

The vibration has to be measured in three planes—that is, z (up-down), x (back-forward) and y (side-to-side)—and for this purpose a triaxial accelerometer is normally used. One plane of vibration is usually dominant.

Common types of transducer fit onto the seat of a vehicle (whole-body vibration) or via an adaptor onto the back of the hand (segmental vibration), as shown in Figure 12.33.

Hand–arm vibration is measured in the three orthogonal vibration directions, x, y and z, and the root sum of the squares of the three single axes is taken.

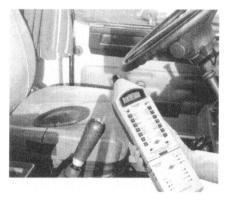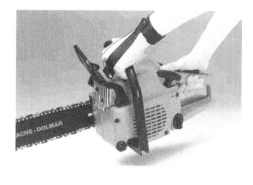

Figure 12.33 Measurement of whole-body and hand-arm vibration
Source: Brüel and Kjær (1989).

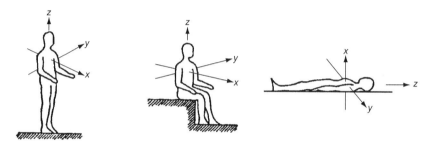

Figure 12.34 Whole-body orthogonal coordinate system
Source: Standards Australia (2001).

Whole-body vibration is determined separately for each of the three orthogonal vibration directions at the point where vibration enters the body—for example, the buttocks or feet (see Figure 12.34).

12.14.3.1 Measurement of whole-body vibration

ISO Standard 2631.1 (ISO, 1997) and AS 2670.1 (Standards Australia, 2001) indicate three criteria for the assessment of whole-body vibration in different situations in the workplace:

- the effects on human health and comfort
- the probability of vibration perception
- the incidence of motion sickness.

The standard does not specify or recommend limits of exposure, but the annexes provide information on the possible effects of vibration on health, comfort and perception.

Annex B is the more important one for the assessment of whole-body vibration with respect to human health. It applies to rectilinear vibration along the x, y and z axes of the body for people in normal health who are exposed regularly to vibration.

Figure 12.35 shows the health guidance caution zones for whole-body vibration. For the assessment of effects on health, two relationships can be used: the average acceleration value (equation B1) and the vibration dose value (equation B2). Exposures below the zones

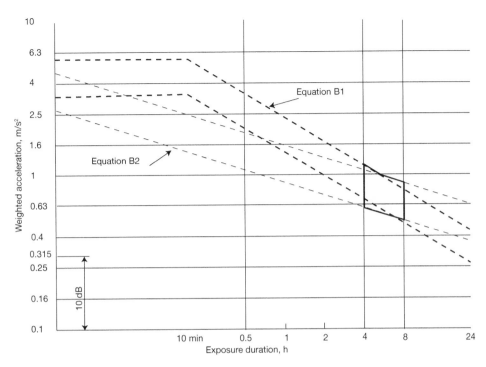

Figure 12.35 Health guidance caution zones
Source: Standards Australia (2001).

are believed unlikely to pose a threat to health. For the zones between the dotted lines caution is indicated, and above the zones health risks are likely. These recommendations are, according to AS 2670.1, mainly based on exposures of four to eight hours.

The caution zone should be viewed as an 'action level' at which intervention to control the exposure is necessary. Exposures in the 'likely health risk zone' would be considered unacceptable under WHS legislation.

For vibration in more than one direction, as is typically the case with whole-body vibration, AS 2670.1 (Standards Australia, 2001) suggests that the effects can be calculated by taking the vector sum, a, of the three weighted acceleration values, a_x and a_y and a_z, as follows:

$$a = \sqrt{[(1{,}4\, a_x)^2 + (1{,}4\, a_y)^2 + a_z^2]} \qquad \text{(Equation 12.5)}$$

The actual exposure time, expressed as a percentage of the total allowed exposure time, is known as the equivalent exposure percentage.

12.14.3.2 Measurement of hand–arm vibration

AS ISO 5439.1-2013 and AS ISO 5439.2-2013 (Standards Australia, 2013b, 2013c) do not provide limits for safe exposure to hand–arm vibration, but they do provide guidelines for its assessment. These standards suggest that the directions of vibrations be measured and reported using two orthogonal coordinate systems: the basicentre system and the

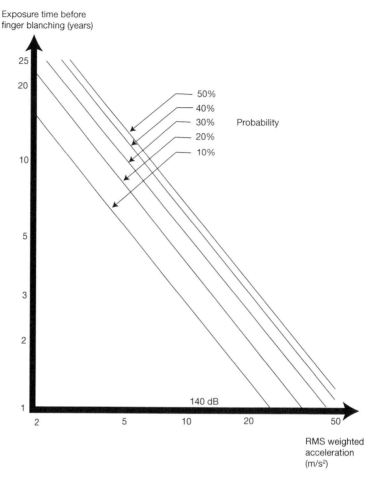

Figure 12.36 Exposure time of percentiles of population groups exposed to vibration suffering mild effects on tip of finger
Source: Brüel and Kjær (1989).

biodynamic system. The basicentre system refers to the tool, while the biodynamic system refers to the hand.

AS ISO 5439.1-2013 and AS ISO 5439.2-2013 (Standards Australia, 2013b, 2013c) provide a chart allowing one to predict when the first signs of VWF in workers will occur on the basis of exposure levels and duration. Figure 12.36 is an example of this chart.

The probability line of Figure 12.36 can be used to assess the long-term effects of eight hour-per-day exposures to hand-arm vibration. For example, the 10 per cent line predicts that exposure to 9 m/s² vibration acceleration will cause 10 per cent of exposed workers to reach stage 1 of VWF (blanching of and tingling sensation in fingertips that tends to disappear with cessation of exposure) in about three years. At a vibration acceleration of 2.5 m/s², it would take about eighteen years for the same percentage of workers to reach stage 1.

A worker's daily exposure to vibration can be calculated from the intensity (magnitude) and duration of the exposure using the equation below.

$$A(8) = a_{hv} \sqrt{\frac{T}{T_0}}$$ (Equation 12.6)

where:
a_{hv} vibration magnitude in m/s^2
T = the actual duration (trigger time) of the exposure
T_0 is the reference duration i.e. the normalised eight-hour period

If the operation is such that the total daily exposure is made up from several exposures, then the total frequency-weighted acceleration may be calculated using the equation:

$$A(8) = \sqrt{A_1(8)^2 + A_2(8)^2 + \ldots}$$ (Equation 12.7)

where:
$A_1(8)$, $A_2(8)$ etc. are the partial vibration exposures from the different tools used

The advantage of the above method is that effects from longer shifts are already taken into account with the calculation of vibration exposure to a normalised eight-hour shift period.

12.14.3.3 Motion sickness
One other kind of vibration that can lead to the special problem of motion sickness is vibration in the frequency range of 0.1 to 0.63 Hz in the vertical z axis. In this range, a significant proportion of unadapted persons will experience discomfort depending on frequency and exposure times.

12.14.4 CONTROL MEASURES

Control of vibration is most important. It usually requires a combination of appropriate tool selection, good work practices and education programs, as well as medical surveillance. It also requires identification of the hazards through vibration measurements and reduction of vibration at the source or transmission. Where vibration is a problem, workers should be warned of the hazards of vibrating tools, and medical supervision should be employed to identify those workers showing early signs of adverse health effects or reversible VWF.
Some control measures for whole-body vibration include:

- ensuring that all on-site road and work-area surfaces are well maintained to minimise rough rides
- ensuring that a traffic management system that incorporates speed limits operates on site
- ensuring that drivers of vehicles are familiar with the on-site road conditions
- insulation of seat and head-rest vibration through springs and dampers
- installation of vibration-dampening seats (suspension seats) that allow correct adjustment for the handling of the vehicle and do not interfere with visibility from the cab
- implementing a seat maintenance system to ensure that suspension seats are regularly checked and maintained in a serviceable condition

- ensuring that only vehicles suitable for the job, and which provide driver comfort, are used
- ensuring that cab layouts are such that drivers do not have to adopt awkward and potentially damaging postures
- ensuring that regular mini-breaks are incorporated in shifts
- where required and practicable, providing special boots with vibration-absorbing soles to protect against vibration through the floor
- limiting the time spent by workers on vibrating surfaces
- ensuring that plant and equipment are well maintained
- mounting machines and plant on vibration-isolating mounting pads.

Some control measures for hand–arm vibration include:

- isolating the vibrations, for example, by using special mountings and/or adjusting the centre of gravity as low as possible
- damping of the vibrations, for example by wearing padded gloves, provided that this does not lead to an increase in clamping force (which results in increased transmission of vibrations).

The adverse effects can also be minimised if the operator's hands are kept warm or the handles of the vibrating tools are warmed in cold work situations.

12.15   REFERENCES

Access Economics 2006, *Listen Hear! The Economic Impact and Cost of Hearing Loss in Australia*, Access Economics, Sydney.

Australian Communications Industry Forum (ACIF) 2013, *Industry Guideline—Acoustic Safety for Telephone Equipment*, ACIF G616, ACIF, Sydney, <www.commsalliance.com.au/Documents/all/guidelines/g616> [accessed 9 October 2018]

Berger, E. 2000, *The Noise Manual*, 5th ed., AIHA, Fairfax, VA.

Brüel, P. and Kjær, V. 1989, *Human Vibration*, BR 0456–12, Brüel & Kjær, Nærum, Denmark.

Eden, D. and Piesse, R. 1991, 'Real world attenuation of hearing protectors', in *Proceedings of the 4th Western Pacific Regional Acoustic Conference*, Brisbane, pp. 508–13.

European Telecommunications Standards Institute (ETSI) 2000, *Acoustic Shock from Terminal Equipment (TE): An Investigation on Standards and Approval Documents*, ETSI TR 101 800 V1.1.1 (2000–07), ETSI, Valbonne, France, <www.etsi.org/deliver/etsi_tr/101800 _101899/101800/01.01.01_60/tr_101800v010101p.pdf> [accessed 2 December 2012]

European Union 2000, *Directive 92/85/EEC—Pregnant Workers*, European Agency for Safety and Health at Work, European Union, Brussels, <https://osha.europa.eu/en/legislation/directives/sector-specific-and-worker-related-provisions/osh-directives/10> [accessed 2 December 2012]

—— 2002, *Directive 2002/44/EC—Vibration*, European Agency for Safety and Health at Work, European Union, Brussels, <https://osha.europa.eu/en/legislation/directives/exposure-to-physical-hazards/osh-directives/19> [accessed 2 December 2012]

Groothoff, B. 2015a, *Human Vibration*, Brüel & Kjær, Sydney.

—— 2015b, *Occupational Noise Management*, Brüel & Kjær, Sydney.

Health and Safety Laboratory 2001, *Advice Regarding Call Centre Working Practices*, LAC 94/1 revised, Health and Safety Executive, London, <www.ispesl.it/dsl/dsl_repository/sch41pdf08marzo06/sche41hse-hela_lac94–1.pdf> [accessed 2 December 2012]

International Organization for Standardization (ISO) 1997, *Mechanical Vibration and Shock—Evaluation of Human Exposure to Whole Body Vibration—Part 1: General Requirements*, ISO 2631–1:1997, ISO, Geneva.

—— 2001, *Mechanical Vibration: Guidelines for the Movement and the Assessment of Human Exposure to Hand-transmitted Vibration*, ISO 5349:2001, ISO, Geneva.

Jastreboff P.J. and Hazell J.W.P. 2004, *Tinnitus Retraining Therapy: Implementing the Neurophysiological Model*, Cambridge University Press, Cambridge.

Kroemer, K.H.E. and Grandjean, E. 1997, *Fitting the Task to the Human: A Textbook for Occupational Ergonomics*, Francis & Taylor, London.

National Acoustics Laboratories (NAL) 1998, *Attenuation and Use of Hearing Protectors*, 8th ed., NAL, Sydney.

Nelson, C.M. and Brereton, P.F. 2005, 'The European vibration directive', *Industrial Health*, vol. 43, pp. 472–9.

Safe Work Australia (SWA) 2010, *Occupational Noise-induced Hearing Loss in Australia: Overcoming Barriers to Effective Noise Control and Hearing Loss Prevention*, SWA, Canberra.

—— 2014, *Occupational Disease Indicators*, July 2014, SWA, Canberra

—— 2015, Media Release: Safer Work Australia Releases Workplace Vibration Guidance Material, 2 October, SWA, Canberra.

—— 2018, *Code of Practice—Managing Noise and Preventing Hearing Loss at Work*, SWA, Canberra.

—— 2019, *Model Work Health and Safety Regulations 2019*, SWA, Canberra.

Standards Australia 1988, Vibration and Shock—Hand-Transmitted Vibration: Guidelines for the Measurement and Assessment of Human Exposure, AS 2763: 1988, SAI Global, Sydney.

—— 2001, Evaluation of Human Exposure to Whole Body Vibration, Part 1: General Requirements, AS 2670.1:2001, SAI Global, Sydney.

—— 2002, Acoustics—Hearing Protectors, AS/NZS 1270: 2002, SAI Global, Sydney.

—— 2004a, Electroacoustics—Sound Level Meters, Part 1: Specifications, AS IEC 61672.1: 2004, SAI Global, Sydney.

—— 2004b, Electroacoustics—Sound Level Meters, Part 2: Pattern Evaluation Tests, AS IEC 61672.2:2004, SAI Global, Sydney.

—— 2005a, Occupational Noise Management, Part 0: Overview, AS/NZS 1269.0: 2005, SAI Global, Sydney.

—— 2005b, Occupational Noise Management, Part 1: Measurement and Assessment of Noise Immission and Exposure, AS/NZS 1269.1: 2005, SAI Global, Sydney.

—— 2005c, Occupational Noise Management, Part 2: Noise Control Management, AS/NZS 1269.2: 2005, SAI Global, Sydney.

—— 2005d, Occupational Noise Management, Part 3: Hearing Protector Program, AS/NZS 1269.3: 2005, SAI Global, Sydney.

—— 2013a, Voice Performance Requirements for Customer Equipment, AS/CA S004: 2013, SAI Global, Sydney.

—— 2013b, Mechanical Vibration—Measurement and Evaluation of Human Exposure to Hand-transmitted Vibration; Part 1: General Requirements, AS ISO 5349.1: 2013, SAI Global, Sydney.

—— 2013c, Mechanical Vibration—Measurement and Evaluation of Human Exposure to Hand-transmitted Vibration; Part 2: Practical Guidance for Measurement at the Workplace, AS ISO 5349.2: 2013, SAI Global, Sydney.

——2014, Occupational Noise Management, Part 4: Auditory Assessment, AS/NZS 1269.4: 2014, SAI Global, Sydney.

——2018, Acoustics—Description and Measurement of Environmental Noise, AS 1055: 2018, SAI Global, Sydney.

Wolfgang, R. and Burgess-Limerick, R. 2014, 'Using consumer electronic devices to estimate whole-body vibration exposure', *Journal of Occupational and Environmental Hygiene*, vol. 11, no. 6, pp. D77–81.

13. Radiation—ionising and non-ionising

Roy Schmid and Geza Benke

13.1 INTRODUCTION

Radiation is energy travelling in the form of either electromagnetic waves or particulates (high-speed particles). Radiation is ionising when it has sufficient energy to remove an electron from an atom of a molecule (Table 13.1).

Table 13.1 Radiation

	Electromagnetic wave	Particle
Ionising	Gamma and X-ray	Alpha, beta and neutrons
Non-ionising	Ultraviolet, visible, infrared, radiofrequency, extremely low frequency radiation	

Electromagnetic waves travel in air and vacuum at the speed (v) of light, 3.0×10^8 m/s, and are characterised by their wavelength (l, or λ, in metres) and frequency (f, in cycles per second, or hertz (Hz)). Wavelength and frequency are inversely proportional.

$$f = v/\lambda \qquad \text{(Equation 13.1)}$$

Radiation made up of high-speed particles travels at a range of speeds, and there are various different particle types (e.g. alpha radiation travels at approximately one-twentieth the speed of light and beta radiation travels at close to the speed of light).

The first half of this chapter discusses ionising radiation, its common sources and its quantification, health effects, measurement and controls. The second half discusses non-ionising radiation, which includes ultraviolet, visible, infrared, radiofrequency and extremely low-frequency radiation and static fields. Lasers are also included as a special application of visible, infrared and ultraviolet radiations.

13.2   IONISING RADIATION

There are two types of ionising radiation: electromagnetic and particulate. Figure 13.1 shows these.

Ionising radiation occurs naturally and also arises from human-made sources. It has many practical uses. Sealed radiation sources and x-rays are used in industrial and medical applications, and unsealed sources are used in medical, research and radiation-monitoring applications. Over-exposure to ionising radiation can cause a range of adverse health effects, including increasing the risk of cancer.

13.2.1   ELECTROMAGNETIC IONISING RADIATION

X-rays and gamma rays are electromagnetic radiation and can be thought of as photons. These types of radiation have the highest energy in the electromagnetic spectrum (see Figure 13.2). Unlike the lower-energy electromagnetic radiations, such as visible light, they have enough energy to cause ionisation of stable atoms.

13.2.2   PARTICULATE IONISING RADIATION

Highly energetic particulate forms of ionising radiation also exist. The most common types used in industrial and biomedical research applications are alpha and beta radiations. Neutron radiation, another form of particulate radiation, is typically used in specialist applications such as physics research, neutron imaging and environmental monitoring. The most common industrial application is moisture-density gauges.

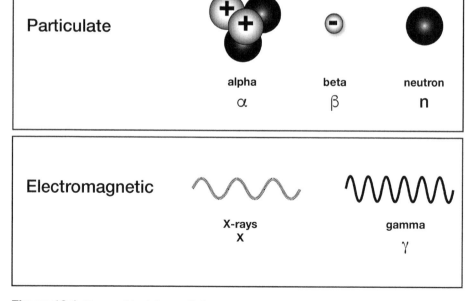

Figure 13.1 Types of ionising radiation

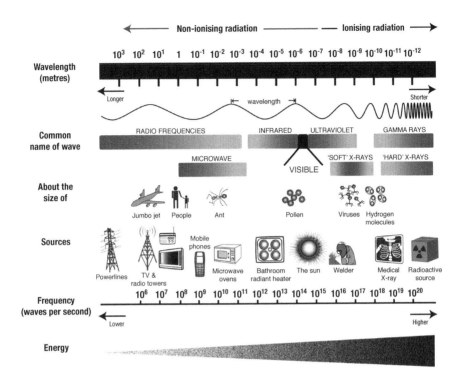

Figure 13.2 Electromagnetic spectrum
Source: Adapted from a diagram by ARPANSA, <www.arpansa.gov.au>.

13.2.3 INTERNATIONAL AND NATIONAL LEGISLATIVE FRAMEWORK

The International System of Radiological Protection (ISRP) has been developed by the International Commission on Radiological Protection (ICRP), which publishes dose limits and many other ionising-radiation safety publications. Many international standards on ionising radiation, such as the Basic Safety Series, are also developed by the United Nations' International Atomic Energy Agency (IAEA).

The ISRP defines three principles:

- *The justification of practice.* No practice involving exposure to ionising radiation should be adopted unless the benefit outweighs the potential or actual harm. In practical terms, when considering the use of ionising radiation in a procedure or activity, the benefits must outweigh the risks.
- *The optimisation of protection (the ALARA principle).* The size of individual doses, the number of people exposed and the likelihood of incurring exposure should be kept as low as reasonably achievable (ALARA), with economic and social factors being taken into account. ICRP has defined dose constraints and reference levels to assist in the application of this principle.
- *Dose limitations.* The total dose administered to any individual from regulated sources in planned exposure situations other than medical exposure of patients should not exceed the recommended dose limits (ICRP, 2007). Planned exposure situations involve the deliberate introduction and operation of radiation sources.

In Australia, the Australasian Radiation Protection and Nuclear Safety Agency (ARPANSA) provides guidance material such as radiation protection advice and guidelines. ARPANSA also regulates Commonwealth entities such as the Australian Nuclear Science and Technology Organisation (ANSTO) and the Commonwealth Scientific and Industrial Research Organisation (CSIRO). State and territory governments regulate non-Commonwealth entities such as hospitals, universities and industry, based on model regulations and recommendations from ARPANSA. Other countries have national regulators.

ARPANSA bases its regulations and recommendations on the ICRP's International System of Radiological Protection, as well as on documentation and data from the IAEA (to which Australia is a signatory). Australia's current dose limits are based on the dose limits recommended by the ICRP.

13.2.4   SOURCES OF IONISING RADIATION

Elements are characterised by the number of *protons* in the nuclei of their atoms. Atomic nuclei also contain *neutrons*. When two atoms of the same element contain different numbers of neutrons, the atoms are called *isotopes*. Examples of naturally occurring isotopes are: carbon-12 (six protons and six neutrons), carbon-13 (six protons and seven neutrons) and carbon-14 (six protons and eight neutrons). Some isotopes are stable and some are unstable—for example, carbon-12 and carbon-13 are stable and carbon-14 is unstable. An atom is unstable when there is an imbalance of energy or mass (for example, when it gains an extra neutron and does not have enough energy to bind its neutrons together). Radioactive decay occurs when an unstable atom emits radiation. Unstable atoms are commonly called radioisotopes or radionuclides.

Often, when radioisotopes decay, they do so in multiple stages, with a number of radioactive daughter products arising before a stable end-form is reached. This is called a *decay chain*.

Radioisotopes can occur naturally in the environment or can be created using equipment such as an x-ray tube, a cyclotron or a nuclear reactor.

13.2.4.1 Background radiation from natural sources
All human beings are exposed to natural background radiation.

- *Cosmic radiation* comes from outer space and the sun, but the earth's atmosphere acts as a shield, so that at sea level the cosmic radiation level is very low.
- *Naturally occurring radioactive material (NORM)* is distributed in the soil, air and water. It ends up in human beings, in food, in building materials and in other objects. For example, potassium-40 (gamma emitter) is naturally elevated in bananas because they have a high total potassium level. Uranium-238 and thorium-232 (mainly gamma and alpha emitters) are found in varying concentrations in soil and rocks.

Another example of elevated NORM levels is seen in the Darling Scarp, in south-western Western Australia. The geology is particularly rich in the alpha-radiation emitters uranium-235 and thorium–232, which results in an average background radiation dose from terrestrial sources that is twice as high as in other areas of Australia.

Some other areas of the world have high natural levels of NORM. Brazil's black beach sands contain a range of naturally occurring radioisotopes. These can result in an average background radiation dose from terrestrial sources up to 100 times higher than elsewhere (Gonzales and Anderer, 1989).

The naturally occurring radioisotope radon-222 gas (alpha emitter) and its short-lived radioactive daughter products are generated when radium-226 within building materials and soil decays. Radon gas can accumulate inside poorly ventilated buildings, especially in areas below the ground (basements or cellars), and can readily be breathed in and its progeny particulates deposited.

Radon is recognised by the IARC (1988) as a cause of lung cancer. In 2017, the ICRP re-evaluated its estimates of lung risk for radon and its progeny, almost doubling the risk. Radon effective dose coefficients are now calculated using biokinetic and dosimetry models with specific radiation and tissue weighting factors. As a result, radon is now seen as a significant contributor to our natural background radiation dose (ICRP, 2017).

Radon levels are generally much lower in Australia than in many other countries for a range of reasons, including geology, climate (no snow cover) and building styles (fewer residential basements). There are only a few areas where radon levels are likely to build up. These include underground and uranium mines and some caves in Victoria, Tasmania and New South Wales that are popular tourist sites. These do not significantly increase the dose for casual visitors, but could significantly raise doses for cave-tour guides (ARPANSA, 2018; Solomon et al., 1996). An Australian action level for residential indoor radon is 200 Bq/m³, corresponding to an effective dose of approximately 10 mSv per year (APRANSA, 2018).

13.2.4.2 Background radiation from human-made sources

A human-made source of ionising radiation can be generated by either concentrating naturally occurring radioisotopes or by using special apparatus such as an x-ray tube or a cyclotron.

Large-scale releases of ionising radiation, such as occurred in the Chernobyl nuclear accident, the atomic bombing of Japan and various nuclear tests and accidents, have added to the background radiation to which all humans are exposed, though the additional exposure from these sources is small (UNSCEIR, 2008). Note that there are some naturally occurring nuclear events.

NORM may be found in minerals, oil and gas. Although it may initially occur at low levels, the refined product, the process line and waste may contain higher levels of radio-activity. Examples include thorium-232 and uranium-238 in the mining and processing of mineral sands, and radium-226 (and associated radon daughters) in oil and gas production (Cooper, 2005). Contamination of equipment used to process NORM can become a problem for disposal and personal exposure during its operation and at demolition.

13.2.4.3 Categories of human-made radiation sources

There are three physical forms of human-made ionising radiation:

- *Radiation apparatus, or a radiation generator*, is equipment that generates radiation. Examples include x-ray machines, cyclotrons and neutron generators. (ICRP now uses the term radiation generator.)

- *Sealed sources and sealed-source apparatus* refer to any quantity of radioisotope whose physical form is so enclosed as to prevent the escape of any of the radioisotope (but not the radiation). Some examples are:
  - electron capture detectors in gas chromatographs, which may use nickel-63 (beta emitter)
  - density, level and thickness gauges, which use a variety of radioactive sources such as americium-241 (alpha emitter), caesium-137 and cobalt-60 (both gamma emitters)
  - bore-hole loggers which use americium-241 in beryllium to cause fission, producing neutrons which are then used to measure moisture
  - smoke detectors, which may use foil containing americium-241 (alpha emitter).
- *Unsealed sources* are usually in a liquid or powder form, and may readily escape into the environment if they are not carefully contained. They are most often used for nuclear medicine and scientific research. Some examples are:
  - iodine-125 (gamma emitter) for labelling peptides in research
  - sulfur-35 (beta emitter) for labelling steroids in research
  - technetium-99m (gamma emitter) for imaging studies in nuclear medicine
  - phosphorus-32 (high-energy beta emitter) and phosphorus-33 (low-energy beta emitter) for labelling DNA in research.

It is useful to classify sources encountered within the workplace based on their form, because these forms influence the type of hazard presented by the source.

13.2.5 EXPOSURE CATEGORIES

Everybody is exposed to background radiation, but ionising radiation is widely used in medical applications and research, as well as industry. For this reason, the ICRP defines three exposure categories: public, medical and occupational.

- *Public exposure* is primarily from background sources such as those discussed above, and varies based on the local environment and dietary habits. The annual average dose from public exposure to background radiation in Australia (excluding medical and work-related sources) is 1500 µSv/year plus an increase in radon exposure (ARPANSA, 2012b, 2015, 2018).
- *Medical exposure* is experienced by all persons who are treated using ionising radiation. Radioisotopes may be used for treating cancer, because at high doses their ionising radiation is effective at destroying rapidly dividing cells, and because it can be directed to selectively irradiate specific tissues or organs. For example, a beam of external gamma rays emitted by a cobalt-60 source can be aimed at the site of an internal tumour, or a small iodine-125 source can be implanted in the tongue to treat tongue cancer. Radioisotopes and x-ray generators are used in a wide range of diagnostic and treatment procedures, such as the injection of technicium-99m (gamma emitter) for brain scans, or iodine-131 (gamma emitter) for thyroid scans.
- *Occupational exposure* occurs among people who are required to work with sources of ionising radiation. Work-related doses vary depending on the work. Table 13.2 shows a comparison of work-related annual doses for workers in various occupations in

Table 13.2 Typical average work-related radiation doses for workers in various occupations in Australia

Occupation	Average whole-body dose (μSv/year)
Radiologists in large hospitals	108
Dentists in private practice	12
Uranium miners	1125
Users of radioactive tracers in research	31
Undergraduate students in tertiary education	19
Note: the occupational exposures are in addition to background radiation (Morris, Thomas and Rafferty, 2004)	
Background radiation	
Australia – average natural radiation (incorporating an increase in radon exposure of 300 μSv/year)	1800
World – Natural radiation	1000 to 13000

Source: ARPANSA (2015, 2018).

Australia. The majority of these occupational doses are significantly below the annual average dose from public exposure in Australia.

Air crew are considered to be occupationally exposed to ionising radiation because their cosmic radiation exposure increases as the atmospheric density decreases at altitude. Air crew thus have a significantly higher exposure over the course of a year than their grounded counterparts. Crews who routinely fly domestic routes receive about double the annual average background dose of the general public, while crews on international routes, who fly higher and stay in the air longer, receive more than three times this dose (ARPANSA, 2012b).

13.2.6 PROPERTIES OF IONISING RADIATION

13.2.6.1 Interaction with matter

Alpha and beta particles, neutrons, gamma rays and x-rays are known as ionising radiation because they have sufficient energy to ionise the medium through which they travel. These types of radiation remove electrons from atoms to produce positive ions.

The charge carried by *alpha* and *beta* particles allows them to interact with target atoms. Alpha particles have a double positive charge and beta particles have a negative charge. When charged particles come close to one another, they are either repulsed if they have the same type of charge or attracted if they have opposite charges. Whenever these attractions or repulsions occur, energy is transferred from alpha and beta radiations to target atoms, causing ionisation.

- Gamma rays and x-rays are electromagnetic waves and do not have any mass or charge, so they cannot directly ionise other atoms via interaction with charged fields. However, through various complex interactions, they can transfer their energy to an atom's orbital electrons and thereby ionise the atom.
- Neutrons have mass but they have no charge. Like gamma rays and x-rays, they do not cause ionisation by direct interaction with an atom's charged electron fields. Instead, they cause ionisation indirectly, by interacting with the nucleus of the atom. By this process, neutrons can cause stable materials to become radioactive. Neutrons are the only type of ionising radiation able to do this.

13.2.6.2 Penetrating vs ionising ability

The penetrating and ionising abilities of the different types of ionising radiation vary. Penetration ability refers to the distance a type of radiation can travel, and ionising ability refers to the degree of ionisation it can cause as it travels. The greater the penetrating ability the greater the risk of interaction with the human body becomes, and as ionising ability increases so does biological damage.

- *Alpha particles* (helium nuclei) consist of two protons and two neutrons tightly bound together. They are heavy and slow moving (around one-twentieth the speed of light), and have a positive charge of 2. They cause a lot of ionisation over a short distance and usually travel only a few centimetres in air. Their large mass and charge mean that alpha particles interact easily with the matter through which they pass, but each interaction reduces the particles' energy until they no longer have enough energy to cause ionisation.
- *Beta particles* are electrons that travel at close to the speed of light. They have a very low mass and a negative charge of 1. Because they have half the charge and much less mass than alpha particles, they cause less ionisation over the same travel distance and typically travel further before they use up their energy. Beta particles are emitted from the nucleus during radioactive decay with a spectrum of energies ranging from close to zero to a maximum energy that is characteristic of the radioisotope. Low-energy beta radiation such as that from tritium (hydrogen-3) travels less distance in air than most alpha particles. Medium-energy beta particles typically travel up to 3 metres in the air, but the distance varies according to their energy. Phosphorus-32 is a high-energy beta emitter, and the highest-energy beta particles it emits may travel up to 7 metres in air.
- *Neutron particles* are fast-moving neutrons with a spectrum of energies and speed. They have a variable ionising capability which is dependent on their energy. They may be more ionising than alpha and beta particles, gamma rays or x-rays. They may travel many metres in air before stopping completely. They are more penetrating than alphas and betas, as they do not interact with the charged fields of atoms as they pass through matter.
- *X-rays and gamma rays* are the least ionising of all types of ionising radiations. They are generally the most penetrating because they do not interact with the charged fields

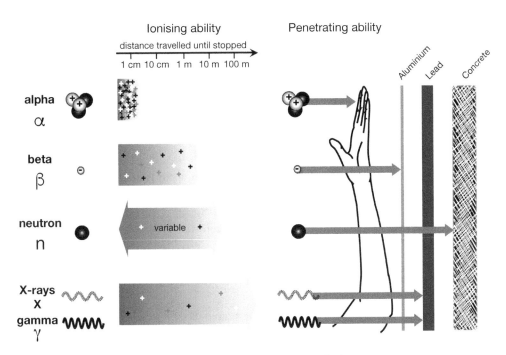

Figure 13.3 Penetrating and ionising ability of ionising radiation

of atoms as they pass through matter. They will travel very large distances through matter—they typically travel many metres in air.

Penetrating and ionising abilities for all types of ionising radiation are summarised in Figure 13.3.

13.2.7 EXTERNAL AND INTERNAL HAZARD

Unlike many other hazards, ionising radiation can present a risk when it is inside the body, when it is outside the body and even when it is at some distance from the body. The degree of internal and/or external hazard is different for the different types of radiation because it depends on their penetrating and ionising ability.

13.2.7.1 External hazard

An external radiation hazard is present when a source of ionising radiation is located outside the body. A person may be irradiated without being aware of it, and without coming into direct contact with the source. The greatest external radiation hazards are x-rays, gamma rays and neutrons because of their ability to travel large distances and penetrate matter, including the human body. Alpha particles and low-energy beta particles, such as those from tritium, do not represent a significant external radiation hazard, even if they are close to the body, as they are unable to penetrate the outer layer of skin. Medium- and high-energy beta particles do present an external hazard, as they are more energetic and penetrating—for example, phosphorus-32 may cause skin burns and eye damage.

13.2.7.2 Internal hazard

Alpha and beta radiation represent the greatest internal radiation hazard when they are emitted inside the body because they are capable of causing intense ionisation in a local area. X-rays, gamma rays and neutrons cause less ionisation, but because they are more penetrating a large proportion of the radiation will pass out of the body without causing ionisation.

Radioisotopes can enter the body and cause an internal radiation exposure by three different exposure pathways:

- *Ingestion* is the most common means, and can occur if items in the work area become contaminated. This contamination can ultimately end up being transferred to the hands and then to items put into the mouth.
- *Inhalation* of radioactive vapours, gases, aerosols or dusts can occur depending on the chemical form of the radioisotopes or the processes in which the radioisotopes are used. For example, unbound radioactive iodine presents a particular danger owing to its volatility, while some equipment and processes, such as centrifuges, can generate a radioactive mist.
- *Skin uptake* occurs when radioactive materials penetrate intact skin under favourable conditions. Skin uptake rates are determined by contact time, the chemical form of the radioisotope and whether the radioisotope is mixed with other chemicals, such as organic solvents. Where there is a wound, the uptake rate is higher.

13.2.8   QUANTIFYING RADIATION

To quantify ionising radiation, it is necessary to know the:

- activity
- radiological half-life
- energy.

13.2.8.1 Activity

The activity tells us how much radiation is being emitted. The international standard (SI) unit for activity is the becquerel (Bq), which is equivalent to one disintegration, or nuclear transformation, per second. The non-SI unit of activity is the curie (Ci), which is still in active use. The relationship between the becquerel and the curie is shown in Equation 13.2.

$$1\ Ci = 3.7 \times 10^{10}\ Bq \qquad \text{(Equation 13.2)}$$

13.2.8.2 Radiological half-life

The radiological half-life of a radioisotope describes its rate of decay. As radiation is emitted, the number of unstable atoms decreases exponentially, so the activity of a set amount of radioactive material decreases with time. This pattern can be used to predict how long the decay process might take, as shown in Figure 13.4.

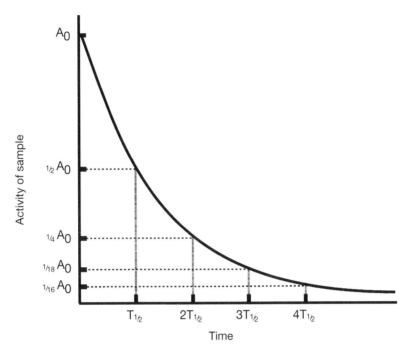

Figure 13.4 The pattern of radioactive decay

Radiological half-life ($T_{1/2}$) is the time taken, on average, for half of a given amount of radioisotope to undergo radioactive decay. The relationship between activity and radiological half-life is defined by equations 13.3 and 13.4.

$$A_t = A_0 e^{-\lambda t}$$

$$\text{Or } A_t = \frac{A_0}{2^n} \qquad \text{(Equation 13.3)}$$

where:
A_t = activity at time elapsed (t)
A_0 = original activity at $t = 0$
λ = decay constant
n = number of half-lives
$n = t / T_{1/2}$
$T_{1/2}$ = radiological half-life

$$T_{1/2} = \frac{\ln 2}{\lambda} \text{ or } = \frac{0.693}{\lambda} \qquad \text{(Equation 13.4)}$$

where:
t = time elapsed
λ = decay constant

If the starting activity of the radioisotope (A_0) is known, the half-life can be used to determine how much radiation will still be present after a given time—that is, for how long the radioisotope will continue emitting radiation. Half-life varies depending on type of radioisotope. For example, phosphorus-32 has a half-life of 14.29 days, while radium-226 has a half-life of 1600 years. A rule of thumb is that after ten half-lives the radiation emitted by the material will be about one thousand times lower than it was at the start.

13.2.8.3 Energy

The SI unit for energy is the electron volt (eV). All types of ionising radiation have either a specific energy or a spectrum of energies. The amount of energy affects the relative penetrating and ionising ability of each ionising radiation type. Typical energies for various types of ionising radiation are shown in Table 13.3.

Table 13.3 Energies of ionising radiation

Type of radiation	Typical energy range
Alpha	3–9 MeV
Beta*	0–3 MeV*
Gamma**	10 keV and 10 MeV
X-rays***	A few eV to several MeV
Neutrons	0–10 MeV

*   Each radioisotope emits beta radiation with a spectrum of energies. The most probable energy
      is approximately one-third of the maximum energy.
**   Have a characteristic energy.
*** Have a spectrum of energy.
Source: Derived from Simon (2007).

13.2.9   *QUANTIFYING EXPOSURE AND DOSE*

To measure the amount of ionising radiation absorbed and the amount of harm it is likely to inflict, it is necessary to understand the following concepts:

- absorbed dose and exposure
- equivalent dose and radiation weighting factor
- effective dose and tissue weighting factor
- biological half-life and effective half-life
- committed dose.

13.2.9.1 Absorbed dose and exposure

Absorbed dose is the primary quantity used to define the energy deposited in matter due to radiation exposure. Historically, exposure was defined as the amount of ionisation that *gamma* or *x-rays* produce *in air*. Exposure was formerly measured in roentgens (R); now the SI unit coulomb per kilogram of air (C/kg) is used. However, in the latest version of the Basic Safety Series, exposure is defined as the state of being subject to irradiation (IAEA, 2011).

Absorbed dose (D) is a measure of the energy absorbed per unit mass of *any matter*. For a given type of ionising radiation of defined energy, the absorbed dose increases with the density of the absorbing matter. The SI unit is joule per kilogram (J/kg), or gray (Gy). The gray is a large unit, so normally milligray (mGy), microgray (µGy) and nanogray (nGy) are used.

Absorbed dose indicates how much energy was absorbed, but not how much ionisation has occurred. Knowing the amount of ionisation, however, is crucial to quantifying the biological damage caused by the radiation exposure.

13.2.9.2 Equivalent dose and radiation weighting factor

The equivalent dose is an average measure of the dose or doses received by a particular mass of tissue from all types of radiation.

Different types of radiation have different relative biological effectiveness, owing to their different ionising abilities. The ICRP defines radiation weighting factors (w_R) to account for these differences (see Table 13.4).

Table 13.4 Radiation weighting factors*

Radiation type	Radiation weighting factor, w_R
Gamma and x-ray	1
Beta	1
Alpha	20
Neutrons	A continuous curve as a function of neutron energy*

* Based on ICRP (2007). For a full list of radiation weighting factors and the energy curve for neutrons, refer to ICRP (2007).

Equivalent dose (H_T) is the sum of the absorbed dose times the radiation weighting factor.

$$H_T = \Sigma \, (w_R \, D) \qquad \text{(Equation 13.5)}$$

The radiation weighting factor is a dimensionless number (i.e. it has no unit), and it is based on experimental data. Like the absorbed dose, the equivalent dose is measured in J/kg, but to distinguish it from absorbed dose its unit is given the name 'sievert' rather than 'gray'.

13.2.9.3 Effective dose and tissue weighting factor

Equivalent dose limits apply to a tissue or an organ. When considering the risk of whole-body health effects such as cancer, it is important to know that some tissues and organs are more likely to be damaged by ionising radiation than others. Tissue weighting factors (w_T) represent the relative amount of damage likely to be caused to various types of tissue based on their sensitivity (Table 13.5). Effective dose (E) is the sum of the equivalent doses multiplied by the tissue weighting factors and is also expressed in sievert (Sv).

$$E = \Sigma\ (H_T\, w_T) \qquad\qquad \text{(Equation 13.6)}$$

Tissue weighting factors are defined by the ICRP after examination and modelling of data from many different human epidemiological studies as well as other research, such as animal studies. It is important to stay abreast of the current ICRP recommendations, as the tissue weighting factors can change, based on new evidence.

Table 13.5 Tissue weighting factors*

Organ/tissue	Number of tissues	Tissue weighting factor, w_T	Total contribution
Lung, stomach, colon, bone marrow, breast, remainder tissues**	6	0.12	0.72
Gonads (testes and ovaries)	1	0.08	0.08
Thyroid, oesophagus, bladder, liver	4	0.04	0.16
Bone surface, skin, brain, salivary glands	4	0.01	0.04

*   Based on ICRP (2007).
**   Remainder tissues are adrenals, extrathoracic tissue, gall bladder, heart, kidneys, lymphatic nodes, muscle, oral mucosa, pancreas, prostate, small intestine, spleen, thymus, uterus/cervix.

13.2.9.4 Biological half-life and effective half-life

The biological half-life of a radioisotope is the time it takes the body to eliminate half of an intake of radioisotope by natural biological processes such as urination. The effective half-life combines the radiological half-life and the biological half-life. The effective half-life is the time taken for half of the radioisotope to be removed from the body by both radioactive decay and biological processes.

13.2.9.5 Committed dose, dose coefficient and limits of intake

The committed dose is the exposure that will be received from a radioisotope that has entered the body. Once a radioisotope is inside the body, it will irradiate the tissues until it either decays (radiological half-life) or is excreted (biological half-life). To allow calculation of the committed dose, the ICRP has recommended coefficients or doses per unit of intake of a radioactive substance. Different dose coefficients exist for different pathways of exposure, and they take into account both the physical half-life of the radioisotope (radiological half-life) and its biological retention rates (biological half-life). The committed dose is calculated for a 50-year period for adults and a 70-year period for children, and then—per ICRP recommendations—assigned to the year in which the intake occurred. Both committed equivalent doses and committed effective doses can be calculated.

ICRP dose coefficients can also be used to derive the secondary operational limit called the annual limit of intake (ALI) for inhalation or for ingestion for occupational exposure (ICRP, 2011). The ALI is the amount of a radioisotope that, when taken into the body, would produce a committed effective dose equal to the annual effective dose limit (see section 13.2.11). As ALIs are calculated limits, it is important to understand the underlying assumptions when using an ALI from another source such as a website or an older national standard. The ALI must be based on the appropriate dose limit and the latest ICRP publications.

13.2.10 HEALTH EFFECTS OF IONISING RADIATION

The health effects of ionising radiation fall into two main categories:

- *Harmful tissue reactions* (also called deterministic effects) occur when enough cells have been killed or damaged to affect the function of an organ or tissue. These reactions are a consequence of the death or malfunction of cells following high doses. In its 2007 recommendations, the ICRP noted that no tissues show loss of function when exposed to either a single acute dose or a fractionated annual dose in the absorbed dose range up to 100 mGy (ICRP, 2007).
- *Stochastic effects* may or may not occur, but their likelihood increases with the exposure. These effects are due mainly to damage to the DNA, and the resulting effects are primarily cancer and hereditary disorders.

Additional effects that need to be considered are the risk to the unborn child and diseases other than cancer.

13.2.10.1 Radiation sickness and death

A short-term dose of approximately 1 Gy may result in radiation sickness (ICRP, 1990). This manifests as symptoms of nausea, vomiting, rapid pulse and fever within a few hours of the exposure. Doses of this magnitude are not common, and historically have usually been associated with accidents.

Death from a short-term dose of ionising radiation occurs only at very high whole-body doses. For example, a short-term dose of 3–5 Gy to the whole body results in damage to the bone marrow and death within a few months. Doses of 5–15 Gy to the whole body can cause damage to the gastrointestinal tract and lungs, and death within a few weeks. Short-term whole-body doses exceeding 15 Gy damage the central nervous system and cause death within a few days (ICRP, 1990).

13.2.10.2 Skin effects

Skin effects of radiation exposure include hair loss, reddening of the skin, burns and cancer. Reddening and damage to the outer layer of skin, called erythema, occur with short-term absorbed dose to the skin of around 3–5 Gy. The initial reddening lasts only a few hours, but may be followed several weeks later by a wave of deeper and more prolonged reddening (ICRP, 1990). Erythema is unlikely in modern workplace settings, where ionising radiation is well controlled.

13.2.10.3 Effects on the eyes

Cataracts can form on the lens of the eye when ionising radiation breaks down the dividing cells in the lens epithelium, although this is unlikely where ionising radiation exposure is well controlled. Cataracts may be induced by a single, relatively large short-term dose or by a series of smaller doses—particularly of beta radiation. The threshold is now thought to be 0.5 Gy, as opposed to the 5 Gy that was defined as the threshold in the past (ARPANSA, 2011).

13.2.10.4 Reproductive system and hereditary effects

Ionising radiation can seriously affect the reproductive system in both sexes, as the reproductive cells are sensitive to radiation. In males, a single acute dose of 0.15 Sv can cause temporary sterility, and a single acute dose of 3.5–6 Sv may cause permanent sterility. In females, a single acute dose of 2.5–6 Sv may cause permanent sterility (ICRP, 2007). Doses high enough to result in sterility would also cause other health effects such as radiation sickness, but are unlikely in settings where exposure to ionising radiation is well controlled.

In addition to direct effects such as sterility, there has been concern about whether radiation can cause hereditary damage in the form of genetic disorders being passed to future generations. The summary of the current state of knowledge is that there is no direct evidence of heritable disease having been caused by exposure of human parents to radiation, but there is compelling evidence of this from experimental studies of animals (ICRP, 2007).

13.2.10.5 Effects on the unborn child

The unborn child (embryo and foetus) is more susceptible to ionising radiation than an adult. The effect of such radiation depends entirely on the stage of development. Early in a pregnancy, a high short-term dose may cause death of the foetus and a miscarriage, but absorbed doses of less than 100 mGy are very unlikely to cause any damage. When the organs are being formed, exposure to ionising radiation may cause deformities in that organ, though the current evidence appears to indicate that the threshold for malformations is about 100 mGy (ICRP, 2007). In the later stages of a pregnancy, the IQ of the unborn child may be adversely affected by exposure to high doses of ionising radiation; doses below 300 mGy pose a very low risk of damage (ICRP, 2007). Radiation exposure must be controlled carefully, even in the most modern facilities, to ensure that unborn children are not affected. Cancer risk from exposure in utero is similar to the risk of an irradiation in early childhood—that is, three times the population risk.

13.2.10.6 Cancer

The association between radiation exposure and cancer is based mostly on data from situations where relatively high exposures have occurred (e.g. in Japan, among atomic bomb survivors). The association between low doses of radiation and cancer is complicated and difficult to prove because of the relatively high rate of cancer within the general population. This is a stochastic effect, where the risk of cancer increases as the radiation dose increases. Some cancers known to be induced by exposure to ionising radiation include:

- leukaemia in atomic-bomb survivors
- thyroid cancer due to exposure to radioactive iodine

- lung cancer in uranium miners as a result of inhaling radioactive dust
- bone sarcoma in workers who applied radium-containing paint to the faces of luminous clocks and dials. The workers used their lips to get a fine point on their brush and ingested radium in the process.

13.2.10.7 Diseases other than cancer

There has been some evidence in highly exposed groups (e.g. Japanese atomic bomb survivors) that ionising radiation may be a risk factor for non-cancer diseases such as heart disease, stroke, digestive disorders and respiratory disease. However, the data have been judged insufficient to extrapolate any of these effects to lower dose exposures—that is, below 100 mSv (ICRP, 2007, p. 57).

13.2.11 DOSE LIMITS (EXPOSURE STANDARDS)

Exposure standards that apply to exposure to ionising radiation from external and internal sources are called dose limits and are defined in Table 13.6. Dose limits for use in Australia are established by ARPANSA based on international recommendations and are adopted into legislation by each state or territory (see Section 13.2.3). Dose limits are defined for occupational exposure as well as for members of the public. An arbitrary safety factor is applied to the public dose limits to take account of the sick, the elderly and the very young. These groups are thought to be more sensitive to radiation than workers exposed to it on the job.

Table 13.6 Dose limits for planned exposure to ionising radiation

Application	Dose limit	
	Occupational	Public
Effective dose	20 mSv per year, averaged over five consecutive years with no more than 50 mSv in any single year*	1 mSv in a year** The limit for the unborn child is 1 mSv from the declaration of pregnancy
Annual equivalent dose:		
Lens of the eye***	20 mSv***	15 mSv
Skin (averaged over 1 cm² of skin)	500 mSv	500 mSv
Hands and feet	500 mSv	—

*   Refer to the ARPANSA publication for all of the provisos.
**   A higher value is allowed in special circumstances, provided the average over five years does not exceed 1 mSv.
*** The dose limit for the lens of the eye has been significantly reduced by the ICRP based on epidemiological data.
Note: No statement has been made about the public dose limit for the lens of the eye.
Sources: ARPANSA (2002, 2011, 2016).

The effective dose limit has been set to minimise the risk of cancer and other effects. The additional equivalent dose limits, for the lens of the eye, the skin and the hands and feet, have been set to ensure that individual tissue thresholds are not exceeded.

13.2.12 DETECTING IONISING RADIATION

Ionising radiation monitors can be categorised according to:

- what they measure:
  - count rate
  - dose rate
  - dose
- the principle of detection:
  - gas-filled detectors
  - scintillation detectors
  - semiconductor detectors
  - thermoluminescent detectors.

13.2.12.1 Types of measurement
Instruments for monitoring ionising radiation provide three main measurements:

- *Count rate* (counts per second). This does not mean disintegrations per second (Bq) unless the device has had appropriate calibration, so count rate monitors typically provide a relative measure of the radiological activity that is present. They are particularly useful for conducting contamination surveys and finding sources.
- *Dose rate* (dose per unit time). Users need to know the type of radiation dose for which the meter is calibrated. Dose-rate monitors are particularly useful for estimating potential exposures for different exposure scenarios. They are also useful for checking surface dose rates for radioisotope waste disposal and for labelling packages for transport.
- *Equivalent and effective doses.* These measures are provided by dosimeters.

Some monitors are multi-functional—for example, they provide both a count rate and a dose rate—while many digital dosimeters measure dose rate. Some examples are shown in Figure 13.5.

13.2.12.2 Types of instrument
Instruments can be classified based on the principle of detection they utilise:

- *Gas-filled detectors* have a gas-filled chamber of known volume with conducting electrodes connected to an electronic circuit to collect the charges created by ionisation events that occur within the chamber. The three main types in common use are Geiger–Müller (GM) detectors, ionisation chambers and proportional counters. In an ionisation chamber, one ionisation event gives rise to one pulse that can be detected. In a GM detector, each ionisation event is multiplied, which causes a cascade of pulses. The proportional counter has some multiplication for each event but does not produce a cascade. This makes the GM more sensitive in general than both the ionisation chamber and the proportional counter. The downside of the GM detection method is

Figure 13.5 Instruments for measuring ionising radiation

that the electronics can detect only one cascade at a time, which can result in overload or dead time in high-radiation areas. The proportional counter can be set up to distinguish between ionisation events caused by alpha and beta particles.

- *Scintillation detectors* contain a material that emits light when it absorbs ionising radiation energy. This light can then be detected using an appropriate electronic device, such as a photomultiplier tube. The scintillant material is often a crystal such as sodium iodide or a powder such as zinc sulfide. The type of scintillant material has to be matched to the type of radiation that needs to be detected. A variant of the scintillation detector is a laboratory-based device in which the scintillant is a liquid that is physically mixed with a sample.
- *Semiconductor detectors* use a semiconductor material that generates a charge when exposed to gamma radiation. Coupled with the correct electronics, this type of detector can be used to analyse the energy spectrum of the radiation.
- *Thermoluminescent detectors* use a thermoluminescent material that changes to a metastable excited state when exposed to ionising radiation. When this material is warmed, the material releases detectable light.

13.2.12.3 Choice of a monitoring instrument
- *Energy response curves.* No single instrument is suitable for measuring all types and all energies of radiation; hence it is important to consider the type of radiation and the energy response curve for any detector in relation to the energy of the radiation that needs to be measured. The energy response curve should be available from the manufacturer. The *GM detector* is most responsive at lower energy levels, but only for a limited energy range. The *scintillation counter* is less responsive, but to a wider range of radiation energies, and the type of radiation detected is highly dependent on the choice of scintillant material. Finally, the *ionisation chamber* is fairly insensitive to an even

wider range of energies. It is important to know the type of radiation to be measured, and its energy, so as to choose the most appropriate detector.

- *Limit of detection and sensitivity.* The required limit of detection of a monitor needs to match its intended use—for example, a count rate meter for detecting contamination in a research laboratory needs a much lower limit of detection than a dose rate meter.

The characteristics of the monitor can affect its sensitivity and limit of detection. Having a large surface area on a detector can increase the sensitivity of the monitor, but increasing the volume of the detector can decrease the sensitivity of the monitor as a result of averaging effects, especially where the radiation is emitted in beams or the source size is small. Saturation of the detector can lead to a false low response. The response time of the detector is also important, especially when measuring pulsed x-ray fields.

13.2.12.4 Calibration and performance checking

Performance checking should be undertaken every time a radiation measuring instrument is used. The battery should be checked for sufficient energy and the detector should be inspected for damage. The detector should be checked against a suitable source (preferably of the same radiation being monitored) and its response noted. This should be logged and referred to over time as a benchmark for any deterioration in detector performance.

Calibration is a more in-depth examination of the performance of the radiation monitor, and requires a specialist calibration service such as ARPANSA. It should be undertaken at regular intervals and any time the radiation monitor is repaired, knocked, damaged or otherwise altered.

13.2.12.5 Monitoring techniques

The technique used for monitoring ionising radiation depends on the aspect of ionising radiation being examined. The main techniques are:

- area surveys with direct read-out monitor—contamination survey and area survey
- area monitoring with wipe testing
- external exposure assessment—passive and active dosimetry
- internal exposure assessment—personal contamination surveys, air monitoring, external monitoring and bioassay.

Area surveys with direct read-out monitor

Area monitoring is necessary to ensure that appropriate control strategies are implemented to minimise exposure. Area monitoring for ionising radiation must be done at various heights over a wide area because ionising radiation can diffract and scatter. An instrument measuring count rate or dose rate is most commonly used. The area to be monitored should be mapped out on a grid pattern and measurements taken at each grid point. For the sake of personal safety, choose an instrument with an adequate response time and commence monitoring at the farthest point from the ionising radiation source. Using this method of monitoring, it is possible to quickly establish areas where no person should enter or where exposure time should be limited.

Surfaces should be checked for contamination before and after any procedure that uses unsealed sources of ionising radiation, or if the integrity of a sealed source is questioned.

Area monitoring with wipe testing

Wipe testing is used to identify surfaces contaminated with radioisotopes. It is particularly useful when:

- the radioisotope has very low energy, such as tritium, as wipe testing is the only effective way to detect surface contamination
- the contamination is very dilute on the surface
- there is a high radiation background that may interfere with the survey instrument—for example, where the integrity of a sealed source is being checked or surfaces in a radiation-source storage area are being tested for contamination.

Wipe testing is done by wiping a moistened filter paper across the surface to be tested. This concentrates any radioactive contamination on the filter paper. The filter paper can then be monitored using a survey instrument (usually a calibrated count rate meter), or counted using laboratory instrumentation such as a liquid scintillation counter. Wipe testing can also be done to determine whether radioactive contamination is fixed or removable, and will provide information about whether clean-up or shielding is required.

External exposure assessment

Dosimetry is the measurement of the radiation dose to either the whole body or certain parts of the body.

Passive dosimetry can be performed by a number of monitors. Typical passive dosimeters are thermoluminescent detector badges (TLD badges) or optically stimulated luminescence (OSL) monitors. Other older forms of passive dosimeters that utilise other detection methods, such as a charged quartz fibre or film badge dosimeters, are also available. A TLD badge can be used to monitor for beta, gamma and x-ray radiation, while the OSL monitor (ARPANSA, 2018) is suitable for gamma and x-rays.

The whole-body dose is assessed by using a single badge at the point of expected highest exposure (usually the waist or the chest) for a period of four to twelve weeks. At the end of this period, the badge is sent away for analysis. If a lead apron is worn, then the badge should be positioned beneath it at the waist or chest to ensure that it represents the actual dose. Passive dosimeters and analytical services are available from ARPANSA and several other suppliers.

Equivalent doses to parts of the body can be measured using small TLD badges. For example, the dose to the hands can be measured using a TLD that is wrapped around the fingers. A person's radiation dosimetry results should be recorded and provided to the regulatory authority. In Australia, ARPANSA maintains a dose register of personal annual radiation doses: the Australian National Radiation Dose Register.

Passive monitoring results indicate what dose of radiation has been received but do not provide any real-time data to help choose interventions or controls.

Active electronic dosimeters can be used to measure dose or dose rate, and can provide immediate feedback on dose rates associated with specific tasks. Dosimeters of this type often contain very small GM tubes. Practitioners planning to use an active electronic dosimeter should ensure that it will detect the types of radiation that are present and that it complies with the standards specified by the International Electrotechnical Commission

(Voytchev et al., 2011). To avoid spurious results electronic dosimeters should be kept away from other electronic devices like mobile phones.

Internal exposure assessment

- *Personal contamination surveys.* Whenever radioisotopes are handled, it is necessary to routinely monitor the hands, clothes and body to check for contamination. This type of contamination can lead to intake of the radioisotope and receipt of a committed dose. Personal contamination surveys are usually done using a direct read-out instrument of the kind used for an area contamination survey. Where contamination is likely, a clean person (someone who has not recently had direct contact with radioisotopes) should survey a person who is potentially contaminated. The surfaces of laboratory coats, cuffs, hands and feet are the usual sites of personal external contamination. Many common laboratory contamination monitors have a stand or clip that allows researchers to use them in hands-free mode to check the hands during experiments.
- *Air monitoring.* Monitoring for airborne contamination should be undertaken where there is a risk of inhaling the radioisotope, such as a radioactive dust or radioactive iodine (because it can sublime—that is, change directly to vapour). Monitoring for airborne radioactive material requires specialist methods and equipment, and the H&S practitioner will probably need to seek expert advice.
- *External monitoring.* The intake of gamma ray- or x-ray-emitting radioisotopes can often be estimated directly on the basis of external monitoring of either the whole body, or specific organs or tissues. For example, when using iodine radioisotopes, the thyroid gland can be monitored with a sensitive scintillation detector to ascertain the dose of radioactive iodine. Whole-body monitoring is usually done only when someone suspects they have received a large dose of radioisotope internally. It is performed using a similar monitor to that used for thyroid monitoring, and detects only gamma and x-ray radiation. Only a few facilities in Australia operate instruments for whole-body monitoring. ARPANSA, ANSTO or a local health authority would be a useful first point of contact for locating such an instrument.
- *Bioassay.* Indirect radioisotope intake can be gauged indirectly by measuring the amount in a biological tissue or product, such as urine, blood, faeces, hair or sweat. Common bioassay methods involve monitoring urine for radioisotopes such as tritium (beta emitter), sulfur-35 (beta emitter) and sometimes carbon-14 (beta emitter). Urine samples can be analysed using an instrument such as a laboratory liquid scintillation counter. If workers routinely use radioisotopes that can be assessed via urine monitoring, it is useful to establish a baseline count before the work using the radioisotope commences.

13.2.13 CONTROLLING RADIATION HAZARDS

Anyone working with ionising radiation should aim to minimise the potential for exposure. This should always be done in the planning stages of the work, in line with the three principles defined in the International Radiological System of Protection (see section 13.2.3). Before a decision to use ionising radiation is made, the benefits need to be weighed carefully against the potential harm. If the use is justified, then appropriate controls should be used.

13.2.13.1 Minimising external exposure

The primary methods for controlling or minimising external exposure to radiation are:

- *shielding:* using a barrier to absorb the radiation
- *distance:* increasing the distance between the person and the radiation source
- *time:* handling the radiation source as briefly as possible.

Shielding

Shielding is the most important control measure to reduce or eliminate external exposure to ionising radiation. The shielding required for each type of radiation depends on its properties:

- *Alpha radiation* does not penetrate far in air, so does not require a specific shield (a sheet of paper is thick enough to stop alpha particles).
- *Beta radiation:*
  - can penetrate a reasonable distance in air
  - needs low-density types of shielding such as Perspex (approximately 1 centimetre thick) or aluminium (several millimetres thick)
  - needs to be shielded with the correct material. Do NOT use high-density shielding such as lead. X-rays known as Bremsstrahlung radiation are generated when beta particles are slowed down too quickly in high-density materials.
- *Gamma and X-rays:*
  - can penetrate a long distance in air
  - are best controlled with high-density shielding. Relatively effective shields can also be made from thicker slices of less dense materials such as brick, concrete, sand or gravel.
  - need thicker shielding as their energy increases.
- *Neutrons:*
  - can penetrate moderate to long distances in air
  - need shielding with specific elements such as boron and cadmium, or with materials containing high levels of hydrogen—for example, paraffin wax, water and concrete.

Distance

If it is not practical to shield a radiation source, or if the shielding does not stop all the radiation, then increasing the worker's distance from the source should be considered. This can be done by using remote handling devices such as tongs or mechanised devices. The intensity of ionising radiation is inversely proportional to the square of distance from the source, so a worker twice as far from the source receives one-quarter the radiation exposure.

Time

If shielding and distance do not completely stop ionising radiation, then reducing the exposure time should be considered.

13.2.13.2 Minimising internal exposure

The main methods for minimising internal exposure are:

- choice of work practices
- use of contamination controls
- multiple layers of containment
- managing the risk of inhalation.

Choice of work practices

Work practices should be chosen that minimise the number of steps in the radioisotope-handling process. Complex processes provide more opportunities for contamination to occur: the chance of contamination with a multi-step chemical process is much higher than when using a simple dilution.

Use of contamination controls

Minimising the amount of contamination reduces the potential for a radioisotope to enter the body. Cleanliness is very important in avoiding internal uptake of radioisotopes. This includes measures such as changing gloves frequently, wearing gloves that do not absorb radioactive materials and, if using surgical gloves, wearing two pairs.

Workers carrying out experiments involving radiation should monitor their clothing, hands and work space frequently, and any spills should be cleaned up without delay. Routine area contamination surveys should be done and the results recorded to ensure that the general work environment remains uncontaminated.

Multiple layers of containment

Multiple layers of containment should be used to minimise the risk and the consequences of spillages. This should be done for both storage and handling. An example of good practice is to work within a tray lined with absorbent material to contain a spill.

Managing the risk of inhalation

The risk of inhaling radioactive material can be managed by choosing work practices that minimise aerosol production, minimise use of volatile chemical forms of a radioisotope and avoid the use of dry powder forms.

13.3   NON-IONISING RADIATION

Non-ionising radiation is electromagnetic radiation of a wavelength greater than 100 nm that does not have sufficient energy to ionise the matter with which it interacts. It includes ultraviolet, visible, infrared, radiofrequency and extremely low-frequency radiation. With such a wide range of wavelengths, frequencies and therefore energies, non-ionising radiation has the potential to cause various adverse health effects in humans in certain body locations.

13.3.1 *ULTRAVIOLET RADIATION*

13.3.1.1 Properties

Ultraviolet radiation is the highest-energy form of non-ionising radiation and exists in three bands, from highest to lowest energy (Figure 13.6):

- *far, short or UV-C*—wavelengths 100–280 nm and frequencies around 10^{16} Hz. Wavelengths below 180 nm are absorbed by air and are therefore of little biological significance. This is why the wavelength range of UV-C is often listed as 180–280 nm.
- *middle, erythemal or UV-B*—wavelengths 280–315 nm and frequencies around 10^{15} Hz.
- *near, long or UV-A*—wavelengths 315–400 nm and frequencies around 10^{14} Hz.

13.3.1.2 Sources

The most common source of ultraviolet radiation is the sun. Other sources capable of providing significant exposure in the occupational setting include arc sources and specialised lamps. Sometimes ultraviolet radiation is produced as an unwanted side-effect, as in plasma torches, gas and electric arc welding. Ultraviolet lamps are used for applications such as:

- tanning and dermatology (UV-A and UV-B)
- black lights for non-destructive testing (UV-A)
- scientific research (all UV)
- photo curing of inks and plastic (UV-A and UV-B)
- photoresist processes (all UV)
- germicidal uses (UV-C and UV-B).

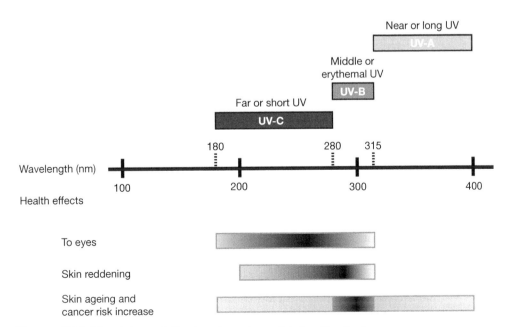

Figure 13.6 Ultraviolet radiation bands and possible health effects

Some lasers also emit ultraviolet radiation. Lasers are discussed further in section 13.3.3.

Ultraviolet light sources with emissions below 250 nm can interact with the workplace atmosphere to produce ozone, oxides of nitrogen and phosgene.

13.3.1.3 Quantification and health effects

The skin and eyes may be affected by exposure to ultraviolet radiation. At one time, wavelengths below 315 nm were collectively known as 'actinic radiation'—that is, radiation that can induce biological effects—but those health effects are also observed in UV-A at substantially higher doses. Health effects can result from occupational exposures in the absence of effective controls.

- *Reddening (or erythema)* results from over-exposure of skin in the middle and far ultraviolet range (200–315 nm), with the greatest sensitivity occurring at 295 nm. Exposure to near ultraviolet alone requires far higher levels to induce erythema; however, exposure to near and middle ultraviolet together intensifies the response.
- Chronic exposure to ultraviolet light—especially middle ultraviolet—increases *skin ageing* and the *risk of developing skin cancer*.
- The cornea and conjunctiva of the eye strongly absorb middle and far ultraviolet radiation. The resulting condition, photo-keratoconjunctivitis, is generally known as 'welder's flash' because it often occurs after welding. Wavelengths above 295 nm penetrate the cornea and are absorbed by the lens, increasing the risk of cataracts.

Individuals who have had an eye lens replaced or exposed to photosensitising agents are at additional risk from ultraviolet exposure.

Exposure to ultraviolet radiation incident on the skin or eyes is measured in joules (J) per square metre (J/m^2). The maximum allowable exposure varies according to the wavelength of the radiation—that is, its relative spectral effectiveness—with the most harmful wavelength being 270 nm. Where a person is exposed to a range of ultraviolet wavelengths simultaneously (in light from a broad-spectrum or incoherent source), the spectrally weighted effective irradiance (E$_{eff}$) is measured or calculated to determine the exposure and compare with exposure limits. Effective irradiance (E$_{eff}$) is the surface exposure dose rate and has units of watt per square metre (W/m^2). Since 1 watt equals 1 J/s, effective irradiance can be used to calculate the time taken to reach a certain maximum permissible exposure to ultraviolet radiation.

Measuring ultraviolet radiation is a task usually undertaken by a professional with specific expertise in the area, using the correct equipment and considering the geometry of exposure. An H&S practitioner needing an exposure assessment of a lamp for ultraviolet radiation should consider the International Electrotechnical Commission (IEC, 2006) lamp risk group.

13.3.1.4 Control methods

Exposure to ultraviolet radiation can be reduced by shielding, distance and time:

- *Hats, eye protection, clothing and sun-shades* serve as shields against sunlight. Clothing and hats chosen should have a formal rating for ultraviolet protection. Wide-brimmed hats and high-protection-factor sunscreens are also useful, but may not provide sufficient protection.

- *Protective glasses* for outdoor or general use, or face shields and glasses for welding, should meet the relevant international or Australian standards. Where shields are required for industrial sources, polycarbonate or methyl methacrylate plastics strongly absorb most ultraviolet radiation. However, where a high-intensity source exists, the radiation may not be absorbed fully. Ultraviolet radiation can also reflect from shiny surfaces.
- Exposures to artificial or industrial sources of ultraviolet radiation should be controlled by *engineering solutions* such as light-tight cabinets and enclosures, and ultra-violet-light-absorbing glass and plastic shielding. *Shields, curtains and barriers* are used to shield against UV light from welding processes.
- *Distance from the source* should be maximised, since the intensity of non-ionising radiation falls off rapidly as this distance increases.
- The *time of exposure* to the source should be limited wherever possible. For outdoor workers, avoiding working outside without proper protective equipment during the middle of the day can also help to reduce exposure.

13.3.2 *VISIBLE AND INFRARED RADIATION*

13.3.2.1 Properties

Visible and infrared radiations are less energetic forms of non-ionising radiation than ultraviolet. They can be divided into bands from highest to lowest energy (Figure 13.7):

- *visible light*—wavelengths 400–780 nm and frequencies around 10^{14} Hz
- *near infrared or IR-A*—wavelengths 780–1400 nm and frequencies around 10^{14} Hz
- *middle infrared or IR-B*—wavelengths 1400–3000 nm and frequencies around 10^{14} Hz
- *far infrared or IR-C*—wavelengths 3000 nm–1 mm and frequencies around 10^{11}–10^{14} Hz.

13.3.2.2 Sources

Visible light comes mainly from the sun, but lacks the intensity to harm the eyes unless a person stares directly at the sun for a sustained period. There are also human-made visible

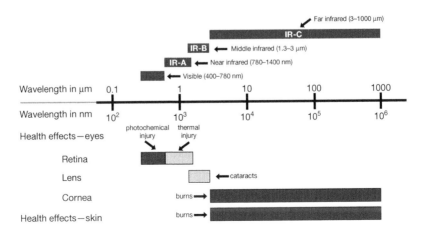

Figure 13.7 Visible and infrared radiation bands and possible health effects

light sources such as incandescent lamps, lasers and gas discharge sources that produce very intense visible light. Intense visible light may also be a by-product of industrial processes such as welding. Most artificial sources used in industry, consumer, scientific and medical applications emit visible radiation that is not dangerous. Usually, the eye is protected by the natural tendency to blink or avert the gaze in response to intense visible light.

Infrared radiation from the sun is felt as warmth. Transfer of infrared radiation also occurs from any object that is at a higher temperature than another receiving object. Intense sources of infrared radiation in the workplace and the home are many, and include heating devices such as furnaces, ovens, infrared lamps and some lasers. In general, most high-intensity broadband sources, such as incandescent lamps, produce negligible levels of IR-C compared with emissions of visible light and other infrared bands. Where substantial IR-C exposure is present (e.g. steel furnaces) it may contribute significantly to heat stress (ICNIRP, 2006).

13.3.2.3 Quantification and health effects

The eye is the most vulnerable organ to visible and near infrared radiation because the wavelengths from these sources are focused by the lens onto the retina, where thermal or photochemical damage may result. Photochemical injury to the retina is most likely due to long exposures to visible light, peaking at 440 nm. Thermal injury to the retina is more common with short exposures to near infrared radiation. The lens is most vulnerable to middle infrared, and the formation of cataracts is common in glass-blowers ('glass-blower's cataract') and furnace workers unless eye protection is worn. Far infrared is absorbed at the surface of the eye, and may cause superficial burns to the cornea or skin; however, skin is not usually at risk unless the source is very intense and pulsed, because the pain reflex limits the duration of exposure.

To injure the eye or skin, visible and infrared radiation must be transmitted to and absorbed by the tissues. Control therefore requires knowledge of the spectral distribution of radiation from the source and the total irradiance, measured at the eye or on the skin. At least five types of injury may be caused by visible and infrared radiation:

- thermal injury of the retina (UV-A, visible, IR-A), which requires very high-intensity sources
- photochemical injury of the retina (UV-A, visible [blue light])
- thermal injury of the lens (IR-A, IR-B)
- thermal injury (burns) to the skin and cornea (visible, IR-C)
- photosensitisation of the skin, including side-effects of medication.

The amount of photochemical injury depends on dose and exposure duration. An extremely bright light viewed for a short time will have the same effect as a bright light viewed for a longer period. These photochemical exposures are accumulative over a working day; however, the amount of thermal injury to a body tissue will depend on whether the heat absorbed can be conducted away from the exposed site quickly enough. If not, the temperature rise will damage the tissue. Sub-threshold thermal exposures are not cumulative over a working day.

Assessing exposure for small sources of visible-light and near-infrared radiation requires knowledge of the spectrally weighted total irradiance (E, W/m²), measured at the location of the eye. Irradiance measurements are made using a spectrally weighted radiometer matching the applicable action spectrum. Once total irradiance is known, the appropriate exposure standard can be applied and a maximum permissible duration of exposure can be established.

Assessing exposure for large sources of visible-light and near-infrared radiation requires information about the radiance (or 'brightness') of the source and not the irradiance at the eye. Radiance is measured with a radiometer and is independent of distance from the source. A large source creates an image on the retina that changes in proportion to the irradiated power on the retina. This means the power level of incident radiation and risk of injury remain in proportion. Allowed exposures to large optical sources are expressed as radiance of the source in watts per square metre per steradian—that is, $W/(m^2.sr)$.

The measurement of infrared radiation is a task usually undertaken by a professional with specific expertise in the area (see section 13.4). Particular consideration should be given to evaluating intense infrared sources that are not visibly bright, as these sources could be easily gazed upon, putting the eye at risk of thermal injury.

H&S practitioners often measure visible light to determine whether a workplace has sufficient illumination. A photometer is used to measure quantities such as luminance (brightness as perceived by a standard human observer) or illuminance (the light falling on a surface) in lux (lx) weighted to the response of the human eye (see Chapter 15).

13.3.2.4 Control methods

Exposure to visible and infrared radiation can be controlled by shielding, distance and time:

- Certain materials can be used to *shield* against or attenuate both visible and infrared radiation. Protective glasses for outdoor or general use, or face shields and glasses for welding, should meet the relevant international or Australian standards for eye and face protection. Glasses or goggles are used to protect the eye and typically are made from plastics such as polycarbonate. Sources of infrared radiation should be shielded close up, using reflective materials such as aluminium. Shields, curtains and barriers are used to protect people from welding processes.
- *Distance* from the source should be maximised, since the intensity of non-ionising radiation falls off rapidly with distance.
- Limiting the *time of exposure* to the source will reduce photochemical injury.

13.3.3 LASERS

13.3.3.1 Properties

LASER is an acronym for light amplification by stimulated emission of radiation. Lasers currently produce light in the ultraviolet, visible or infrared region of the electromagnetic spectrum. Laser light differs from ordinary electromagnetic radiation in that the light beam is coherent in space (all the waves are in phase) and time (all the waves are of the same frequency). As a result, the beam has very little divergence and is able to be transmitted over large distances while retaining a relatively high level of energy per unit area. In other words, a laser beam remains very intense over long distances. This is in contrast to ordinary light, the intensity of which falls off with the square of distance. Lasers produce a single or narrow wavelength band of light.

13.3.3.2 Sources

Lasers have a wide range of industrial and commercial applications, including for alignment, range-finding in construction, printing, drilling, welding, cutting, advertising and entertainment. Lasers are also used for surgery and ophthalmic procedures, and as pointers in lectures and business presentations.

13.3.3.3 Health effects

The eyes and skin are the body parts most susceptible to damage from lasers. The intensity of the beam influences the extent of damage, while the emitted wavelength determines the site of damage. A powerful laser that is not appropriately controlled may cause serious injury before the body's natural aversion response (blinking or turning away) is triggered. Reflections from mirrors, equipment and other surfaces, including watch faces, are also a significant hazard. A laser may cause photo-chemical effects (chemical reactions, after-images), thermal effects (burns) and non-linear effects.

Non-linear effects are caused by a rapid heating and thermal expansion of biological tissue. The high irradiance produced by some lasers delivers a lot of energy to the biological target in a very short time. The target tissues experience such a rapid rise in temperature that the liquid components of their cells are converted to gas and the cells rupture.

13.3.3.4 Control methods

The laser classification system is almost internationally consistent and indicates a laser's level of risk along with the measures needed for safe use (Table 13.7). Products that contain (embedded) lasers are also classified according to the risk they present. Such 'laser products' should be handled and dismantled with caution, as the lasers within may be more hazardous (i.e. a higher class) when exposed than when properly encased.

Laser pointers are increasing in efficiency and power, with many capable of dazzling and causing an after-image (flash blindness) if the beam enters the eye. This dazzling may have catastrophic consequences for operators of machinery, vehicles or aircraft.

Laser pointers of power above 1 mW are prohibited from public use in some countries. They are prohibited imports into Australia (Australian Customs and Border Protection Service, n.d.) and controlled or prohibited as potential weapons in many jurisdictions (e.g. NSW Police Force, 2009). Teachers and public speakers should therefore not exceed a class 1 or 2 laser pointer.

13.3.4   RADIOFREQUENCY RADIATION

13.3.4.1 Properties

Radiofrequency (RF) radiation covers the frequency range from 300 Hz to 300 GHz and has corresponding wavelengths from approximately 1000 km to 1 mm (ARPANSA, 2002). This type of electromagnetic radiation exists whenever there is a current and voltage flowing.

Table 13.7 Summary of laser classes, characteristics and control measures

Laser class	Risk	Characteristics	Specific control measures
1	No risk to eyes or skin	Safe under most circumstances including when viewed with optical instruments.	None
1M	No risk to naked eyes No risk to skin	These lasers emit in the range 302.5 to 4000 nm, but may be hazardous if optics are used in the beam.	i–iii
2	No risk to eyes or skin for short exposure times, including viewing through optical instruments	These lasers must emit visible radiation 400–700 nm as eye and skin protection is afforded by normal aversion responses (e.g. blinking) including where optical instruments are used for intra-beam viewing.	None
2M	No risk to eyes or skin for short exposure times	As for class 2, except that viewing may be more hazardous if optics are used in the beam.	i–iii
3R	A risk to eyes No risk to skin	These low-power lasers (<5 mW) emit in the range 302.5–10^6 nm, where direct intra-beam viewing is potentially hazardous.	i–iii
3B	Medium to high risk to eyes Low risk to skin	As for class 3R, except that direct intra-beam viewing is hazardous. Viewing diffuse surface reflections is normally safe.	i–vii
4	High risk to eyes High risk to skin Fire hazard	These lasers have power of 500 mW and above. They cause eye injury, and are capable of producing hazardous diffuse reflections, skin injuries and may also be a fire hazard.	i–viii

Explanation of control measures:

i	Training		v	Warning signs on enclosures
ii	Beam stop or attenuator		vi	Remote interlock to laser
iii	Avoid specular reflections		vii	Key control of laser
iv	Protective eyewear		viii	Protective clothing

Source: Derived from AS 2211.1 (Standards Australia, 2004) by the author.

13.3.4.2 Sources

Microwaves and communications devices are the major sources of radiofrequency radiation. There are also natural sources, such as the Earth and the cosmos, whose RF radiation covers a range of frequencies but is of very low intensity. Human-made sources, many of which may be found in workplaces, include:

- radar
- RF induction heaters
- electronic article-surveillance systems, radiofrequency identification systems
- microwave ovens
- mobile telephones, wi-fi
- television transmitters
- FM and AM radio transmitters.

Exposure to sources such as radar devices would only be high if a person is very close to it. Medical uses of radiofrequency radiation include diathermy, microwave treatment and magnetic resonance imaging (MRI). These sources are in the MHz and GHz range and can potentially expose patients to high-intensity levels.

13.3.4.3 Quantification and health effects

The human body responds differently to different electromagnetic field frequencies, resulting in different health effects and safety limits. Health effects from radiofrequency radiation may broadly be grouped as either thermal or non-thermal. Thermal effects are well established and occur when human tissue is irradiated, especially with frequencies between 1 and 10,000 MHz. In much the same way as water absorbs heat in a microwave oven, body tissues absorb and are heated by RF radiation. Different water content means that tissues are not heated uniformly, and some (e.g. the brain, lens, testes and the unborn child) are more sensitive to heat than others. The heating effect also depends on frequency. For example, localised warming occurs in the ankles at 30–80 MHz. Above 10 GHz, radiation is absorbed at the skin surface, increasing the likelihood of cataracts and skin burns.

Non-thermal effects include the induction of electrical currents in tissues, shocks and burns, and the microwave hearing effect whereby people report a sense of buzzing, clicking or popping in their ears. Shocks and burns can be caused by contact with the source of RF radiation. There is uncertainty about possible carcinogenic effects (Sanders, 1996); however, reviews of current studies suggest that any risk of cancer from exposure to RF radiation, if it exists, is very low. 'It is the opinion of ICNIRP that the scientific literature published since the 1998 guidelines has provided no evidence of any adverse effects below the basic restrictions and does not necessitate an immediate revision of its guidance on limiting exposure to high frequency electromagnetic fields' (ICNIRP, 2009).

Exposure standards for RF radiation are defined as 'basic restrictions' and take different forms depending on the section of the radiofrequency spectrum under discussion. Basic restrictions are quantities that are impractical to measure. Thus 'reference levels' are quantities that can be measured to assess compliance with basic restrictions. The relationship between basic restrictions and reference levels and the RF ranges and health effects where each applies are summarised in Table 13.8. Measurement of radiofrequency radiation and interpretation of the results against reference levels is best undertaken by a radiation safety professional (see section 13.4).

Measuring non-ionising radiation is often complex, and it is easy to obtain false readings. It is best to engage a health physicist or occupational hygienist with specific expertise in the type of radiation and with access to the appropriate instrumentation. For some non-ionising radiation, the instrumentation should be able to measure both

Table 13.8 Relationship between basic restrictions and reference levels for radiofrequency radiation

Frequency range	Basic restriction—limit factors that protect against known health effects	Reference level—practical measurable quantities	Health effects
3 kHz–10 MHz	Current density (rms), J measured at a point in space and time	Instantaneous electric, E and/or magnetic, H fields (rms), or instantaneous contact current, I	Induction of electrical currents in body tissue. Inhomogeneous deep penetration of power into the human body
100 kHz–6 GHz	Specific absorption rate, SAR—average over whole body, and/or head and torso Alternatively, specific absorption rate, SAR—spatial peak in limbs	Time averaged electric, E and magnetic, H fields (rms) and/or induced limb currents for legs and arms (10–110 MHz) or contact point currents (100 kHz–110 MHz)	Between 30 and 300 MHz: the wavelengths are similar to that of human body or body parts. Most of the field energy is absorbed. This can lead to whole-body heat stress and excessive localised temperature rises.
300 MHz–6 GHz	Specific absorption, SA—spatial peak in head	Instantaneous electric, E and/or magnetic, H fields (rms) or equivalent power flux density, S	Microwave hearing effect (clicking) may be observed between 200 MHz and 3 GHz.
10 MHz–6 GHz	Specific absorption rate, SAR — instantaneous spatial peak in head and torso	Instantaneous electric, E and/or magnetic, H fields (rms) or equivalent power flux density, S	
6–300 GHz	Power flux density, S, instantaneous and averaged over time	Time averaged and instantaneous electric, E and/or magnetic, H fields (rms)	Excessive heating in surface tissue. Potential skin burns with high exposure.

Units:
Current (I): amperes (A)
Current density (J): amperes per square metre (A/m^2)
Magnetic field strength (H): amperes per metre (A/m)
Electric field strength (E): volts per metre (V/m)
Magnetic flux density (B): tesla (T)
Specific absorption rate (SAR): (the rate at which energy is imparted to human tissue) watts per kilogram (W/kg)
Specific absorption (SA): joules per kilogram (J/kg)
Power flux density (S): watts per square metre (W/m^2)
rms: root mean square
Source: derived from ARPANSA (2002).

electric and magnetic field strengths in the near field (< 3x wavelength), for comparison to worker and general public exposure standards.

13.3.4.4 Control methods

Exposure to radiofrequency fields is best controlled by avoidance (including distance) and shielding at source.

Leakage of RF radiation should be limited at the source, using shielding materials to absorb or contain the radiation. While electric fields are easier to shield, the magnetic field component penetrates most materials. Choosing such shielding is a specialist task, as inappropriate shielding can enhance the radiofrequency field and lead to higher exposures (see section 13.4).

In the case of communications technology such as mobile phone antennas and radar devices, the RF radiation cannot be suppressed. Care must be taken to prevent contact between the human body and radiofrequency fields by limiting access to transmitters. Conductive suits can partially shield the user if fully enclosed. Such suits have the disadvantage of high thermal load and limited visibility. In addition, any opening in the suit will result in an enhanced local field.

Care should also be taken to prevent access to objects like tools that may cause high levels of contact current and thus pose a risk of shocks. Ideally, these objects should be electrically grounded. Using electrically insulating gloves when handling metal tools can prevent 'startle' currents.

13.3.5　EXTREMELY LOW FREQUENCY RADIATION

13.3.5.1 Properties

Extremely low frequency (ELF) radiation has frequencies up to 300 Hz and wavelengths above 1000 kilometres.

13.3.5.2 Sources

The ELF region of the electromagnetic spectrum is associated with the generation, distribution and use of electricity. Electricity supply in Australia is at 50 Hz (6 km). The presence of electric charges gives rise to electric fields, which are measured in volts per metre (V/m). The motion of the electric charges (the current) gives rise to magnetic fields, measured expressed in tesla (T), millitesla (mT) or gauss (G; 10 000 G = 1 T). Electric fields are easily shielded with common materials, but magnetic fields pass through such materials. Both types of fields are strongest close to the source and diminish with distance.

13.3.5.3 Quantification and health effects

The health effects of ELF radiation depend on the component electric and magnetic fields. A WHO task group (WHO, 2007) concluded that there were no substantive health issues related to ELF electric fields at levels generally encountered by members of the public. External ELF magnetic fields induce electric fields and currents in the body which, at very high field strengths (well above the 100 μT general public exposure limit), can cause nerve and muscle stimulation and changes in nerve-cell excitability in the central nervous system (CNS).

Both electric and magnetic fields are known to interfere with cardiac pacemakers at levels of exposure that may not otherwise cause adverse health effects. There is conjecture surrounding the possible carcinogenic effects of extremely low-frequency radiation and static electric and magnetic fields, but this link remains unproven (WHO, 2007).

Measurement of ELF radiation is a task for a radiation specialist (see section 13.4).

13.3.5.4 Control methods

Control of ELF radiation also requires specialist expertise (see section 13.4). In power-transmission applications, it is impractical to shield ELF radiation. Instead, attempts are made to control the electric field strength at ground level by locating high-voltage transmission lines high in the air and with a corridor of land around them.

13.3.6 STATIC FIELDS

13.3.6.1 Properties

Static magnetic and electric fields occur where frequency is 0 Hz, and are characterised by magnetic and electric field strengths that do not change over time.

13.3.6.2 Sources

The Earth has a static magnetic field generated by the electric current flowing in its core. Human-made static magnetic fields are usually stronger, and occur around direct current (DC) devices, electric trains, nuclear magnetic resonance spectrometers, aluminium production plants, powder coating, galvanisation and in some welding. MRI can expose the patient from 0.2 to 7 T.

Static electric fields (also known as electrostatic fields) occur where there are charged bodies. Friction can separate charges and generate strong static electric fields that create a spark on discharge.

13.3.6.3 Quantification and health effects

The health effects of static electric fields are poorly understood. Known effects are currently limited to discomfort from spark discharges. However, it is recommended that electric field strength (E) be controlled in order to limit both currents on the body surface and induced internal currents. Field strength should also be controlled in order to prevent safety hazards such as spark discharge and high-contact currents on metal objects.

Static magnetic fields are likely to cause health effects only when there is movement within the field. A person moving within a field above 2 T can experience sensations of vertigo, nausea, a metallic taste and perceptions of light flashes. Static magnetic fields also affect implanted metallic devices such as cardiac pacemakers.

Magnetic field strength (H), measured in amperes per metre (A/m), and magnetic flux density (B), measured in tesla (T), are used to quantify a magnetic field.

13.3.6.4 Control methods

Protection from static electric fields entails grounding objects carrying current and prescribing protective suits similar to those used in the presence of radiofrequency radiation.

Static electric fields in isolated applications may be shielded using an earthed conducting enclosure (a Faraday cage) and static magnetic fields may be shielded using magnetic shielding. Protection is also gained by maximising distance from the source and minimising time spent in close proximity to it. Small metal objects need to be kept away from strong static magnetic fields, which can turn them into missiles.

13.4   THE ROLE OF THE H&S PRACTITIONER WITH REGARD TO RADIATION HAZARDS

13.4.1   IONISING RADIATION

H&S practitioners may become involved with ionising radiation when a source is present in their workplace. Their role is generally one of identifying sources of ionising radiation, and assisting in complying with any routine requirements such as registration, licensing and managing radiation dose badges for radiation workers within their organisations. They may undertake basic surveys for contamination, but other monitoring methods and interpretation of the results can be complex and can require specialist expertise.

If a source has very high activity or high energy, or generates beams of ionising radiation, a specialist will be needed to advise on correct placement of shielding and recommend other appropriate controls.

13.4.2   NON-IONISING RADIATION

H&S practitioners may become involved with non-ionising radiation when a source is present in their workplace. Their role is generally one of identifying and characterising sources of non-ionising radiation, assisting in complying with any routine requirements such as registration, licensing and managing radiation dose badges for radiation workers within their organisations. They may also be involved in:

- reviewing new workplace plans, equipment or arrangements
- identifying potential hazards associated with equipment, processes or environments, like a new high-intensity radiation source, ultraviolet lamp or laser
- dealing with concerns of the workforce
- arranging the appropriate specialist to assess and address issues related to non-ionising radiation
- prescribing simple control measures, such as hats and sunscreen for protection from ultraviolet radiation outdoors. Some controls should be carefully chosen to avoid increasing the risk of exposure to non-ionising radiation.
- ensuring radiation workers undertake the appropriate training
- reviewing procedures and processes.

13.4.3   SPECIALIST EXPERTISE

Specialists such as health physicists and occupational/industrial hygienists are often engaged to measure and prescribe controls for both ionising and non-ionising radiation.

These specialists should be members of relevant national or international professional societies. In part, specialists have access to special equipment required for surveys that determine the risk to workers.

It should be remembered that the electromagnetic compatibility (EMC), that gives rise to the CE Mark for radiation immunity of equipment does not indicate health effects on people. Human exposure should be determined by appropriate radiation surveys and personal monitors.

13.5 REFERENCES

Australian Customs and Border Protection Service n.d., *Prohibited and Restricted Imports*, <www.customs.gov.au/site/page4369.asp> [accessed 4 December 2012]

Australian Radiation Protection and Nuclear Safety Agency (ARPANSA) 2002, *Radiation Protection Standard—Maximum Exposure Levels to Radiofrequency Fields—3kHz to 300GHz*, ARPANSA, Melbourne.

—— 2011, *Monitoring, Assessing and Recording Occupational Radiation Doses on Mining and Mineral Processing*, ARPANSA, Melbourne, <www.arpansa.gov.au/sites/default/files/legacy/pubs/rps/rps9_1.pdf> [accessed 18 August 2019]

—— 2012a, *Alpha Particles*, ARPANSA, Melbourne, <www.arpansa.gov.au/radiation protection/Basics/alpha.cfm> [accessed 18 August 2019]

—— 2012b, *Cosmic Radiation Exposure When Flying*, Fact Sheet 27, ARPANSA, Melbourne, <www.arpansa.gov.au/radiationprotection/Factsheets/is_cosmic.cfm> [accessed 4 December 2012]

—— 2012b, *Understanding Radiation*, ARPANSA, Melbourne, <www.arpansa.gov.au/RadiationProtection/basics/understand.cfm> [accessed 4 December 2012]

—— 2015, *Ionising Radiation and Health*, ARPANSA, Melbourne, <www.arpansa.gov.au/understanding-radiation/radiation-sources/more-radiation-sources/ionising-radiation-and-health> [accessed 10 October 2018]

—— 2016, *Radiation Protection in Planned Exposures, Radiation Protection Series C-1*, ARPANSA Melbourne.

—— 2018, *New Dose Coefficients for Radon Progeny: Impact on Workers and the Public*, ARPANSA Advisory Note, <www.arpansa.gov.au/sites/g/files/net3086/f/new-dose-coefficients.pdf> [accessed 5 October 2018]

Australian Radiation Protection and Nuclear Safety Agency Radiation Health Committee 2011, *Statement on Changes to Occupational Dose Limit for Lens of the Eye*, ARPANSA, Melbourne, <www.arpansa.gov.au/pubs/rhc/LensofEye_stat.pdf> [4 December 2012]

Cooper, M.B. 2005, *Naturally Occurring Radioactive Materials (NORM) in Australian Industries: Review of Current Inventories and Future Generation*, Radiation Health and Safety Advisory Council, ARPANSA, Melbourne, <www.arpansa.gov.au/pubs/norm/cooper_norm.pdf> [accessed 4 December 2012]

Gonzalez, G.A. and Anderer, J. 1989, 'Radiation versus radiation: Nuclear energy in perspective—a comparative analysis of radiation in the living environment', *IAEA Bulletin*, vol. 31, no. 2, pp. 21–31, <www.iaea.org/Publications/Magazines/Bulletin/Bull312/31205642131.pdf> [accessed 4 December 2012]

International Agency for Research on Cancer (IARC) 1988, *Man-made Mineral Fibres and Radon*, IARC, Lyon.

International Atomic Energy Agency 2011, *Radiation Protection and Safety of Radiation Sources: International basic safety standards: Interim Edition No. GSR Part 3*, <www-pub.iaea.org/MTCD/publications/PDF/p1531interim_web.pdf> [accessed 12 December 2012]

International Commission on Non-Ionizing Radiation Protection (ICNIRP) 2006, 'ICNIRP statement on far infrared radiation exposure', *Health Physics*, vol. 91, no. 6, pp. 630–45, <www.icnirp.org/documents/infrared.pdf> [accessed 4 December 2012]

—— 2009, 'ICNIRP statement on the "Guidelines for limiting exposure time-varying electric, magnetic, and electromagnetic fields (up to 300 GHz)"', *Health Physics*, vol. 97, no. 3, pp. 257–8, <www.icnirp.org/documents/StatementEMF.pdf> [accessed 4 December 2012]

International Commission on Radiological Protection (ICRP) 1990, '1990 recommendations of the International Commission on Radiological Protection, ICRP publication 60', *Annals of the ICRP*, vol. 31, no. 1–3, pp. 100–5.

—— 2007, 'The 2007 recommendations of the International Commission on Radiological Protection, ICRP publication 103', *Annals of the ICRP*, vol. 37, nos 2–4.

—— 2011, *ICRP Database of Dose Coefficients: Workers and Members of the Public*, version 3, [software], ICRP, Ottawa, <www.icrp.org/page.asp?id=145> [accessed 4 December 2012]

——2014, *Radiological Protection Against Radon Exposure*, *Annals of the ICRP*, vol. 43, no. 3.

—— 2017, *Occupational Intakes of Radionuclides: Part 3. Annals of the ICRP*, vol. 43, nos 3–4.

International Electrotechnical Commission (IEC) 2006, *Photobiological Safety of Lamps and Lamp Systems*, IEC 62471: 2006, IEC, Geneva.

Morris, N.D., Thomas, P.D. and Rafferty, K.P. 2004, *Personal Radiation Monitoring Service and Assessment of Doses Received*, ARPANSA, Melbourne, <www.arpansa.gov.au/pubs/technicalreports/tr139.pdf> [accessed 4 December 2012]

NSW Police Force 2009, *Laser Pointers: Questions and Answers*, NSW Government, Sydney, <www.police.nsw.gov.au/about_us/structure/operations_command/firearms/laser_pointers/laser_pointers_-_questions_and_answers>.

Sanders, R.D. 1996, 'Biological effects of radiofrequency radiation', in *Non-ionizing Radiation: Proceedings of the Third International Non-Ionizing Radiation Workshop, 22–26 April, Baden, Austria*, ICNIRP, Munich, pp. 245–54.

Simon, S.L. 2007, 'Introduction to radiation physics and dosimetry', *Radiation Epidemiology Course*, US National Cancer Institute, Bethesda, Maryland, <radepicourse2007.cancer.gov/content/presentations/slides/SIMON1_slides.pdf> [accessed 4 December 2012]

Solomon, S.B., Langroo, R., Peggie, J.R., Lyons, R.G. and James, J.M. 1996, *Occupational Exposure to Radon in Australian Tourist Caves: An Australia-wide Study of Radon Levels*, Australian Radiation Laboratory, Melbourne, <www.arpansa.gov.au/pubs/technicalreports/arl119tx.pdf> [accessed 5 December 2012]

Standards Australia 2004, Safety of Laser Products, Part 1: Equipment Classification, Requirements and User's Guide, AS 2211.1: 2004/IEC 60825.1, Standards Australia, Sydney, pp. 21–2, 101.

United Nations Scientific Committee on the Effects of Ionising Radiation (UNSCEIR) 2008, *Sources and Effects of Ionizing Radiation*, vol. 1, United Nations, New York.

Voytchev, M., Ambrosi, P., Behrens, R. and Chiaro, P. 2011, 'IEC standards for individual monitoring of ionising radiation', *Radiation Protection Dosimetry*, vol. 144, nos 1–4, pp. 33–6.

World Health Organization (WHO) 2007, *Electromagnetic Fields and Public Health: Exposure to Extremely Low Frequency Fields*, Fact Sheet no. 322, WHO, Geneva, <www.who.int/mediacentre/factsheets/fs322/en/index.html> [accessed 5 December 2012]

14. The thermal environment

Ross Di Corleto and Jodie Britton

14.1   INTRODUCTION

The workplace is seldom an ideal environment, and can often expose the individual to physical extremes of temperature. This chapter looks at the impact of these parameters on the individual, and outlines assessment protocols and guidelines for their management.

The iron ore mines in the Pilbara region of northern Western Australia and the diamond mines of the Canadian North West Territories have one thing in common: extremes of temperature and what can often be an inhospitable working environment. The predicted influence of global climate change will no doubt further exacerbate working in these extreme environments, with all regions of the planet expected to become warmer in the future. The forecasted increase in environmental heat is likely to affect large populations of people, causing both acute and chronic health impacts with the average global temperature expected to increase between 1.8 and 4.0°C by the year 2100 (Kjellstrom et al, 2010). The human body functions best in a moderate climate, and variations that can result in a lowering or increase of the core body temperature can lead to serious physiological injury or illness. While the identification of such conditions is often obvious, their control and successful correction are not always straightforward owing to the myriad variations that can be involved. The three key areas where variation can occur are the environment, the task and the individual. By taking a systematic approach to the investigation, such confusion can often be overcome and practical controls identified and implemented.

14.2   WORK IN HOT ENVIRONMENTS

Workers in hot environments, around furnaces, smelters, boilers or out in the sun, can be subjected to considerable thermal stress. Because of natural climatic conditions and outdoors lifestyles and work styles, many environments have a high potential for heat-related work illnesses. The WHS practitioner should be able to recognise the physical factors contributing to heat stress and how the body responds to them, and be familiar with control procedures for these adverse factors. This section provides a basic introduction to the concepts of heat stress and its management. More comprehensive discussion in this area may be found in specialist texts and guides such as the Australian Institute of Occupational Hygienists' *A Guide to Managing Heat Stress: Developed for Use in the Australian Environment.* (Di Corleto, Firth and Maté, 2013), from which sections of this chapter have been adapted.

14.2.1   *HEAT STRESS AND HEAT STRAIN*

It is important at the outset to define two key terms associated with work in the thermal environment. The combined effect of external thermal environment and internal metabolic heat production constitutes the *thermal stress* on the body. The response to the thermal stress from bodily systems such as the cardiovascular, thermoregulatory, respiratory, renal and endocrine systems constitutes the *thermal strain*. Thus environmental conditions, metabolic workload and clothing, individually or combined, can create *heat stress* for the worker. The body's physiological response to that stress—for example, sweating, increased heart rate and elevated core temperature—is the *heat* strain (Di Corleto, Firth and Maté, 2013).

14.2.2   THE HEAT BALANCE EQUATION

In order for the body to maintain thermal equilibrium and avoid illness or injury, a thermal balance must be maintained. This is represented by the following equation.

$$H = M \pm C \pm R \pm K - E \qquad \text{(Equation 14.1)}$$

where:
H = net heat accumulation by the body
M = metabolic heat output
C = convective heat input or loss (can be positive or negative)
R = radiant heat input or loss (can be positive or negative)
K = conductive input or loss (can be positive or negative)
E = evaporative cooling by sweating (can only be negative)

During any activity, the body automatically attempts to maintain a constant working temperature range by balancing out the heat gain and heat loss. Working creates metabolic heat, and that heat is carried by the blood to the surface of the skin. The work causes the heart to pump faster, and so carries the blood faster to the surface. The body dissipates heat through the skin via vasodilation, whereby heat is transferred from the blood to the air surrounding the skin surface against a temperature gradient via radiation, convection and the cooling mechanism provided by evaporation of sweat off the skin.

14.2.3   ACCLIMATISATION

Workers exposed to repetitive bouts of work in hot environments (either artificial or natural) eventually become acclimatised, which ultimately reduces heat strain. Key physiological responses to acclimatisation are a lower heart rate, an earlier onset of sweating and dilute sweat content. There are different rates of physiological change and adaptation in the acclimatisation process, with some occurring more rapidly than others:

- The *first stage* of the acclimatisation process usually involves the cardiovascular processes of the body, such as heart rate decreases, which can occur in the first four to five days in physically fit individuals. Plasma volume expansion of up to 16 per cent can occur over the first three to five days (Pryor, Minson and Ferrara, 2018).
- The *intermediate phase* occurs when cardiovascular stability has been assured and surface and internal body temperatures are lower. Usually, 75–80 per cent of optimum stability can be achieved by day 8 (Pandolf, 1998), although some research carried out in northern Australia (Brake and Bates, 2001) suggests that about 70–80 per cent will be achieved after about seven to ten days.
- The *third phase* (>15 days) sees a decrease in the salt content of sweat and urine, and other compensations to conserve body fluids and restore electrolyte balances. Usually, 93 per cent of optimum is achieved by day 18 and 99 per cent by day 21 (ACGIH, 2000).

It is very important to note that employees who have been on extended leave, new employees and contract workers from a cooler climatic location will not be acclimatised, and this must be taken into consideration when scheduling work in a hot environment. Generally, new workers in hot environments must be permitted one to two weeks to acclimatise. Additionally, as acclimatisation is obtained to the level of the heat exposure present, a person will not be able to fully acclimatise to a sudden higher level of exposure such as those experienced during a heat wave.

The rate of decay of this heat acclimatisation has been suggested to occur such that one day of acclimatisation is lost for every two days spent without working in the heat (Casa, 2018). The first heat adaptations to decline are usually those that were achieved first, such as the cardiovascular adaptations (Garrett, Rehrer and Patterson, 2011).

14.2.4 THE BASIC FORMS OF HEAT ILLNESS

If environmental or personal work factors prevent the body from maintaining heat balance because the air temperature or humidity is too high, there is a high radiant heat load, the worker is constricted by insulating clothing, the worker's personal factors (age, health, drugs—illicit or prescribed), acclimatisation or the level of hydration, the body experiences physiological heat strain, with a range of different symptoms and illnesses that are dependent on the degree of heat stress. The following conditions of medical importance range from least to most hazardous:

- *Behavioural disorders.* Chronic or transient simple physical heat fatigue often occurs in workers from colder climates who are unacclimatised to continuously hot weather. This can often manifest as a change in demeanour, irritability, tiredness and lethargy, impaired judgement, loss of cognitive function and poor concentration. The relationship between performing manual work in heat and a subsequent reduction in cognitive function has been shown to have a significant link to workplace safety performance (Ganio et al., 2011; Knapik et al., 2002; Xiang et al., 2014). Lifestyle changes (suitable clothing, midday resting), avoiding strenuous work during the heat of the day and acclimatisation are appropriate measures to take.
- *Heat rash (or prickly heat).* This usually occurs as a result of continued exposure to humid heat during which the skin remains continuously wet from unevaporated sweat. This can often result in blocked glands, itchy skin and reduced sweating. In some cases, prickly heat can lead to lengthy periods of disablement (Donoghue and Sinclair, 2000). Where conditions encourage the occurrence of prickly heat—for example, in damp situations in tropical or deep underground mines—control measures may be important to prevent onset. Keeping the skin as clean, cool and dry as possible to allow recovery is generally the most successful approach.
- *Heat cramps.* These are characterised by painful spasms in one or more muscles. Heat cramps may occur in persons who sweat profusely in heat without replacing their salt losses, or unacclimatised personnel with higher levels of salt in their sweat. Resting in a cool place and oral replacement of electrolytes will rapidly alleviate cramps (Casa et al., 2008). Use of salt tablets is undesirable. Counselling by medical staff or a healthcare practitioner should be sought to ensure that workers maintain a balanced intake

of electrolytes, with meals if required. Note that heat cramps occur most commonly during heat exposure, but can also occur some time after heat exposure (Di Corleto, Coles and Firth, 2003).

- *Fainting (or heat syncope).* Exposure of fluid-deficient persons to hot environmental conditions can cause a major shift in the body's remaining blood supply to the skin vessels, in an attempt to dissipate heat. This ultimately results in an inadequate supply of blood being delivered to the brain, one consequence of which is fainting. Fainting may also occur without a significant reduction in blood volume as a result of wearing restrictive or confining clothing, or postural restrictions and changes.

- *Heat exhaustion.* While serious, heat exhaustion is initially a less medically severe heat injury than heat stroke, although it can become a precursor to heat stroke. Heat exhaustion generally is characterised by clammy, moist skin, weakness or extreme fatigue, nausea, headache, no excessive increase in body temperature and low blood pressure with a weak pulse. Without prompt treatment, collapse is inevitable. Heat exhaustion occurs most often in persons whose total blood volume has been reduced by dehydration (i.e. depletion of body water as a consequence of deficient water intake), but can also be associated with inadequate salt intake even when fluid intake is adequate. Individuals who have a low level of cardiovascular fitness and/or are not acclimatised to heat have a greater potential to suffer heat exhaustion, sometimes recurrently. This is particularly important where self-pacing of work is not practised. Note that in workplaces where self-pacing is practised, both fit and unfit workers tend to have a similar frequency of heat exhaustion (Brake and Bates, 2001). Lying down in a cool place and drinking an electrolyte supplement will usually result in rapid recovery of the victim of heat exhaustion, but a physician should be consulted prior to resumption of work. Heat exhaustion from salt depletion may require further medical treatment under supervision (Glazer, 2005).

- *Heat stroke.* This is a state of thermoregulatory failure and is the most serious of the heat illnesses. Heat stroke is usually characterised by hot, dry skin, rapidly rising body temperature, collapse, loss of consciousness and convulsions. Without prompt and appropriate medical attention, including the removal of the victim to a cool area and applying a suitable method for reduction of the rapidly increasing body temperature (exceeding 40°C), heat stroke can be fatal. Immediate cooling is necessary to reduce the body's core temperature. It has been recommended that whole-body immersion into an ice bath should be the method of choice when quick reduction of core temperature is needed (Casa et al., 2007). Caution needs to be taken to ensure that over-cooling of hyperthermic individuals does not occur (Gagnon et al., 2010). *A heat-stroke victim is a medical emergency and needs immediate and experienced medical attention.*

- *Chronic illness.* While the acute forms of illness associated with heat exposure are well known, there is increasing evidence that chronic exposure to heat can result in other illnesses in the long term. Studies have shown potential increase in susceptibility to kidney stones (Atan et al., 2005; Borghi et al., 1993) and kidney disease (Jimenez et al., 2014). Chronic dehydration has been linked with cardiopulmonary disorders, gastrointestinal dysfunction (El-Sharkawy, Sahota and Lobo, 2015) and bladder cancer (Jones and Ross, 1999).

14.2.5 FACTORS INFLUENCING HEAT STRESS

The working body gains heat from several sources:

- *Muscular activity from the work.* As the muscles of the body undertake work and oxygen is consumed, heat is released, which increases core temperature.
- *Conductive and convective heat from working in hot environmental conditions.* In some cases, heat is transferred to the body when hot objects are handled. Cool air can cool the body directly. If air temperature is hotter than body temperature, heat flows from the hotter air to the cooler skin surface. Air speed is also very important in workplace cooling because it influences evaporation rate and convective cooling. The humidity of the air is also able to affect evaporation rate. High humidity has less capacity to absorb moisture, resulting in poorer evaporation of sweat off the skin and less consequent cooling. Normal resting skin temperature is approximately 33°C.
- *Radiant heat from nearby or distant hot bodies.* These radiate heat in the infrared region, which passes through air (or vacuum) unobstructed. The infrared energy is absorbed by the body of the worker, by equipment in the workplace and by surrounding materials. This can be a major factor contributing to heat stress.

It is important to note that, in addition to environmental factors and metabolic workload, there are several personal factors that may exacerbate a worker's physiological response to working in heat; these include age, health, diet, hydration, medication and acclimatisation.

14.2.6 FLUID INTAKE AND THIRST

The importance of adequate fluid intake and the maintenance of correct bodily electrolyte balance cannot be over-emphasised. Maintaining heat balance in conditions of heat stress demands the production and evaporation of enough sweat to cool the body and assist with the balance of the heat gain from the environment and metabolism. Continuous production of sweat is influenced by the upper limit of fluid absorption from the digestive tract.

Fluid absorption is dependent on gastric emptying and intestinal absorption. Gastric emptying is the process in which the fluids in the stomach pass into the small intestine, where they are absorbed into the bloodstream. These processes are affected by a number of factors, including the volume, temperature, calorie content and osmolality of the fluid, as well as exercise intensity (Casa et al., 2000). Bariatric surgery is another factor that must be taken into consideration when assessing adequate fluid absorption and subsequent hydration in an industrial setting, with the challenges around sufficient fluid intake for those post bariatric surgery often posing a challenge. Physiological effects from dehydration may commence at 1.5–2 per cent change in total body weight (Casa et al., 2000; Hunt, Stewart and Parker, 2009), while a net fluid loss of 5 per cent or more in an occupational setting is considered severe dehydration.

A linear relationship has been shown to exist between colour, specific gravity and osmolality of urine (Shirreffs, 2003). Consequently, urine specific gravity (U_{sg}) is increasingly being used as a measure of dehydration in sporting and industrial settings. The methodology usually employed utilises refractometers, and self-testing via hydration specific urine test strips is now becoming popular. The US National Athletic Trainers'

Association (NATA) recommends that, 'Fluid replacement should approximate sweat and urine losses, and at least maintain hydration at less than 2% body weight reduction' (Casa et al., 2000). In its guideline, the American College of Sports Medicine recommends drinking 0.4–0.8 L/h of fluid during exercise, depending on the size of the individual and the level of work/exercise being undertaken (Sawka et al., 2007). Research has shown that body-weight loss levels of 2 per cent or more can be regarded as indicating that an individual is in the early stages of dehydration (Casa et al., 2000; Cheuvront and Sawka, 2005; Ganio et al. 2007; Sawka et al., 2007). The relationship between urine specific gravity and hydration level is illustrated in Table 14.1.

Table 14.1 NATA index of hydration status

	Body weight loss (%)	Urine specific gravity (USG)
Well hydrated	<1	<1.010
Minimal dehydration	1–3	1.010–1.020
Significant dehydration	3–5	1.021–1.030
Severe dehydration	> 5	> 1.030

Source: adapted from Casa et al. (2000).

Thirst is an inadequate indicator of the need for fluid replacement in an occupational environment. The sensation of thirst lags behind the loss of fluid, and most individuals are dehydrated to some degree. Workers should be encouraged to drink small amounts of water frequently rather than larger quantities infrequently. The water should be cool (10–15°C) and be available close to the workplace. In some cases, it may be desirable to flavour the water to make it more palatable, but low-sugar flavouring should be used.

It should be noted that high solute levels reduce the rate of water absorption in the gastrointestinal system. Alcohol is a diuretic, and hence increases water loss. While caffeine can increase non-sweat body fluid losses (urine production) in some individuals, recent studies (Armstrong et al., 2007; Killer, Blannin and Jeukendrup, 2014; Roti et al, 2006; Seal et al., 2017) have shown that moderate caffeine intake does not have the diuretic effect first thought and can add to the overall fluid intake of an individual. In situations of severe fluid loss, electrolyte replacement may also be required. In most situations, the diet provides sufficient salt to maintain electrolyte requirements for acclimatised individuals. With unacclimatised workers in a high heat stress scenario, a deficiency in electrolytes can occur even when large volumes of fluid are consumed. In such situations, salt tablets should not be used for electrolyte replacement, as commercially available electrolyte replacement drinks can be used sparingly to fulfil this role. Table 14.2 lists the advantages and disadvantages of a number of drinks commonly used for fluid replacement.

Table 14.2 Analysis of fluid replacement

Beverage type	Uses	Advantages	Disadvantages
Tea/coffee	Before, during and after work	Provide energy. Palatable.	Excessive quantities, which result in high levels of caffeine, may result in nervousness, insomnia and gastrointestinal upset.
Sports drinks	Before, during and after work	Provide energy. Aid electrolyte replacement. Palatable.	May not be correct mix. Excessive use may exceed salt replacement requirement levels. Low pH levels may affect teeth. Elevated carbohydrate content may contribute to excess calorie consumption and weight gain.
Fruit juices	Recovery	Provide energy. Palatable. Low in sodium.	Not absorbed rapidly.
Carbonated drinks	Recovery	Palatable. Variety of flavours. Provide potassium. Low sodium. Quick 'fillingness'.	Cause belching. 'Diet' drinks supply no energy. Risk of dental cavities. Some may contain caffeine and excessive intake is undesirable. Elevated carbohydrate content may contribute to excess calorie consumption and weight gain.
Water and mineral water	Before, during and after exercise	Palatable. Most obvious fluid. Readily available. Low sodium.	Not as good for high-output events of 60 minutes or more. No energy provided.
Milk	Before and after recovery	Contains sodium. Provides some energy. Use with fruit, cereal.	Slowly absorbed. Contains fat. Not suitable during an event.

Source: Adapted from Pearce (1996).

14.2.7 EDUCATION

Education and training are key components in any health-management program. In relation to heat stress, it should be conducted for all personnel likely to be involved with:

- hot environments
- physically demanding work at elevated temperatures
- the use of impermeable protective clothing.

Any combination of the above conditions will further increase the risk of a heat-related illness. The education and training should encompass the following:

- mechanisms of heat exposure
- potential heat exposure situations
- recognition of predisposing factors
- the importance of fluid intake
- the nature of acclimatisation
- effects of alcohol and drug (illicit and prescription) use in hot environments
- early recognition of symptoms of heat illness
- prevention of heat illness
- first aid treatment for heat-related illnesses
- self-assessment
- management and control
- medical surveillance programs and the advantages of employee participation in such programs.

Training of all personnel in heat stress management should be recorded on their personal training record.

14.2.8 SELF-ASSESSMENT

Self-assessment is a key element in the training of workers potentially exposed to heat stress. With the correct knowledge about signs and symptoms, individuals will be in a position to identify the onset of a heat illness in the earliest stages and take appropriate action. This may simply involve taking a short break and a drink of water, which in most cases should take only a matter of minutes. This brief intervention can help significantly in preventing the onset of the more serious heat-related illnesses, particularly when workers are also allowed to carry out tasks at their own pace.

14.2.9 ASSESSMENT OF THE HOT THERMAL ENVIRONMENT

Numerous factors can affect the heat stress associated with a particular task or environment, and no single factor can be assessed in isolation. A structured assessment protocol is the recommended approach, with the flexibility to address a variety of situations.

The use of a heat stress index alone to determine heat stress and the resultant heat strain is not recommended. Each situation requires an assessment that will incorporate the many parameters that may impact on an individual working in hot conditions.

In effect, a risk assessment must be carried out that includes additional observations such as workload, worker characteristics and personal protective equipment, as well as measurement and calculation of the thermal environmental conditions. This process may involve a variety of heat stress indices, including but not limited to wet bulb globe temperature, predicted heat strain, basic effective temperature and thermal work limit.

Wet bulb globe temperature (WBGT) uses air temperature (T_a), globe temperature (T_g) and a natural wet bulb temperature (T_{nwb}). These parameters are incorporated into one of two formulae for either indoor or outdoor measurements:

Indoor: $\quad\quad$ WBGT = 0.7 T_{nwb} + 0.3 T_g $\quad\quad\quad\quad\quad\quad\quad\quad\quad\quad$ (Equation 14.2)

Outdoor: $\quad\quad$ WBGT = 0.7 T_{nwb} + 0.2 T_g + 0.1 T_a $\quad\quad\quad\quad\quad\quad\quad$ (Equation 14.3)

The basic effective temperature (BET) is used predominantly in the underground coal-mining industry, and combines dry bulb and aspirated wet bulb temperatures with air velocity. It should be noted that caution should be exercised when utilising BET in situations where values above 31°C may be encountered, since the index is less accurate above this value (Hanson et al., 2000).

Predicted heat strain (PHS) is a rational index (i.e. it is an index based on the heat balance equation—see section 14.2.2). It estimates the required sweat rate and the maximal evaporation rate, utilising the ratio of the two as an initial measure of 'required wettedness'. This required wettedness is the fraction of the skin surface that would have to be covered by sweat in order for the required evaporation rate to occur. The evaporation rate required to maintain a heat balance is then calculated (Di Corleto, Firth and Maté, 2013).

The thermal work limit (TWL) was developed in the underground mining industry in Australia by Brake and Bates (2002). TWL is defined as the limiting (or maximum) sustainable metabolic rate that hydrated, acclimatised individuals can maintain in a specific thermal environment, within a safe deep body core temperature (<38.2°C) and sweat rate (<1.2 kg/hr).

14.2.9.1 Assessment protocol

A recommended method of assessment is as follows:

- *Stage 1:* Conduct a basic heat stress risk assessment incorporating a simple index, and a basic review of the task and the environment.
- *Stage 2:* If a potential problem is indicated from the initial step, then progress to a second level rational index (e.g. PHS, TWL) to undertake a more comprehensive investigation of the situation and general environment. Ensure that factors such as temperature, radiant heat load, air velocity, humidity, clothing, metabolic load, posture and acclimatisation are taken into account.
- *Stage 3:* Where the calculated allowable exposure time is less than 30 minutes, or there is an involvement of high-level personal protective equipment, employ some form of physiological monitoring.

The first two stages involve measuring the environmental parameters, measuring metabolic work rate factors and/or estimating workload factors from task observation. The third stage looks at the individual's physiological response to the exposure. Technological methods for measuring deep core temperature include swallowing a continuous radio-transmitting temperature sensor. However, for the average H&S practitioner, such techniques or other physiological measurements have limitations.

14.2.10 STAGE 1: BASIC RISK ASSESSMENT

The first level of assessment utilises a basic observational risk assessment in conjunction with a simple index such as apparent temperature (AT) or WBGT. It is important for the initial assessment to involve a review of the work conditions, the task and the personnel involved. Risk assessments may be carried out using checklists or pro-formas designed to prompt the assessor to identify potential problem areas. The method may range from a short checklist through to a more comprehensive calculation matrix that will produce a numerical result for comparative or priority listing. Also now available are mobile phone and tablet applications to aid in conducting a basic thermal risk assessment (BTRA).

A BTRA, such as that used in the AIOH guidelines (Di Corleto, Firth and Maté, 2013) and illustrated in Table 14.3, is a simple first approach. The table incorporates a number of heat stress-related factors that could influence an individual. These factors are given a numerical value and weighted according to their potential influence. The values are then used in a simple calculation that yields a numerical value, which may be used to assess the potential risk according to a predetermined scale. This approach encourages the individual or team assessing the situation to review a number of parameters and not focus solely on one measure, such as air temperature. The simple AT index assists with the final result by adding a level of environmental measurement and objectivity.

The AT index provides a basic, convenient measure for heat stress evaluations as in its simplest form only air temperature and humidity are required to estimate the final temperature. The original simple version of the AT is used to try to keep the qualitative assessment easy to use.

Two simple temperature measurements will provide the necessary information:

1   Measure the dry air temperature in the workplace with a thermometer (e.g. mercury in glass or thermocouple junction), shielded from radiant energy. A sling psychrometer that comprises a dry bulb and wet bulb thermometer is often used for this purpose, as it will also allow the calculation of humidity (Figure 14.1).

Figure 14.1 A modern wet bulb globe temperature (WBGT) instrument

2   Measure the humidity at the same point in the workplace, which can readily be accomplished with the psychrometer or a hygrometer. The wet bulb is a standard thermometer in which the temperature-sensing element is enclosed in a white cotton wick that is wetted by immersion in distilled water. This temperature reading will take into account the ability of the air to cool by evaporation and the air movement that aids cooling by sweat evaporation from the skin.

The dry bulb temperature is aligned with the corresponding relative humidity to determine apparent temperature in Table 14.3. Numbers in parentheses refer to skin humidity above 90 per cent and are only approximate.

Table 14.3 Apparent temperature: temperature–humidity scale

Dry bulb temperature (°C)	Relative humidity (%)										
	0	10	20	30	40	50	60	70	80	90	100
20	16	17	17	18	19	19	20	20	21	21	21
21	18	18	19	19	20	20	21	21	22	22	23
22	19	19	20	20	21	21	22	22	23	23	24
23	20	20	21	22	22	23	23	24	24	24	25
24	21	22	22	23	23	24	24	25	25	26	26
25	22	23	24	24	24	25	25	26	27	27	28
26	24	24	25	25	26	26	27	27	28	29	30
27	25	25	26	26	27	27	28	29	30	31	33
28	26	26	27	27	28	29	29	31	32	34	(36)
29	26	27	27	28	29	30	30	33	35	37	(40)
30	27	28	28	29	30	31	33	35	37	(40)	(45)
31	28	29	29	30	31	33	35	37	40	(45)	
32	29	29	30	31	33	35	37	40	44	(51)	
33	29	30	31	33	34	36	39	43	(49)		
34	30	31	32	34	36	38	42	(47)			
35	31	32	33	35	37	40	(45)	(51)			
36	32	33	35	37	39	43	(49)				
37	32	34	36	38	41	46					
38	33	35	37	40	44	(49)					
39	34	36	38	41	46						
40	35	37	40	43	49						
41	35	38	41	45							
42	36	39	42	47							
43	37	40	44	49							
44	38	41	45	52							
45	38	42	47								
46	39	43	49								
47	40	44	51								
48	41	45	53								
49	42	47									
50	42	48									

Source: Steadman (1979).

14.2.11 *WORKED EXAMPLE OF A BASIC THERMAL RISK ASSESSMENT*

An example of the application of the basic thermal risk assessment would be as follows:

> A fitter is working on a pump out in the plant at ground level that has been taken out of service the previous day. The task involves removing bolts and a casing to check the impellers for wear, approximately 2 hours of work. The pump is situated approximately 25 metres from the workshop. The fitter is acclimatised, has attended a training session and is wearing a standard single layer long shirt and trousers, is carrying a water bottle, and a respirator is not required. The work rate is light, there is a light breeze and the air temperature has been measured at 30°C, and the relative humidity at 70 per cent. This equates to an apparent temperature of 35°C (see Table 14.3).

An example of the application of the basic thermal risk assessment using this information is given in Table 14.4.

Examples of work rate include:

- *light work:* sitting or standing to control machines; hand and arm work assembly or sorting of light materials
- *moderate work:* sustained hand and arm work such as hammering, handling of moderately heavy materials
- *heavy work:* pick and shovel work, continuous axe work and carrying loads upstairs.

- *Subtotal A* = 9. This is a general measure of the working environment (excluding temperature) and some key influencing factors associated with heat stress.
- *Subtotal B* = 2. This evaluates the metabolic load on the individual (i.e. the intensity of work being undertaken in the task).
- *Subtotal C* = 3. Here the assessment incorporates the actual environmental temperatures using the apparent temperature index. As this is an important factor in the scenario, it is given added weight by using a multiplication factor rather than addition.

The formula used to arrive at the numerical assessment of risk is (A + B) × C; hence the total = (9 + 2) × 3 = 33.

- If the total is *less than 28*, then the risk from thermal conditions is low to moderate.
- If the total is *28 to 60*, there is a risk of heat-induced illnesses occurring if the conditions are not addressed. Further analysis of heat stress risk is required.
- If the total *exceeds 60*, then the onset of a heat-induced illness is very likely and action should be taken to implement controls as soon as possible.

As the total lies between 28 and 60, there is a risk of heat-induced illness if the conditions are not addressed, and more comprehensive analysis of heat stress risk or implementation of controls is required.

Table 14.4 Worked example of basic thermal risk assessment

Hazard type	Assessment point value							
	0		**1**		**2**		**3**	
Sun exposure	Indoors	☐	Shade	☑	Part shade	☐	No shade	☐
Hot surfaces	Neutral	☑	Warm on contact	☐	Hot on contact	☐	Burn on contact	☐
Exposure period	<30 min	☐	30 min–1 hour	☐	1–2 hours	☑	>2 hrs	☐
Confined space	No	☑					Yes	☐
Task complexity			Simple	☐	Moderate	☑	Complex	☐
Climbing, up/down stairs or ladders	None	☑	One level	☐	Two levels	☐	>Two levels	☐
Distance from cool rest area	<10 m	☐	<50 m	☑	50–100 m	☐	>100 m	☐
Distance from drinking water	<10 m	☑	<30 m	☐	30–50 m	☐	>50 m	☐
Clothing (permeable)			Single layer (light)	☑	Single layer (mod)	☐	Multiple layer	☐
Understanding of heat strain risk	Training given	☑					No training given	☐
Air movement	Strong wind	☐	Moderate wind	☐	Light wind	☑	No wind	☐
Resp. protection (-ve pressure)	None	☑	Disposable half face	☐	Rubber half face	☐	Full face	☐
Acclimatisation	Acclimatised	☑					Unacclimatised	☐
				3		6		0
SUB-TOTAL A								9

	2		**4**		**6**	
Metabolic work rate*	Light	☑	Moderate	☐	Heavy	☐
SUB-TOTAL B						2

	1		**2**		**3**		**4**	
Apparent temperature	<27°C	☐	>27°C ≤ 33°C	☐	>33°C ≤ 41°C	☑	>41°C	☐
SUB-TOTAL C								3

TOTAL = A plus B	Multiplied by	C		=		33

* Examples of work rate
Light work: sitting or standing to control machines; hand and arm work assembly or sorting of light materials.
Moderate work: sustained hand and arm work such as hammering, handling of moderately heavy materials.
Heavy work: pick and shovel work, continuous axe work, carrying loads up stairs.

14.2.12 *STAGE 2 OF THE ASSESSMENT*

When stage 1 of an assessment indicates that the conditions may be unacceptable, as in the above example, relatively simple and practical control measures should be considered based on the outcomes identified in the BTRA. Where these are unavailable, a more detailed assessment is required. This stage usually involves a more extensive measurement survey of the environment, including humidity, air velocity, clothing, posture, globe temperature and metabolic load. These additional data would then be used in a higher-level heat stress index such as the rational index ISO 7933: 2004 *Ergonomics of the Thermal Environment: Analytical Determination and Interpretation of Heat Stress Using Calculation of the Predicted Heat Strain* (ISO, 2004) or other rational indices. Assessments at this level require technical interpretation, and assistance may be required from a suitably qualified specialist such as an occupational hygienist. The number of calculations required with rational indices necessitates the use of pre-programmed instrumentation, a computer program or a calculation spreadsheet. Annex E2 (ISO, 2004, p. 24) details an example of code that may be used to develop a computer program for performing predicted heat strain model computations. There are also mobile phone and tablet applications available and some equipment providers include software to calculate PHS with their instruments. This allows calculation of predicted body core temperatures for specific environmental conditions and workloads for a number of task or work phases, and may be used as a guide in the development of heat stress controls. It should always be taken into consideration that the aforementioned tools are guides only and should not be used for the development of absolute safe/unsafe limits.

Figure 14.2 shows a simple, practical portable device for WBGT measurements being used to measure temperatures in a hot area of an aluminium smelter.

Figure 14.2 Use of a portable WBGT in an indoor industrial setting

A heat-stress risk assessment checklist should also be used as part of the stage 2 process to ensure that a comprehensive assessment is undertaken. An example of a suitable checklist is presented in Table 14.5.

Table 14.5 Heat stress risk assessment impacts

Assessment parameter	Impact
Dry bulb temperature	Elevated temperatures will add to the overall heat burden.
Globe temperature	Will give some indication of the radiant heat load.
Air movement—wind speed	Poor air movement will reduce the effectiveness of sweat evaporation; high air movements at high temperatures (>42°C) will add to the heat load.
Humidity	High humidity is also detrimental to sweat evaporation.
Hot surfaces	Can produce radiant heat as well as result in contact burns.
Metabolic work rate	Elevated work rates can potentially increase internal core body temperatures.
Exposure period	Extended periods of exposure can increase heat stress.
Confined space	Normally results in poor air movement and increased temperatures.
Task complexity	Will require more concentration and manipulation.
Climbing, ascending, descending—work rate increase	Can increase metabolic load on the body.
Distance from cool rest area	Long distances may be disincentive to leave hot work area or seen as time-wasting.
Distance from drinking water	Prevents adequate rehydration.

Employee condition	Impact
Medications	Diuretics, some anti-depressants and anti-cholinergics may affect the body's ability to manage heat.
Chronic conditions—such as heart or circulatory	May result in poor blood circulation and reduced body cooling.
Acute infections—such as colds, influenza, fevers	Will affect the way the body handles heat stress—thermoregulation.

Table 14.5 Heat stress risk assessment impacts *continued*

Employee condition	Impact
Acclimatisation	Poor acclimatisation will result in poorer tolerance of heat—less sweating, more salt loss.
Obesity	Excessive weight will increase the risk of a heat illness.
Age	Older individuals (>50) may cope less well with the heat.
Fitness	A low level of fitness reduces cardiovascular and aerobic capacity.
Alcohol in last 24 hours	Will increase the likelihood of dehydration.
Chemical exposure factors	**Impact**
Gases, vapours and dusts soluble in sweat	May result in chemical irritation/burns and dermatitis.
Impermeable clothing	Significantly affects the body's ability to cool.
Respiratory protection (negative pressure)	Will affect the breathing rate and add an additional stress on the worker.
Increased workload owing to personal protective equipment PPE	Items such as self-contained breathing apparatus (SCBA) will add weight and increase metabolic load.
Restricted mobility	Will affect posture and positioning of employee.

14.2.13 STAGE 3: INDIVIDUAL HEAT STRESS MONITORING

In some circumstances, rational indices cannot provide the necessary information to guide the assessment of the exposed work group, and the use of individual physiological monitoring may be required. This may include situations of high heat stress risk or where the rational index indicates that the exposure time is limited to less than 30 minutes or where the individual's working environment cannot be assessed accurately. A common example is work involving the use of encapsulating suits.

Instruments for personal heat stress monitoring do not measure the environmental conditions leading to heat stress; rather, they monitor the physiological indicators of heat strain—usually body temperature and/or heart rate. Temperature may be measured by a number of routes, including oral, rectal, aural (ear canal or tympanic), oesophageal, skin and internal telemetry. Each of these methods (measuring heart rate and body temperature) has advantages and drawbacks. The method chosen must not only provide the required data but also be acceptable to the individual being monitored.

Figure 14.3 shows a personal heat stress monitor that employs an ingestible core temperature capsule, which transmits physiological parameters to an external data-logging

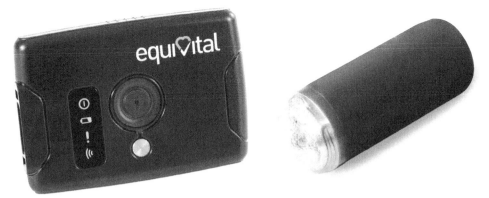

Figure 14.3 Personal heat stress monitor (left) data-logging sensor; (right) ingestible capsule

sensor or laptop computer. It is small, lightweight, can be slipped into a shirt pocket or worn on a belt and is fitted with an audible indicator to warn of stressing conditions. Thresholds are not fixed for heart rate or temperature, but are established as the physiological measurements are taken.

Such instruments provide a logged time-history graph that may be downloaded to a computer for further analysis. Several of these also have the ability to be able to connect to a mobile phone or tablet via an application, with the data being able to be viewed in real time as the task at hand is being performed. Depending on the circumstances, physiological monitoring can be involved and complex. When a risk assessment deems it necessary, a competent person with proven technical skills and experience in heat stress and/or human physiology must undertake the assessment. H&S practitioners involved in industries where heat stress is a continuing problem may find this approach very useful. A good example is the stripping of asbestos while dressed in impermeable plastic clothing, a practice not at all suitable for tropical and subtropical climates.

14.2.14 PRE-PLACEMENT HEALTH ASSESSMENT

Pre-placement health assessment screening should be considered for identifying those susceptible to heat injury, or for tasks involving high heat stress exposures. The standard ISO 12894 *Ergonomics of the Thermal Environment: Medical Supervision of Individuals Exposed to Extreme Hot or Cold Environments.* (ISO, 2001) provides guidance for medical supervision of individuals exposed to extreme heat. Health assessment screening should consider the worker's physiological and biomedical aspects, and provide an interpretation of job fitness for the tasks to be performed. Specific indicators of heat intolerance should be included.

Some workers may be more vulnerable to heat stress than others. They include but are not limited to individuals who:

- are dehydrated
- are unacclimatised to workplace heat levels
- are physically unfit
- have low aerobic capacity, as measured by maximal oxygen consumption
- carry excessive weight

- are more than 50 years old
- suffer from diabetes
- suffer from hypertension
- suffer from heart, circulatory or skin disorders
- suffer from thyroid disease
- are anaemic
- use medications that impair temperature regulation or perspiration.

Workers with a history of renal, neuromuscular or respiratory disorders, previous head injury or fainting spells, or previous susceptibility to heat illness may also be at risk (Hanson and Graveling 1997, Brake, Donoghue and Bates, 1998). Those individuals who are at greater risk may be excluded from certain work conditions or have more frequent medical checks.

Workers with short-term disorders or illnesses, such as colds or flu, diarrhoea, vomiting, lack of sleep and hangover, should also be considered at risk. These acute disorders will limit their ability to tolerate heat stress and hence make them more susceptible to heat illness.

14.2.15 SETTING LIMITS FOR HEAT EXPOSURE

As can be seen from the preceding section, a number of variables associated with individual wellbeing can affect workers' response to heat. When climatic variables such as humidity and air speed, the clothing/PPE required for the job and the workload itself are taken into account, the equation for estimating heat loads becomes extremely complex. For this reason, the practice of selecting a single parameter such as air temperature as a work/no-work limit is dangerous. It is impossible to select any one temperature that is suitable for work for all individuals and tasks. Heavy work at temperatures not normally classed as elevated can produce a significant heat stress risk equal to low workloads at higher temperatures. A temperature of 35°C at 15 per cent humidity is more tolerable than one of 30°C at 95 per cent humidity. Considering individual health and fitness variations reveals how difficult it would be to select a single valid temperature for limiting work. Using the 'one temperature' principle, it is quite possible to over-protect some workers and under-protect others within the same group. Management of heat stress requires investigating all variables in the workplace rather than only one easily measurable parameter.

14.2.16 CONTROL MEASURES AGAINST HEAT STRESS

A number of controls may be utilised in the workplace so environmental and task factors can be manipulated to reduce heat stress. Ideally, controls should be prioritised based on the hierarchy of controls (HoC). From the data collected in the risk assessment and measurement stages, controls may be determined. Some examples of these potential controls are listed in Table 14.6.

Worker training and selection are also important. Training content should include:

- heat acclimatisation
- appropriate levels of physical fitness

- adherence to a liquid replacement schedule
- maintenance of electrolyte balance in body fluids, especially for non-acclimatised workers
- training of supervisors and workers in the recognition of various heat illnesses
- information for workers on the impact of drugs, alcohol and obesity on heat illnesses
- screening workers for heat intolerance (particularly previous episodes)
- seasonal factors relating to climate.

Application of these guidelines should ensure that workplaces are free of heat stress-inducing conditions. Illness from heat stress is totally preventable.

Table 14.6 Heat exposure control examples

Elimination

Utilise remote-controlled equipment such as mini-diggers, automated hydroblasting equipment etc.

Conduct the task at cooler parts of the day.

Allow hot equipment or vessels to cool before working on or nearby.

Engineering controls

Erect shade or barriers when radiant heat sources are involved.

Use insulating and/or clad equipment.

Improve ventilation by using force draft fans with or without chiller units.

Use cranes, forklifts, manual handling aids to reduce metabolic work rate.

Establish cooled rest areas in close vicinity to the task.

Change the emissivity of the hot surface.

Dehumidify air to increase evaporative cooling from sweating.

Eliminate sources of water vapour from leaks in steam lines or standing water evaporating from floors.

Administrative

Use extra human resources or mechanisation to reduce exposure time for each worker.

Establish appropriate work/rest regimes.

Simplify tasks into smaller components.

Pre-plan tasks to include heat as a risk.

Ensure cool, palatable water is readily available close by.

Supply electrolyte replacement where appropriate.

Personal protective equipment

Provide specialised vortex air-cooled or ice/phase change vests for some continuous-demand tasks.

14.3   WORK IN COLD CLIMATES

Concerns relating to work in cold conditions occur predominantly in countries or regions where snow falls during winter and outdoor work must continue. Cold is also relevant to work in freezer plants and cold-storage facilities, and for a few outdoor occupations in winter. Efforts in the prevention of cold-related health problems are directed towards the maintenance of body heat.

During exposure to cold environments, the body responds by constricting blood vessels in the skin. When heat is continually lost and body temperature decreases, muscles begin to involuntarily contract (shiver) to produce heat. The body's 'thermometer' is a pea-sized region of the brain called the hypothalamus. This area regulates heat loss and heat gain mechanisms such as changes in skin blood flow and initiation of sweating and shivering. As a result of severe cold, nerve reactions become impeded and fingers and hands lose dexterity, thus presenting additional safety hazards.

14.3.1   HEALTH EFFECTS OF EXPOSURE TO EXTREMES OF COLD

Local injuries are generally the most common forms of cold injuries. These involve the impact of low temperatures on tissue in the exposed extremities. Localised effects include frostbite and chilblains, which result when insufficient blood reaches the extremities and the fluids around cells freeze. This causes tissue damage and can occur on any superficial tissues on the body. Frostbite can occur when very cold objects are handled or when cold air is passed over the skin for a period of time. The initial stages are sometimes also referred to as 'frostnip'. A sensation of cold is followed by numbness. Frostbite occurs in three degrees: freezing; freezing with blistering or peeling; and freezing with tissue death.

Actual freezing is not always necessary to cause serious tissue injury. Injury can also occur from prolonged local cooling at temperatures well above freezing. Trench foot was common in the trenches during World War I, and immersion foot was described in World War II. Both conditions can occur as a result of prolonged exposure in damp or wet conditions at temperatures from 0°C to 10°C. The feet become cold and swollen, and may be painful, itchy or numb. Damage occurs to the nerves and muscles, and gangrene develops in severe cases (Burton and Edholm, 1955).

Generalised effects of exposure to severe cold include uncontrollable shivering accompanied by slowing heart rate and a decrease in blood pressure. Similarly to heat stress, exposure to extreme cold is known to have an impact on mood (tension, depression, anger) as well as fatigue and cognitive performance. Cold-induced cognitive decline has also been linked to workplace safety (Muller et al., 2012). Speech may become slurred and incoherent, with drowsiness, irregular breathing and cool skin noted as body core temperature decreases to around 30°C. Serious problems occur at a body temperature below 30°C. Thermogenesis ceases, heat loss becomes pronounced and the respiratory rate decreases markedly. Popular remedies such as drinking alcohol to keep warm can be dangerous because they exacerbate heat loss by dilating surface blood vessels.

14.3.2 ASSESSMENT OF THE COLD THERMAL ENVIRONMENT

Assessment of cold environments can be undertaken using a similar approach to that for hot environments; however, the effects of cold should be evaluated from two different perspectives:

- the impact of local cooling on the extremities of the body (e.g. hands, feet and face), employing approaches such as the Wind Chill Index (WCI)
- the impact of the cooling effect on the body overall, utilising thermal indices based on the heat balance equations, such as the Required Clothing Insulation Index (IREQ).

The main factors contributing to cold injuries are humidity, wind contact with cold bodies, improper clothing and general state of health. Assessment of localised cooling also requires the consideration of parameters such as convective cooling (wind chill), conductive cooling, extremity cooling and airway cooling.

Wind chill, when the wind blows away insulating layers of air near the body, can make cold conditions feel bitterly cold. The wind-chill factor is significant for individuals working in cold because threshold limit values (TLVs®) (Table 14.7) are based on workload and wind speed. Not all countries have standards for cold exposure, but the TLV® table, developed in North America, can be applied.

The IREQ is used in the assessment of general body cooling. It is defined as 'the resultant clothing insulation required to maintain the body in thermal equilibrium under steady state conditions when sweating is absent and peripheral vasoconstriction is present' (BOHS, 1996, p. 56). It incorporates the effects of elements of the heat balance equation, which include:

- air temperature
- mean radiant temperature
- relative humidity
- air velocity
- metabolic rate.

An important aspect of the IREQ is that it can also be utilised to identify and evaluate controls and improvements in work planning for tasks under cold environmental conditions. As with the heat rational indices such as PHS, any of the parameters of the heat balance equation may be changed, and the relative impact on the IREQ can then be assessed.

ISO 11079: 2007 (ISO, 2007) provides more comprehensive detail about the principles and application of the index. The risk assessment and management of work in cold environments can follow similar principles to those employed for heat stress. In both cases, the strategy of a three-stage protocol is readily applied:

1 *Observation:* basic thermal risk assessment or checklist as per annex A in ISO 15743:2008.
2 *Analysis:* determination of wind cooling effects or incorporating a cold stress index such as the IREQ.
3 *Expertise:* utilisation of individuals with specific competencies such as occupational health-care professionals and occupational hygienists and specialised monitoring equipment (ISO, 2008).

Table 14.7 Threshold limit values as work/warm-up schedule for a four-hour shift

Air temp °C (approx.), sunny sky	No noticeable wind		8 km/h wind speed		15 km/h wind speed		25 km/h wind speed		35 km/h wind speed	
	Max work (mins)	No. of breaks	Max work (mins)	No. of breaks	Max work (mins)	No. of breaks	Max work (mins)	No. of breaks	Max work (mins)	No. of breaks
−26 to −28	(Norm. breaks)	1	(Norm. breaks)	1	75	2	55	3	40	4
−29 to −31	(Norm. breaks)	1	75	2	55	3	40	4	30	5
−32 to −34	75	2	55	3	40	4	30	5	EWO*	
−35 to −37	55	3	40	4	30	5	EWO*			
−38 to −39	40	4	30	5	EWO*					
−40 to −42	30	5	EWO*							
−43 and below	EWO*									

Notes: Schedule applies to any four-hour work period with moderate to heavy work activity, with warm-up periods of 10 minutes in a warm location and with an extended break (e.g. lunch) at the end of the four-hour work period in a warm location.

Examples of wind movement are: 8 km/h: light flag moves; 15 km/h: light flag is fully extended; 25 km/h: newspaper sheet is lifted; 35 km/h: snow blows and drifts.

TLVs® apply only for workers who are dressed in dry clothing.

*EWO = emergency work only

Source: Adapted from Occupational Health and Safety Division, Saskatchewan Department of Labor, Canada. Reproduced by permission of the American Conference of Governmental Industrial Hygienists.

14.3.3 *PREVENTION AND CONTROL MEASURES*

In formulating measures to reduce the risk of hypothermia and cold injury, the following are relevant factors:

- Skin exposure at equivalent chill temperatures of –32°C or less needs to be prevented.
- Wet clothing should be changed for exposures at temperatures below –2°C.
- Special care needs to be taken if working with evaporative liquids such as alcohol, petrol or solvents that may spill on the hands (Parsons, 2003).
- Acclimatisation to cold conditions can have a beneficial, though small, effect.
- Cold air is often very dry, and insidious dehydration via water loss through the skin must be prevented by consumption of warm, non-alcoholic drinks.
- Salt balance is best controlled by normal dietary means.
- Engineering controls can include:
    - provision of wind shields outdoors, or against circulated air indoors in freezer rooms
    - provision of local heating, hot air jets, radiant heating if bare hands have to be used
    - avoiding metal tools
    - avoiding seats for low temperature work (< –1°C)
    - use of powered equipment to reduce physical workload
    - heated shelters for recovery or, if possible, for working in.
- Administrative controls can include:
    - staying within the TLVs® for cold work (this involves measuring air temperature and wind speed and minimising work in cold temperatures)
    - maintaining a schedule of rest and liquid refreshment
    - having an adequate workforce
    - using the least cold part of the day for the coldest work (i.e. work with the highest exposure potential)
    - instructing workers to recognise and act on adverse effects of cold
    - avoiding long shifts or excessive overtime in the cold.

PPE against cold primarily takes the form of adequate clothing. Work in cold environments is one of the few work situations where PPE is the first line of defence. Layers of clothing provide a number of insulating air layers. The clothing must be permeable to sweat—inner layers of cotton are ideal. Particular attention must be paid to the hands, the feet and the head—a large heat emitter.

14.4   REFERENCES

American Conference of Governmental Industrial Hygienists (ACGIH) 2000, *Threshold Limit Values for Chemical Substances and Physical Agents and Biological Exposure Indices*, 6th ed., Supplement, ACGIH®, Cincinatti, OH.

Armstrong, L.E., Casa, D.J., Maresh, C.M. and Ganio, M.S. 2007, 'Caffeine, fluid–electrolyte balance, temperature regulation, and exercise-heat tolerance', *Exercise Sport Science Review*, vol. 35, pp. 135–40.

Atan, L., Andreoni, C., Ortiz, V., Silva, E.K., Pitta, R., Atan, F. and Sougi, M. 2005, 'High kidney stone risk in men working in steel industry at hot temperatures', *Urology*, vol. 65, pp. 858–61.

Borghi, L., Meshi, T., Amato, F., Novarini, A., Romanelli, A. and Cigala, F. 1993, 'Hot occupation and nephrolithiasis', *Journal of Urology*, vol. 150, pp. 1757–60.

Brake, D.J. and Bates, G.P. 2001, personal communication with the author.

—— 2002, 'Limiting metabolic rate (thermal work limit) as an index of thermal stress', *Applied Occupational and Environmental Hygiene*, vol. 17, no. 3, pp. 176–86.

Brake, D.J., Donoghue, A.M. and Bates, G.P. 1998, 'A new generation of health and safety protocols for working in heat', in *Proceedings of Queensland Mining Industry Health and Safety Conference: New Opportunities, Yeppoon, 30 August–2 September*, pp. 91–100.

British Occupational Hygiene Society (BOHS) 1996, *The Thermal Environment*, Technical Guide no. 12, 2nd ed., H and H Scientific Consultants, Leeds, UK.

Burton, A.C. and Edholm, O.G. 1955, *Man in a Cold Environment—Physiological and pathological effects of exposure to low temperatures*, Edward Arnold, London

Casa, D.J. (ed.) 2018, *Sport and Physical Activity in the Heat*, Springer, Dordrecht, pp. 42–3.

Casa, D.J., Armstrong, L.E., Hillman, S.K., Montain, S.J., Reiff, R.V., Rich, B.S., Roberts, W.O. and Stone, J.A. 2000, 'National Athletic Trainers' Association position statement: Fluid replacement for athletes', *Journal of Athletic Training*, vol. 35, no. 2, pp. 212–24.

Casa, D.J., Ganio, M.S., Lopez, R.M., McDermott, B.P., Armstrong, L.E. and Maresh, C.M. 2008, 'Intravenous versus oral rehydration: Physiological, performance, and legal considerations', *Current Sports Medicine Reports*, vol. 7, no. 4, pp. S41–9.

Casa, D.J., McDermott, B.P., Lee, E.C., Yeargin, S.W., Armstrong, L.E. and Maresh, C.M. 2007, 'Cold water immersion: The gold standard for exertional heatstroke treatment', *Exercise and Sport Science Reviews*, vol. 35, no. 3, pp. 141–9.

Cheuvront, S.N. and Sawka, M.N. 2005, 'Hydration assessment of athletes', *Sports Science Exchange*, vol. 18, no. 2, pp. 1–8.

Di Corleto, R., Coles, R. and Firth, I. 2003, 'The development of a heat stress standard for Australian conditions', in *Australian Institute of Occupational Hygienists Inc. 20th Annual Conference Proceedings*, AIOH, Melbourne.

Di Corleto, R., Firth, I. and Maté, J. 2013, *A Guide to Managing Heat Stress: Developed for Use in the Australian Environment*, AIOH, Melbourne.

Donoghue, A.M. and Sinclair, M.J. 2000, 'Miliaria rubra of the lower limbs in underground miners', *Occupational Medicine*, vol. 50, no. 6, pp. 430–3.

El-Sharkawy, A.M., Sahota, O. and Lobo, D.N. 2015, 'Acute and chronic effects of hydration status on health', *Nutrition Reviews*, vol. 73, supp. 2, pp. 97–109.

Gagnon, D., Lemire, B.B., Casa, D.J. and Kenny, G.P. 2010, 'Cold-water immersion and the treatment of hyperthermia: Using 38.6°C as a safe rectal temperature cooling limit', *Journal of Athletic Training*, vol. 4, no. 5, pp. 439–44.

Ganio, M.S., Armstrong, L.E., Casa, D.J. et al. 2011, 'Mild dehydration impairs cognitive performance and mood of men', *British Journal of Nutrition*, vol. 106, pp. 1535–43.

Ganio, M.S., Casa, D.J., Armstrong, L.E. and Maresh, C.M. 2007, 'Evidence based approach to lingering hydration questions', *Clinics in Sports Medicine*, vol. 26, no. 1, pp. 1–16.

Garrett, A.T., Rehrer, N.J. and Patterson, M.J. 2011, 'Induction and decay of short-term heat acclimation in moderately and highly trained athletes', *Sports Medicine*, vol. 41, p. 757.

Glazer, J.L. 2005, 'Management of heatstroke and heat exhaustion', *American Family Physician*, vol. 71, no. 11, pp. 2133–40.

Hanson, M.A., Cowie, H.A., George, J.P.K., Graham, M.K., Graveling, R.A. and Hutchison, P.A. 2000, *Physiological Monitoring of Heat Stress in UK Coal Mines*, IOM Report TM/00/05, Institute of Occupational Medicine, Edinburgh.

Hanson, M.A. and Graveling, R.A. 1997, *Development of a Code of Practice for Work in Hot and Humid Conditions in Coal Mines*, IOM Report TM/97/06, Institute of Occupational Medicine, Edinburgh.

Hunt, A.P., Stewart, I.B. and Parker, T.W. 2009, 'Dehydration is a health & safety concern for surface mine workers', in *International Conference on Environmental Ergonomics, Boston, 2–7 August*, International Society for Environmental Ergonomics, London.

International Organization for Standardization (ISO) 2001, *Ergonomics of the Thermal Environment: Medical Supervision of Individuals Exposed to Extreme Hot or Cold Environments*, ISO 12894: 2001, ISO, Geneva.

—— 2004, *Ergonomics of the Thermal Environment: Analytical Determination and Interpretation of Heat Stress Using Calculation of the Predicted Heat Strain*, ISO 7933: 2004, ISO, Geneva.

—— 2007, *Ergonomics of the Thermal Environment: Determination and Interpretation of Cold Stress When Using Required Clothing Insulation (IREQ) and Local Cooling Effects*, ISO 11079: 2007, ISO, Geneva.

—— 2008, *Ergonomics of the Thermal Environment: Cold Workplaces—Risk Assessment and Management*, ISO 15743: 2008, ISO, Geneva.

Jimenez, C.A.R., Ishimoto, T., Lanaspa, M.A., Rivard, C.J., Nakagawa, T. et al. 2014, 'Fructokinase activity mediates dehydration-induced renal injury', *Kidney International*, vol. 86, pp. 294–302.

Jones PA. and Ross, R.K. 1999, 'Prevention of bladder cancer', *New England Journal of Medicine*, vol. 340, pp. 1424–6.

Killer, S.C., Blannin, A.K. and Jeukendrup, A.E. 2014, 'No evidence of dehydration with moderate daily coffee intake: A counterbalanced cross-over study in a free-living population', *PloS one*, vol. 9, no. 1, p. e84154.

Kjellstrom, T., Butler, A.J., Lucas, R.M. and Bonita, R. 2010, 'Public health impact of global heating due to climate change: Potential effects on chronic non-communicable diseases', *International Journal of Public Health*, vol. 55, pp. 97–103.

Knapik, J.J., Canham-Chervak, M., Hauret, K., Laurin, M., Hoedebecke, E., Craig, S. and Montain, S.J. 2002, 'Seasonal variations in injury rates during US Army basic combat training', *Annals of Occupational Hygiene*, vol. 46, no. 1, pp. 15–23.

Muller, M.D., Gunstad, J., Alosco, M.L., Miller, L.A., Updegraff, J., Spitznagel, M.B. and Glickman, E.L. 2012, 'Acute cold exposure and cognitive function: evidence for sustained impairment', *Ergonomics*, vol. 55, no. 7, pp. 792–8.

Pandolf, K.B. 1998, 'Time course of heat acclimation and its decay', *International Journal of Sports Medicine*, vol. 19, Supp. 2, pp. S157–60.

Parsons, K. 2003, *Human Thermal Environments*, 2nd ed., Taylor & Francis, London.

Pearce, J. 1996, 'Nutritional analysis of fluid replacement beverages', *Australian Journal of Nutrition and Dietetics*, vol. 43, pp. 535–42.

Pryor, J.L., Minson, C.T. and Ferrara, M.S. 2018, 'Heat acclimation', in D.J. Casa (ed.), *Sport and Physical Activity in the Heat*, Springer, Dordrecht, pp. 33–58.

Roti, M.W., Casa, D.J., Pumerantz, A.C., Watson, G., Judelson, D.Q., Dias, J.C., Ruffin, K. and Armstrong, L.E. 2006, 'Thermoregulatory responses to exercise in the heat: Chronic caffeine intake has no effect', *Aviation, Space & Environmental Medicine*, vol. 77, no. 2, pp. 124–9.

Sawka, M.N., Burke, L.M., Eichner, E.R., Maughan, R.J., Montain, S.J. and Stachenfeld, N.S. 2007, 'ACSM position stand: Exercise and fluid replacement', *Medicine & Science in Sports & Exercise*, vol. 39, no. 2, pp. 377–90.

Seal, A.D., Bardis, C.N., Gavrieli, A., Grigorakis, P., Adams, J.D., Arnaoutis, G., Yannakoulia, M. and Kavouras, S.A. 2017, 'Coffee with high but not low caffeine content augments fluid and electrolyte excretion at rest', *Frontiers in Nutrition*, vol. 4, p. 40.

Shirreffs, S.M. 2003, 'Markers of hydration status', *European Journal of Clinical Nutrition*, vol. 57, supp. 2, pp. s6–9.

Steadman, R.G. 1979, 'The assessment of sultriness. Part 1: A temperature humidity index based on human physiology and clothing science', *J. Applied Meteorology*, vol. 18, pp. 861–73.

Xiang, J., Peng, B.I., Pisaniello, D. and Hansen, A. 2014, 'Health impact of workplace heat exposure: An epidemiological review', *Industrial Health*, vol. 52, pp. 91–101.

15. Lighting

Martin Jennings and So Young Lee

15.1   INTRODUCTION

Life on Earth would not be possible without light. Humans are light-dependent beings and while all our senses are important, our sense of vision is recognised as being our most important link with our surroundings. Approximately 80 per cent of all information to be processed by the human brain is received by way of the eyes, and this has a major influence on the way we function. The sense of sight depends upon light; this may be natural light (sunlight), sunlight reflected by the moon (moonlight) or artificial light.

This chapter looks at the basics of lighting. It discusses the eye and vision, the eye's response to different light levels, terminology used and some of the most familiar types of lamps and lighting. It also discusses light in the work environment, where good lighting is essential for health, safety and productivity. Good lighting in the design of a workplace refers to the quality as well as the quantity of light. In this regard, the ergonomics of lighting are described briefly to show how workers' operating performance levels improve in a well-lit environment.

Lastly, this chapter discusses the physiological effects of light and the impact of light on the health of individuals. In recent years, a considerable volume of literature has been published on the response of organs such as the pineal gland to light and its impact on the neuroendocrine system. Light has been shown to be a major factor in entraining the body's circadian rhythms, and is also a factor in some disorders as well as being a therapeutic agent for others. These considerations are all relevant to the design of a well-lit working environment, and it is hoped that by the end of this chapter the reader will be looking at light in a different way.

15.2   LIGHT

In 1665, Isaac Newton placed a prism in the path of a sunbeam and observed the spectrum it cast on the far wall of a darkened room. Newton's experiment changed forever how people thought about light. We now know that in the electromagnetic spectrum, visible light forms the part ranging from approximately 380 nm to 770 nm, with violet light having the shortest and red the longest wavelengths. Immediately below the visible spectrum is the ultraviolet region of the electromagnetic spectrum, while the infrared region lies immediately above it. Light may be natural, from sources such as the sun, or it may be artificial, from sources such as lamps. There are other sources of natural light, such as chemiluminescence (emission of light as a result of a chemical reaction), as seen in glow-worms; phosphorescence (in which a phosphorescent material re-emits the radiation it absorbs) can sometimes be seen in the sea from plankton. However, these phenomena are of little relevance in the workplace.

15.2.1 WHY IS LIGHTING IMPORTANT?

Aside from its indispensable role in vision, light affects the functioning of the neuroendocrine system, which in turn affects other systems and organs of the body. For example, it has been suggested that exposure to light at night may increase the risk of breast cancer by disrupting the normal nocturnal production of melatonin by the pineal gland, which in turn could increase the release of oestrogen by the ovaries (Hansen, 2001).

In the context of this book, good lighting—that is, of an appropriate quality and adequate quantity—makes all work tasks easier, whether in industrial or office settings. Appropriate lighting, without glare or shadows, can reduce eye fatigue and headaches. It enables workers to see moving machinery and other safety hazards. It also reduces the chance of accidents and injuries from 'momentary blindness' while the eyes adjust to brighter or darker surroundings.

The ability to 'see' at work depends not only on lighting, but also on other factors, such as:

- the time it takes to focus on an object: fast-moving objects are hard to see
- the size of an object: very small objects are hard to see
- the location of the object in the visual field: objects are more difficult to see if too close or at the periphery of the visual field
- brightness: too much or too little reflected light makes objects hard to see
- contrast between an object and its immediate background: too little contrast makes it hard to distinguish the two.

Light possesses wavelike properties, which enable it to be focused, reflected and refracted. These properties provide for the sense of vision, as light is *reflected* off surfaces, *refracted* by the lens of the eye and *focused* to form an image on the retina.

15.3 THE EYE AND VISION

In relation to vision, the two main functions of the eye are:

- acting as an optical instrument to collect light waves from the environment and project them as images onto the retina
- functioning as a sensory receptor that responds to the images formed on the retina by sending information by way of the visual nerve to the visual areas of the brain.

In addition, the eye also sends information from light falling on the retina to non-visual parts of the brain, to synchronise neuroendocrine activity.

15.3.1 VISION

Light rays striking the eye first enter the cornea, which has a high degree of curvature and a refractive index of 1.33, compared with 1.00 for air. These two factors in combination cause the light rays to bend as they enter the eye. The rays then pass through the lens of the eye, which has a refractive index of 1.39, and this causes further bending of the rays. As a result of this process, the light rays are brought to a focus on the retina at the back of the eye (see Figure 15.1).

The *lens* is normally a flat elastic capsule, but it assumes a more spherical shape when acted upon by the ring-shaped ciliary muscles that encircle it. This process is known as *accommodation*. The ciliary muscles relax when the eye is focused on distant objects. This is why people working with computer screens are advised to take regular breaks and relax their eyes—for example, by gazing out the window.

The *retina* contains two sets of visual photoreceptor cells, the *rods* and the *cones*, so called because of their shape. The cones are concerned with colour vision and fine detail (photopic vision); the rods are much more sensitive to low levels of illumination than the cones but have less visual acuity. The human eye has three types of cones to sense light in three respective bands of colour. The biological pigments of the cones have maximum absorption values at wavelengths of about 420 nm (blue), 534 nm (bluish-green) and 564 nm (yellowish-green). Their sensitivity ranges overlap to provide vision throughout the visible spectrum, with maximum efficacy at 555 nm (yellow-green). The *fovea* is the part of the retina located in the centre of the macula lutea of the retina, and is a region where only cones are found. The cones are particularly densely packed in the fovea; hence it is the region of the highest visual acuity or most detailed vision—about 40 times as sharp as that at the retinal border. The photoreceptor cells convert light energy into nerve impulses that travel via the optic nerve to the visual area of the brain, where the image is formed.

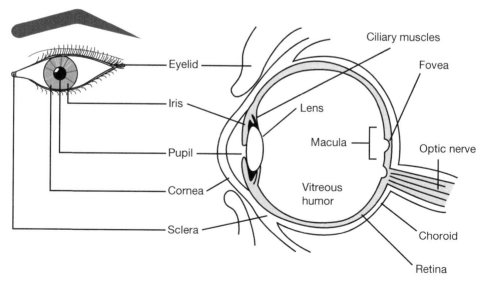

Figure 15.1 The human eye
Based on information from US National Eye Institute/National Institutes of Health.

The function of the rods is to enable vision in low light conditions, known as *scotopic vision*. Rod cells are most sensitive to wavelengths of light around 498 nm (green-blue) and are insensitive to wavelengths longer than about 640 nm (red). Most people have experienced the Purkinje shift or effect, when the eyes transition from predominantly photopic (cone-based) to scotopic (rod-based) vision. The Purkinje shift usually occurs around dusk. As this happens, the maximum sensitivity of the eye shifts towards the blue end of the spectrum at very low illumination levels (*Chambers Dictionary of Science and Technology*, 2007). The transition vision is termed *mesopic vision*, where colour appears to vanish from the scene and all objects seem to become black and white. This has a practical application in the workplace. Under conditions where both the photopic and scotopic systems need to be active, red lighting was thought to provide a solution. For example, submarines and flight decks on aircraft carriers are dimly lit to preserve the night vision of the crew members working there, but the control room must be lit to allow crew members to read instrument panels or maps. Using red lights allows the cones to receive enough light to provide photopic vision (for reading or viewing the control panel). Because the rods are not saturated by bright light and are not sensitive to long-wavelength red light, however, the crew members remain dark-adapted, ready and able to view the external environment at night (Poston, 1974).

More recently, a third set of photoreceptors has been found in the eye: a small population of intrinsically photosensitive retinal ganglion cells (ipRGCs). These cells contain melanopsin and are connected to the non-visual parts of the brain (Hattar et al., 2002). They show peak response to blue light with a wavelength of 460–480 nm. This action spectrum of ipRGCs is referred to as melanopic. It appears that melanopsin cells play a vital role in the entrainment of the body's circadian rhythms by providing light input to the suprachiasmatic nucleus, located in the hypothalamus. The neurons in the hypothalamus control the body's circadian rhythms.

The discovery of ipRGCs is significant to any consideration of the lit environment. Traditionally, lighting industry guidelines followed several scientific principles for efficacy (energy efficiency), light quantity (illumination levels), light quality (colour temperature, colour rendition, glare and so on), and lighting uses (ambient, task or accent lighting). The discovery of ipRGCs introduced a new dimension of considerations for lighting or display designs—that is, how to minimise the adverse effect of artificial lights, via ipRGC photo-transduction, on mental and physical health while maximising visual functions and energy efficiency (Cao and Barrionuevo, 2015).

The spectral responsivity of ipRGCs is of particular interest to lighting designers. Since the discovery of the ipRGC, the lighting industry has responded by producing design guidelines for creating melanopic lighting environments. Until recently, lighting designers had no real idea of how to design lighting schemes that would not only provide effective lighting to support visual tasks, but also provide non-visual biologically effective illumination. The first steps are now underway. In Germany, the DIN SPEC 67600 (2013–04) Biologically Effective Illumination—Design Guidelines specifically address this.

Similarly, a number of lighting manufacturers are currently looking at ways of optimising the spectral power distributions of their products to produce biologically effective white light—that is, white light with an abundance of blue light centred on the melanopic peak wavelength (Jennings, 2016). This is termed *human-centric lighting*.

15.3.2 THE EYE'S RESPONSE TO DIFFERING LIGHT LEVELS

The level of illumination has a critical effect on the ability of the eye to focus. At low light levels, contrast and image sharpness are diminished, making it more difficult to see. Focusing requires more effort when illumination is poor—especially in older workers.

The size of the pupils is controlled by the muscles of the iris, enlarging or reducing the opening to regulate the brightness of the image projected on the retina. Effort is needed to change the size of the opening, but not to maintain it: when the pupil reaches the required diameter, the muscles of the iris relax. For this reason, repeated viewing of objects that have different brightness levels is more tiring than viewing objects of uniform brightness (Aronoff and Kaplan, 1995).

In workplaces, a crucial aspect of safety is avoiding sudden changes from very bright to dimly lit areas or vice versa. Immediately after a person enters a dark room, the eye initially is unable to see anything. Objects then become discernible, usually within five minutes. This is due to the initial rapid response by the cones in the retina, followed by the slower response of the rods. The rods adapt to a greater degree, however, although this process may take as long as 25–30 minutes and continue slowly for several hours. When a person moves from dark to light, the reverse process occurs, with a temporary loss of vision occurring until the eyes adapt to the sudden increase in the level of illumination. Table 15.1 illustrates the extent of the difference between sunlight and artificial lighting.

Table 15.1 Typical natural and artificial light levels

Environment	Lux
Typical bright sunny day in summer outside	100,000
Typical overcast sky in summer	20,000
Sunny winter's day outside, temperate climate	10,000
Dull winter's day	3000
Good workplace lighting	1000
Good street lighting	40
Full moon	0.25
Starlight	0.01

Historically, occupational conditions such as miner's nystagmus were thought to be caused by working in poorly lit conditions or, more accurately, by trying to maintain foveal fixation under poor light. Typically, this occurred in middle age after years of underground work. Symptoms commenced with headaches and vertigo, followed by a pendular or jerking type of nystagmus defined in the *Oxford English Dictionary* (2007) as a rapid involuntary oscillation of the eyeball, usually lateral. This is seen as a flicking backwards and forwards when the eye is deviated, blepharospasm (abnormal blinking) and poor fixation under reduced ambient lighting (Brennan, 1987).

15.4   TYPES OF LAMPS AND LIGHTING

A lamp is a type of energy transformer that transforms electrical energy into electro-magnetic (EM) radiation in the visible region of the EM spectrum. Forster (2012) provides a good review of this subject.

15.4.1   *INCANDESCENT LAMPS*

When solids and liquids are heated above 1000 K (approximately 726°C), they emit radiation as visible light. (K, or kelvin, is the base SI unit of thermodynamic temperature, defined as the fraction 1/273.16 of the thermodynamic temperature of the triple point of water.) This property is termed *incandescence*. One of the oldest ways of generating light is by bringing a solid body to incandescence by heating. A hot object emits a broad spectrum of light. The frequency and wavelength at which most of the light is emitted depend on the temperature of the object. The hotter the object, the shorter the wavelength of the energy emitted. That is, the hotter the object, the greater the energy emitted. Observing a piece of metal in a very hot flame will demonstrate this relationship. When the metal starts to heat up, it will initially glow a dull red colour. The glow becomes a brighter red, then yellow as it becomes hotter, continuing until the metal glows blue and eventually white. These colour changes are due to the increase in frequency, causing the wavelength to become shorter, thus emitting more energy.

This principle is used in incandescent light bulbs, where a fine filament—typically tungsten wire—is heated by passing an electrical current through it. The temperature of the filament reaches approximately 2700 K, above which the evaporation of the filament becomes excessive. The glass bulb is usually filled with an inert gas at low pressure, to prevent loss of tungsten from evaporation and oxidation and to reduce the blackening of the inside of the bulb. At 2700 K, the majority of radiation is emitted as heat in the infrared region of the EM spectrum, rather than as light.

Figure 15.2 shows the spectral distribution of seven different light sources. This diagram compares the spectral distribution of natural light (blue sky) to the distribution of light from an incandescent lamp, a compact fluorescent lamp (CFL), a cool white and a warm white LED and a metal halide lamp. Note that the incandescent lamp's curve lies predominantly in the infrared region of the spectrum.

Incandescent lamps are extremely energy inefficient—only around 10 per cent of the energy is emitted as visible light. Since 2009, the traditional pear-shaped incandescent bulb has been phased out of use in Australia (see Energy Rating, 2009), and all regular light bulbs have been required to meet minimum energy performance standards (MEPS). Incandescent light bulbs that meet the new standards—for example, high-efficiency halogen bulbs—continue to be available. In quartz halogen lamps, halogen gases such as iodine are used to fill the lamp. This permits the tungsten filament to operate at a higher temperature so more of its energy is available as visible light.

In 2018, the Australian government agreed to further improve lighting energy effi-ciency regulation by phasing out inefficient halogen lamps and introducing minimum standards for LED lamps in Australia and New Zealand that are consistent with European Union (EU) standards. The phase out will remove remaining incandescent light bulbs and

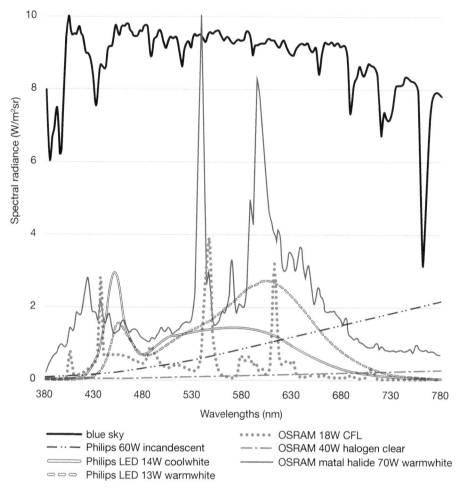

Figure 15.2 Spectral distribution of light sources (1 m distance excepting blue sky, using Specbos 1211UV, od2.5 filter)
Source: Lee (2018).

a range of halogen light bulbs from the Australian market, where an equivalent LED light bulb is available (refer to <www.energyrating.gov.au> for further details).

15.4.2 FLUORESCENT LAMPS

Fluorescent lamps are low-pressure mercury lamps, available as hot-cathode and cold-cathode versions. The hot cathode is the type normally found in offices and workplaces, while the cold-cathode type is used mainly for signage and advertising. The central element in a fluorescent lamp is a sealed glass tube containing a small quantity of mercury and an inert gas—typically argon—under very low pressure. The glass tube also contains a phosphor powder coating on the inside of the glass. Different phosphors are selected by the manufacturer, depending on the specific frequencies of the light they emit. When a current is passed through the tube, the mercury atoms become excited and emit UV radiation, which strikes the phosphors on the inside of the glass tube. The phosphors absorb the radiation and then re-emit it at a different wavelength within the visible spectrum. Fluorescent lamps are much more efficient converters of electrical energy than incandescent lamps.

Claims have arisen that the flickering sometimes perceived with fluorescent lighting is linked to health concerns such as photosensitive epilepsy and migraine; however, the health concern most often linked with fluorescent lamps is of mercury exposure in the event of a lamp being damaged.

15.4.3 ELECTRIC DISCHARGE LAMPS

An alternative and more efficient form of lighting is achieved by passing an electric current through a gas. This excites the atoms and molecules of the gas, causing it to emit radiation, with the spectral distribution characteristic of the specific gas used. The most commonly used metals are mercury and sodium vapour, as they emit useful visible radiation. Discharge lamps are often classified as either high- or low-pressure lamps. In a low-pressure lamp with mercury or sodium vapour as an active ingredient the metal vapour is mixed with an inert gas—often neon or argon. High-pressure lamps are referred to as high-intensity discharge (HID) lamps. They include mercury vapour, metal halide, high-pressure sodium and xenon short-arc lamps. Compared with fluorescent and incandescent lamps, HID lamps are highly efficient because they produce a large quantity of light in a small package. They generate light by striking an electrical arc across tungsten electrodes housed inside a specially designed inner fused quartz or fused alumina tube. This tube is filled with both gas and metal halides. The gas aids in the starting of the lamps; the metals produce the light once they are heated to a point of evaporation.

15.4.4 LIGHT-EMITTING DIODES

A diode allows an electric current to flow in one direction only. Light-emitting diodes (LEDs) are diodes that emit light. Almost any conductive materials will form a diode when placed in contact with each other. When current passes through the diode (within a semiconductor chip, for example), the atoms in one of the materials are excited to a higher energy level. The atoms in the first material possess excess energy, which is released in shedding electrons to the second material within the chip. During this release of energy, light is generated. The efficiency (measured in lumens per watt) and effectiveness (as shown by their relative brightness) of LEDs have resulted in their usage being extended to other applications, including commercial and household settings.

The colour of the LED light is a function of the materials and processes used in making the chip. Red and green LEDs have been around for several decades, but were initially used only in electronic equipment such as watches or calculators. It was only in the 1990s that blue LEDs were invented. This advance finally enabled manufacturers to create white LED light. White LED light allows companies to create smartphone and computer screens, as well as light bulbs that last longer and use less electricity than any bulb invented before.

Table 15.2 shows a range of artificial light sources.

Table 15.2 Classification of artificial light sources (measured by Specbos 1211UV)

Type of light sources	Typical types of shape	Workplaces used	Example of photo of light source	Examples of emission characteristics (luminance, CCT, L_B)
Incandescent	Globe	Gradually disappearing		Philips 60W 52330 cd/m^2 2614K 12.9 W/m^{2}sr
Fluorescent and CFL	Coil, tube	Various aspects in office areas or industrial background lightings		OSRAM 18W CFL Dulux 47070 cd/m^2 6533 K 41.95 W/m^{2}sr
Halogen	Globe/MR	Various aspects		OSRAM 40W 6376 cd/m^2 2770 K 1.848 W/m^{2}sr
Discharge lamps: metal halides	Bulb, tube	Retail shops, sports arenas, factories, etc.		OSRAM 70w Warm white 225700 cd/m^2 3161 K 112.3 W/m^{2}sr
LEDs	Various	Various places and IT devices (most common used)		Philips 14W Cool daylight 93040 cd/m^2 6038 K 74.97 W/m^{2}sr
				Philips 13W Warm white 135300 cd/m^2 3048 K 42.29 W/m^{2}sr

15.5   TERMS AND UNITS OF MEASUREMENT

The following terms are taken from Australian Standard AS 3665 Simplified Definitions of Lighting Terms and Quantities. Such terms are used universally, and can be found in other international sources (CIE, 2004; Zumtobel Lighting GmbH, 2018).

15.5.1   *LUMINOUS FLUX*

Luminous flux—symbol ϕ unit lumen (lm)—is the quantity of light energy emitted per second in all directions. One lumen is the luminous flux of a uniform point light source with luminous intensity of 1 candela and contained in 1 unit of solid angle (or one steradian). A solid angle can be regarded as a cone and is measured in steradians. It is the area of the segment of a sphere subtended, or covered, by the light from a point source at the centre of the sphere. This concept is shown in Figure 15.3 for a sphere of 1 m radius. Since the surface area of the sphere is $4\pi r^2$, the luminous flux of the point light source is 4π lumens.

15.5.2   *LUMINOUS INTENSITY*

Luminous intensity—symbol I, unit candela (cd)—is the concentration of luminous flux emitted in a specified direction. Luminous intensity is the ability to emit light in a given direction, or the luminous flux that is radiated by the light source in a given direction within a given spatial angle (measured in degrees). If the point light source emits ϕ lumens into a small spatial angle ß, the luminous intensity is $I = \phi/ß$.

1 candela = 1 lumen per steradian

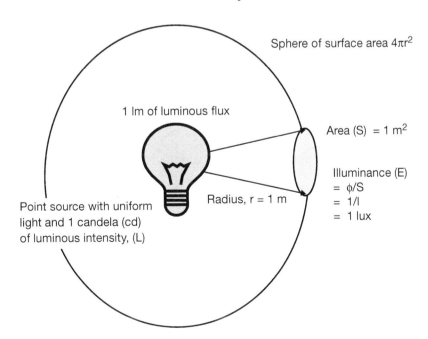

Figure 15.3 Relationship between luminous intensity, luminous flux and illuminance

15.5.3   *ILLUMINANCE*

Illuminance—symbol E, unit lux (lx)—is the luminous flux density at a surface (luminous flux incident per unit area, lumens/m²). It describes the amount of light that reaches a given surface. If ϕ is the luminous flux and S is the area of the surface, then the illuminance E is determined by $E = \phi/S$. One lux is the illuminance of 1 m² surface area uniformly illuminated by 1 lm of luminous flux. Figure 15.4 illustrates this definition.

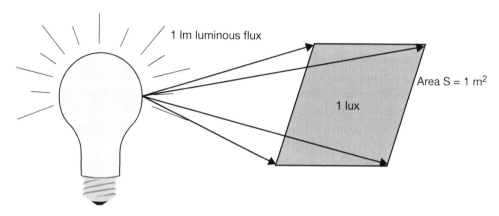

Figure 15.4 The relationship between lumen and lux

15.5.4   *LUMINANCE*

Luminance—symbol L—is the luminous intensity of 1 cm² (or 1 m²) of the surface area of a given light source. Mathematically, $L = I/S$, where I is the luminous intensity and S is the area of the light source surface perpendicular to the given direction. The unit of luminance is cd/m² or cd/cm² (see Figure 15.5).

The lumen (lm) is the photometric equivalent of the watt, weighted to match the eye response of the 'standard observer'. Yellowish-green light (555 nm) receives the greatest

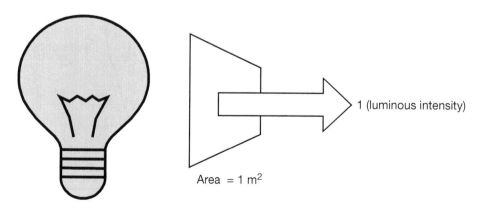

Figure 15.5 Luminance, the intensity of light emitted in a given direction by a unit area

weight because it stimulates the eye more than blue (450 nm) or red light (660 nm) of equal radiometric power. One watt at 555 nm is equal to 683.0 lumens.

To put this into perspective, the human eye can detect a flux of about 10 photons per second at a wavelength of 555 nm. This corresponds to a radiant power of 3.58×10^{18} W (or joule/s). Similarly, the eye can detect a minimum flux of 214 and 126 photons per second at 450 nm and 650 nm, respectively.

Lighting design and evaluation are concerned with the light entering the eye, the surface being viewed (e.g. the work) and the source of the light. This is known as the lighting triangle, and is shown in Figure 15.6.

The distance from the eye to the surface, and the distance from the light source to the surface, are the critical factors. The distance from the light source to the surface (d) and the angle (θ) at which the light reaches the surface determine the illuminance, E, or the amount of light received by the surface. Illuminance is described by the equation: $E = (I \cos \theta)/d^2$. The smaller the angle (θ) for any given light source of luminous intensity (I), and the smaller the distance (d), the higher the illuminance and the better the visibility of the task (Longmore, 1980).

15.5.5 SPECTRAL RADIANCE

Spectral radiance—symbol L_λ—is the radiance of a surface in watts (W) per square metre, steradian taking into account the wavelength (in nm) and the spatial angle of light. The unit of spectral radiance is W/m²·sr·nm. The value of integrated spectral radiance can be used to measure the effective radiance (L_B) of the light source, especially blue light source.

15.5.6 SPECTRAL IRRADIANCE

Spectral irradiance—symbol E_λ—is the irradiance of a surface in watts per square metre taking into account the wavelength (in nm) of light. The unit of spectral irradiance is

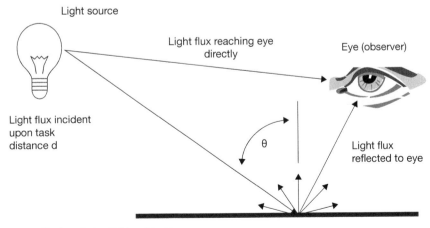

Figure 15.6 The lighting triangle

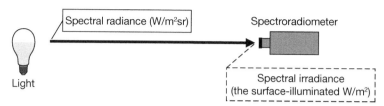

Figure 15.7 Spectral radiance and spectral irradiance
Source: Lee 2018

W/m^2·nm. The value of integrated spectral irradiance can be used to measure the effective irradiance (E_B) of the light source, especially blue-weighted light source.

15.6   MEASUREMENT OF LIGHT

15.6.1   LIGHT INSTRUMENTS

15.6.1.1 Photometers

Photometers are photoelectric instruments that use special optic filters to reconstruct the exact response of the human eye to light intensity. They consist of a photocell that converts light energy into an electric current. A light-measuring cell in the meter (photo diode) converts incident light into an electronic signal that is read and displayed in lux on the meter's liquid crystal display (LCD). In older instruments, this is connected to a moving coil meter which displays the current in lux.

Ideally, the photometer (sometimes referred to as a lux meter or a light meter) should have a measurement range of 0.001–100 000 lux, although a range of 0.1–20 000 lux is acceptable for most workplace applications. The photo cell should be connected to the meter by a cable long enough to ensure that the meter can be read without the user casting a shadow over the cell. Instruments fitted with colour (spectral response) correction filters to limit the sensitivity to ultraviolet (UV) and infrared (IR) radiation are preferred because their spectral responsivity has been matched to that of the human eye. If the receptor is not colour corrected, the appropriate correction factor (usually supplied by the manufacturer) should be applied. The photoreceptor should also be fitted with cosine correction devices to take account of the effects of light falling on it at oblique angles.

All meter types should be calibrated regularly for accuracy. The intervals for calibration of photometer instruments depend on the type of photoreceptor used in the instrument and should be observed in accordance with the manufacturer's recommendations. Before readings are taken, photocells should be exposed to the approximate illuminance to be measured until the reading becomes stabilised. Care should be taken not to cast a shadow on the photocell while taking a reading. An example of a photometer is shown in Figure 15.8.

15.6.1.2 Luminance meter

A luminance meter, as shown in Figure 15.9, is a device used in photometry that can measure the luminance in a particular direction and with a particular solid angle.

Figure 15.8 Extech
LT505 Pocket Light
Meter and Extech Pocket
UV505 UV-AB Light
Meter
(Images courtesy of
Extech Instruments, a FLIR
company)

Figure 15.9 Minolta
Luminance Meter 1°
(Photo courtesy of Konica
Minolta, Inc. All rights
reserved.)

The simplest devices measure the luminance in a single direction while imaging luminance meters measure luminance in a way similar to the way a digital camera records colour images.

15.6.1.3 Spectroradiometers

A spectroradiometer is an instrument for measuring the spectral irradiance ($W/m^2.nm$ $W/m^2/micrometer$). It is able to measure both the wavelength and amplitude of the light emitted from a light source. An example of a spectroradiometer is shown in Figure 15.10.

Spectroradiometers are ideal for measuring spectral energy distribution of a light source, determining not only the radiometric and photometric quantities, but also the colorimetric quantities of light. These instruments record the radiation spectrum of the light source and calculate parameters such as chromaticity and luminance. Dispersion of light is usually accomplished in a spectroradiometer by means of prisms or diffraction gratings.

15.6.2 *LIGHTING FOR THE WORKPLACE*

The visual demands of the different tasks in a given work area will determine the lighting required. The International Labour Office's *Encyclopaedia of Occupational Health and Safety* (ILO, 2012) recommends the illumination levels for types of office tasks, activities and interiors (Table 15.3).

This sets out general principles and recommendations for the lighting of building interiors for performance and comfort. It applies primarily to interiors in which specific visual tasks are undertaken and takes into account both electric lighting and daylight. The recommendations aim to achieve a visual environment in which essential task details are easy to see and factors that may cause visual discomfort are either excluded or appropriately controlled. It should be noted that the recommendations do not apply to lighting for physiological purposes.

Figure 15.10
Spectroradiometer: Specbos 1211UV, mainly used for radiometric, photometric and colorimetric measurements.
(Photo courtesy of JETI Technische Instrumente GmbH. All rights reserved.)

Table 15.3 ILO recommended levels of illumination for particular applications

Location/task	Typical recommended level of maintained illuminance (lux)
General offices	500
Computer workstations	500
Rough work	300
Medium work	500
Fine work	750
Instrument assembly	1,000
Jewellery assembly/repairs	1,500
Hospital operating theatres	50,000

Source: ILO (2012).

Workplace lighting or illumination affects the safety, task performance and visual environment of the workers. Although a poorly lit workplace can result in eyestrain or headaches, workplace lighting traditionally has been considered a matter of ergonomics rather than health. For this reason, it has not traditionally been regarded as a major health issue; however, there is now evidence that challenges this view. This will be discussed later in the chapter.

Other relevant factors to consider in the visual environment include:

- avoiding excessive illuminance variations
- absence of direct glare from lamps, luminaires or windows
- an appropriate luminance distribution on interior surfaces
- use of suitable colours on the main interior surfaces
- control of flickering lights
- use of light sources with suitable colour characteristics.

15.6.2.1 General

Before measurements are taken, the light meter should be calibrated by covering the light sensor and recording the measured value, which should be zero. Measurement of illuminance of a workplace lit by artificial lighting is made either after dark, or with daylight excluded from the interior. Some light sources, such as gas-discharge lamps (e.g. mercury or sodium vapour), should be switched on at least 30 minutes before the measurement to allow the light output to stabilise.

15.6.2.2 Measurement of task illuminance

To measure illuminance at a workstation or equipment operator's position, the following procedure should be followed:

- Use at least four measurement points at each sampling location to obtain a representative average illuminance value.

- Take measurements on the plane (horizontal, vertical or inclined) in which the visual task is performed while the worker is in their normal position, even if this results in shadow on the sensor. Calculate the average illuminance at the task position as the arithmetic mean and compare the average illuminance to the recommended level for the task, as shown in Table 15.3.

15.6.2.3 Measurement of illuminance of an interior

The average illuminance in a space is determined by measuring the illuminance at each of a series of points, set out in a regular pattern, then calculating the arithmetic mean of those values. Accuracy is dependent on the number and spacing of measurement points and the relative location of the points with respect to the luminaires. The following criteria should be applied:

- The selected measurement areas chosen should collectively represent all areas of both the lighting layout and the physical environment, but they do not necessarily need to cover the whole space to be measured.
- Where there is a regular array of lighting, the measurement area should cover an entire pattern.
- The measurement area should be divided into a number of squares.
- The illuminance should be measured at the centre of each square, at the height of the working plane. *Note:* The use of a tripod to support the meter at the correct height and in a horizontal position may be useful.
- The maximum distance between measurement points is 1 metre.
- No measurement point should be closer than 1 metre from a wall.
- For areas with a floor area greater than 25 square metres, the minimum area of measurement is 9 square metres.
- The minimum number of luminaires to be included in the measurement area is four.
- For spaces with non-regular lighting layouts or different room conditions, several measurement areas need to be selected to give a representative set of measurements.
- For areas with a floor area less than 25 square metres, the minimum number of measurement points is nine, and the whole area should be measured.

15.6.2.4 Uniformity of illuminance

Where a *general lighting system* is used, a fairly high degree of uniformity of illuminance in the working plane is required, as the aim of such systems is to allow a particular task or series of tasks to be performed anywhere within the space. Therefore, for general lighting systems, the uniformity of illuminance (the ratio of the minimum illuminance to the average illuminance) should be ≥ 0.5 over the space covered or included by the general lighting system.

The uniformity of illuminance should be ≥ 0.7 over the *task area*. The illuminance on the area immediately surrounding the task area should preferably be no greater than the illuminance on the task area, and any change in illuminance should be gradual.

15.6.2.5 Measurement of blue light

Light can cause photochemical damage to the retina. Although all visible light as well as UV-A can contribute to photochemical injury, light in the wavelength range of

400 nm–500 nm (violet, blue and blue-green) is most detrimental (AIHA, 2012). This is termed the *blue light hazard*. To assess the blue light hazard, it is necessary to characterise the radiant exposure of workers by exposure level (how much) and duration (how long) workers are exposed to blue light in their workplaces and to determine the part of the retina most sensitive to blue light exposure. The exposure assessment should focus on the visual field of a worker. In the work environment, tasks vary and quantification of light exposure in the visual field of all tasks is difficult. Workers generally work under various exposure conditions (e.g. types of light sources, working distance and durations) and the potential risk of retinal injury depends on a person's field of view in their work space. This is the occupational visual field (OVF) and should be considered when identifying the blue light hazard in the workplace (Piccoli et al., 2004).

Significant sources of blue light include blue LED arrays, intense white light sources (such as projection lamps, floodlights, microscope lights, welding arcs) and strong sunlight (AIHA, 2012). Blue light is also found in nail curing lamps used in beauty salons. Blue light is measured differently to other types of light. Lee and colleagues (2016) have described the application for a simple LED spot source and explored the exposure methodology for two commercially available nail-curing lamps used in beauty salons.

The assessment of the blue light hazard requires the use of a radiometer, or preferably a spectroradiometer (Figure 15.10), rather than a photometric instrument (lux or luminance meter). The spectral radiance is the radiant intensity per exposed area, which is generally used to evaluate the retinal photochemical damage from a light source. Figures 15.8, 15.9 and 15.10 show the light-measuring instruments used in assessing the radiometric and photometric output of light sources. There is no simple or reliable conversion factor between photometric and radiometric quantities.

Both ACGIH® and ICNIRP provide equations for measuring effective radiance (W/m^2·sr) and radiance dose (J/m^2·sr) for blue light exposure. For viewing durations less than 10,000 seconds (2.8 hours) in a day, the effective radiance dose must not exceed 10^6 J/m^2, and for periods greater than 10,000 seconds the radiance limit is 100 W/m^{2}sr (ACGIH, 2018a; ICNIRP, 2013).

There are nuances in these limits, and aspects of practical measurement not immediately obvious to those seeking to assess workplace risk according to the guidelines.

Table 15.4 Equations for blue light exposure to normal eyes (non-aphakic) by ICNIRP guidelines

Formulae	Details
$L_B = \sum_{300}^{700} L_\lambda \cdot B(\lambda) \cdot \Delta\lambda$ $D_B = L_B \cdot t = \sum_{300}^{700} L_\lambda \cdot t \cdot B(\lambda) \cdot \Delta\lambda$	L_B: Effective radiance of blue light (W/m^2·sr) D_B: Effective blue light radiance dose (J/m^2·sr) L_λ: Spectral radiance (W/m^2·sr·nm) $B(\lambda)$: Blue light hazard function $\Delta\lambda$: wavelength interval (nm) t: Exposure duration (seconds)
$D^{EL}_B = 10^6$ J/m^{2}sr	D^{EL}_B: Exposure limit of the radiance dose (0.25s ≤ t < 10000s)
$L^{EL}_B = 100$ W/m^{2}sr	L^{EL}_B: Exposure limit of the radiance (t > 10000s)

Measurements should be conducted in the occupational visual field (Piccoli et al., 2004), with special consideration of the averaging angle of acceptance. According to ICNIRP guidelines (ICNIRP, 2013) the acceptance averaging angle is time dependent (from 0.01 to 0.1 radians; 0.6 to 6 degrees). Depending on the actual task and behaviour, however, a larger averaging angle can be used, provided that any part of the retina—particularly the macula—is not exposed beyond the radiance dose limit.

Damage to the peripheral retina has relatively little adverse impact on visual function, as evidenced by laser treatment for diabetic retinopathy (Palanker, Blumenkranz and Marmor, 2011). If the macula is exposed beyond the radiance dose limit, however, then visual acuity may be adversely affected, or it may contribute to macular degeneration. This means that the risk should be assessed for any blue light source within the visual field that may be imaged on the macula. The product of this particular radiance and the viewing time limits the risk for short exposures (up to 2.8 hours). This would apply to most real-life exposure scenarios. If it were certain that the worker was constantly exposed to a blue light source within the acceptance angle for more than 2.8 hours, the radiance limit would apply. In any case, the averaging is assumed to be an arithmetic mean.

15.6.2.6 Using photometric evaluation

The most common method of measuring light is to use a photometer to measure the level of illuminance. Moreover, standards related to light in the workplace refer to levels of illuminance. Yet illuminance alone is not the only factor that should be measured when evaluating lighting in the workplace. There are differences between illuminance and luminance levels in the workstation, shown in Figures 15.11. The values of illuminance on the keyboard illuminated by overhead ceiling lamps and sunlight from the window are similar (490 lx in Figure 15.11a and 500 lx in Figure 15.11b); however, the values of luminance in Figure 15.11b are around 100 to 1000 times higher than in Figure 15.11a (Piccoli et al., 2004).

Similarly, levels of radiance and irradiance vary depending on the types of light sources and/or conditions of lighting. Photometric evaluations must therefore consider various factors, such as types of light sources or work environments in the occupational visual field (OVF). When occupational hygienists evaluate lighting in the work environment, all these factors need to be considered.

15.7   LIGHT SOURCE COLOUR PROPERTIES

Colour is not an inherent property of objects, but more a human perception facilitated by light. Light sources have two colour-related properties: the apparent colour of the light they emit (colour appearance) and the effect the light has on the colours of surfaces (colour rendering). An application of these properties with which most people will be familiar is the use of lighting in a supermarket to enhance the red colour of meat in the butchery section. This is typically very different from the lighting used in the fresh produce section (ASSIST, 2010).

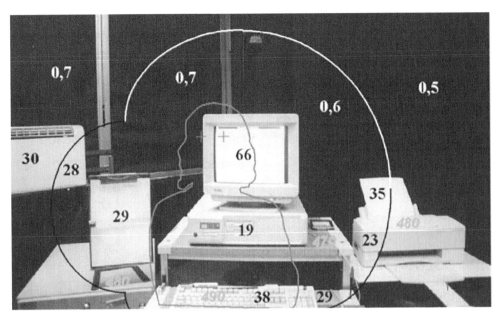

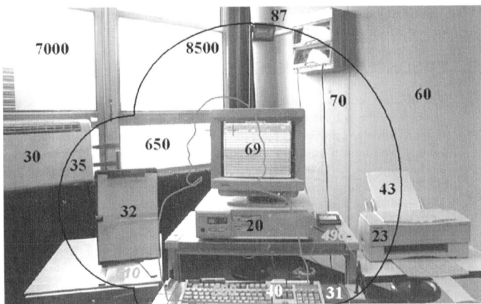

Figures 15.11a and b Differences between luminance and illuminance at a workstation
Source: Piccoli et al. (2004).

15.7.1 COLOUR APPEARANCE

The colour appearance of near-white light sources is normally defined in terms of their correlated colour temperature (CCT), expressed in Kelvin (K). The higher the CCT, the cooler the appearance of the source. For example, the reddish-yellow flame of a candle has a CCT of about 1900 K, the ordinary incandescent lamp about 2800 K, and cool bluish-white southern-sky daylight has a CCT of over 6500 K (Standards Australia, 2006, s.7.2). The CCTs of lamp types have been grouped into three classes, as shown in Table 15.5.

Table 15.5 Lamp colour appearance groups

Colour appearance group	Colour appearance	CCT (K)
1	Warm	< 3300
2	Intermediate	$3300 \leq 5300$
3	Cool	> 5300

Source: International Commission on Illumination, 2004.

15.7.2 COLOUR RENDERING

Colour rendering is the ability of a light source to render the colours of an object as similarly as possible to the way they appear in an ideal or natural light source such as daylight. The colour-rendering properties of a light source can be measured and described according to the CIE Colour Rendering Index (CRI) system (Standards Australia, 2006).

15.8 ERGONOMICS OF LIGHTING

The objective of good lighting design is to apply visual ergonomics to optimise the perception of visual information, provide conditions conducive to task performance, maintain safety and provide an acceptable level of visual comfort. Ergonomically designed lighting takes human capability into account, and generally results in improvements in productivity.

There are four parameters that influence the nature of visual information processed, and therefore a worker's visual performance (Clarke, 1989):

• task characteristics, including size of object, distance, texture, colour, contrast, motion and time factors
• the worker's own visual ability, which depends on ophthalmic condition, age, adaptation, depth and colour perception
• lighting characteristics, including illuminance, uniformity, glare and flicker
• work-space factors such as postural constraints, safety requirements, other physical constraints and psychological factors.

It is often possible to compensate for a deficit in one or more of these factors by enhancing one or more of the others. An obvious consequence of this is that the application of visual ergonomics can increase the number of options available for providing an acceptable visual environment.

15.9 THE BIOLOGICAL EFFECTS OF LIGHT

Each of the effects of light upon mammalian tissue may be classified as direct or indirect, depending on whether its immediate cause is a photochemical reaction occurring within that tissue or a neuroendocrine signal generated by a photoreceptor (Wurtman, 1967).

15.9.1 *THE DIRECT EFFECTS OF LIGHT*

Exposure to sunlight is known to be harmful. This is especially true for mid-ultraviolet erythemal radiation (290–320 nm), which can cause sunburn to appear within a few hours of exposure. Chronic exposure to sunlight for several hours a day over a period of some years can cause permanent changes in skin structure—for example, skin atrophy, the formation of keratin plaques and, in susceptible individuals, the appearance of squamous-cell carcinomas (Wurtman, 1975). One of the best-known direct effects of sunlight on humans is the stimulation of vitamin D synthesis in the skin and subcutaneous tissue. Not only is vitamin D essential for healthy bone development (conditions such as rickets and osteomalacia have been linked to a deficiency), but there is some evidence that the vitamin can also help protect against certain forms of cancer. Research by Grant (2018) suggests that Vitamin D produced in response to sunlight prevents cancer, including breast, colon and prostate cancers.

Sunlight has been used by physicians in the treatment of herpes, psoriasis and hyper-bilirubinaemia (jaundice). Treatment of the two former conditions entails the use of photosensitisers; that of the latter involves direct illumination of the entire body to lower the level of plasma bilirubin.

Photosensitivity is an abnormally high reactivity in the skin or eyes to UV radiation or natural sunlight. It may be induced by ingestion, inhalation or skin contact with certain substances known as photosensitisers. Symptoms will vary with the amount of UV radiation, the type and amount of photosensitiser, skin type, and the age and sex of the person exposed. Photosensitisation of the skin and eyes can be caused by exposure to specific industrial chemicals. The skin can be affected by dermal exposure, and volatile fumes can affect inhalation and the eyes.

In certain occupations, the risk from exposure to particular photosensitising chemicals and solar UV radiation is severe. For example, exposure to tar and sunlight can cause pre-cancerous and cancerous skin lesions. Exposure to coal tar fumes can cause simultaneous inflammation of the conjunctiva and cornea. Ingestion or topical application of particular medications or prescribed drugs may cause photosensitivity in some individuals. Chemical-induced photosensitivity can occur in anybody, is usually dose related and may not happen on first exposure (ASCC, 2008). Photosensitisation has been observed in workers exposed to creosote, which can induce phototoxic or photoallergic reactions, sometimes accompanied by general symptoms such as depression, weakness, headache, slight confusion, vertigo, nausea, increased salivation or vomiting. Exposure to creosote and sunlight may be linked to a significantly elevated risk of lip and skin cancer observed in cohort studies of Swedish and Norwegian wood impregnators and Finnish round timber workers (WHO, 2004). For a good account of photosensitisation, and a list of industrial chemicals, drugs, plants and miscellaneous other substances that are known to cause photosensitisation, the reader is referred to Appendix 2 of the ASCC's (2008) *Guidance Note for the Protection of Workers from the Ultraviolet Radiation in Sunlight.*

15.9.2   THE INDIRECT EFFECTS OF LIGHT

Light exerts an indirect effect on various metabolic, hormonal and organic functions by way of the eyes. Circadian rhythms are entrained by the light–dark cycle, which is sometimes described as a *zeitgeber* (time-giver), a stimulus that 'sets' the circadian clock. The term 'circadian' was devised by Halberg (1959, cited by Aschoff 1965), derived from the Latin *circa* (about) and *dies* (day) (*Oxford English Dictionary*, 2007). Furthermore, free-running circadian rhythms—in which the sleep cycle is not synchronised to environmental cues and oscillates on a period other than 24 hours—seem to be influenced by light intensity. The free-running period is longer than 24 hours in dim light and shorter in bright light (Aschoff, 1965). Another interesting aspect of the interaction of light and circadian rhythms is that most totally blind people have free-running circadian rhythms. This condition causes recurrent insomnia and daytime sleepiness when the rhythms drift out of phase with the normal 24-hour cycle (Sack et al., 2000).

Budnick, Lerman and Nicolich (1995) demonstrated that exposure to high levels of bright light (i.e. 6000 to 12,000 lux) on at least half of a worker's night shifts over three months was effective in altering the worker's circadian rhythm pacemakers. Additionally, the effectiveness of light in setting a diurnal rhythm is a function of the light's wavelength (colour), with peak sensitivity between 460 nm and 484 nm in all vertebrates studied so far, including humans. Circadian photoreception is mediated primarily by melanopsin (a vitamin-A photopigment) contained in intrinsically photosensitive retinal ganglion cells (ipRGCs) distributed in a network across the inner retina (SCENIHR, 2012).

15.9.3   IMPLICATIONS FOR OCCUPATIONAL HEALTH

From the above, it can be seen that workers such as shift workers or those whose health and wellbeing is adversely affected by travelling over several time zones—for example, regular travellers or commercial air crew—may benefit from exposure to high levels of light to re-establish their circadian rhythms. There is also evidence that exogenous melatonin can help mitigate the adverse effects of shift work or jet lag (Arendt et al., 1997). The term 'chronobiotic' has been applied to the use of melatonin; it refers to an agent that can cause phase adjustment of the body clock.

It was predicted long ago that women working a non-day shift would have a higher risk of developing breast cancer than their day-working counterparts. This prediction was based on the idea that electric light at night (LAN) might account for a portion of the high and rising incidence of breast cancer worldwide. Studies such as those of Davis, Mirick and Stevens (2001) suggest that exposure to light at night may increase the risk of breast cancer by suppressing the normal nocturnal production of melatonin by the pineal gland, which in turn could increase the release of oestrogen by the ovaries. This hypothesis has been extended more recently to include prostate cancer. On the basis of limited human evidence and sufficient evidence in experimental animals, in 2007 the International Agency for

Research on Cancer (IARC) classified 'shift work that involves circadian disruption' as a probable human carcinogen, group 2A (Stevens et al., 2011).

In addition to the diurnal circadian rhythms of melatonin, there are also circannual (yearly) rhythms, and melatonin production seems to follow seasonal variations. Moreover, this has been linked to clinical variables in patients with depression (Wetterburg, 1990). The US National Institute of Mental Health and the National Academy of Sciences have studied how exposure to bright light can offset the negative effects of a type of depression called *seasonal affective disorder (SAD)*, which appears to affect a large number of adults who become depressed during the dark days of winter (Clarke, 1989) and start to feel better in the spring. The exact cause of SAD is unknown, but since it is more common during winter and in the higher latitudes of the Northern Hemisphere, some doctors believe that it can arise when a lack of sunlight alters serotonin levels. Daylight suppresses secretion of melatonin, and hence the shorter daylight hours of winter encourage prolonged secretion of melatonin, which may be a contributory factor for those susceptible to SAD. In the early 1990s, scientists started to study the efficacy of light therapy in the treatment of SAD. Terman and colleagues (1989) first noted the anti-depressant impact of bright, artificial light on SAD patients. Now, in severe cases of SAD, doctors recommend bright-light therapy—the controlled use of artificial light that mimics the sunlight spectrum. Daily sessions may range in duration from 20 to 60 minutes, depending on the severity of symptoms. Light therapy taken in the morning seems to be most effective in resetting the circadian rhythm. However, SAD is likely to be an occupational health problem only for workers who are 'light deprived', such as those working in polar regions (Arendt, 2012).

The American Conference of Governmental Industrial Hygienists (ACGIH®) has established threshold limit values (TLVs®) for visible light, to avoid retinal injuries from exposure to very intense light sources, such as welding arcs (ACGIH, 2018a). The International Commission on Non-Ionizing Radiation Protection (ICNIRP) also provides guidance for integrated effective spectral radiance and radiance dose in the visual field.

ACGIH and ICNIRP guidelines require specialised instrumentation to assess, are complex and are difficult to apply. Given that individuals can readily protect themselves through constricting the size of the pupil, blinking or even turning away from a bright light source, these TLVs® are unlikely to be used by most H&S practitioners.

Table 15.6 shows types of artificial light sources using levels of applicable eye damage provided by the ACGIH® TLVs®.

In a statement on the circadian, neuroendocrine and neurobehavioural effects of light, the ACGIH® (2018b) states that it does not consider it practical to develop TLVs® to protect against light-induced changes in circadian rhythms; instead, it makes the following recommendations:

- Shift work is best addressed by optimal planning of work schedules rather than exposure limits such as TLVs®.
- The colour palette of computer displays should be adjusted to reduce the short wavelength content or dim computer screens for evening work.
- The lighting conditions in occupational settings should provide the safest and most alerting environment possible, while maintaining typical visual function. Work environments should therefore incorporate high intensity, blue-enriched (high

Table 15.6 Example sources of non-laser optical radiation and applicable TLVs®

Source type*	Arc sources	Discharge lamps	Fluorescent lamps and LEDs	Thermal sources	Germicidal lamps
	Arc welding; arc lamps; xenon-arc searchlights		White-light and 'black-light' fluorescent lamps; Visible or UV-A LEDs	Hot and molten metals; gas welding; incandescent lamps; IR LEDs	Low-pressure mercury discharge lamps; UV-B and UV-C lamps and LEDs
Ultraviolet See UV TLV®	◆◆	◆	◆		◆◆
Blue-light See LNIR Section 1	◆◆	◆◆	◆		
IR cornea/lens See LNIR Section 2	◆	◆		◆◆	
Infrared retina See LNIR Section 3	f	f		◆	
Retinal thermal See LNIR Section 4	M				

◆◆ – Likely
◆ – Possible
f – Applicable when filtered lamp blocks visible emission
M – Only if magnified source size (e.g. searchlight or projection optics)
* A special type of diode emitter, the super-luminescent diode (SLD), although not a laser, should be assessed with the laser TLV®.

melanopic) light during both the day and especially at night, given the high risk of sleepiness-related accidents and injuries. In occupational settings, there are potentially conflicting needs—for example, at a hospital during the night when patients sleep but staff are awake, the ward environment should be optimised for patient sleep with low-intensity, blue-depleted (low-melanopic) light while the staff environment (nursing station, break rooms) should enhance alertness with high-intensity, blue-enriched (high-melanopic) light. These more complex environments need careful consideration of the spectrum, location and use of the light, but these issues are likely to be solved through the lighting design process.

- Worker complaints related to new installations of high-intensity, high-brightness LED lighting fixtures frequently relate to discomfort glare because of poor luminaire design or installation, not the blue-enriched spectrum of the light. Consulting good lighting guides may be helpful.

From this, it is evident that there is a coherent relationship between light and health in humans. For many workers, the majority of their waking hours are spent under artificial illumination, which provides less than 10 per cent of the light received under a shady tree on a sunny day (Clarke, 1989). Furthermore, the spectral distribution of this illumination is very different from that of natural light. It follows that the level of activity of the pineal gland, and hence melatonin secretion, will be affected by this lighting regime. Given the role of melatonin in various neuroendocrine functions, it was suggested even before this relationship was discovered that prolonged exposure to artificial light and the concomitant lack of exposure to natural light may be contributory factors in sick building syndrome (Robertson et al., 1989; Stone, 1992). Workers have been shown to prefer more daylight than artificial light at their workstations (Coyne, Bradley and Cowling, 1992). From an occupational health perspective, the following points should be considered in relation to the design of a workplace:

- Buildings should be built so as to maximise the use of natural light.
- Individuals should be allowed to control their lighting environment to meet their own requirements.
- All workers should have access to natural light at their workstation.
- Windows should allow full transmission of light.
- Workers should be encouraged to take 'light breaks' and get out of artificially lit environments during their lunch periods.
- Where artificial lighting must be used, preference should be given to use of 'daylight' tubes rather than 'warm white' or 'cool white' tubes.

5.9.3.1 Workers at risk of potential eye injury from exposure to blue light

Humans perceive light through photoreceptors in the eyes and the range of visible wavelengths lies between 390 nm and 700 nm (Starr, Evers and Starr, 2006). Blue light passes through the cornea to reach the retina, that is, the innermost area of an eyeball. The retina has seven thin layers of cells that can sense light. These can be damaged by short, high-energy wavelengths that comprise blue light. There are two types of retinal damage from blue light exposure: photochemical and photothermal damage.

- *Photochemical damage* is the injury to photoreceptors caused by high energy, short wavelengths in the range 400 nm to 550 nm. Certain artificial light sources, such as LEDs in workplaces, could potentially contribute to this. This is termed the 'blue light hazard' (Sliney, Bitran and Murray, 2012).
- *Photothermal damage* is the thermal (burning) retinal injury caused by UV and near-infrared radiation (IR) in the range 400 to nearly 1400 nm. Extremely intense infrared sources could induce the damage. However, the cornea and crystalline lens absorb most UV, thereby limiting thermal damage to retinal cells from general light sources such as LEDs (Sliney, Bitran and Murray, 2012).

15.9.4   MAKING FULL USE OF DAYLIGHT

Natural lighting is most effective for improving illumination. Maximising the use of daylight improves morale and reduces energy costs. Examine the workplace layout, material flow and workers' needs, then consider the following options for making effective use of daylight:

- Provide skylights—for example, by replacing roof panels with translucent ones.
- Equip the workplace with additional windows.
- Place machines near windows.
- Move work requiring more light nearer to windows.

Before planning and installing windows and skylights, do the following:

- Consider the height, width and position needed for windows or skylights. More light is available when the window is placed high on a wall.
- Install shades, screens, louvres, canopies or curtains on the windows and skylights to protect the workplace from external heat and cold while taking advantage of the natural light.
- Orient skylights and windows away from direct sunlight to obtain constant but less bright light.
- Direct skylights and windows towards the sun if variations in levels of brightness throughout the day do not disturb workers.
- While low-emissivity (low-E) windows will help to conserve energy by preventing heat loss from the building during hot weather and will draw heat from the sun in cool weather, they significantly reduce the quantity and quality of transmitted light (Salares and Russell, 1996).

Lighting is a subject area in which there is much work to be done. There is now significant evidence indicating a strong correlation between lighting and human behaviour and mood, but there is still inadequate recognition of the importance of lighting in the design of buildings. There are some exceptions, however, and a good example is the design of lighting in Melbourne City Council House 2 (CH2), constructed in 2004 to accommodate council staff. The design remit was not only to conserve energy and water but to improve the wellbeing of building occupants through the quality of the internal environment, including lighting (Altomonte, 2006). CH2 maximises the penetration of natural light within the building, reducing the requirement for artificial lighting, by:

Table 15.7 Potential health effects of shorter wavelength EMR or visible light

Potential health effects			Wavelengths/ peak	Typical light sources	Work-related
Impact	**Disorder**	**Exemplification**			
Eye	Photochemical damage (blue light hazard)	Photo-retinitis; possibly age-related macular degeneration (AMD)	Blue light (380–500 nm)/ potential hazardous wavelength: 441 nm)	Blue light sources (e.g. white LED, CFL, metal-halides, welding arcs, dental curing lamps, nail lamps, etc.)	Outdoor workers, welders, medical officials such as dental professionals, etc.
	Cornea/lens	Cataract	UVR (100–400 nm)	Sun, UV lamps	Outdoor workers
Skin	Skin cell	Melanoma, skin cancer	UVR (100–400 nm)	Sun, UV lamps	Outdoor workers
Circadian rhythm disruption	Biological rhythm, retinal ganglion cell damage	Metabolic syndrome, obesity	Blue light (380–500 nm, potential hazardous wavelength: 450–470 nm)	Sun, white or blue light sources (e.g. cool white LEDs)	Night shift workers, office workers, medical officials, etc.
	Melatonin levels	Sleep disorder			
	Cancers	Breast, prostate, brain cancers		White light sources (e.g. cool white LEDs or metal halides)	Night shift workers

Sources: SCENIHR (2012); SCHEER (2018).

- locating windows at the highest point of the curved concrete ceilings
- having an external 'light shelf' on the northern windows that, while protecting the windows from the direct rays of the sun, also bounces natural light into the building
- having movable timber shutters that remain open to catch the morning sun and close in the afternoon glare
- carefully positioning and using blinds on the northern windows. These windows are divided into upper and lower sections, each with its own blind. The upper blinds are needed only when the sun is low in the sky in winter. The lower blind is a partial blind (900 mm high) shielding against the direct rays of the sun while still letting in natural light (City of Melbourne, 2012).

In the absence of exposure standards, or of universally agreed guidelines, consideration should be given to the five principles of healthy lighting, promulgated by Canada's International Commission on Illumination CIE (2004):

1 The daily light dose received by people in Western [i.e. industrialised] countries might be too low.
2 Healthy light is inextricably linked to healthy darkness.
3 Light for biological action should be rich in the regions of the spectrum to which the non-visual system is most sensitive.
4 The important consideration in determining light dose is the light received at the eye, both directly from the light source and reflected off surrounding surfaces.
5 The timing of light exposure influences the effects of the dose.

Traditionally, most H&S practitioners have confined their efforts regarding lighting to ensuring that workplaces had adequate illumination for work to be done safely and efficiently. This chapter has demonstrated to the reader that there is much more to the lighting environment than simply measuring light levels. H&S practitioners now need to be aware that the lighting environment directly affects the health not only of workers, but of all members of the community. The onus is on occupational hygienists in particular to ensure that the impact of the lit environment is considered an integral part of the design of any facility, work regime or work environment.

15.10   REFERENCES

Alliance for Solid State Illuminations Systems and Technologies (ASSIST) 2010, *ASSIST Recommends . . . Guide to Lighting and Colour in Retail Merchandising*, Lighting Research Centre, New York.

Altomonte, S. 2006, *Lighting and Physiology*, City of Melbourne, Melbourne, <www.melbourne.vic.gov.au/SiteCollectionDocuments/ch2-lighting-physiology-technical-paper.pdf> [accessed 14 August 2019]

American Conference of Governmental Industrial Hygienists (ACGIH) 2018a, 'Documentation for light and near infra-red radiation', in *Threshold Limit Values for Chemical Substances and Physical Agents and Biological Exposure Indices*, ACGIH®, Cincinnati, OH.

—— 2018b, 'Notice of intent to establish—Appendix A: Statement on the occupational health aspects of new lighting technologies—circadian, neuroendocrine and neurobehavioural effects of light', in *Threshold Limit Values for Chemical Substances and Physical Agents and Biological Exposure Indices*, ACGIH®, Cincinnati, OH.

American Industrial Hygiene Association (AIHA) 2012, *Blue Light Hazard Quick Reference Sheet*, AIHA Non-Ionizing Radiation Committee, <www.aiha.org/get-involved/Volunteer Groups/Documents/NONIONRAD-BlueLightquickreferenceguideDec2012.pdf> [accessed 14 August 2019]

Arendt, J. 2012, 'Biological rhythms during residence in polar regions', *Chronobiology International*, vol. 29, no. 4, pp. 379–94, <www.ncbi.nlm.nih.gov/pmc/articles/PMC 3793275/pdf/CBI-29-379.pdf> [accessed 14 August 2019]

Arendt, J., Skene, D.J., Middleton, B., Lockley, S.W. and Deacon, S. 1997, 'Efficacy of melatonin treatment in jet lag, shift work, and blindness', *Journal of Biological Rhythms*, vol. 12, no. 6, pp. 604–17.

Aronoff, S. and Kaplan, A. 1995, *Total Workplace Performance: Rethinking the Office Environment*, WDL Publications, Ottawa.

Aschoff, J. 1965, 'Circadian rhythms in man', *Science,* vol. 148, no. 3676, pp. 1427–32, <https://mechanism.ucsd.edu/teaching/F11/philbiology2011/aschoff.circadianrhythms inman.1965.pdf> [accessed 28 July 2018]

Australian Safety and Compensation Council (ASCC) 2008, *Guidance Note for the Protection of Workers from the Ultraviolet Radiation in Sunlight*, ASCC, Canberra, <www.safeworkaustralia.gov.au/system/files/documents/1702/guidancenote_protection ofworkers_ultravioletradiationinsunlight_2008_pdf.pdf> [accessed 9 October 2018]

Brennan, D.H. 1987, 'Non-ionising radiation', in P.A.B. Raffle, W.R. Lee, R.I. McCallum and R. Murray (eds), *Hunter's Diseases of Occupations*, Hodder & Stoughton, London.

Budnick, L.D., Lerman, S.E. and Nicolich, M.J. 1995, 'An evaluation of scheduled bright light and darkness on rotating shift workers: trials and limitations', *American Journal of Industrial Medicine*, vol. 27, no. 6, pp. 771–82.

Cao, D. and Barrionuevo, P.A. 2015, 'The importance of intrinsically photosensitive retinal ganglion cells and implications for lighting design', *Journal of Solid State Lighting*, vol. 2, no. 10, <https://journalofsolidstatelighting.springeropen.com/track/pdf/10. 1186/s40539-015-0030-0> [accessed 9 October 2018]

Chambers Dictionary of Science and Technology 2007, Chambers Harrap, Edinburgh.

CIE, *see* International Commission on Illumination (Canada).

City of Melbourne 2012, *Natural Lighting Opportunities*, City of Melbourne, Melbourne, <www.melbourne.vic.gov.au/sitecollectiondocuments/ch2-natural-lighting-opportunities.pdf> [accessed 9 October 2018]

Clarke, G. 1989, 'Lighting the workplace for people', in *Proceedings of the 25th Annual Conference of the Ergonomics Society of Australia, 26–29 November 1989*, Ergonomics Society of Australia, Canberra.

Coyne, S., Bradley, G. and Cowling, I. 1992, 'The influence of daylight on office workers in Brisbane', *Journal of Occupational Health and Safety—Australia and New Zealand*, vol. 8, no. 1, pp. 13–20.

Davis, S., Mirick, D.K. and Stevens R.G. 2001, 'Night shift work, light at night, and risk of breast cancer', *Journal of the National Cancer Institute*, vol. 93, no. 20, pp. 1557–62.

Energy Rating 2009, *Lighting Phase Out*, <www.energyrating.gov.au/products/lighting/phaseout> [accessed 9 October 2018]

Forster, R. 2012, 'Types of lamps and lighting', in *Encyclopaedia of Occupational Health and Safety*, International Labour Organization, Geneva, <www.iloencyclopaedia.org/part-vi-16255/lighting> [accessed 9 October 2018]

Grant, W.B. 2018, 'A review of the evidence supporting the vitamin D cancer prevention hypothesis in 2017', *Anticancer Research*, vol. 38, pp. 1121–36.

Hansen, J. 2001, 'Light at night: Shiftwork, and breast cancer risk', *Journal of the National Cancer Institute*, vol. 93, no. 20, pp. 1513–15.

Hattar, S., Liao, H.-W., Takao, M., Berson, D.M. and Yau, K.-W. 2002, 'Melanopsin-containing retinal ganglion cells: Architecture, projections, and intrinsic photosensitivity', *Science*, vol. 295, no. 5557, pp. 1065–70.

IARC Monographs Working Group on the Evaluation of Carcinogenic Risks to Humans 2007, 'Shiftwork', in *Monographs on the Evaluation of Carcinogenic Risks to Humans Volume 98: Painting, Firefighting, and Shiftwork, 2–9 October 2007*, International Agency for Research on Cancer, Lyon, <https://monographs.iarc.fr/wp-content/uploads/2018/06/mono98-8.pdf> [accessed 9 October 2018]

International Commission on Illumination (CIE) (Canada) 2004, *Research Roadmap For Healthful Interior Lighting Applications*, Government of Canada, Ontario.

International Commission on Non-Ionizing Radiation Protection (ICNIRP) 2013, 'Guidelines on limits of exposure to broad-band incoherent optical radiation', *Health Physics*, vol. 105, no. 1, pp. 74–96.

International Labour Organisation (ILO) 2012, 'General hazards: Lighting', in *Encyclopaedia of Occupational Health and Safety*, ILO, Geneva, http://www.iloencyclopaedia.org/part-vi-16255/lighting. [accessed 9 October 2018]

Jennings, A.M. 2016, 'What is the blue light hazard? A review of recent developments', in *Proceedings of AIOH 34th Annual Conference, Gold Coast, 3–7 December 2016*.

Lee, S.Y., Pisaniello, D., Gaskin, S. and Piccoli, B. 2016, 'Characterising blue light exposure: Methodological considerations and preliminary results', in *Proceedings of AIOH 34th Annual Conference, Gold Coast, 3–7 December 2016*.

Longmore, J. 1980, 'Light', in H.A. Waldron and J.M. Harrington (eds), *Occupational Hygiene*, Blackwell, Oxford.

Oxford English Dictionary 2007, 6th ed., Oxford University Press, Oxford.

Palanker, D.V., Blumenkranz, M.S. and Marmor, M.F. 2011, 'Fifty years of ophthalmic laser therapy', *Archives of Ophthalmology*, vol. 129, no. 12, pp. 1613–19.

Piccoli, B., Soci, G., Zambelli, P.L. and Pisaniello, D. 2004, 'Photometry in the workplace: The rationale for a new method', *Annals of Occupational Hygiene*, vol. 48, pp. 29–38.

Poston, A.M. 1974, *A Literature Review of Cockpit Lighting*, Fort Belvoir Defense Technical Information Center, Fort Belvoir, VA.

Robertson, A.S., McInnes, M., Glass, D., Dalton, G. and Sherwood Burge, P. 1989, 'Building sickness: Are symptoms related to the office lighting?', *The Annals of Occupational Hygiene*, vol. 33, no. 1, pp. 47–59.

Sack, R.L., Brandes, R.W., Kendall, A.R. and Lewy, A.J. 2000, 'Entrainment of free-running circadian rhythms by melatonin in blind people', *The New England Journal of Medicine*, vol. 343, no. 15, pp. 1070–7.

Salares, V. and Russell, P. 1996, 'Energy efficient windows, lighting and human health', in *Indoor Air '96, Proceedings of the 7th International Conference on Indoor Air Quality and Climate, Nagoya, Japan*, International Society of Indoor Air Quality and Climate, Santa Cruz, CA.

SCENIHR [European Union Scientific Committee on Emerging and Newly Identified Health Risks] 2012, *Health Effects of Artificial Light*, <https://ec.europa.eu/health/scientific_committees/emerging/docs/scenihr_o_035.pdf> [accessed 8 October 2018]

Scientific Committee on Health, Environmental and Emerging Risks (SCHEER) 2018, *Final Opinion on Potential Risks to Human Health of Light Emitting Diodes (LEDs)*, <https://ec.europa.eu/health/sites/health/files/scientific_committees/scheer/docs/scheer_o_011.pdf> [accessed 8 October 2018]

Shang, Y.-M., Wang, G.-S., Sliney, D.H., Yang, C.-H. and Lee, L.-L. 2014, 'White light-emitting diodes (LEDs) at domestic lighting levels and retinal injury in a rat model', *Environmental Health Perspectives,* vol. 122, no. 3, pp. 269–76.

Sliney, D.H., Bitran, M. and Murray, W. 2012, *Infrared, Visible, and Ultraviolet Radiation*, John Wiley and Sons, New York.

Standards Australia 2006, AS/NZS 1680.1: 2006, Interior and Workplace Lighting Part 1: General Principles and Recommendations, SAI Global, Sydney.

Starr, C., Evers, C.A. and Starr, L. 2006, *Biology: Concepts and Applications*, Thomson, Brooks/Cole, Pacific Grove, CA.

Stevens, R.G., Hansen, J., Costa, G., Haus, E., Kauppinen, T., Aronson, K.J., Castano-Vinyals, G., Davis, S., Frings-Dresen, M.H.W., Fritschi, L., Kogevinas, M., Kogi, K., Lie, J-A., Lowden, A., Peplonska, B., Pesch, B., Pukkala, E., Schernhammer, E., Travis, R.C., Vermuelen, R., Zheng, T., Cogliano, V. and Straif, K. 2011, 'Considerations of circadian impact for defining "shift work" in cancer studies: IARC Working Group report', *Occupational and Environmental Medicine*, vol. 68, no. 2, pp. 154–63.

Stone, P.T. 1992, 'Fluorescent lighting and health', *Lighting Research and Technology*, vol. 24, no. 2, pp. 55–61.

Terman, M., Terman, J.S., Quitkin, F.M., McGrath, P.J., Stewart, J.W. and Rafferty, B. 1989, 'Light therapy for connective disorder: A review of efficacy', *Neuropsychopharmacology*, vol. 2, no. 1, pp. 1–22.

Wetterburg, L. 1990, 'Lighting: Nonvisual effects', *Scandinavian Journal of Work Environmental Health*, vol. 16, suppl. 1, pp. 26–8.

World Health Organization (WHO) 2004, *Coal Tar Creosote*, <http://apps.who.int/iris/bitstream/handle/10665/42943/9241530626.pdf?sequence=1> [accessed 9 October 2018]

Wurtman, R.J. 1967, 'Effects of light and visual stimuli on endocrine function', in L. Martini and W.F. Ganong (eds), *Neuroendocrinology, Vol. II*, Academic Press, New York, pp. 19–59.

—— 1975, 'The effects of light on man and other mammals', *Annual Review of Physiology*, vol. 37, pp. 467–84.

Zumtobel Lighting GmbH 2018, *The Lighting Handbook,* 6th ed., <www.zumtobel.com/PDB/teaser/EN/lichthandbuch.pdf> [accessed 9 October 2018]

16. Biological hazards

Margaret Davidson, Ryan Kift and Sue Reed

16.1   INTRODUCTION

Biological hazards in the workplace have been a topic of study, discussion and publications for many centuries. Notable early researchers and their works include Bernadino Ramazinni's (2001) eighteenth-century treatise on occupational diseases, *De Morbis Artificum Diatriba* [Diseases of Workers], John Tyndall's (1888) *Essays on the Floating-Matter of the Air: In Relation to Putrefaction and Infection*; and Thomas Oliver's (1902) *Dangerous Trades: The Historical, Social, and Legal Aspects of Industrial Occupations as Affecting Health*. Biological hazards such as viruses, bacteria and allergens exert a significant burden on worker health and wellbeing, as well as impacting the economy. Between 2014 and 2015, infectious and parasitic disease caused 290 serious workers' compensation cases (0.2 per cent of all claims), while injuries and illnesses associated with biological factors ranged from 606 cases in 2000–01 down to 360 cases in 2012–13 (Safe Work Australia, 2018). Industries with elevated risk of biological exposures—particularly those relating to micro-organisms—include health care, agriculture, waste management, forestry and food production (Safe Work Australia, 2018; Viegas et al., 2017). Consideration should also be given to previously unquantified biological hazards, whether these are atmospheres such as space (Lang et al., 2017), evolving ecosystems due to climate change, such as permafrost melt and anthrax (Charlier et al., 2017), or emergent industries such as medicinal and recreational cannabis production (Davidson et al., 2018; Green et al., 2018).

We are experiencing a rapid growth in our understanding of microbiological communities, both in our bodies and in the environment, and how they interact, courtesy of molecular-based, next-generation sequencing (NGS) technologies. Access to microbial analytical tools advanced rapidly over the second half of the twentieth century. The technology has stimulated an explosion in the field of *metagenomics*. *Meta*, translated from the Greek, means the higher study (transcendent), and *genome* is the total genetic material in an organism, translating as the higher study of genes at the community, as opposed to individual species, level. New sequencing technologies can simultaneously identify the collection of microbial genomes in an environment/sample (microbiome) without the need for culturing in the lab. A *microbiome* is simply a collection of genetic material, microbiota, in a location, be it human, animal or environmental. While these techniques are being used more commonly in research and some specialised commercial laboratories, it may be a few years before they became commonly available as test methods.

NGS advances have helped researchers to overcome reliance on culture-based analysis, one of the biggest obstacles to studying the microbial communities. It is estimated that less than 1 per cent of micro-organisms are estimated to be culturable in the lab (Schloss and Handelsman, 2005). We must not lose perspective on factors such as dose–response, and the ability of just one viable cell to cause an infection. Deoxyribonucleic acid (DNA) may be present from either dead or living cells, and we need methods to distinguish between genetic material from living and dead cells.

In recent times, outbreaks of infectious diseases such as measles, Ebola virus, pandemic influenza, Japanese encephalitis, Legionellosis and zoonotic diseases such as Brucellosis and Q Fever have increased the public's awareness of the importance of risk-management planning and preparedness for biological hazards. There are still many areas in which knowledge of biological hazards is limited, particularly regarding the way we monitor and estimate health risk associated with exposure to infectious (communicable) particles such as bacteria, viruses, prions, protozoa, non-infectious biological hazards such as plant pollens, insect dander, myco- and bacterial toxins, flour dust, enzymes and the many other substances that can trigger an immune response. The assessment of biological hazards remains a challenging area in the field of occupational hygiene, about which this chapter presents an introductory overview.

16.2   BIOLOGICAL AGENTS

Biological hazards encompass a wide array of agents, including potentially infectious, bacteria, archaea, viruses, fungi, protozoa and parasitic organisms, as well as non-living structural components and products of microbes, plants, animals and insects, including bacterial toxins and cell-wall chemicals (endotoxin, exotoxins), fungal spores, toxins (mycotoxins) and fragments (hyphae), prions (infectious proteins), animal and insect dander, dust mites, plant fibres, pollen, grain, cotton and organic dusts (Oppliger, 2014; Safe Work Australia, 2011; Viegas et al. 2017).

Exposure to biological agents can occur through multiple pathways, such as ingestion of toxins and infectious bacteria, inhalation of proinflammatory biological aerosols, puncture/injection of viruses from sharps, vector transfer of arboviruses by mosquitoes and ticks and dermal contact with allergenic pollens and saps. In occupational hygiene, we look for the potential for worker exposure to biological hazards that may be present on surfaces, as well as in soil, air, water and even human tissues and biological materials.

Aerobiology is a specialist area that studies biological aerosols (bioaerosols); these are defined as 'airborne particles that are living or originate from living organisms' (Macher, 1998, p. 1). The following sections provide a brief introduction to a number of biological hazards: bacteria, fungi, allergens, viruses, protozoans and prions.

Zoonoses are diseases that are transmitted between animals and humans, and encompass all infectious (communicable) biological agents.

16.2.1   BACTERIA

Bacteria are single-celled organisms with no nucleus or membrane-bound organelles. Bacteria may have structure that include flagella (motility), glycocalyces (protective outer coating), pilli (reproduction) and fimbrae (attachment and protective biofilm/slime layer). Bacteria typically reproduce asexually through division (binary fission) (Tortora, Funke and Case, 2016). Some bacteria, such as *Bacillus anthracis* (anthrax), can produce hardy endospores that assist them to survive in inhospitable environments (Liu et al., 2004), which makes them difficult to eradicate and control.

Bacteria can be categorised into gram negative or gram positive, based on a gram stain that targets the cell wall, which identifies the presence of lipopolysaccharide (LPS) or peptidoglycan (PG) in their outer cell wall (Bauman, 2012). Gram-positive bacteria typically stain purple and gram-negative bacteria, with LPS in their outer wall, stain pink. Some bacteria, such as *Neisseria* species and environmental isolates, can produce a gram-variable stain that is both pink and purple (Tortora, Funke and Case, 2016).

Discussion of gram-positive and gram-negative bacteria may arise when reviewing organic dust exposures and bioaerosols, particularly in relation to the respiratory conditions associated with LPS exposure. LPS, which is referred to as endotoxin, is a proinflammatory microbial constituent. Industries with potential for high endotoxin exposures include agriculture, forestry and fisheries (Donham and Thelin, 2016), laboratory workers (Stave, 2018;), waste management (Barker et al., 2017), textiles (Paudyal et al., 2015) and food production (Farokhi, Heederik and Smit, 2018). Conversely, endotoxin exposure has also been linked to reduced risk of occupational cancers (Khedher et al., 2017). Some bacteria species produce pathogenic exotoxins that may interfere with nerve-cell function (botulinum and tetanus) or can alter cellular function in the gastrointestinal tract (cholera, diphtheria and staphylococcal food poisoning) (Bauman, 2012; Popoff and Poulain, 2010).

Cyanobacteria (blue-green algae), a group of water-borne photosynthetic bacteria that are some of the Earth's oldest micro-organisms (Yates et al., 2016), can become problematic when they rapidly reproduce, creating 'blooms' and producing intracellular cyanotoxins that can be harmful to humans and livestock (hepatoxins, neurotoxins etc.). To date, no human deaths have been associated directly with their toxins, but ingestion may cause gastroenteritis, and direct contact with the bacteria and toxins can cause skin rashes and eye irritation (NHMRC, 2018).

16.2.2 FUNGI

Fungi are single or multicellular organisms with cell walls composed predominantly of chitin. There are over 81,000 known species, which vary in size and shape. Common characteristics include that they are all non-mobile and non-photosynthetic, and capable of both sexual and asexual reproduction. Fungi may be classified as yeasts that are single celled and reproduce by budding or spores, or unicellular or multicellular moulds that reproduce by spores or gamete formation (Tortora, Funke and Case, 2016). Many species of fungi, such as mushrooms, are edible; however, there are also poisonous types that are indistinguishable from the non-poisonous varieties. Eating a single mushroom of some species can be fatal, and ingestion of hallucinogenic mushrooms may also lead to renal failure (Austin et al., 2019; McKenzie, 2012).

Direct fungal growths on human and animal hosts are called *mycoses*, while diseases associated with ingestion, dermal contact and inhalation of toxic fungal metabolites are called *mycotoxicoses* (Bennett and Klich, 2003). Moulds of occupational health significance include *Aspergillus*, *Penicillium*, *Cladosporium*, *Fusarium*, *Alternaria* and *Stachybotrys* species (Dutkiewicz et al., 2011; Semple, 2010). Moulds are also sources of mycotoxins, microbial volatile organic compounds (MVOCs), allergenic mycelia (hyphae) and cell-wall constituents such as β-(1,3)-D-glucans (Viegas et al., 2015). Fungal spores typically range between 2 and 50 μm in size, and are bigger than bacteria but usually smaller than pollens (Madelin and Madelin, 1995).

Mycotoxins produced by fungi can be toxic to animals and humans. The following are some of the main mycotoxins of occupational health significance:

- Aflatoxins, produced by *Aspergillus* species, are a hepatotoxic carcinogen.
- Fumonisins, from *Fusarium* species, are carcinogenic and neurotoxic.
- Ochratoxins, from *Aspergillus* species, cause renal disease.
- Trichothecenes from multiple fungi genera, including *Fusarium, Myrothecium, Phomopsis, Stachybotrys* and *Trichodermachartarum*, are immunotoxins and cause gastroenteritis.
- Citrinin, from *Penicillium* and *Aspergillus* species, are nephrotoxic, hepatotoxic and cytotoxic.
- Patulin, from *Penicillium* species, are allergenic (Bennett and Klich, 2003; Samson, 2015).

β–1,3-D-glucans are a cell-wall component of fungi, plants and some bacteria, but there is conflicting information about their associated health effects (Donham and Thelin, 2016; Viegas et al., 2017). Research has indicated that β–1,3-D-glucans may trigger an immune response such as asthma or high blood pressure (Oluwole et al., 2018; Viegas et al., 2017; Zhang et al., 2015). Further research is required to determine the specific health effects associated with β–1,3-D-glucans exposure.

16.2.3   ALLERGENS

Allergens include a range of biological agents that can cause allergic alveolitis, contact dermatitis and occupational asthma. Occupational allergens include flour dust in bakeries and mills, enzymes used in washing powder and food production, latex from rubber gloves, animal proteins in laboratory animal-handling facilities, wood dust, pollens, animal dander and hair, plant materials and pollens, fungal spores and hyphae, bacteria, protozoa, bird droppings, feathers, insects and microbial components such as endotoxin or cyanotoxin (Donham and Thelin, 2016; Macher, 1999).

16.2.4   VIRUSES

Viruses are found in all environments, in the home and the occupational environments, and also form part of our personal microflora, referred to as the *human virome*. Viruses are comprised of protein and nucleic acid, either DNA or ribonucleic acid (RNA). They are very small (20–450 nm), and can only replicate inside other living cells (Cowan, 2012). Interest in airborne exposure to viruses has increased alongside the development of new technology which enables viruses to be identified more easily; however, monitoring for airborne viruses in Australia is only conducted for sampling of the vector (mosquitoes) or in sentinel chicken flocks and not for the virus itself (Knope et al., 2016). Limitations of monitoring include the need for diagnostic markers and databases to identify viruses, and sufficient samples for identification (Zou et al., 2016). More information on viruses is presented in Table 16.1 in section 16.3.1.

16.2.5 OTHER BIOLOGICAL HAZARDS

Other biological hazards include protozoan pathogens, heterotrophic single-celled organisms capable of independent movement that include:

- *Giardia lamblia* (*G. intestinalis*, *G. duodenalis*) and *Cryptosporidium parvum* and *C. hominis*, which cause gastroenteritis when infectious cysts are ingested
- the malaria parasite *Plasmodium ovale*, *P. malariae*, *P. knowlesi*, *P. vivax* and *P. falciparum*, which are a concern in deployed defence personal and overseas travellers (Heymann, 2015). Australia was declared malaria free in 1981 (non-endemic), and the greatest risk for people is contracting the disease while travelling overseas (Russell, 2018).
- amoeba such as *Entamobeba histolytica*, a human parasite that is transmitted through ingestion of contaminated food or water, as well as contact with contaminated surfaces.

Algae, which are photosynthetic eukaryotic organisms, include aquatic diatoms and dinoflagellates, both of which can produce neurotoxins that humans may be exposed to through consumption of contaminated fish and shellfish (Tortora, Funke and Case, 2016). They should not be confused with blue-green algae, which are in fact bacteria.

Prions are microscopic self-replicating proteins that cause neurodegenerative diseases such as classical Creutzfeldt-Jakob Disease (CJD) and variant CJD (vCJD) in humans, and bovine spongiform encephalopathy (BSE) in animals. These diseases are always fatal. Prions are infectious proteins that replicate by acting as a template that causes normal cellular proteins to reshape into prion proteins. Spasmodic cases of classical CJD have been recorded in Australian since 1970; however, there have been no reported cases of vCJD. Australia is considered a negligible-risk country for BSE (NSW Health, 2015).

Last but not least are the macroscopic biological hazards such as spiders, snakes, predatory and stinging aquatic organisms, and even farm animals that can cause crush injuries and even occasionally death. There are also parasitic worms (helminths), tapeworms, liver flukes and hookworm that can be contracted when working with animals, handling soil and waste products and working in outdoor environments (CDC, 2016).

16.3 OCCUPATIONAL DISEASE

Occupational diseases associated with biological hazards can be grouped into infectious diseases, respiratory diseases, dermal/skin conditions and cancer. *Infectious* diseases are transferred from one human or animal to another, including zoonotic diseases passed between humans and animals (Tortora, Funke and Case, 2016). *Respiratory* diseases are complex responses by the body to various agents; they may be allergic and non-allergic diseases, along with dermatoses (skin diseases) stimulated by irritation and biological (immune) responses. In the longer term, the body may respond to a biological hazard by developing cancer.

Workers in a range of occupations are at significant risk of acquiring an occupational disease through exposure to biological hazards if exposure controls are inadequate (Viegas et al., 2017):

- health care
- social assistance
- veterinary medicine
- waste management
- biomedical research
- agriculture forestry and fishing.

It must be noted that, unlike most chemicals and physical agents, biological agents are omnipresent in our working, social and home environments, and we are continuously being exposed. Consideration needs to be given to military and charitable aid workers deployed in foreign countries, who may be exposed to various exotic infectious diseases (Rapp et al., 2014) and to the targeting of government agencies in bioterrorism attacks (Edmonds et al., 2016).

16.3.1 INFECTIOUS DISEASE

Infectious diseases, also called communicable diseases, may be transmitted through direct human or animal contact, accidental auto-inoculation (needlestick), inhalation of contaminated aerosols and/or contact with contaminated objects (fomites) or ingestion of contaminated food or water (Figure 16.1).

Bites from disease vectors such as mosquitoes or ticks represent a significant health risk to outdoor workers because Australia has many endemic mosquito- and tick-borne viral diseases, including Ross River virus, Murray Valley encephalitis and Barmah

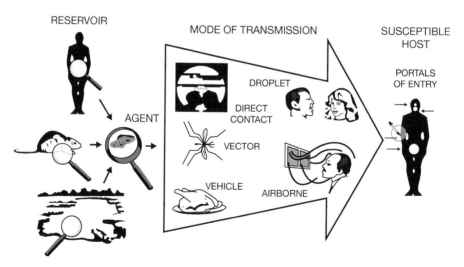

Figure 16.1 Chain of infection
Source: CDC (2016).

Forest virus, which can cause severe illness or even death. Sporadic cases of malaria, dengue and Japanese encephalitis also occur in northern Australia (Russell and Doggett, 2012). Australia uses a mandatory reporting system of notifiable infectious diseases, and further information and a full list of notifiable diseases are available from the National Notifiable Disease Surveillance System (Department of Health, 2018). In Australia, cases of workplace-related infectious diseases have occurred, including brucellosis, human immunodeficiency virus (HIV), leptospirosis, mycosis, tuberculosis, measles, legionellosis, Q fever, Hendra virus, cellulitis (bacterial skin infection) and viral hepatitis (Safe Work Australia, 2006, 2011). A list of infectious diseases that have been reported in Australia is presented in Table 16.1.

Legionnaires' disease is a pneumonia caused by the *Legionella* species. Workers who may be at risk include those exposed to aerosols from cooling towers and decorative fountains, as well as to water sprayed to cool people or water sprays being used to control dust levels or for cleaning, which potentially contain *Legionella pneumophila*. Smokers and the immuno-compromised are at higher risk of contracting *Legionella* illness.

Q fever is a zoonotic, flu-like illness caused by the organism *Coxiella burnetii*. Workers most at risk of exposure to Q fever include those exposed to animal placental material or urine of infected animals; examples of those who may be at risk of contracting Q fever include veterinarians, abattoir workers and others who handle livestock. Q fever is the most reported of six zoonoses on the Australian National Notifiable Disease Surveillance System (NNDSS). Q fever can be controlled effectively by the uses of a vaccine, which must be given before exposure to the bacteria (Australian Meat Processor Corporation, 2018).

For further information on communicable diseases, consult the US *Control of Communicable Diseases Manual*, published by the American Public Health Association or the Australian Department of Health's Communicable Diseases website (Department of Health, 2018). Pathogen safety data sheets (PSDSs) can also be obtained from the Canadian Public Health Agency (CPHA, 2018) as well as the GESTIS Biological Agents Database, <www.dguv.de/ifa/gestis/gestis-biostoffdatenbank/index-2.jsp>.

16.3.2 RESPIRATORY DISEASE

Respiratory diseases associated with exposure to biological agents are divided into allergic and non-allergic conditions (Table 16.2). An allergic response can be defined as a respiratory disease that is triggered by activation of the adaptive (active) immune system by a foreign substance, while a non-allergic respiratory disease is caused by mechanical blockage, irritation or innate (passive) immune-system response (Donham and Thelin, 2016).

16.3.2.1 Allergic responses

Occupational asthma, resulting from workplace exposure to allergens and starting in adulthood, is estimated to account for 1 in 6 adult-onset asthma cases (Torén and Blanc, 2009) and is under-reported in Australia (Crewe et al., 2016; Hoy and Brims, 2017). Allergic reactions occur when the human body identifies a foreign material (antigen) and an immune response is stimulated. The body forms antibodies, which bind to the antigen, making them harmless. On occasion, however, the body overreacts (hyper responds), leading to an allergic reaction/hypersensitivity disease.

Table 16.1 Infectious diseases in Australian occupational environments

Disease	Agent	Reservoir	Transmission	High-risk occupations
Viral				
Arboviruses				
Barmah Forest virus	Alphavirus	Mosquitoes and humans	Mosquitoes	Outdoor workers on South Coast of NSW
Dengue	Flavivirus (DEN1, DEN2, DEN3 or DEN4)	Human–*Aedes aegypti* cycle		Outdoor workers in northern Queensland
Japanese encephalitis	Flavivirus	Mosquito (*Culex annulirostiris*), wild and domestic birds and pigs		Outdoor workers in northern Australia
Kunjin virus infection	Flavivirus	Mosquito (*Culex annulirostris*) and water birds		Outdoor workers in northern Australia
Murray Valley encephalitis	Flavivirus	Mosquitoes and water birds		Outdoor workers in northern Australia
Ross River fever	Alphavirus	Mosquitoes and humans		
Rabies virus				
Australian bat lyssavirus infection	Lyssavirus	Insectivorous bats, flying foxes	Bites or scratches from infected animals	Bat carers, wildlife workers and veterinarians

Disease	Agent	Reservoir	Transmission	High-risk occupations
Blood-borne viruses				
Hepatitis A	Picornaviridae family RNA virus	Humans	Faecal/oral transmission (Hep A), contact with infected blood and other bodily fluids, sexual intercourse and mother-to-child (i.e. vertical transmission)	Health-care workers, sex industry and waste-management workers
Hepatitis B	Hepadnavirus			
Hepatitis C	Flavivirus			
Human immunodeficiency virus infection	HIV 1, HIV 2			
Hendra virus infection				
Hendra virus disease	Henipavirus	Horses, flying foxes, dogs and cats	Contact with infected fruit bats or horses in eastern Australia	Horse industry workers, veterinarians and government inspectors
Influenzas and para-influenzas				
Seasonal influenza	Influenza virus	Humans	Contact with contaminated people and objects, or inhalation of aerosolised viruses	Health-care workers, educators
Swine flu	H1N1v virus	Swine and humans	Contact with infected animals, less often contact with infected humans	Swine handlers, government inspectors, veterinarians, abattoir workers

Table 16.1 Infectious diseases in Australian occupational environments *continued*

Disease	Agent	Reservoir	Transmission	High-risk occupations
Menangle virus infection	Rubulavirus	Swine, humans and fruit bats	Contact with stillborn piglets and bats in NSW	Swine handlers, veterinarians and government inspectors
Newcastle disease	Avulavirus	Birds	Contact with infected birds	Poultry workers, lab staff, veterinarians
Bacterial				
Anthrax	*Bacillus anthracis*	Domestic and wild animals	Inhalation, ingestions or contact of broken skin with diseased animals, hides, hair or offal	Veterinarians, wool sorters, abattoir workers, farmers and stock handlers. Also at risk are security and postal workers in the event of bioterrorism
Brucellosis	*Brucella suis*	Feral pigs in Queensland	Contact with infected tissues and bodily fluids, inhalation of contaminated aerosols	Feral animal hunters, agricultural workers and veterinarians. *Brucella abortis* was eradicated from Australia in 1989
Campylobacteriosis	*Campylobacter jejuni, C. upsaliensis*	Wild birds and domestic animals and birds	Contact with infected animals and objects, contaminated water or faecal/oral transmission	Veterinarians, agricultural workers, child-care, waste-management and health-care workers

Disease	Agent	Reservoir	Transmission	High-risk occupations
Erysipeloid	*Erysipelothrix rhusiopathiae*	Fish, shellfish, mammals and poultry	Contact with contaminated meat	Fishermen, abattoir workers, farmers and butchers
Haemophilus influenza	*Haemophilus influenzae*	Humans	Inhalation of respiratory droplets from infected person. Rarely, contact with infected mucal discharges	Health-care and child-care workers
Legionellosis/ Legionnaires' disease	*Legionella pneumophila, L. longbeachae, L. micdadei, L. bozemanni*	Soil, water, mulch, wood chips and hot springs	Inhalation of contaminated water and dust aerosols	Office and health-care workers, horticulturalists, agricultural workers
Leptospirosis (Weil's disease)	*Leptospira interrogans*	Wild and domestic animals including rats, cows and pigs	Contact with infected animal urine/ flesh, ingestion of contaminated water or soil	Farmers, veterinarians, waste management and abattoir workers, sugar cane and banana farmers and miners
Listeriosis	*Listeria monocytogenes*	Wild and domestic animals, sewage, silage and birds	Ingestion of contaminated food and other materials	Agricultural workers, veterinarians and lab workers
Methicillin-resistant *Staphylococcus aureus* (MRSA)	*Staphylococcus aureus*	Swine, dogs, cats, horses, humans and wild animals	Contact with infected animals or people	Farmers, animal handlers, veterinarians, health-care and abattoir workers

Table 16.1 Infectious diseases in Australian occupational environments *continued*

Disease	Agent	Reservoir	Transmission	High-risk occupations
Mycobacterial infections				
Tuberculosis	*Mycobacterium tuberculosis*	Humans	Inhalation of infectious droplets	Health-care and child-care workers, prison wardens
Non-tuberculosis	*M. avium-intracellulare* *M. kansasii* *M. scrofulaceum* *M. fortuitum* *M. marinum* *M. chelonae*	Groundwater, dust and soil	Rarely determined, possibly inhalation or ingestion of soil, dust or water, *M. marinum* skin inoculation	Recreational industry, agriculture, forestry and fishing workers
Pasteurellosis	*Pasteurella multocida*	Livestock and domestic pets	Animal scratches	Agricultural workers and veterinarians
Pertussis (whooping cough)	*Bordetella pertussis*	Humans	Inhalation of contaminated aerosols or contact with infected mouth or nose secretions	Health-care and child-care workers
Psittacosis	*Chlamydia psittaci*	Birds, rarely cats, dogs, goats or sheep	Inhalation of dust infected with bird faeces, or contact with eye or nasal secretions of infected animals. Rarely human to human	Veterinarians, pet shop workers, breeders

Disease	Agent	Reservoir	Transmission	High-risk occupations
Q fever	*Coxiella burnetii*	Sheep, goats	Inhalation of contaminated aerosols	Abattoir workers, researchers, veterinarians, farmers
Salmonellosis	*Salmonella species*	Domestic and wild animals, birds and reptiles	Faecal/oral transmission, ingestions of contaminated foods	Child-care, health-care, hospitality and agricultural workers
Shigellosis	*Shigella sonnei*	Humans	Faecal/oral transmission, ingestion of contaminated food, water or milk	Child-care and health-care workers, laboratory workers
Escherichia coli infection	*Escherichia coli* (enterohaemorrhagic, enteropathogenic, enterotoxigenic or enteroinvasive)	Soil, dusts, silage, domestic and wild animals, reptiles, birds and fish	Faecal/oral transmission, ingestion of contaminated food and water. Contact with infected animals	Health-care, child-care, laboratory and agricultural workers
Spotted fevers (Queensland tick typhus and Flinders Island spotted fever)	*Rickettsia australis*, *R. honei*	Ticks and marsupials (suspected)	Tick bites	Agriculture, forestry and fisheries workers, veterinarians and horticulturalists

Table 16.1 Infectious diseases in Australian occupational environments *continued*

Disease	Agent	Reservoir	Transmission	High-risk occupations
Fungal				
Cryptococcal infections	*Cryptococcus neoformans var. neoformans*, *var. gattii***	Bird droppings* *Eucalyptus sp.***	Inhalation of fungal spores	Forestry workers, horticulturalists, veterinarians and agricultural workers
Dermatophytosis (tinea)	*Trichophyton tonsurans*, *T. verrucosum***	Humans*, cattle**	Skin to skin, contact with contaminated objects or surfaces	Veterinarians, abattoir and agricultural workers
Histoplasmosis	*Histoplasma capsulatum*	Soil and bats	Inhalation of spores	Bat handlers and people who enter bat caves
Protozoan				
Cryptosporidiosis	*Cryptosporidium parvum*	Humans, cattle and domestic animals	Faecal/oral transmission, ingestion of contaminated food or water	Health-care, child-care and recreational industry workers, agricultural workers and veterinarians
Giardiasis	*Giardia lamblia*	Humans, animals and contaminated waters	Faecal/oral transmission, contacted with contaminated items or ingestion of contaminated water	Health-care and child-care workers, recreation industry, waste management, agricultural and forestry workers, veterinarians

Disease	Agent	Reservoir	Transmission	High-risk occupations
Toxoplasmosis	*Toxoplasma gondii*	Felines; intermediate hosts include domestic and wild animals and birds	Ingestion of contaminated food, unpasteurised milk, contaminated soil; blood transfusion, organ transplant and maternal transmission	Laboratory, agricultural and abattoir workers, veterinarians, hunters and agricultural workers
Parasitic				
Echinococcosis (hydatid disease)	*Echinococcus granulosus*	Canine/domestic and wild animal cycle. Sheep are main intermediate host	Hand-to-mouth transfer of eggs from dog faeces	Farmers (particularly sheep) and hunters

Sources: Harries and Lear (2004); Heinsohn (2011); Heymann (2015); Jordan et al. (2011); McLeod et al. (2011); McCormack and Allworth 2002; Morrell and Stratman (2011); NOHSC (1990); ; Russell et al. (2009); Russell and Doggett (2012); Smith et al. (2011); Biosecurity Australia (2012); Victorian Department of Health (2018).

Table 16.2 Respiratory diseases from exposure to non-communicable biological hazards

Respiratory disease	Agent	High-risk occupations
Non-allergic		
Non-allergic asthma, non-allergic rhinitis/ mucous membrane irritations (MMI), chronic bronchitis, chronic airflow obstruction, organic toxic dust syndrome (ODTS)	Fungi, bacteria actinomycetes, endotoxin, ß-(1,3)-D glucans, peptidoglycan, mycotoxin	Agriculture, forestry and fishing workers, waste treatment workers, composting and recycling workers, textile and food production workers, horticulturalists, metal workers and machinists, veterinarians, zookeepers, laboratory workers, construction workers, archaeologists and biofuel production workers
Allergic		
Allergic asthma, allergic rhinitis, hypersensitivity pneumonitis (allergic alveolitis/farmer's lung)	Fungi, thermophilic mycelial bacteria, *Mycobacterium immunogenum* MVOCs, spores, allergens (dust mite, plant, insect and animal proteins, pollens etc.), endotoxin, ß-(1,3)-glucans	Waste treatment, composting and recycling workers, biomedical researchers, enzyme production workers and lab animal tenders, health-care workers, bakers, industrial (detergent and biopesticide manufacturing) workers, pet shop owners, agricultural, forestry and fishing workers, wood processors/furniture makers and horticulturalists

Sources: adapted from Donham and Thelin (2016); Liebers, Raulf-Heimsoth and Brüning (2008).

There are four types of allergic/hypersensitivity reactions, which have been described by Bogaert et al. (2009) and Coico and Sunshine (2015):

- *Type 1:* immediate hypersensitivity (allergic reactions) is a rapid reaction (15–30 minutes) mediated by immunoglobulin E (IgE). Diseases include asthma, eczema, conjunctivitis, allergic rhinitis and in severe cases anaphylaxis.
- *Type 2:* cytotoxic hypersensitivity, which occurs when antigens bind to cells, affecting a variety of organs and tissues. Type 2 is mediated primarily by IgM and IgG, including anaphylactic shock such as the response to penicillin and rheumatic fever.
- *Type 3:* immune-complex hypersensitivity occurs three to ten hours after exposure. Examples of immune-complex diseases include hypersensitivity pneumonitis (HP).

- *Type 4:* cell-mediated/delayed hypersensitivity is associated with many infectious diseases, including leprosy, toxoplasmosis and tuberculosis, as well as contact dermatitis expressed as papular lesions. Unlike Types 1 to 3, Type 4 is cell mediated, and not antibody mediated. The response time is 48–72 hours.

The body's response to inhaled allergens can range from mild rhinitis to the more severe allergic asthma and allergic alveolitis. The intensity of the response is dependent on intrinsic (internal) and extrinsic (external) factors, such as genetic susceptibility, smoking status, early childhood or living on a farm, among many others (Reynolds et al., 2013). Common biological causes of occupational asthma and allergic rhinitis include animal proteins, arthropods, seafood, mites, flour, pollens, plant proteins and derived products, amylases from the bacteria *Bacillus subtilis* (subtilisins) and wood dusts (Dao and Bernstein, 2018).

- *Allergic rhinitis* is one of the most common medical conditions worldwide (Adelman, Casale and Corren, 2012). Symptoms of allergic rhinitis (hayfever) are associated with a blocked or runny nose, sneezing, itchy, sore eyes and a sore throat. Common triggers include pollen, fungi, dust mite, pet dander and cockroaches (Dao and Bernstein, 2018). Agents that maybe be an issue in workplaces include grain dust, latex, α-amylase in flour, biological enzymes, fish and seafood proteins and wood dusts (Greiner et al., 2011).
- *Allergic asthma* is an airway-obstructive condition that can have many names. Work-aggravated asthma (WAA) is triggered by occupational exposures in people with pre-existing sensitivities, while occupational asthma is reserved for new adult-onset cases directly attributed to a specific agent or exposure in the workplace (Dao and Bernstein, 2018). Together, the two clinical definitions form the group of respiratory diseases known as work-related asthma (WRA). Symptoms of allergic asthma may occur outside of work hours, often at night, making identification of the sources more difficult. An atopic worker with a history of asthma or eczema may be more likely than other workers to develop asthma. High-risk industries include:
  - agriculture and forestry
  - waste management
  - wood production
  - baking
  - biotechnology and laboratory workers
  - fish handlers and seafood processors
  - production of biological enzymes used in detergents and food production (Crewe et al., 2016; El-Zaemey et al., 2018).

  An employer who is hiring for a position that may entail exposure to workplace allergens (e.g. wood-working, animal handling) may consider undertaking pre-employment screening to identify workers with a history of hypersensitivity reactions to common environmental antigens, and take appropriate measures to ensure access to appropriate medicines.
- *Hypersensitivity pneumonitis (HP)* (allergic alveolitis) is a complex lung disease caused by an immune response to inhaled organic bioaerosols (<5 μm) which can reach the alveoli (Selman, Pardo and King, 2012). Symptoms include shortness of breath or a bronchitis-like illness, which usually is caused by microbial contamination of plant matter stored in a damp environment. Symptoms of acute HP are very similar to those

of the non-immunogenic organic dust toxic syndrome (ODTS), and must be clinically distinguished from this more common disease (Selman et al., 2012). HP in various industries may be called farmer's lung, bagassosis, mushroom lung, coal worker's lung, aspergillosis, malt worker's disease, wood pulp worker's disease, cheese washer's disease, greenhouse lung, wine grower's lung, chicken fancier's disease (Bønløkke, Cormier and Sigsgaard, 2010) and even bagpiper's lung (King et al., 2017).

16.3.2.2 Non-allergic responses

Certain bioaerosols can act as irritants, cause airway blockages or stimulate the innate immune system, even when they do not produce an allergic response in the worker. One of the most common symptoms, mucous membrane irritation (MMI), which may appear consistently during the work shift, may take the form of irritation and inflammation of the upper respiratory tract—conjunctivitis, sinusitis, rhinitis, pharyngitis, laryngitis and tracheitis. There may also be a chronic cough or bronchitis (Rusca et al., 2008; Schlosser et al., 2009). Important non-allergenic diseases include non-allergic asthma, inhalation fever and ODTS.

Non-allergic asthma is diagnosed frequently in farmers or other workers with high bioaerosol exposure, which results in a marked reduction of lung function across the period of a work shift (Linaker and Smedley, 2002). Although similar to allergic asthma in presentation, non-allergic asthma can result in a relatively quick decline in lung function and is not accompanied by production of IgE antibodies (Linaker and Smedley, 2002).

Inhalation fevers (flu-like illnesses) are caused by the inhalation of microbially contaminated aerosols from agricultural dust, air humidifiers or other aerosols contaminated with *Legionella* bacteria. The following are examples of inhalation fevers:

- Humidifier fever can occur in buildings with humidification systems or air-conditioning. Symptoms appear within four to twelve hours of exposure, and the illness is self-limiting (resolves without treatment), with recovery occurring in a matter of days.
- Pontiac fever results from exposure to *Legionella pneumophila* and other *Legionella* species. This is a variant of legionellosis that is mild, self-limiting and does not result in pneumonia (Tortora, Funke and Case, 2016).

Toxic alveolitis, known as ODTS, occurs as a result of exposure to high levels of organic dust and endotoxin (Donham and Thelin, 2016), fungal spores and fragments, including mouldy hay, silage and corn (Madelin and Madelin, 1995). The symptoms of OTDS include fever, shivering, dry cough, chest tightness, dyspnoea, headache, muscular and joint pain, fatigue, nausea and general malaise, which typically disappear after 24 hours but may persist for up to seven days (Douwes, Eduard and Thorne 2008). Many occupations have been associated with increased risk of ODTS, including farmers, veterinarians, biofuel workers and waste management workers, including those involved in composting (Basinas, Elhom and Wouters, 2017; Madsen et al., 2012; Pearson et al., 2015).

16.3.3   CANCER

Biological agents classified as carcinogens by the International Agency for Research on Cancer (IARC) are listed in Table 16.3.

Table 16.3 Carcinogenic biological agents

IARC category	Definition	Agents
Group 1	Carcinogenic to humans	Virus Hepatitis C Hepatitis B Epstein-Barr virus Bacteria *Helicobacter pylori* Fungi Aflatoxins (*Aspergillus* spp.) Protozoa Malaria parasite *Schistosoma haemotobium* Plants Some wood dusts Animals Leather based dusts

Source: IARC (2018).

16.4 REGULATIONS AND STANDARDS PERTAINING TO BIOLOGICAL HAZARDS

There are a number of areas in which biological hazards in Australia are either regulated or subject to Australian or international standards, as well as their general duties of care under applicable OHS/WHS legislation.

16.4.1 SECURITY-SENSITIVE BIOLOGICAL AGENTS

The Department of Health and Ageing administers the legislation relating to security-sensitive biological agents (SSBAs). The *National Health Security Act 2007 (as amended 2016)* and the associated *National Health Security Regulations 2008 (as amended 2013)* describe regulated biological agents based on their potential for use as biological weapons.

16.4.2 GENETICALLY MANIPULATED ORGANISMS

The Australian Office of the Gene Technology Regulator (OGTR) administers the legislation relating to genetically modified (GM) biological materials. The *Gene Technology Act 2000* (as amended 2016) and the *Gene Technology Regulations 2001 (as amended 2016)* aim to 'protect the health and safety of people, and the environment, by identifying risks posed by or as a result of gene technology, and by managing those risks through regulating certain dealings with genetically modified organisms' (Ley, 2015, p. 1). As well as regulating genetically manipulated organisms (GMOs), the OGTR also sets requirements for laboratory construction and procedures to be used when handling GM biological materials (OTGR, 2018).

16.4.3 DEPARTMENT OF AGRICULTURE AND WATER RESOURCES

The Department of Agriculture and Water Resources (DAWR) regulates the importation of biological materials from overseas. It also authorises and inspects quarantine-approved premises (QAP) within Australia in order to prevent the introduction of new biological agents into the country. QAP may be used to hold or work on particular imported biological materials, including plants, animals and other products. The DAWR (previously DAFF, AQIS) administers the *Biosecurity Act 2015, Export Control Act 1982* and *Imported Food Control Act 1992*.

16.4.4 AUSTRALIAN STANDARD/NEW ZEALAND STANDARD 2243.3: SAFETY IN LABORATORIES

This standard is relevant to the operation of biological laboratories in Australia and New Zealand, including those used for research, teaching and pathology among others—particularly labs that are not regulated by other means (e.g. by the OGTR). AS/NZ 2243.3 sets out facility construction and practices requirements for laboratory, animal, plant and invertebrate containment facilities. This standard classifies micro-organisms into risk groups (RG) according to their pathogenicity (Table 16.4). Laboratories are divided into four levels of containment, depending on the work to be conducted (Table 16.5). The majority of biological laboratories in Australia are classified as physical containment (PC) level 2.

Table 16.4 Adapted from AS/NZS 2243.3 Definition of micro-organism risk groups

Risk group level	Work undertaken
RG1	Unlikely to cause human or animal disease
RG2	May infect an individual worker, but treatment is likely to be available and spread to the wider community is unlikely
RG3	May pose a significant risk to an individual worker, but further spread to the community is most likely able to be controlled
RG4	May cause life-threatening human or animal disease and may spread in the community or environment. Treatment may not be available

Table 16.5 Adapted from AS/NZS2243.3 Definition of physical containment levels for laboratories

Physical containment level	Work undertaken
PC1	Low-hazard, risk group (RG) 1 micro-organisms, e.g. undergraduate teaching laboratory
PC2	Research or diagnostic laboratory, RG2 micro-organisms
PC3	Research or diagnostic laboratory, RG3 micro-organisms
PC4	High-risk specialist work with RG4 micro-organisms

Risk groupings of organisms relate directly to the containment levels of the associated laboratories. In other words, RG1 organisms should be handled in a PC1 laboratory, and so on. AS 2243.3 gives other useful information for laboratories, such as standards for spill clean-up, transportation of biological materials, use of specialist equipment (centrifuges, biological safety cabinets etc.), useful disinfectants for various organisms and waste disposal.

16.4.5 *NOTIFIABLE DISEASE SURVEILLANCE*

In Australia, there are a number of notifiable diseases for which reporting is mandatory, which are listed in the *National Health Security (National Notifiable Disease List) Instrument 2018*, and include potential occupationally acquired diseases such as Q fever, leptospirosis, Hendra and malaria.

16.5 OCCUPATIONAL EXPOSURE STANDARDS

The number of occupational exposure standards (OES) for biological hazards is extremely limited, and all relate to chemical, non-viable constituents such as wood dust, bacterial endotoxin or gaseous emissions rather than living micro-organisms such as a count of viable bacteria or fungi cells. The establishment of limits is complicated by the highly variable nature of exposures to viable micro-organisms; airborne concentrations can vary by species of micro-organisms and orders of magnitude, depending on ventilation of the building, its occupancy and activity, as well as season, climate, construction and age of the building and outdoor air intake. The measurement of viable airborne micro-organisms is complicated further by the limited duration of sampling available (30 seconds to three minutes), intended to prevent overloading of sampling media. When considered using any exposure limits for biological hazards, the following should be taken into account (Gorny, 2007; Walser et al., 2015):

- In relation to infectious organisms, the dose can vary greatly between humans, based on their individual health and genetics.
- There is insufficient understanding of the dose–response relationship of exposures to non-infectious organisms in the workplace.
- For some agents, there is no pathological difference between an occupationally or domestically acquired infectious disease.
- The lack of standardisation in sampling and analytical methods makes critical review of historical studies and meta-analysis difficult.

The use of biological markers to estimate microbial exposure, such as cell-wall components, has gained popularity. The Dutch Expert Committee on Occupational Safety (2010) has recommended exposure limit of 90 EU/m^3 for bacterial endotoxins, and researchers are undertaking epidemiological studies on workplace exposures to other microbial constituents such as muramic acid (gram-positive bacteria), ergosterol (fungi) and 3-OHFA (gram-negative bacteria) (Davidson et al., 2018). For biological agents other than bacteria and fungi, occupational exposure limits have been established for:

- nuisance dusts, wood, cotton, cellulose and grain dusts (GESTIS, 2018; Safe Work Australia, 2018)
- volatile compounds such as ammonia and hydrogen sulfide, which can be produced by biological organisms (Safe Work Australia, 2018)
- subtilisins, enzymes obtained from *Bacillus subtilis* (ACGIH, 2018).

The GESTIS (2018) International Limit Values database is a valuable resource for obtaining information, and searches can be readily made for micro-organisms of interest to download safety data sheets (SDS) and check risk classifications (GESTIS, 2018). It is hard to identify the point at which exposure to bioaerosols becomes a health hazard (Umbrell, 2003), and for the majority of biological agents, their role in the initiation and development of occupationally acquired illnesses is still poorly understood. Intrinsic (personal) and extrinsic (external) factors, such as host susceptibility, and gene–environment interactions also play a role in the induction and severity of occupational diseases (Kwo and Christiani, 2017). The new techniques, such as Next Generation Sequencing/metagenomics, are similar in principle to those being used to study human biology (genomics, transcriptomic, epigenetics) to better understand the relationship between biological agents and disease (Mardis, 2017).

16.6   SAMPLING AND ANALYSIS OF BIOLOGICAL HAZARDS

There is no perfect method for the investigation of biological hazards, and occupational hygienists and researchers frequently use combinations of traditional and novel sampling and analytical methods from a multitude of scientific disciplines, including occupational hygiene, microbiology, food technology, veterinary and environmental science and engineering.

It is recommended that the occupational hygienist consult with a microbiologist, mycologist or communicable disease expert before undertaking monitoring for any biological hazard. A comprehensive discussion of all the available sampling and analytical methods is beyond the scope of this chapter, and readers are encouraged to consult key texts published by professional bodies, such as:

- *NIOSH Manual of Analytical Methods*, 'Sampling and characterisation of bioaerosols' (Lindsley et al., 2017)
- *Field Guide for the Determination of Biological Contaminants in Environmental Samples* (Hung, Miller and Dillon, 2005)
- *Recognition, Evaluation and Control of Indoor Mold* (Pezant, Weekes and Miller, 2016)
- *Sampling for Biological Agents in the Environment* (Emanuel, Roos and Niyogi, 2008).

16.6.1   SAMPLING STRATEGY AND STUDY DESIGN

The most important consideration when designing a sampling strategy is how the data will be analysed so that results are meaningful, especially if there is no occupational exposure standard for the agent/s of interest. The main factors influencing the selection of methods for sampling and analysis of biological agents will relate to the purpose of the investigation. Reasons for monitoring of biological agents vary, and include:

- regulatory and compliance investigations for infectious agents in response to reports of notifiable conditions—Q fever, HIV or Hendra virus—or for non-infectious biological hazard with WES (wood dust, cotton etc).
- complaint investigation, source identification, staff reassurance, exposure documentation (sick building syndrome (SBS) investigations), clusters of occupational diseases (occupational asthma, HP), building certification (Green Star etc.)
- risk assessment data collection
- epidemiological and research studies; identification of no-adverse-effect levels; research for exposure standards (endotoxin, 3-OHFA, peptidoglycan, muramic acid or mycotoxins)
- remediation assessments; determining the adequacy of remediation work such as flood damage and potential for mould contamination
- assessment of engineering controls; determining whether controls are working effectively, such as decontamination HAZMAT gear used in emergency response to bioterrorism events.

In Australia, a common practice when sampling for mould is to collect outdoor samples as a background reference for comparison with indoor proliferation of fungi, and changes in the microbiological profile and relative abundance of species present. Another approach (which is not recommended) is to monitor two or more indoor locations: an area of concern versus a reference environment—for example, a flood-damaged area versus another office or room within a building with a separate ventilation system. Key points to consider when developing a sampling plan include:

- whether appropriate sampling and accessible analysis techniques are available
- the nature and potential concentration of the biological agent, to prevent over- or under-estimation of exposure owing to poor sampling methodology
- the size distribution and/or environmental matrix of the biological agent, to ensure selection of suitable sampling devices and media
- the sampling duration (time, continuous, periodic, random or worst-case), equipment placement (personal or area), collection method (active or passive) and sample type (aerosol, bulk or surface)
- the cost, suitability and availability of sampling equipment
- the cost and availability of accredited analytical facilities
- any potential constraints that analytical methods may place on sample collection, such as transportation requirements and minimum detection limits
- the technical expertise required of field and laboratory personnel
- any potential sources of cross-contamination.

16.6.1.1 Sample collection

Sampling methods may include the collection of airborne contaminants, surface (swabs, RODAC plates, tape lifts, spore traps), bulk materials (soil, wash water etc.) or biological samples (blood, urine and nasal secretions), as shown in Figure 16.2. There are Australian and international standardised methods for the sampling of bulk materials, such as water and food, but unfortunately, currently there is no similar breadth of standard sampling techniques in Australia for exposure to airborne or surface biological agents in the workplace.

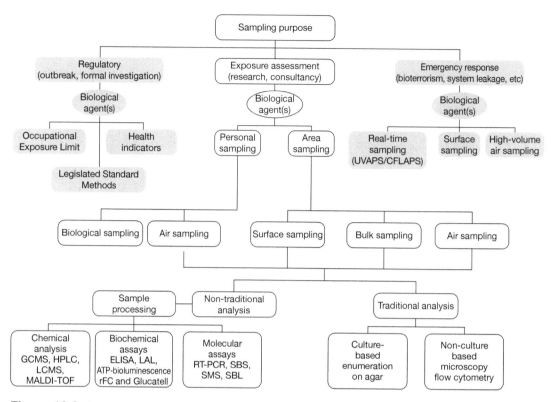

Figure 16.2 Overview of selected sampling and analytical methods for assessing biological hazards

The technology for monitoring and analysing samples for biological hazards is developing rapidly, and researchers are constantly inventing and refining sampling devices and analytical techniques to improve reliability, collection efficiency and sensitivity. More recent developments include the NIOSH two-stage personal bioaerosol sampler (Lindsley et al., 2017).

Willeke and Macher (1999) list the three key performance principles that should be considered when reviewing and selecting sampling methodologies: inlet sampling efficiency (ability to remove atmospheric particulates); particle removal efficiency (ability of media to effectively retain aerosol); and biological recovery efficiency, or ensuring that collected particles maintain the chemical, physical and biological integrity of the environmental aerosols. Table 16.6 presents some of the more common commercially available sampling devices.

16.6.1.2 Impactors

Impactors are commonly used for indoor air-quality assessments, clean-room testing and industrial operations (Haig et al., 2016). Commercially available impactors range from slit to single-stage or multi-stage (cascade) impactors, which enable size-selective sampling of viable and culturable micro-organisms in different aerosol fractions (Lindsley et al., 2017; Reponen et al., 2011).

Slit impactors direct the air stream through a rectangular slit onto agar, which undergoes incubation for culturable organisms (slit-to-agar sampler) or onto glass slides or

Table 16.6 Sampling devices

Device	Media	Method	Aerosol	Comments
Impactors				
Single-stage	Agar	Area	Viable micro-organisms or spores (agar, tape and gelatin filters)	Relatively inexpensive, and equipment and analytical facilities readily available for viable micro-organism analysis in Australia
Cascade	Agar or filters	Area and personal		Slit impactors can give real-time indication of temporal variation in bioaerosols
Slit	Agar or tape strip		Allergens, dusts, microbial cell-wall components (filters)	Collection and impaction on agar and gelatin filters can stress micro-organisms, making them viable but non-culturable; selective for more resilient organisms. May require short sampling periods of one to three minutes in heavily contaminated atmospheres. Cascade samplers enable size-selective measurement of biological agents, and both personal and area samplers are commercially available. Gelatin filters have short sampling periods of 30 minutes, after which filters become brittle. Micro-organisms may become viable and non-culturable through sampling process. Use of different filter media alters recovery of cell-wall components like endotoxin
Impingers				
AS PCR	Liquid	Area	Viable and countable micro-organisms	Samples can undergo a wide variety of culture- and non-culture-based analyses.
Rotating cup	Liquid	Area		
AGI–30	Liquid	Area	Cell-wall components, MVOCs and toxins	Subject to inlet sampling losses
Biosampler	Liquid	Area		
Midget	Liquid	Personal	Viable and countable micro-organisms	Require short sampling periods of less than 30 minutes owing to sample media evaporation. Glass presents personal injury risk with midget impingers. Bulky and cumbersome to wear

Table 16.6 Sampling devices *continued*

Device	Media	Method	Aerosol	Comments
Filtration IOM	Filters	Area Personal	Dust, microbial and cell-wall components, spores, MVOC, toxins and allergens	Can be adapted for full work-shift sampling. Best suited for microbial indicators such as cell-wall components (endotoxin, muramic acid etc.) or direct counts owing to sampling stress such as desiccation on viable organisms
Button				Gelatin filters for viable micro-organisms have short sampling periods, maximum 30 minutes
Cyclone Versa Trap Microspore cassettes	Custom cassettes			Particle removal efficiency and biological recovery efficiency will vary widely for various filter materials
Other				
NIOSH personal bioaerosol sampler	Filter and centrifuge tubes	Personal	Inhalable dust	Highly suited to culture independent analytical approaches
Vertical elutriators	Liquid	Area	Organic dust (<15 µm diameter) and endotoxin	Vertical elutriators are standard method in cotton and textile industry in relation to byssinosis
Electrostatic precipitators	Liquid	Area	Vegetative cells	Experimental technology
Wetted wall cyclones	Liquid	Area	Vegetative cells	Experimental technology, may not be efficient for hydrophobic bacteria or fungal spores. Subject to impaction and rehydration stresses
Settling plates	Agar	Area	Vegetative cells and spores	Unsuitable for spore sampling. Highly subject to environmental conditions, and low repeatability
HEPA vacuum	Filter sock	Surface	Viable cells, dust, cell-wall components, toxins, etc.	Not representative of airborne exposures

Sources: Cage et al. (1996); Engelhart et al. (2007); Görner et al. (2006); Jensen et al. (1992); Kulkarni, Baron and Willeke (2011); Madelin and Madelin (1995); Reponen et al. (2011); Spurgeon (2007); Yamamoto et al. (2011); Yao and Mainelis (2007).

moving tapes for microscopic analysis of pollen and spore traps (Ruzer and Harley, 2013). Commercially available devices include the BioSlide direct-to-slide sampler and the Hirst volumetric spore sampler from Burkard Scientific, which uses tape.

The single and cascade impactors work by directing an air stream onto a plate containing agar, which is subsequently incubated and the cultivated micro-organisms counted, or onto a filter that can be analysed by a variety of methods, including direct microscopy analysis (counts) and the extraction and biochemical/chemical analysis of cell-wall constituents. Particles in the airstream collect on the agar based on their velocity, with heavier/large particles with greater momentum collected on the top stages and smaller particles on the lower stages (Figure 16.3).

Commonly used agars include non-selective nutrient or tryptic soy agar (TSA) for bacteria (Figure 16.4a) and malt extract agar (MEA) for fungi (Atlas, 2010); other selective agars, such as Endo, eosin methylene blue (EMB) and MacConkey agars, can be used. Differential media such as blood agar can be used to help presumptively identify bacteria through their growth morphology and ability to lyse red blood cells (Figure 16.4b). Some agars can be both differential and selective; examples include EMB agar on which *E. coli* develop a gold sheen (Figure 16.4c), and chromogenic agars, which are used to select for and presumptive identification of specific micro-organisms such as *Salmonella* colonies that form mauve colonies when cultured on chromagar (Figure 16.4d). However, chromogenic agars are expensive compared with other types, as well as being prone to false positive results. Presumptive samples must be verified by more advanced identification methods.

The advantage of using impactors such as the Andersen microbial sampler, SAS microbial air sampler or BioSampler are that they are readily accessible and easy to operate and calibrate, and the instruments, media and sample analysis all are relatively inexpensive compared with other methods, such as endotoxin or MVOC analysis. The disadvantage of the impactors is that results may not be truly representative of the microbial community because the impaction of micro-organisms onto solid surfaces can render them viable but non-culturable and various species may go undetected. Sampling times need to be very short (30 seconds to three minutes) in heavily contaminated environments because the agar plates quickly become overloaded. The majority of impactors are not designed, and hence are unsuitable, for personal sampling. Due to loading, bioaerosol concentrations

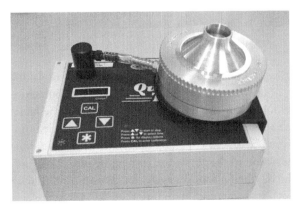

Figure 16.3 Commercial impactors
(a) SKC BioStage single-stage impactor; (b) Marple cascade impactor

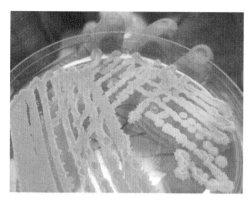
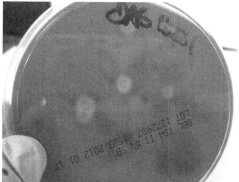
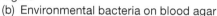

Figure 16.4 General, selective and differential agars
(a) *Bacillus* spp. on TSA   (b) Environmental bacteria on blood agar

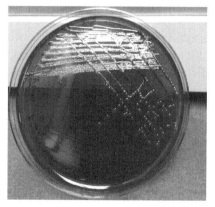
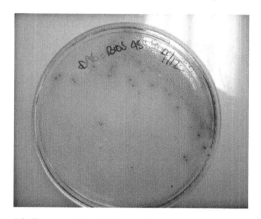

(c) *E. coli* on EMB agar   (d) Presumptive *Salmonella* spp. on
                                              chromogenic agar

may be under-estimated because they are based on plate counts of individual colonies, which may grow from an aggregate of cells or spores rather than the individual cells. It is estimated that only a fraction of the microbial organisms can be cultured in the lab (Schloss and Handelsman, 2005), which is why the advent of culture-independent approaches enables better understanding of complex microbial communities.

16.6.1.3 Impingers

Impingers historically have been used to sample a variety of biological aerosols, including viable and countable bacteria, fungi, viruses, pollens and cell-wall components. They collect air contaminants by drawing an air stream through a thin glass tube and aspirating into a liquid medium, which separates out large particles. The liquid medium can be plated directly onto agar, examined by microscope and analysed either biochemically or using molecular techniques. The action of the air stream impinging into the liquid aids in the separation of aggregates, which, when plated on agar, can provide a more representative count than impactors can. Most impingers, including the SKC BioSampler and the AGI-30, are designed for area sampling and have large, cumbersome pumps (Figure 16.5a). Midget impingers (Figure 16.5b) designed for personal sampling have been around since

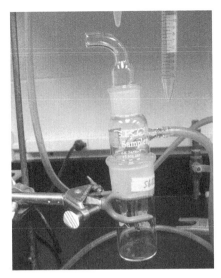
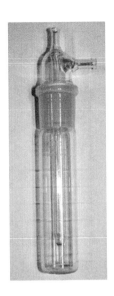

Figure 16.5 Commercially available impingers
(a) SKC BioSampler
(b) Midget impinger

the 1930s, but could be considered a hazard due to their glass construction. Multi-stage impingers are also available, but their glass construction makes it difficult to accurately predict the mass median diameters collected by each stage (Vincent, 2007).

Limitations of impingers for biological hazards monitoring include short sampling time owing to rapid evaporation of media (maximum 30 minutes); the fact that the hydrophobic particles will bounce and escape the system (Hung, Miller and Dillon, 2005); and the effects of moderate changes in flow rate and liquid quantity on sample collection (Dart and Thornburg, 2008). Impingers may be subject to inlet and internal losses as well as re-aerosolisation of organisms at lower concentrations (Han and Mainelis, 2012; Kesavan, Schepers and McFarland, 2010). ViaTrap mineral oil can be used as a collection liquid to reduce evaporation, allowing for eight-hour sampling periods, but such oils do not plate well and are better suited for microscopic analysis.

16.6.1.4 Filtration
Biological aerosols can be collected using the methods for particulate sampling: AS 2985: 2009 (respirable dust) and AS 3640: 2009 (inhalable dust) (Standards Australia, 2009a, 2009b). These methods are suitable for the collection of non-viable bioaerosol samples such as pollen, allergens, spores, endotoxins and DNA/RNA, but not necessarily viable cells to undergo culturing in the lab. This is because microbial cells can be sub-lethally injured during sampling, making them viable but non-culturable and causing both under-estimation of bacterial loadings and sampling bias for hardier organisms such as the endospore-forming *Bacillus* species. Choice of collection media is also critical. Gelatin filters have been recommended for collection of viable micro-organisms, but they become brittle and fragile during sampling and have a maximum sampling period of 30 minutes, depending on the environmental conditions. Filter materials must be tested for background endotoxin and genetic materials, as well as for recovery efficiency of various bioaerosols such as endotoxin or β-D-glucans (Davidson et al., 2004; Hung, Miller and

Figure 16.6
Filtration devices

Dillon, 2005). Figure 16.6 displays different sampling heads that have been used for bioaerosol sampling. Metal sampling heads, such as IOM cartridges or button samples, are ideal because they can be heat treated to remove background contamination. This is important when sampling for endotoxin and genomic DNA material, as they may not be removed fully from devices using standard cleaning methods such as autoclaving and/or alcohol rinsing (Davidson et al., 2005). There are now some single-use filter sampling devices, such as spore traps, which are commercially available for the collection of fungal spores and which also reduce the risk of sample cross-contamination.

16.6.1.5 Other samplers

Other commercially available samplers included wetted-wall, centrifugal and electrostatic samplers. Wetted-wall cyclones operate on the same principle as particle-sampling cyclones except that the internal surfaces are coated with suitable liquid to wash particles into the collection chamber (Reponen et al., 2011; Vincent, 2007). The hand-held, battery-operated Reuter Centrifugal Sampler (RCS) has been commercially available for a number of years, and has been adopted in Canada as a reference sampling method for fungal contamination assessments (Vincent, 2007). The operating principle is a centrifugal fan (impeller), which draws air into the sampling head and directs the stream onto an agar strip fitted to the internal walls. The strip is then incubated and the number of colonies counted to estimate the bioaerosol loading (Semple, 2010).

Real-time monitors that operate by measuring the fluorescence emitted by particles are commercially available. Real-time monitors can be linked into existing alarm systems as engineering controls for biohazard containment or clean-room efficiency (Fennelly et al., 2018). A personal PCR-based detection unit has also been developed to assess bioaerosol exposures in real time (Agranovski et al. 2017), as well as wearable sensors to detect biological threat agents (Ozanich, 2018).

16.6.2   SAMPLE STORAGE AND TRANSPORTATION

Before sampling commences, the technician collecting must consult with their analytical laboratory regarding special storage and transport requirements, which will change depending on sample type and endpoint analysis. The basic principles for sample storage and transport are to protect and stabilise samples from environmental disturbances (sunshine, humidity, desiccation, heat, dust and so on). Consideration needs to be given to temperature fluctuations and storage time, or cell growth in favourable conditions. Freeze/thaw cycling of endotoxin causes sample losses and should be avoided (Hung, Miller and Dillon, 2005), while collection and storage methods for metagenome studies must be optimised based on sample type to avoid confounding relative abundance estimates and community profiles (Quince et al., 2017). Workplace atmosphere guideline for the measurement of airborne micro-organisms and endotoxins, as well as local standard methods published by governmental and other professional agencies, should be consulted.

16.6.3   SAMPLE ANALYSIS

The selection of analytical methods will be driven primarily by the agent of interest, regulatory requirements—that is, *Legionella* spp. in cooling towers or blood-borne viruses—as well as budget and access to analytical services. A wide variety of approaches can be harnessed, as illustrated in Figure 16.7.

In recent years, there has been a shift away from culture-based analysis to culture-independent approaches to the study of microbial diversity with metagenomics, whole

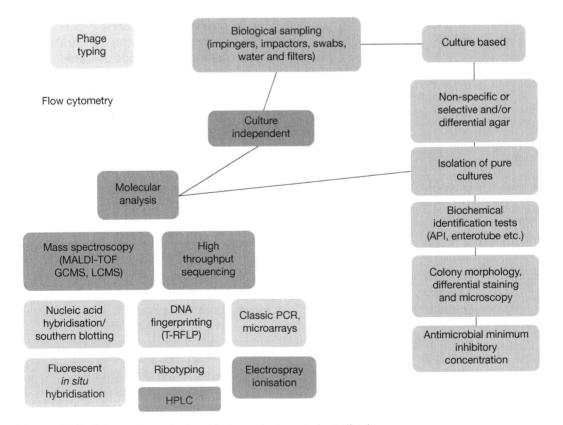

Figure 16.7 Culture-dependent and independent analysis methods

genome sequencing and real-time and quantitative PCR (Mbareche et al., 2017). Access to these instruments/techniques can be difficult in remote areas, however, and collection and analysis protocols are yet to be standardised both locally and internationally.

Genomic sequencers are providing greater insight into the complex microbial community that inhabits the body and environment, including non-culturable micro-organisms such as archaea (prokaryotic bacteria with no peptidoglycan in the cell wall) and viruses (Blais-Lecours, Perrott and Duchaine, 2015). Given the flexibility in terms of the sampling approaches that can be used with culture-independent analysis, it is anticipated that these methods will continue to increase in popularity (Schloss and Handelsman, 2005). Approaches are being developed to distinguish live from dead micro-organisms in communities using specialised dyes, ribonucleic acid (RNA) and protein-based methods (Emerson et al., 2017). Table 16.7 outlines the various methodologies that can be used for analysis of bioaerosols, but for more comprehensive overviews please consult your local analytical laboratory.

Table 16.7 Analytical methodologies for biological agents

Category	Methods	Endpoint	Comments
Culture based	Plate counts, most probable number, membrane filtration, turbidity IDEXX	Viable bacteria, fungi and viruses	Most common method in Australia at present. Accredited labs readily accessible in all states and territories. Most inexpensive method. Australian standard methods available.
Countable	Microscopic counts, flow cytometry	Total bacteria, fungi, spores and pollen	Accredited labs readily available for microscopy work. Flow cytometry primarily a research method.
Chemical	GC/MS and GC/MSMS, LC/MSMS, HPLC, TLC, MALDI-TOF	MVOCs, lipopolysaccharide (gram negative), muramic acid (gram positive), ergosterol (fungi) 16s rDNA	Limited number of labs in Australia perform these analyses, which are mainly used for epidemiological and exposure research at present.
Molecular	Next generation (massively parallel) sequencing. Single-cell sequencing. Real-time PCR, Western blot, ribotyping pulsed-field gel electrophoresis	RNA and DNA of viruses, fungi, bacteria, cytokines and allergens	Expensive and laboratories offering these services commercially are limited in Australia, but are steadily becoming more accessible. Analysis of next-generation sequencing data can be complex and requires specialised bioinformatics software.

Category	Methods	Endpoint	Comments
Biochemical	Bacteria and fungi ID kits (API, Enterotube, IDEXX etc.)	Bacteria and fungi	Commercial labs offer bacteria/fungi identification and ELISA analyses.
	ELISA Western Blot	Proteins/allergens	
	LAL & rFC assay	β-D-glucans, fungal extracellular polysaccharides, endotoxin, peptidoglycan	Endotoxin and β-D-glucans analysis of air samples not readily accessible in Australia, and mainly used for research at present. Kits are available, but a large number of variables must be addressed when using. Refer to BS EN 14031: 2003 or ISO 29701: 2010 for more information.

16.7 CONTROL

16.7.1 FOCUS ON INFECTIOUS/COMMUNICABLE DISEASES

For infectious/communicable diseases, the intention is to break the chain of infection by disrupting the agent–host–environment interaction, often referred to as the *epidemiological triad* (Figure 16.8). The *agent* is the infectious organism that is required for disease to occur, the *host* is the human who contracts the disease and the *environment* includes the external/extrinsic biological, physical and socioeconomic factors that can influence the host–agent interaction (CDC, 2012).

Control or prevention of the spread of infectious disease can be achieved by targeting the host (immunisation, diet), agent/pathogen (genetic modification), environment (sanitation, removal water sources) or vector (sterilisation of organisms, retractable sharps for injections). Figure 16.9 presents a worked model of control and prevention interventions for a mosquitoes-borne virus, using the Barmah Forest virus as an example.

The AIHA (2006) has produced a guideline for risk management of pandemic influenza in the workplace; for other communicable diseases, you should refer to the American Public Health Association's (APHA) *Control of Communicable Diseases Manual* (Heymann, 2015), which sets out methods of control for significant public health pathogens. For health-care settings, in Australia the National Health and Medical Research Council (NHMRC, 2010) has published a comprehensive guideline on the prevention and control of infection in health care, which was under review in 2018. Other key guidelines for control of infectious micro-organisms such Legionellosis include the enHealth *Guidelines for Legionella Control* (Department of Health, 2016) and AS3666: 2011 Air-Handling and Water Systems of Buildings—Microbial Control (Standards Australia, 2011).

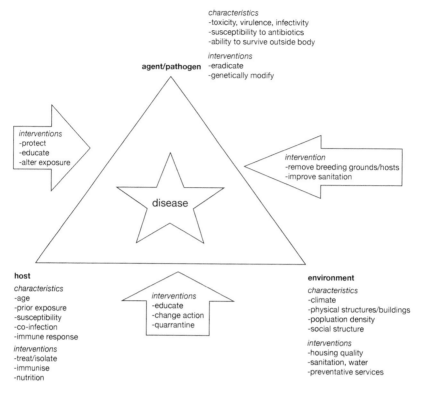

Figure 16.8 Control of infectious disease
Source: University of Ottawa (2014).

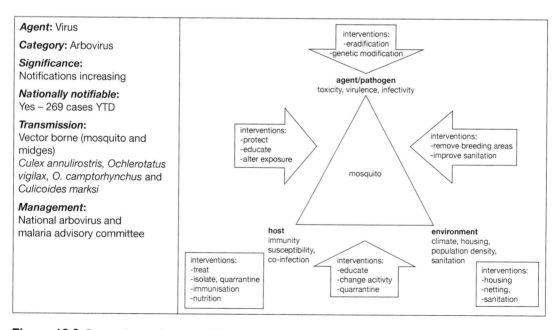

Figure 16.9 Prevention and control of Barmah Forest virus
Source: Adapted from University of Ottawa (2014).

16.7.2 NON-INFECTIOUS BIOLOGICAL HAZARDS

When controlling biological hazards, the standard principles of the hierarchy of control, beginning with elimination as part of a risk management approach, are prescribed in AS/NZS ISO45001: 2018, Occupational Health and Safety Management Systems—Requirements with Guidance for Use (Standards Australia, 2018). Administrative controls and PPE are considered the lowest-order controls, and should not be used alone; rather, they should be combined with administrative and engineering controls such as time limits on work shifts to prevent fatigue, or use of positive/negative pressure environments depending on the situation/agent. Table 16.8 provides examples of the different types of controls that may be applied in various industries.

Table 16.8 Potential controls for biological hazards

Hierarchy of control	Examples
Elimination	Elimination of contaminated water sources Elimination of vectors, such as mosquitoes
Substitution	Substituting high-dust animal feed with a low-dust variety Substituting a high-risk bacteria with a low-risk organism in a laboratory Substitute latex gloves with nitrile Undertaking in vitro experiments instead of using an animal model
Engineering	Use of biological safety cabinets when handling hazardous micro-organisms in a laboratory Use of fully enclosed and air-conditioned tractor cabs to protect the operator from outside dusts Use of approved sharps containers to dispose of needles or other sharp items Using single-use equipment to remove the risk of transferring an infectious disease in medical centres, beauty salons or tattoo parlours Maintenance of negative or positive air pressure, via ventilation, in spaces where biological hazards may be handled or naturally present
Administrative	Development of a vaccination program to reduce the risk of infection from a identified pathogen—for example, Hepatitis A or B Use of a documented biological waste segregation and disposal system—for example, in a hospital, laboratory or home health-care setting Scheduling work to limit exposure during high-exposure activities Implementation of an induction or training activities and appropriate supervision Use of biohazard signage and barricading Frequent changing of machine oil in sumps to prevent bacterial growth Instigation of routine maintenance schedule for mechanical ventilation systems to ensure biological material does not accumulate
PPE	Appropriate approved respiratory protection Safety glasses, goggles or face shield as appropriate Lab coat, gown, overalls or Hazmat suit Gloves

16.8   EMERGING DISEASE ISSUES IN AUSTRALIA

Emerging infectious diseases (EIDs) of significance are:

- zoonotic diseases, such as Q fever, Henda virus, avian influenza and Ebola
- arboviruses, such as Ross River virus, Dengue fever, Barmah Forest virus, Nipah virus, Chikungunya virus—which means 'to become contorted' (One Health, 2018), Japanese encephalitis, drug-resistant malaria and Zika virus (McCloskey et al., 2014).

Although it must be stated that with ever increasing globalisation, and as witnessed in the Zika virus, H1N1 Influenza and SARs public health emergencies of international concern (PHEIC) and the more recent Ebola epidemic, no nation is immune to the potential for outbreaks from exotic pathogens, especially viruses (Heymann et al., 2015). A quick perusal of the WHO outbreaks page readily displays the vast array of outbreaks that may be underway at any one time around the globe.

Factors that are influencing the emergence and re-emergence of infectious diseases include climate change, political, social and environmental instability and anti-microbial resistance. Climate change is resulting in shifting geographical boundaries for infectious diseases vectors such as ticks and mosquitoes, enabling their movement into previously unreported regions. Increased global temperatures impact on the re-emergence of pathogens previously in stasis in colder climates, especially temperature-resistant pathogens such as smallpox, influenza and anthrax that can survive for long periods in the environment—even in ice (Sköld, 2017). It sounds like the plot of a Hollywood movie, but viable 1918 influenza viruses have been extracted from bodies buried in the Arctic permafrost (Stevens et al., 2004), and an anthrax outbreak in Scandinavian nomadic herders was linked to *Mycobacterium anthracis* originating from reindeer that died in 1941 (Sköld, 2017). The more conducive temperatures also promote survival and proliferation of pathogens in their environmental niches, thereby increasing their outbreak potential (Watts et al., 2015).

Heymann (in Heymann et al., 2015, p. 1884) states:

> Health security—essentially the protection from threats to health—is recognised as one of the most important non-traditional security issues . . . this is security that comes from access to safe and effective health services, products, and technologies which are not always available in conflict areas.

Increased incidence of parasitic helminth infections, tuberculosis, measles and Ebola have been recorded in geographical regions of increased conflict and instability (Ismail et al., 2016; Jaffer and Hotez, 2016). Polio outbreaks have also been reported in conflict regions from which the disease had previously been eradicated (Akil and Ahmad, 2016), while Leishmaniosis has been dubbed the disease of guerrilla warfare due to its association with conflict zones (Berry and Berrang-Ford, 2016). The reason for the resurgence in politically unstable areas relates to limited access to anti-virals, vaccines and medicine, under-nutrition and limited public health infrastructure. Overcrowding and insanitary conditions associated with waste disposal in refugee centres can increase the risk of infectious disease outbreaks (Bowles, 2016; Heymann et al., 2015).

Anti-microbial resistance (AMR) is a significant threat to infectious disease control in addition to the personal choice by some people not to vaccinate against preventable diseases. AMR is the ability of micro-organisms to develop immunity to anti-microbial agents such as antibiotics, anti-virals and anti-malarials. This immunity renders standard medical treatments ineffective and results in persistent infections in hosts. The rise in AMR has been attributed to the over-prescribing of antibiotics, patients not finishing their course of treatment, poor infection-control practices, a lack of hygiene and sanitation, limited research into new treatments, and the over-use of antibiotics in agriculture, both for preventing infection and as growth promoters in animal production. The prevalence of AMR is increasing both in Australia and internationally, with resistance increasing at a pace that is currently exceeding the development of new anti-microbial drugs. The Australian government released the first National Anti-microbial Resistance Strategy in 2015 to guide the response to the threat of antibiotic misuse and resistance. Current initiatives include:

- regulatory restrictions on the prescription and use of anti-microbials
- surveillance activities
- hand hygiene and anti-microbial stewardship (AMS) programs
- strict requirements to manage pathogen levels along the food production and processing chain
- education for prescribers on the judicious use of antibiotics, and research into new products
- approaches to prevent and respond to AMR (Australian Government, 2015).

16.9   REFERENCES

Adelman, D.C., Casale, T.B. and Corren, J. 2012, *Manual of Allergy and Immunology*, Wolters Kluwer/Lippincott Williams & Wilkina, Philadelphia, PA.

Agranovski, I.E., Usachev, E.V., Agranovski E. and Usacheva, O.V. 2017, 'Miniature PCR based portable bioaerosol monitor development', *Journal of Applied Microbiology*, vol. 122, no. 1, pp. 129–38.

Akil, L. and Ahmad, H.A. 2016, 'The recent outbreaks and re-emergence of poliovirus in war and conflict-affected areas', *International Journal of Infectious Diseases*, vol. 49, pp. 40–6.

American Conference of Governmental Industrial Hygienists (ACGIH) 2018, *Threshold Limit Values for Chemical Substances and Physical Agents and Biological Exposure Indices*, ACGIH®, Cincinnati, OH.

American Industrial Hygiene Association (AIHA) 2006, *The Role of the Industrial Hygienist in a Pandemic*, AIHA, Cincinnati, OH.

Atlas, R.M. 2010, *Handbook of Microbiological Media*, CRC Press, Washington, DC.

Austin, E., Myron, H.S., Summerbell, R.K. and Mackenzie, C.A., 2019, 'Acute renal injury cause by confirmed Psilocybe cubensis mushroom ingestion', Medical mycology case reports, 23, pp. 55–7.

Australian Government 2015, *National Antimicrobial Resistance Strategy 2015–2019*, Commonwealth of Australia, Canberra, <www.amr.gov.au/resources/importance-ratings-and-summary-antibacterial-uses-human-and-animal-health-australia>.

Australian Meat Processor Corporation (AMPC) 2018, 'FAQs: Australian Q Fever Register', AMPC, <www.qfever.org/faqs> [accessed 28 June 2018]

Barker, S.F., O'Toole, J., Sinclair, M.I., Keywood, M. and Leder, K. 2017, 'Endotoxin health risk associated with high pressure cleaning using reclaimed water', *Microbial Risk Analysis*, vol. 5, pp. 65–70.

Basinas, I., Elholm, G. and Wouters, I. 2017, 'Endotoxins, glucans and other microbial cell wall agents', in C. Viegas, S. Viegas, A. Gomes, M. Täubel and R. Sabino (eds), *Exposure to Microbiological Agents in Indoor and Occupational Environments*, Springer, Dordrecht, pp. 159–90.

Bauman, R.W. 2012, *Microbiology with Diseases by Body System*, Benjamin Cummings, Boston.

Bennett, J.W. and Klich, M. 2003, 'Mycotoxins', *Clinical Microbiology Reviews*, vol. 16, no. 3, pp. 497–516.

Berry, I. and Berrang-Ford, L. 2016, 'Leishmaniasis, conflict, and political terror: A spatio-temporal analysis', *Social Science & Medicine*, vol. 167, pp. 140–9.

Biosecurity Australia 2012, *Animal Pests and Diseases*, Department of Agriculture, Fisheries and Forestry, Canberra, <www.daff.gov.au/animal-plant-health/pests-diseases-weeds/animal> [accessed 8 December 2012]

Blais-Lecours, P., Perrott, P. and Duchaine, C. 2015, 'Non-culturable bioaerosols in indoor settings: Impact on health and molecular approaches for detection', *Atmospheric Environment*, vol. 110, pp. 45–53.

Blais-Lecours, P., Veillette, M., Marsolais, D. and Duchaine, C. 2012, 'Characterization of bioaerosols from dairy barns: Reconstructing the puzzle of occupational respiratory diseases by using molecular approaches', *Applied and Environmental Microbiology*, vol. 78, no. 9, pp. 3242–8.

Bogaert, P., Tournoy, K.G., Naessens, T. and Grooten, J. 2009, 'Where asthma and hypersensitivity pneumonitis meet and differ: Noneosinophilic severe asthma', *American Journal of Pathology*, vol. 174, no. 1, pp. 3–13.

Bønløkke, J.H., Cormier, Y. and Sigsgaard, T. 2010, 'Agricultural environments and the food industry', in S.M. Tarlo (ed.), *Occupational and Environmental Lung Diseases*, John Wiley & Sons, Hoboken, NJ.

Bowles, D.C. 2016, 'Climate change-associated conflict and infectious disease', in M. Bouzid (ed.), *Examining the Role of Environmental Change on Emerging Infectious Diseases and Pandemics*, IGI Global, Hershey, PA, pp. 68–88.

Cage, B.R., Schreiber, K., Portnoy, J. and Barnes, C. 1996, 'Evaluation of four bioaerosol samplers in the outdoor environment', *Annals of Allergy, Asthma and Immunology*, vol. 77, no. 5, pp. 401–6.

Canadian Public Health Agency (CPHA) 2018, *Pathogen Safety Data Sheets*, Government of Canada, <https://www.canada.ca/en/public-health/services/laboratory->biosafety-biosecurity/pathogen-safety-data-sheets-risk-assessment.html>.

Centers for Disease Control (CDC) 2012, *Principles of Epidemiology in Public Health Practice, Third Edition: An introduction to applied epidemiology and biostatistics. Centers for Disease Control and Prevention*, <https://www.cdc.gov/csels/dsepd/ss1978/lesson1/section8.html> [accessed 10 June 2019]

—— 2016, 'About parasites', Centers for Disease Control, Atlanta, GA, <www.cdc.gov/parasites/about.html> [accessed 12 December 2018]

Charlier, P., Claverie, J.-M., Sansonetti, P., Coppens, Y., Augias Anaïs, Jacqueline, S., Rengot, F. and Deo, S. 2017, 'Re-emerging infectious diseases from the past: Hysteria or real risk?', *European Journal of Internal Medicine*, vol. 44, pp. 28–30.

Coico, R. and Sunshine, G. 2015, *Immunology: A Short Course*, John Wiley & Sons, New York.

Cowan, M.K. 2012, *Microbiology: A Systems Approach*, 3rd ed., McGraw-Hill, New York.

Crewe, J. et al. 2016, 'A comprehensive list of asthmagens to inform health interventions in the Australian workplace', *Australian and New Zealand Journal of Public Health*, vol. 40, no. 2, pp. 170–3.

Dao, A. & Bernstein, D.I. 2018, 'Occupational exposure and asthma', *Annals of Allergy, Asthma & Immunology*, vol. 120, no. 5, pp. 468–75.

Dart, A. and Thornburg, J. 2008, 'Collection efficiencies of bioaerosol impingers for virus-containing aerosols', *Atmospheric Environment*, vol. 2, no. 4, pp. 828–32.

Davidson, M., Reed, S., Markham, J. and Kit, R. 2004, 'Filter performance in endotoxin sampling and quantification in the Australian agricultural environment', in *22nd Annual Conference Proceedings, Fremantle, WA*, AIOH, Melbourne.

—— 2005, Guide to Environmental Endotoxin Sampling, in *23rd Annual Conference Proceedings, Terrigal NSW*, AIOH, Melbourne.

Davidson, M.E. et al. 2018, 'Personal exposure of dairy workers to dust, endotoxin, muramic acid, ergosterol and ammonia on large-scale dairies in the high plains western United States', *Journal of Occupational Environmental Hygiene*, vol. 15, no. 3, pp. 182–93.

Department of Health 2016, *Guidelines for Legionella Control*, Commonwealth Government, Canberra.

—— 2018, *National Notifiable Diseases Surveillance System*, Commonwealth Government, Canberra, < www9.health.gov.au/cda/source/cda-index.cfm> [accessed 28 June 2018]

Donham, K.J. and Thelin, A. 2016, *Agricultural Medicine: Rural Occupational and Environmental Health, Safety, and Prevention*, 2nd ed., John Wiley & Sons, Hoboken, NJ.

Douwes, J., Eduard, P. and Thorne, P.S. 2008, 'Bioaerosols', in K. Heggenhougen and S. Quah (eds), *International Encyclopaedia of Public Health*, Elsevier, Oxford.

Dutch Expert Committee on Occupational Safety 2010, *Endotoxins: Health-based Recommended Occupational Exposure Limit*, Health Council of the Netherlands, The Hague.

Dutkiewicz, J., Cisak, E., Sroka, J., Wojcik-Fatla, A. and Zajac, V. 2011, 'Biological agents as occupational hazards—selected issues', *Annals of Agricultural and Environmental Medicine*, vol. 18, no. 2, pp. 286–93.

Edmonds, J., Lindquist, H.D.A., Sabol, J., Martinez, K., Shadomy, S., Cymet, T. and Emanuel, P. 2016, 'Multigeneration Cross-Contamination of Mail with Bacillus anthracis Spores', PLoS ONE, vol. 11, no. 4, p. e0152225.

El-Zaemey, S. et al. 2018, 'Prevalence of occupational exposure to asthmagens derived from animals, fish and/or shellfish among Australian workers', *Occupational Environmental Medicine*, vol. 75, pp. 310–16.

Emanuel, P., Roos, J.W. and Niyogi, K. 2008, *Sampling for Biological Agents in the Environment*, ASM Press, Washington, DC.

Emerson, J.B. et al. 2017, 'Schrödinger's microbes: Tools for distinguishing the living from the dead in microbial ecosystems', *Microbiome*, vol. 5, no. 86, doi:10.1186/s40168-017-0285-3.

Engelhart, S., Glasmacher, A., Simon, A. and Exner, M. 2007, 'Air sampling of Aspergillus fumigatus and other thermotolerant fungi: Comparative performance of the Sartorius MD8 airport and the Merck MAS–100 portable bioaerosol sampler', *International Journal of Hygiene and Environmental Health*, vol. 210, no. 6, pp. 733–9.

Farokhi, A., Heederik, D. and Smit, L.A.M. 2018, 'Respiratory health effects of exposure to low levels of airborne endotoxin: A systematic review', *Environmental Health*, vol. 17, no. 1, p. 14.

Fennelly, M.J., Sewell, G., Prentice, M.B. and Sodeau, J.R. (2018), 'Review: The Use of Real-Time Fluorescence Instrumentation to Monitor Ambient Primary Biological Aerosol Particles (PBAP)', *Atmosphere, vol.* 9, no. 1, doi:10.3390/atmos9010001.

GESTIS 2018, *GESTIS International Limit Values*, DGUV, Germany, <http://limitvalue.ifa.dguv.de>.

Görner, P., Fabries, J.-F., Duquenne, P., Witschger, O. and Wrobel, R. 2006, 'Bioaerosol sampling by a personal rotating cup sampler CIP 10-M', *Journal of Environmental Monitoring*, vol. 8, no. 1, pp. 43–8.

Gorny, R. 2007, *Biological Agents: Need for Occupational Exposure Limits (OELs) and Feasibility of OEL Setting*, Department of Biohazards, WHO Collaborating Center, Lodz, Poland, <https://osha.europa.eu/sites/default/files/seminars/documents/en/seminars/occupational-risks-from-biological-agents-facing-up-the-challenges/speech-venues/speeches/biological-agents-need-for-occupational-exposure-limits-oels-and-feasibility-of-oel-setting/Presentation%20by%20Mr%20Gorny.ppt>.

Green, B.J., Couch, J.R., Lemons, A.R., Burton, N.C., Victory, K.R., Navak, A.P. and Beezhold, D.H. 2018, 'Microbial hazards during harvesting and processing at an outdoor United States cannabis farm', *Journal of Occupational and Environmental Hygiene*, vol. 15, no. 5, pp. 430–40.

Greiner, A.N., Hellings, P.W., Rotiroti, G. and Scadding, G.K. 2011, 'Allergic rhinitis', *The Lancet*, vol. 378, no. 9809, pp. 2112–22.

Haig, C.W., Mackay, W.G., Walker, J.T. and Williams, C., 2016, 'Bioaerosol sampling: sampling mechanisms, bioefficiency and field studies', *Journal of Hospital Infection*, vol. 93, no.3, pp. 242–55.

Han, T. and Mainelis, G. 2012, 'Investigation of inherent and latent internal losses in liquid-based bioaerosol samplers', *Journal of Aerosol Science,* vol. 45, pp. 58–68.

Harries, M.J. and Lear, J.T. 2004, 'Occupational skin infections', *Occupational Medicine*, vol. 54, no. 7, pp. 441–9.

Heinsohn, P.A. 2011, 'Biohazards and occupational disease', in V.A. Rose and B. Cohrssen (eds), *Patty's Industrial Hygiene*, 6th ed., John Wiley & Sons, Hoboken, NJ.

Heymann, D.L. 2015, *Control of Communicable Diseases Manual*, 20th ed., APHA, Washington, DC.

Heymann, D.L. et al. 2015, 'Global health security: The wider lessons from the west African Ebola virus disease epidemic', *The Lancet*, vol. 385, no. 9980, pp. 1884-901.

Hoy, R. and Brims, F. 2017, 'Occupational lung diseases in Australia', *The Medical Journal of Australia*, vol. 207, no. 10, pp. 443–8.

Hung, L.L., Miller, J.D. and Dillon, K.H. 2005, *Field Guide for the Determination of Biological Contaminants in Environmental Samples*, 2nd ed., AIHA, Fairfax, VA.

International Agency for Research on Cancer (IARC) 2018, *IARC Monographs*, Lyon, France, <https://monographs.iarc.fr/?s=biological+agents+carcinogens&sa=> [accessed 29 August 2018]

Ismail, S.A. et al. 2016, 'Communicable disease surveillance and control in the context of conflict and mass displacement in Syria', *International Journal of Infectious Diseases*, vol. 47, pp. 15–22.

Jaffer, A. and Hotez, P.J. 2016, 'Somalia: A nation at the crossroads of extreme poverty, conflict, and neglected tropical diseases', *PLOS Neglected Tropical Diseases*, vol. 10, no. 9, p. e0004670.

Jensen, P.A., Todd, W.F., Davis, G.N. and Scarpino, P.V. 1992, 'Evaluation of eight bioaerosol samplers challenged with aerosols of free bacteria', *American Industrial Hygiene Association Journal*, vol. 53, no. 10, pp. 660–7.

Jordan, D. et al. 2011, 'Carriage of methicillin-resistant Staphylococcus aureus by veterinarians in Australia', *Australian Veterinary Journal*, vol. 89, no. 5, pp. 152–9.

Kesavan, J., Schepers, D. and McFarland, A.R. 2010, 'Sampling and retention efficiencies of batch-type liquid-based bioaerosol samplers', *Aerosol Science and Technology*, vol. 44, no. 10, pp. 817–29.

Khedher, B.S, Guida, N.M., Matrat, F., Cenée, M., Sanchez, S., Menvielle, M., Molinié, G., Luce, F.D. and Stücker, I. 2017, 'Occupational exposure to endotoxins and lung cancer risk: Results of the ICARE Study', *Occupational and Environmental Medicine*, vol. 74, no. 9, pp. 667–79.

King, J., Richardson, M., Quinn, A., Holme, J. and Chaudhuri, N. 2017, 'Bagpipe lung—a new type of interstitial lung disease?', *Thorax*, 72(4), pp. 380–2.

Knope, K.E., Kurucz, N., Doggett, S.L., Muller, M., Johansen, C.A., Feldman, R., Hobby, M., Bennett, S., Sly, A., Lynch, S., Currie, B.J., Nicholson, J. and National Arbovirus and Malaria Advisory Committee 2016, 'Arboviral diseases and malaria in Australia, 2012–13: Annual report of the National Arbovirus and Malaria Advisory Committee', *Communicable Disease Intelligence*, vol. 40, no. 1, pp. E17–47.

Kulkarni, P., Baron, P.A. and Willeke, K. (eds) 2011, *Aerosol Measurement: Principles, Techniques, and Applications*, Wiley, Hoboken, NJ.

Kwo, E. and Christiani, D. 2017, 'The role of gene–environment interplay in occupational and environmental diseases: Current concepts and knowledge gaps', *Current Opinion in Pulmonary Medicine*, vol. 23, no. 2, pp. 173–6.

Lang, J.M. et al. 2017, 'A microbial survey of the International Space Station (ISS), *PeerJ*, vol. 5, p. e4029, doi:10.7717/peerj.4029.

Ley, S. 2015, *Gene Technology Amendment Bill*, Parliament of the Commonwealth of Australia, Canberra.

Liebers, V., Raulf-Heimsoth, M. and Brüning, T. 2008, 'Health effects due to endotoxin inhalation (review)', *Archives of Toxicology*, vol. 82, no. 4, pp. 203–10.

Linaker, C. and Smedley, J. 2002, 'Respiratory illness in agricultural workers', *Occupational Medicine*, vol. 52, no. 8, pp. 451–9.

Lindsley, W.G., Green, B.J., Blachere, F.M., Law, B.F., Jensen, P.A. and Schafer, M.P. 2017, 'Sampling and characterisation of bioaerosols', in *NIOSH Manual of Analytical Methods (NMAM)*, 5th ed., NIOSH, Cincinnati, OH.

Liu, H. et al. 2004, 'Formation and composition of the Bacillus anthracis endospore', *Journal of Bacteriology*, vol. 186, no. 1, pp. 164–78.

Macher, J. 1999, *Bioaerosols: Assessment and Control*, ACGIH®, Cincinnati, OH.

Madelin, T.M. and Madelin, M.F. 1995, 'Biological analysis of fungi and associated molds', in C.S. Cox and C.M. Wathes (eds), *Bioaerosols Handbook*, Lewis Publishers, Boca Raton, FL.

Madsen, A.M., Tendal, K., Schlünssen, V. and Heltberg, I. 2012, 'Organic dust toxic syndrome at a grass seed plant caused by exposure to high concentrations of bioaerosols', *Annals of Occupational Hygiene*, vol. 56, no. 7, pp. 776–88.

Mardis, E.R. 2017, 'DNA sequencing technologies: 2006–2016', *Nature Protocols*, vol. 12, p. 213.

Mbareche, H., Brisebois, E., Veillette, M. and Duchaine, C. 2017, 'Bioaerosol sampling and detection methods based on molecular approaches: No pain no gain', *Science of the Total Environment*, vols 599–600, pp. 2095–104.

McCloskey, B., Dar, O., Zumla, A. and Heymann, D.L. 2014, 'Emerging infectious diseases and pandemic potential: Status quo and reducing risk of global spread', *The Lancet Infectious Diseases*, vol. 14, no. 10, pp. 1001–10.

McCormack, J.G. and Allworth, A.M. 2002, 'Emerging viral infections in Australia', *Medical Journal of Australia*, vol. 177, no. 1, pp. 45–9.

McKenzie, R.A., 2012, *Australia's Poisonous Plants, Fungi and Cyanobacteria: a guide to species of medical and veterinary importance*, CSIRO Publishing, Collingwood, Vic.

McLeod, D.S.A. et al. 2011, 'Histoplasmosis in Australia: Report of 16 cases and literature review', *Medicine*, vol. 90, no. 8, pp. 61–8.

Morrell, J. and Stratman, E. 2011, 'Primary care and specialty care delays in diagnosing Trichophyton verrucosum infection related to cattle exposure', *Journal of Agromedicine*, vol. 16, no. 4, pp. 244–50.

National Health and Medical Research Council (NHMRC) 2010, *Australian Guidelines for the Prevention and Control of Infection in Healthcare*, Department of Health, Canberra, <https://nhmrc.gov.au/about-us/publications/australian-guidelines-prevention-and-control-infection-healthcare-2010>.

—— 2018, 'Australian Drinking Water Guidelines 6 2011', Version 3.5 updated August 2018, Australian Government, Canberra, < https://www.nhmrc.gov.au/sites/default/files/documents/attachments/australian-drinking-water-guidelines-may19.pdf>.

National Occupational Health and Safety Commission (NOHSC) 1990, *Occupational Diseases of the Skin*, AGPS, Canberra.

NSW Health 2015, *Creutzfeldt-Jakob Disease (CJD)*, NSW Government, Sydney, <www.health.nsw.gov.au/Infectious/factsheets/Pages/creutzfeldt-jakob-disease.aspx> [accessed 30 August 2018]

Office of the Gene Technology Regulator (OTGR) 2018, Website, Australian Government, Canberra, <www.ogtr.gov.au>.

Oliver, T. 1902, *Dangerous Trades: The Historical, Social, and Legal Aspects of Industrial Occupations as Affecting Health*, John Murray, London.

Oluwole, O. et al. 2018, 'The association between endotoxin and beta-$(1 \rightarrow 3)$-D-glucan in house dust with asthma severity among schoolchildren', *Respiratory Medicine*, vol. 138, pp. 38–46.

One Health 2018, *One Health Initiative*, <www.onehealthinitiative.com/map.php> [accessed 3 September 2018]

Oppliger, A. 2014, 'Advancing the science of bioaerosol exposure assessment', *The Annals of Occupational Hygiene*, vol. 58, no. 6, pp. 661–3.

Ozanich, R. 2018, 'Chem/bio wearable sensors: current and future direction', *Pure and Applied Chemistry*, vol. 90, no. 10, pp. 1605–13.

Paudyal, P., Semple, S., Gairhe, S., Steiner, M.F.C., Niven, R. and Ayres, J.G. 2015, 'Respiratory symptoms and cross-shift lung function in relation to cotton dust and endotoxin exposure in textile workers in Nepal: A cross-sectional study', *Occupational and Environmental Medicine*, vol. 72, no. 12, pp. 870–6.

Pearson, C., Littlewood, E., Douglas, P., Robertson, S., Gant, T.W. and Hansell, A.L. 2015, 'Exposures and health outcomes in relation to bioaerosol emissions from composting facilities: A systematic review of occupational and community studies', *Journal of Toxicology and Environmental Health. Part B, Critical Reviews*, vol. 18, no. 1, pp. 43–69.

Pezant, B., Weekes, D.M. and Miller, J.D. 2016, *Recognition, Evaluation and Control of Indoor Mold*, AIHA, Cincinnati, OH.

Popoff, M.R. and Poulain, B. 2010, 'Bacterial toxins and the nervous system: Neurotoxins and multipotential toxins interacting with neuronal cells', *Toxins*, vol. 2, no. 4, pp. 683–737.

Quince, C., Walker, A.W., Simpson, J.T., Loman, N.J. and Segata, N. 2017, 'Shotgun metagenomics, from sampling to analysis', *Nature Biotechnology*, vol. 35, pp. 833–44.

Ramazzini, B. 2001, '*De morbis artificum* [Diseases of workers]', *American Journal of Public Health*, vol. 91, no. 9, pp. 1380–2.

Rapp, C., Aoun, O., Ficko, C., Andriamanantena, D. and Flateau, C. 2014, 'Infectious diseases related to aeromedical evacuation of French soldiers in a level 4 military treatment facility: A ten-year retrospective analysis', *Travel Medicine and Infectious Disease*, vol. 12, no. 4, pp. 355–9.

Reponen, T., Willeke, K., Grinshpun, S. and Nevalainen, A. 2011, 'Biological particle sampling', in P. Kulkarni, P.A. Baron and K. Willeke (eds), *Aerosol Measurement*, Wiley, Hoboken, NJ.

Reynolds, S.J. et al. 2013, 'Systematic review of respiratory health among dairy workers', Journal of Agromedicine, vol. 18, no. 3, pp. 219–43.

Rusca, S., Charrière, N., Droz, P. and Oppliger, A. 2008, 'Effects of bioaerosol exposure on work-related symptoms among Swiss sawmill workers', *International Archives of Occupational and Environmental Health*, vol. 81, no. 4, pp. 415–21.

Russell, R.C. 2018, *Malaria*, University of Sydney, <http://medent.usyd.edu.au/fact/malaria.htm>, [accessed 10 June 2019]

Russell, R.C., Currie, B.J., Lindsay, M.D., Mackenzie, J.S., Ritchie, S.A. and Whelan, P.I. 2009, 'Dengue and climate change in Australia: Predictions for the future should incorporate knowledge from the past', *Medical Journal of Australia*, vol. 190, no. 5, pp. 265–8.

Russell, R.C. and Doggett, S.L. 2012, *Vector Borne Diseases*, Department of Medical Entomology, University of Sydney, Sydney, <http://medent.usyd.edu.au/fact/fact.htm> [accessed 30 August 2018]

Ruzer, L.S. and Harley, N.H. 2013, *Aerosols Handbook: Measurement, Dosimetry, and Health Effects*, CRC Press, Boca Raton, FL.

Safe Work Australia 2006, *Work-related Infectious and Parasitic Diseases in Australia*, Safe Work Australia, Canberra, <www.safeworkaustralia.gov.au/system/files/documents/1702/workrelated_infectious_parastitic_disease_australia.pdf> [accessed 12 December 2018]

—— 2011, *National Hazard Exposure Worker Surveillance: Exposure to Biological Hazards and the Provision of Controls Against Biological Hazards in Australian Workplaces*, Safe Work Australia, Canberra, <www.safeworkaustralia.gov.au/system/files/documents/1702/nhews_biologicalmaterials.pdf> [accessed 30 August 2018]

—— 2018, *Disease and Injury Statistics by Type, Safe Work Australia*, Canberra, <www.safeworkaustralia.gov.au/statistics-and-research/statistics/disease-and-injuries/disease-and-injury-statistics-type> [accessed 18 December 2018]

Samson, R.A. 2015, 'Cellular constitution, water and nutritional needs, and secondary metabolites', in C. Viegas et al. (eds), *Environmental Mycology in Public Health: An Overview on Fungi and Mycotoxins Risk Assessment and Management*, Elsevier, New York, pp. 5–13.

Schloss, P.D. and Handelsman, J. 2005, 'Metagenomics for studying unculturable micro-organisms: Cutting the Gordian knot', *Genome Biology*, 6(8), p. 229.

Schlosser, O., Huyard, A., Cartnick, K., Yanez, A., Catalan, V. and Do Quang, Z. 2009, 'Bioaerosol in composting facilities: occupational health risk assessment', *Water Environment Research*, vol. 81, no. 9, pp. 866–77.

Selman, M., Mejia, M., Ortega, H. and Navarro, C. 2012, 'Hypersensitivity pneumonitis: A clinical perspective', in R.P. Baughman and R.M.D. Bois (eds), *Diffuse Lung Disease*, 2nd ed., SpringerLink, London.

Selman, M., Pardo, A. and King, T.E. 2012, 'Hypersensitivity pneumonitis: Insights in diagnosis and pathobiology', *American Journal of Respiratory and Critical Care Medicine*, vol. 186, no. 4, pp. 314–24.

Semple, S. 2010, 'Bioaerosols', in C. John, H. Robin and S. Sean (eds), *Monitoring for Health Hazards at Work*, Wiley Blackwell, Oxford, pp. 111–22.

Sköld, P. 2017, 'The health transition: A challenge to Indigenous peoples in the Arctic', in K. Latola and H Savela (eds), *The Interconnected Arctic: UArctic Congress 2016*, Springer, Dordrecht, pp. 107–13.

Smith, T.C., Harper, A.L., Nair, R., Wardyn, S.E., Hanson, B.M, Ferguson, D.D. and Dressler, A.E. 2011, 'Emerging swine zoonoses', *Vector Borne and Zoonotic Diseases*, vol. 11, no. 9, pp. 1125–234.

Spurgeon, J. 2007, 'A comparison of replicate field samples collected with the Bi-Air, Air-O-Cell, and Graesby-Andersen N6 bioaerosol samplers', *Aerosol Science and Technology*, vol. 41, no. 8, pp. 761–9.

Standards Australia 2009a, *AS 2985: 2009 Respirable Dust*, SAI Global, Sydney.

—— 2009b, AS 3640: 2009, Inhalable Dust, SAI Global, Sydney.

—— 2011, AS3666: 2011 Air-Handling and Water Systems of Buildings—Microbial Control, SAI Global, Sydney.

—— 2018, AS/NZS/ISO 45000: 2018, *Occupational Health and Safety Management Systems—Requirements with Guidance for Use*, SAI Global: Sydney.

Stave, G.M. 2018, 'Occupational animal allergy', *Current Allergy and Asthma Reports*, vol. 18, no. 2, p. 11.

Stevens, J., Corper, A.L., Basler, C.F., Taubenberger, J.K., Palese, P. and Wilson, I.A. 2004, 'Structure of the uncleaved human H1 Hemagglutinin from the extinct 1918 influenza virus', *Science*, vol. 303, no. 5665, pp. 1866–70.

Torén, K. and Blanc, P.D. 2009, 'Asthma caused by occupational exposures is common: A systematic analysis of estimates of the population-attributable fraction', *BMC Pulmonary Medicine*, vol. 9, no. 1, p. 7.

Tortora, G.J., Funke, B.R. and Case, L. 2016, *Microbiology: An introduction*, Pearson, London.

Tyndall, J. 1888, *Essays on the Floating-Matter of the Air: In Relation to Putrefaction and Infection*, D. Appleton and Co, New York, <http://bhl-china.org/bhldatas/pdfs/e/essaysonfloating00tynd.pdf>.

Umbrell, C. 2003, 'Mould: Creating a scientific consensus on a hot topic', *The Synergist*, April, pp. 35–8.

University of Ottawa 2014, *Infectious Disease Control*, <www.med.uottawa.ca/sim/data/Infectious_Diseases_e.htm> [accessed 10 December 2018]

Victorian Department of Health 2018, *Blue Book: Guidelines for the Control of Infectious Diseases*, Department of Health, Melbourne.

Viegas, C., Brandão, J., Pinheiro, A.C., Sabino, R., Veríssimo, C. and Viegas, S.D.M. (eds) 2015, *Environmental Mycology in Public Health: An Overview on Fungi and Mycotoxins Risk Assessment and Management*, Elsevier, New York.

Viegas, C., Viegas, S., Gomes, A., Täubel, M. and Sabino, R. 2017, *Exposure to Microbiological Agents in Indoor and Occupational Environments*, Springer, New York.

Vincent, J.H. 2007, *Aerosol Sampling: Science, Standards, Instruments and Applications*, John Wiley and Sons, Chichester.

Walser, S.M. et al. 2015, 'Evaluation of exposure–response relationships for health effects of microbial bioaerosols: A systematic review', *International Journal of Hygiene and Environmental Health*, vol. 218, no. 7, pp. 577–89.

Watts, N. et al. 2015, 'Health and climate change: Policy responses to protect public health', *The Lancet*, vol. 386, no. 10006, pp. 1861–914.

Willeke, K. and Macher, J.M. 1999, 'Air sampling', in J.M. Macher (ed.), *Bioaerosols: Assessment and Control*, ACGIH®, Cincinnati, OH.

Yamamoto, N., Schmechel, D., Chen, B.T., Lindsley, W.G. and Peccia, J. 2011, 'Comparison of quantitative airborne fungi measurements by active and passive sampling methods', *Journal of Aerosol Science*, vol. 42, no. 8, pp. 499–507.

Yao, M. and Mainelis, G. 2007, 'Analysis of portable impactor performance for enumeration of viable bioaerosols', *Journal of Occupational and Environmental Hygiene*, vol. 4, no. 7, pp. 514–24.

Yates, M.V., Nakatsu, C.H., Miller, R.V. and Pillai, S.D. 2016, *Manual of Environmental Microbiology*, 4th ed., ASM, Washington, DC, <https://app.knovel.com/hotlink/pdf/id:kt0110TGI1/manual-environmental/approaches-cyanohab-monitoring>.

Zhang, J. et al. 2015, 'Endotoxin and β-1,3-d-glucan in concentrated ambient particles induce rapid increase in blood pressure in controlled human exposures', *Hypertension*, vol. 66, no. 3, pp. 509–16.

Zou, S., Caler, L., Colombini-Hatch, S., Glynn, S. and Srinivas, P. 2016, 'Research on the human virome: Where are we and what is next', *Microbiome*, vol. 4, p. 32.

17. Postscript

Dino Pisaniello

17.1 EVOLVING AND EMERGING ISSUES IN OCCUPATIONAL HYGIENE

Occupational hygiene is a dynamic area with new challenges, opportunities and directions. One can consider these in terms of *evolving* and *emerging* issues, impacting on the scope of professional practice, relationships with other health professionals and the wider community. Although there is overlap between evolving and emerging issues, the former can be considered those which fall in traditional disciplinary domains but are being shaped by new materials, technologies and applications. This includes nanomaterials introduced in Chapter 7, new forms of lighting mentioned in Chapter 15, and new sensors and video exposure monitoring alluded to in Chapter 9. Emerging issues may be considered those that arise from broader or disruptive contexts, for example, new forms of work, a changing workforce, or climate, data science and often require an interdisciplinary approach.

17.2 EVOLVING ISSUES

With respect to evolving issues, an expanding knowledge base has led to more professional specialisations and overlapping interests. For example, some exposure assessment aspects of occupational health and hygiene are being addressed through an International Society of Exposure Science, formed in 1989. Relatedly, the British Occupational Hygiene Society (BOHS) journal, previously known as the *Annals of Occupational Hygiene*, is now called the *Annals of Work Exposures and Health* and is dedicated to presenting advances in exposure science as it relates to the health of workers, families and communities.

Occupational hygiene has also evolved with the availability of low-cost wearable technology and networked data. The use of mathematical models, apps and exposure databases for risk assessment is becoming more common. New opportunities exist for wide-scale and personalised (or even worker-driven) exposure assessment and physiological measurement. The concept of an 'exposome' has been put forward by Wild (2005) as a complement to the genome, and as it represents the *totality* of exposure, it challenges the traditional limits of exposure assessment to include non-occupational and life course (conception to old age) exposures (Robinson and Vrijheid, 2015). There is an expectation that hygienists will be engaged in exposomics and turn concept into reality (Wild, 2012). The need to characterise and control exposures in critical time periods such as during pregnancy and early life has become more evident. It is now accepted that most chronic disease, including cancer, is not attributable solely to genetic factors, and must be associated with the exposome. A challenge for occupational hygienists is the compilation and interpretation of increasingly large volumes of exposure data of variable quality and purpose.

17.3 EMERGING ISSUES

With respect to emerging issues, advances in systems biology, toxicology and molecular epidemiology have led to new insights into risk factors, including gene-environment interactions. The so-called 'fourth industrial revolution' involving Big Data, robotics, artificial intelligence and the 'internet of things' is also relevant here. Safe Work Australia and CSIRO have recently published a report on workplace safety futures and megatrends (Horton et al, 2018).

The workforce in many developed countries is ageing and becoming more diverse, with migrant workers and international students representing an important and growing segment (Farrow and Reynolds, 2012; Moyce and Schenker, 2018). Temporary and fragmented work as well as screen-based work is more common (Horton et al., 2018). Finally, in an Asian century, rapid industrialisation and globalisation may require new occupational health management and communication approaches including a greater understanding of cultural perspectives and social systems (European Agency for Safety and Health at Work, 2013).

17.4 REFERENCES

European Agency for Safety and Health at Work, 2013, *Diverse cultures at work: ensuring safety and health through leadership and participation*, Luxembourg: Publications Office of the European Union. DOI: 10.2802/14394.

Farrow, A. and Reynolds, F. 2012, 'Health and safety of the older worker', *Occupational Medicine*, vol. 62, pp. 4–11. DOI:10.1093/occmed/kqr148.

Horton J., Cameron, A., Devaraj, D., Hanson, R.T. and Hajkowicz, S.A. 2018. 'Workplace Safety Futures: The impact of emerging technologies and platforms on work health and safety and workers' compensation over the next 20 years', CSIRO, Canberra.

Moyce, S.C. and Schenker, M. 2018, 'Migrant workers and their occupational health and safety', *Annual Review of Public Health*, vol. 39, pp. 351–65, DOI: 10.1146/annurev-publhealth-040617-013714.

Robinson, O. and Vrijheid, M. 2015, 'The pregnancy exposome', *Current Environmental Health Reports*, vol. 2, no. 2, pp. 204–213. DOI: 10.1007/s40572-015-0043-2.

Wild, C. 2005, 'Complementing the genome with an "exposome": the outstanding challenge of environmental exposure measurement in molecular epidemiology', *Cancer Epidemiology, Biomarkers and Prevention*, 14(8), pp. 1847–50, DOI: 10.1158/1055-9965.EPI-05-0456.

—— 2012, 'The exposome: from concept to utility', *International Journal of Epidemiology*, vol. 41, pp. 24–32, DOI: 10.1093/ije/dyr236.

Index